Tools, Concepts and Applications of Biochemical Engineering

Tools, Concepts and Applications of Biochemical Engineering

Edited by **Gerald Cole**

Willford Press

New York

Published by Willford Press,
118-35 Queens Blvd., Suite 400,
Forest Hills, NY 11375, USA
www.willfordpress.com

Tools, Concepts and Applications of Biochemical Engineering
Edited by Gerald Cole

International Standard Book Number: 978-1-68285-221-7 (Hardback)

Printed in China.

Contents

Preface

In my initial years as a student, I used to run to the library at every possible instance to grab a book and learn something new. Books were my primary source of knowledge and I would not have come such a long way without all that I learnt from them. Thus, when I was approached to edit this book; I became understandably nostalgic. It was an absolute honor to be considered worthy of guiding the current generation as well as those to come. I put all my knowledge and hard work into making this book most beneficial for its readers.

Biochemical engineering is an important branch of engineering that has diverse applications in fields like pharmaceutical industry, petrochemical industry, food, water treatment, etc. The chapters included herein elucidate new techniques of biochemical engineering and their applications in a multidisciplinary approach. This book is compiled in such a manner, that it will provide in-depth knowledge about the theoretical aspects of topics like bioprocess analysis, catalysis, fermentation, etc. along with their practical applications. It also presents concepts of bioreactor designing, biochemical reaction engineering, etc. It will be suitable for the students of biochemical engineering, chemical engineering and similar branches of engineering.

I wish to thank my publisher for supporting me at every step. I would also like to thank all the authors who have contributed their researches in this book. I hope this book will be a valuable contribution to the progress of the field.

Editor

From Solar Hydrogen to Desert Development: A Challenging Approach

Hussein K Abdel-Aal*

Professor Emeritus of Chemical Engineering / Petroleum Operations, NRC, Cairo, Egypt (Retired)

Abstract

This paper presents a novel approach for desert development, taking the Arab countries a model in our study. Solutions to make the most from the sun to turn our desert into useful and arable land are proposed. Saline water resources from sea or rejects from desalination plants (brines) are chemically processed to produce partially desalted water along with valuable chemical products. Solar hydrogen is a cornerstone in our system to produce ammonia, which has dual functions in the scheme. Ammonia represents a source of nitrogen for plants. It is used together with carbon dioxide in the chemical separation process of saline water.

For the Arab countries with an average solar intensity flux = 700 Watt/m^2, an average sunshine hours of 3000 annually and only 1% of total land area for solar power generation, it is feasible to produce the equivalent of 184 × 106 tons of hydrogen yearly by water electrolysis. Ammonia gas synthesis takes place, using solar hydrogen and atmospheric nitrogen.

Experimental findings are reported by the author and co-workers. Separation of salt brine (from inland sources or from the sea) was carried out in a gas bubbler using a modified Solvay process to study the conversion of sodium chloride into chemical products (namely sodium carbonate and ammonium chloride). The highest conversion achieved of sodium chloride was 82.2%, and the balance makes partially desalted water. Magnesium chloride is obtained as a by-product. The role of hydrogen as a vector coupled with solar energy for desert development is schematically illustrated.

Keywords: Solar hydrogen; Ammonia; Brine salt; Ammonium chloride; Soda ash; Fertile water

Introduction

As the population continues to grow in many parts of the world, the infrastructure that supports that population must be increased as well. Desert development represents a challenging problem that need to be solved. It offers one potential solution to problems of rural unemployment, urban overcrowding, and food security, if pursued in a sustainable means. World hunger is one of the most pressing problems of our time. To overcome it, the three basic economic resources that contribute to agriculture have to be satisfactorily provided. These are:-

- Arable land
- Water and fertilizers
- Energy

It is towards this objective that this study is carried out to harness the sun for desert development into useful and arable land.

Energy security is a challenge that we all face whether this is due to a lack of resources or a dependence on global markets or more recently the prospect of dwindling resources. Hydrogen offers a solution to this in that it can be produced domestically from available renewable sources of energy, in particular solar energy.

You never change things by fighting the existing reality. To change something, build a new model that makes the existing model obsolete. In this paper, we will explore the usage of hydrogen power as an effective means in desert development. A comprehensive and visionary approach that utilizes solar hydrogen, carbon dioxide and saline water resources is presented. Arab countries are taken as a case study to apply our proposed model, where the vast land resources and the abundant solar energy are very much in favour of this study.

The proposed system encompass three main components, as shown in Figure 1

1- Solar Hydrogen Production

2- Ammonia Synthesis

3- Chemical Separation Process of Saline Water Resources.

Solar hydrogen (H_2), once produced by water electrolysis [1-11] is utilized in the synthesis of ammonia (NH_3), a well-established process. Ammonia plays two distinctive roles in the proposed system:-

- As a source of nitrogen-based fertilizers
- As a chemical agent to be used in the separation of minerals salts for saline water resources.

The proposed separation process is carried out for highly-saline water (brine) using a non-conventional partial desalting method [12-14]. The method utilizes a series of chemical reactions involving the conversion of sodium chloride, the major constituent in saline waters, into sodium bicarbonate and ammonium chloride soluble in partially desalted water making what is called "fertile water". It involves the absorption of ammonia gas first, in the brine solution; to make what is called "ammoniated brine". This is followed next, by bubbling carbon dioxide into this brine.

***Corresponding author:** Hussein K. Abdel-Aal, Professor Emeritus of Chemical Engineering / Petroleum Operations, NRC, Cairo, Egypt

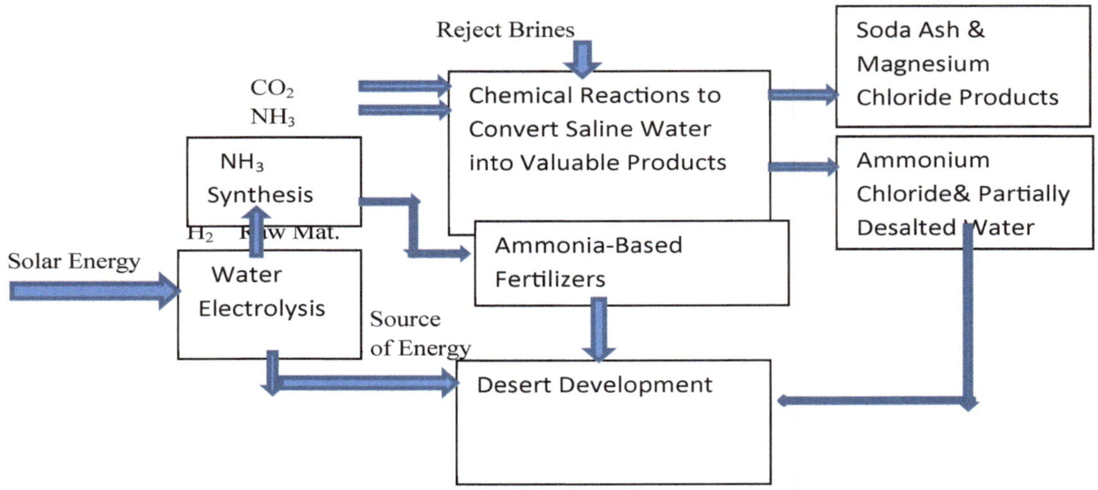

Figure 1: A Demonstration of the Role of Hydrogen Vector Coupled with Solar Energy for Desert Development.

Source of CO_2, Brines and Ammonia

CO_2

Almost all CO_2 emissions (about 96.5%) come from fossil fuels use. When fossil fuels are combusted, the carbon stored in them is emitted almost entirely as CO_2 causing what is known as the green-house effect. While CO_2 could in theory be extracted from the atmosphere, most of the CO_2 collected in processes would be derived from ammonia plants.

Brines

This proposal would be ideal for arid countries that have access to sea shores. Usually because fresh water is scarce, desalination plants are erected near or at sea locations.

It is estimated that the world production of desalination water exceeds 30 million cubic meters per day and the desalination market worldwide is expected to reach $30 billion by 2015 [15]. One of the major economic and environmental challenges to the desalination industry, especially in those countries that depend on desalination for potable water, is the handling of reject brine. Assuming that 50% of the intake sea water to be desalinated, then about 15 million cubic meters/day of rejected brines is produced. The common practice in dealing with these huge amounts of brine is to discharge back into the sea affecting seriously the aquatic life.

Ammonia

Ammonia is a principal raw material for the desalination process of brine along with CO_2 as shown in the diagram. The availability of solar hydrogen makes ammonia synthesis very feasible to produce by the well-established Haber-Bosch process [16].

Role of Solar Energy

Solar energy provides electricity via photovoltaic cells. Sunlight reaching the land surface of our planet can produce the equivalence of 1,600 times the total energy consumption of the world; the amount of solar energy derived from the sun's radiation on just one square kilometer is about 4,000 megawatts. Many of the Arab countries have all practical futuristic possibilities of supplying energy in significant quantity from their vast solar rich deserts, by erecting solar farms on them to generate power which could be used for hydrogen production.

In the wake of the first Gulf War, the US Army assessed Saudi Arabia's solar energy resource potential in a classified effort to determine how oil fires had affected the region. In addition to being a vast petroleum repository, the desert nation was also the heart of the most potentially productive region on the planet for harvesting power from the sun. Sitting in the center of the so-called Sun Belt, the country is part of a vast, rainless region reaching from western edge of North Africa to the eastern edge of Central Asia that boasts the best solar energy resources on Earth [17].

Role of Efficiency in Hydrogen Production

Having solar energy at hand, there is more than one option to generate electricity to be used in the production of hydrogen by water electrolysis.

The following data on efficiencies are cited in order to select the optimum path for hydrogen production [1].

E_1, Solar to EE = 20% (43.5 is the highest)

E_2, Solar to thermal = 80%

$E_{3,}$ Thermal to EE = 40%

$E_{4,}$ Electrolysis = 50-80%, use 70% as an average.

E_5 Hydrogen/EE (Fuel Cell) = >90%

Using these data, let us compare the following options to produce hydrogen from a sustainable energy source (solar energy):-

Option 1: To use E.E produced by PVC for water electrolysis, $E_{overall}$, for Hydrogen production = 0.435×0.7 = 30%

Option 2: To use E.E produced by thermal collector for water electrolysis, $E_{overall} =$ 0.8×0.4×0.7 = 22.4%

Water Electrolysis

The operation of an electrolyser is a trade -off between energy efficiency and productivity [18-21]. There is a minimum theoretical voltage required to split water (about 1.48 volt for the cell to remain thermally neutral), but practically higher voltages are required to overcome inefficiencies in the system. This additional energy is what we call "over-potential".

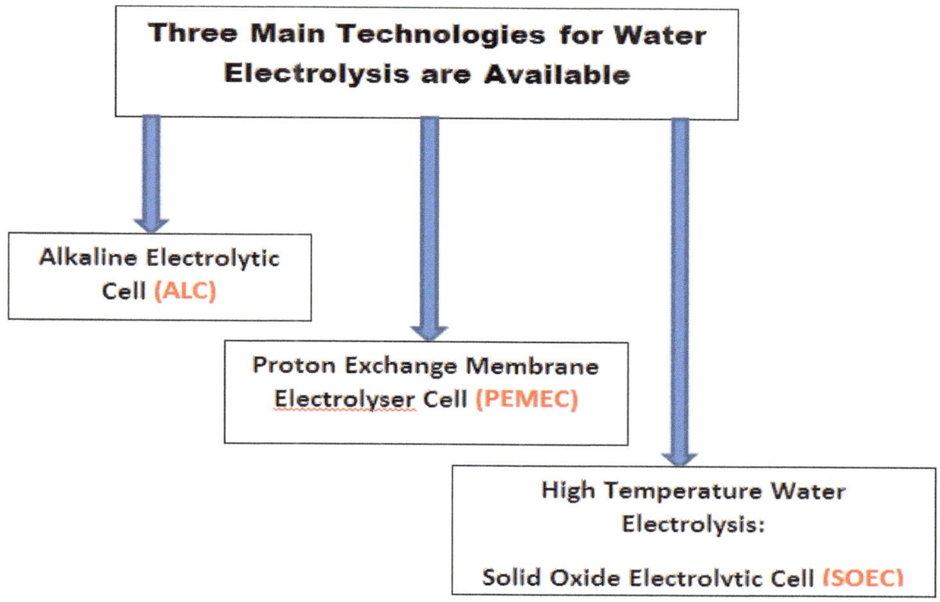

Figure 2: Technology Available for Water Electrolysis.

Type of technology cell	Electrolyte	Operating Temperature °c	Electrode Catalyst
AEC	KOH or NaOH	50-100	Nickel-based
PEMEC	H.P.M	20-100	Platinum/Iridium
SOEC	Ceramic	500-100	Nickel cement

AEC = Alkaline Electrolyte Cell
PEMEC = Proton Exchange Membrane Electrolyser Cell
SOEC = Solid Oxide Electrolytic Cell
H.P.M = Humidified Polymer Membrane

Table1: Technical data for the technology of cells used in water electrolysis

Now, the following scenario applies:-

- As the current density is increased between the electrodes, the over-potential increases. This causes heat to be generated, resulting in loss of efficiency.
- If, however, the current density is reduced, too little gas is produced.

Therefore, a balance has to be established for practical practice in electrolysis. There are a number of ways to reduce over-potential to allow for improved production while maintaining efficiency at the same time. This include:-

- Designing the electrodes in order to maximize the surface (contact area between the electrodes and electrolyte.
- The use of catalytic materials on the electrode surfaces.

Therefore, minimizing the over-potential contributes to lowering the operating costs of an electrolyser, since the cost of electricity represents a substantial element in the running costs. As far as the types of electrolyzers, three main technologies are recognized as shown in Figure 2.

A comparison between the three types of electrolysis cells is presented in Table 1.

High-temperature electrolysis (H.T.E.) or steam electrolysis is a method currently being investigated for water electrolysis with a heat engine. High temperature electrolysis may be preferable to traditional room-temperature electrolysis because some of the energy is supplied as heat, which is cheaper than electricity. Also, electrolysis reaction is more efficient at higher temperatures.

Hydrogen Production

Basis and assumptions

The average solar intensity flux = 700 Watt/m^2

- The average sunshine hours is 3000 annually
- The total land area of the Arab countries considered in the survey for producing non-fossil hydrogen ≈ 11 × 106 Km^2
- Only 1% of total land area is utilized for solar power generation
- The efficiency of solar conversion is taken 10%
- The efficiency of hydrogen production is taken 30%
- One cubic meter of hydrogen produces 3 Kw.hr (thermal)

Calculations

Based on the above, the following calculations are presented:

1st The total annual Kw.hr received by one m^2 = 0.1×0.7Kw × 3000 hr/y = 210 Kw.hr ≈ 200 Kw.hr

2nd The total annual Kw.hr received by land area = 0.01 × 200 ×11× 106 (Km^2) ×106(m^2/Km^2) = 22 × 1012 Kw.hr

3rd The annual hydrogen production = [0.3 × 22 × 1012] /3 = 22 ×1011 m^3 = 22 × 1011 × 83.76 (g/m^3) × 10^{-6}(ton/g)

= 184 × 106 Tons of Hydrogen/year

Ammonia Production

All nitrogen (N_2) fertilizers begin with a source of hydrogen gas (in our case it is the "Solar" hydrogen) and atmospheric nitrogen that are

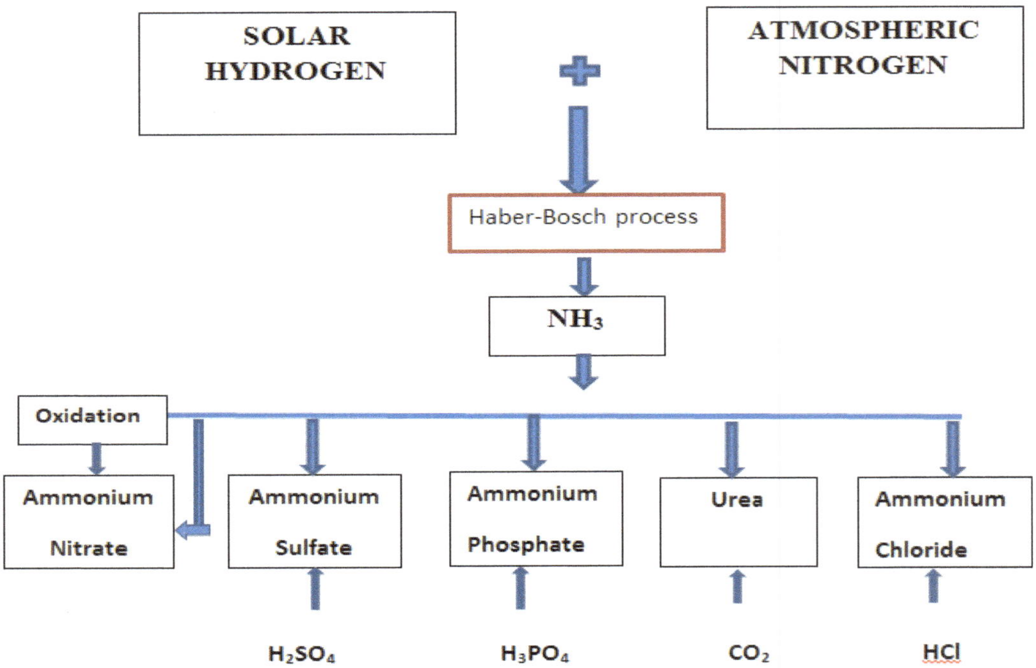

Figure 3: The Chain of Nitrogen Fertilizers produced from Ammonia using Solar hydrogen.

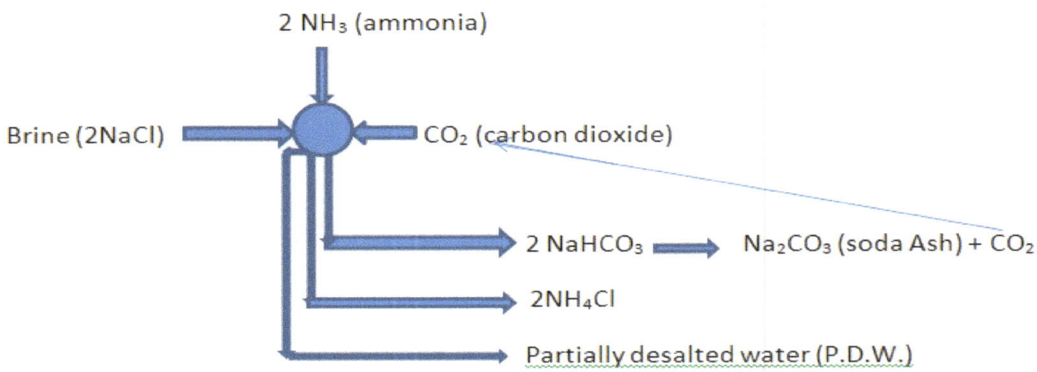

Figure 4: Input and Output for the Chemical Separation Process of Brine.

reacted to form ammonia using the well-known Haber-Bosch process [16].

Nitrogen is essential for many metabolic processes in plants and animals. Perhaps the best-known role of N is in forming amino acids, which are the building blocks for protein. Since the Haber-Bosch process for synthesizing N fertilizer was developed early in the 20th century, its importance in maintaining the global food supply has rapidly grown. It is estimated that half of the food produced now in the world is supported by the use of N_2 fertilizer. After hydrogen and N_2 are combined under conditions of high temperature and pressure to form ammonia, many other important N_2-containing fertilizers can then be made. Urea is the most common N_2 fertilizer, but there are many excellent N_2 fertilizers that can be made from ammonia. For example, some ammonia is oxidized to make nitrate fertilizer. This same conversion of ammonia to nitrate takes place in agricultural soils through the microbial process of nitrification.

Figure 3 represents a diagram for the different types of N_2 obtained from ammonia using solar hydrogen.

Details of the chemical separation process are illustrated in the block diagram, as shown in Figures 4 and 5.

Chemical Separation of Saline Water Resources

The proposed chemical separation process is credited to Abdel-Aal and co-workers [12,13]. The main feed input to this process is highly-saline water resource, which is best represented by the reject brines exit water desalination plants. Ammonia and carbon dioxide are introduced as chemical agents required for executing the underlying reactions, as explained next.

Process outline

As stated earlier, the proposed separation method utilizes a series of chemical reactions involving the conversion of sodium chloride

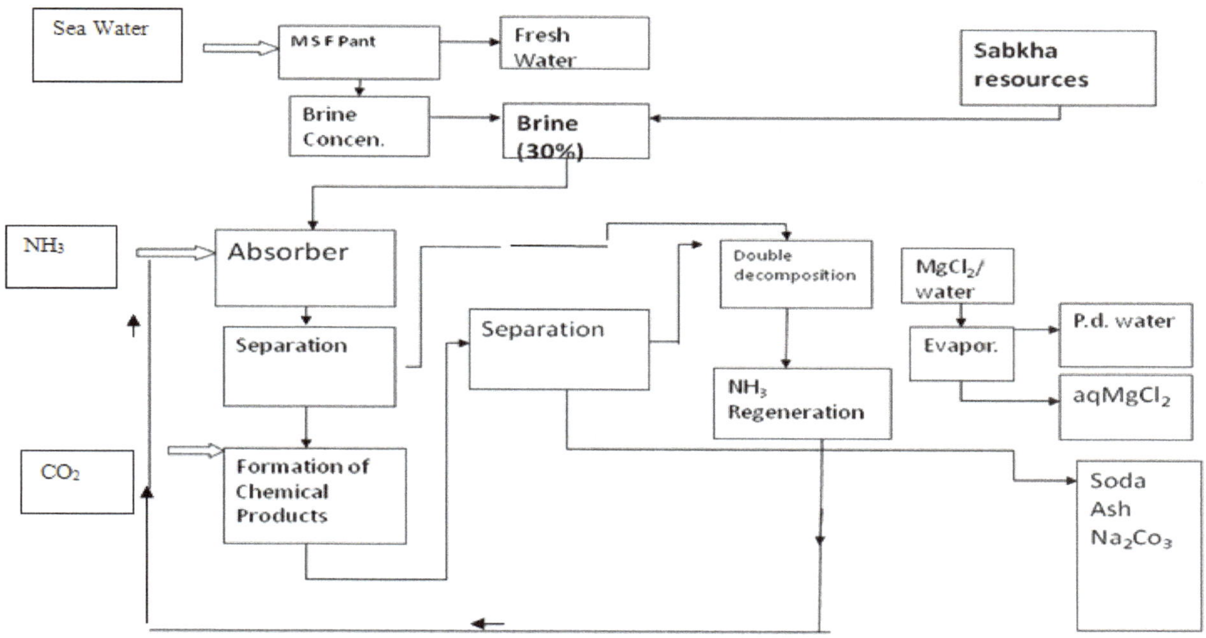

Figure 5: Block Diagram for the Chemical Separation Process.

the major constituent in saline water into sodium bicarbonate and ammonium chloride, by using ammonia and carbon dioxide respectively. The chemical production process involves the following consecutive steps:

1. Treating saturated brine (30 % sodium chloride) with ammonia gas forming what we call "ammoniated brine".
2. Bubbling carbon dioxide gas into the ammoniated brine will trigger a series of chemical reactions. This will lead to the conversion of both sodium cation and chloride anoion respectively into sodium bicarbonate and ammonium chloride.
3. Sodium bicarbonate precipitates under the experimental conditions (slight cooling) and is separated by filtration, leaving ammonium chloride in solution. Soda ash is produced as an end product by the calcination of the bicarbonate.
4. The overall reaction involved could be visualized to take place as follows:-

$NH_4HCO_3^- + NaCl$ à $NaHCO_3 + NH_4Cl$

Partial generation of ammonia gas could be done and recycled back by a double-decomposition chemical reaction with magnesium hydroxide, formed during the ammonia bubbling in the brine.

Ammonia regeneration:

$NH_4Cl + Mg(OH_2)$ à $NH_4OH + MgCl_2$

Reported Experimental Findings

As presented in reference [12,13] the feasibility of the process was investigated experimentally. Some of the results are reported:

a) The Separation of high-salinity synthetic saline water (brines) was carried out in a gas bubbler using a modified Solvay process. The conversion of NaCl in the ammoniated brine into $NaHCO_3$ and NH_4Cl was investigated. The highest conversion of NaCl achieved was 82.2%.

b) The effect of temperature, initial concentrations of NH_3 and NaCl, and the ratio of NH_3/NaCl were studied. An optimum separation temperature of 22°C was found.

c) Indigenous saline resources (such as Sabkha) are abundant in many regions at seashores in countries like Saudi Arabia and Egypt which can make the chemical separation approach very attractive.

Sabkha resources by definition are saline sources containing up to 4-5 mol/L of NaCl. When these resources are subjected to the proposed separation process, partially desalted water is obtained along with other valuable chemical products.

Conclusions

With desert covering a large part of its surface, Arab countries have to develop solutions and propose new strategies to overcome the lack of arable land and potable water. The proposed project begins with a vision of capturing solar energy to produce hydrogen and using it in an innovative way for desert development. The study makes the most from the sun to turn the desert into useful and arable land. Along with solar energy, access to sea water (salt brine) is required; both are abundant for many desert locations.

In the proposed scheme, producing solar hydrogen is considered the cornerstone of our process. Solar energy is used to generate electric power first, for water electrolysis. Once produced, hydrogen is combined with atmospheric nitrogen for the synthesis of ammonia gas. This represents a key source for nitrogen fertilizers.

Ammonia, along with carbon dioxide, assumes an important role to provide what is called "fertile water", in the chemical conversion of highly saline water and brines. Fertile water is defined, in this context as partially-desalted water containing ammonium chloride salt.

Saline-water resources are made available by utilizing rejects from water desalination plants, or using indigenous saline resources (Sabkha).

Solar energy is used to boost salinity of brine to a concentration of about 30% NaCl.

On the other side, other valuable mineral salts are recovered as well. Magnesium chloride, a raw material for the strategic element of magnesium metal, could be recovered.

In the proposed scheme, ammonia is consumed in producing ammonium chloride, as a fertilizer. However, provision could be made to recycle, at least part of the ammonia by regeneration as shown before.

Two options exist, either to have fertile water rich in ammonium chloride; or partially desalted water, ammonium-free to be used for irrigation.

References

1. Ohta T (1979) Solar-Hydrogen Energy Systems. Pergamon Press Ltd.
2. Abdel-Aal HK, Fahim MA, Gashgari MA (1974) Utilization of ConcentratedSolar Energy as a Source of Heat in the Production of Hydrogen.
3. Abdel-Aal GK, Peattie EG (1974) Hydrogen Opportunities in Saudi Arabia, Hydrogen Energy. Part A, Plenum Publishing Corporation, New York, USA 345-369.
4. Abdel-Aal HK (1976) Non-Conventional Energy Trends to Supplement Our Petroleum Reserves. Energy Communications.
5. Abdel-Aal HK, Al-Somait F (1978) Solar Energy Prospects in Saudi Arabia. Energy Communications 4: 271-291.
6. Abdel-Aal HK (1984) From Solar Hydrogen to Ammonium Fertilizers: An Assessment of Large-Scale Production in Saudi Arabia. Hydrogen Energy Progress V, Proceedings of the 5th World Hydrogen Energy Conference, Toronto, Canada 1803-1817.
7. Abdel-Aal HK (1986) Solar Hydrogen Energy Systems: Present State of the Art. Solar Energy Prospects in the Arab World, Pergamon Press 348-353.
8. Abdel-Aal HK, Mohammed MA (1989) Storage of Solar Energy in the Form of Hydrogen: A Pioneering Experiment in Egypt. Energy Sources 11: 95-103.
9. Abdel-Aal HK, Hussein AI (1992) Electrolysis of Saline Water Under Simulated Conditions of P.V. Solar Cell. Volume 1, 9th WHEC, Paris 439-442.
10. Abdel-Aal HK (1992) Storage and Transport of Solar Energy on a Massive Scale:The Hydrogen Option. Int J Hydrogen Energy 17: 875-882.
11. Abdel-Aal HK, Al-Naafa MA (1998) Prospects of Solar Hydrogen for Desert Development in the Arab Countries. Int J Hydrogen Energy 23: 83-88.
12. Abdel-Aal HK, Ibrahim AA, Shalabi MA, Al-Harbi DK (1996) Chemical Separation Process for Highly Saline Water: 1. Parametric Experimental Investigation. I&EC Research 35: 799-804.
13. Abdel-Aal HK, Ibrahim AA, Shalabi MA, Al-Harbi DK (1996) Chemical Separation Process for Highly Saline Water: 2. Systems Analysis and Modeling. I&EC Research 35: 805-810.
14. Hussein Abdel-Aal, Zohdy KM, Maha Abdelkreem, Abdelhamid SM, Bassyouni M (2013) A novel Application in Green Chemical Processing Using Reject Brines and Carbon Dioxide as Raw Materials. Engineering 5: 154-156.
15. Schorr M (2011) Desalination, Trends and Technologies. Corrosion Engineering, Science and Technology 46: 315-316.
16. Shreve NR (1967) Chemical Process industries. McGraw-Hill Company.
17. Middle East at a Glance: Includes Middle Eastern countries and Arab countries of North Africa.
18. Abdel-Aal HK, Hussein IA (1993) Parametric Study for Saline WaterElectrolysis-Part I: Hydrogen Production. Int J Hydrogen Energy 18: 485-489.
19. Abdel-Aal HK, Sultana SM, Hussein IA (1993) Parametric Study for Saline Water Electrolysis - Part II: Chlorine Evolution, Selectivity & Determination. Int J Hydrogen Energy 18: 545-551 (1993).
20. Abdel-Aal HK, Hussein IA (1993) Parametric Study for Saline Water Electrolysis - Part III: Precipitate Formation and Recovery of Magnesium Salts. Int J Hydrogen Energy 18: 553-556.
21. Abdel-Aal HK, Hussein IA (1992) Electrolysis of Saline Water Under Simulated Conditions of P.V. Solar Cell, 9th WHEC, Paris 1: 439-442.

2

A Review Study on Precipitated Silica and Activated Carbon from Rice Husk

Rajesh Ghosh* and Sounak Bhattacherjee

Department of Chemical Engineering, Calcutta Institute of Technology, Uluberia, Howrah, West Bengal, India

Abstract

The present work entitled "Precipitated Silica and Activated Carbon from Rice Husk" was taken up to study the optimum operating conditions at laboratory scale. Precipitation of Silica was carried out using various parameters such as pre-treatment or acid washing, carbonization temperature of acid washed rice husk, activating agent such as sodium hydroxide, time of carbonization, time of heating with activating agent. The activated carbon was obtained from rice husk ash at various concentration of activating agent (Sodium Hydroxide) and different activation temperature. These studies indicate that nano-silica powder can be produced by optimizing the parameters used in the study. Without optimization, one would get different grades of silica with different yields. Similarly, different grades of activated carbon can be obtained by varying the parameters of activating agent, carbonizing temperature after activation and time of carbonization. The article presents a procedure to produce nano-silica powder from rice husk along with wastewater treatment grade of activated carbon.

Keywords: Activated carbon; Pretreatment; Optimum condition; Laboratory scale; Nano-silica powder; Carbonizing temperature

Introduction

Rice husk, an agro waste material, contains about 20% ash which can be retrieved as amorphous, chemically reactive silica. This silica finds wide applications as filler, catalyst support, adsorbent and a source for synthesizing high performance silicon and its compounds. Various metal ions and unburned carbon influence the purity and color of the ash. Controlled burning of the husk after removing these ions can produce white silica of high purity. India produces around 25 million tons of rice husks (widely available waste). 78% of weight as rice, broken rice and bran, rest 22% of weight of paddy as husk. 75% of organic volatile matter and 25% of weight is converted as Rice Husk Ash (RHA) during firing process. Husk contains 17%-20% silica in complex form and RHA contains 85%-95% amorphous silica. RHA is a great environment threat causing damage to the land and the surrounding area in which it is dumped.

The study focuses on samples from the state of Andhra Pradesh Rice Husk (APRH) in the central part of India and the other from Kerala (KRH) the southernmost part of the country. Leaching the husk with acetic and oxalic acids was attempted for the first time and the improvement in properties of the ash was studied. The husk samples were also treated with hydrochloric and nitric acids of different concentrations for comparison. The ashes produced by controlled burning of these samples before and after acid treatment, were characterized for the optical properties in addition to the chemical and physical nature. The APRH ash was found to be inferior to the KRH ash in all properties. Pretreatment of the husks with the organic acids improved the properties of ashes and the effect was comparable to that achieved by mineral acid leaching. Amorphous, reactive and high purity silica with high surface area and pore volume and good optical properties could be prepared from both the husks under specific conditions. Rice husk ash or silica ash, as it is commonly called, is classified as an industrial waste obtained after burning the rice husks. It has approximately 55-97% silica in partly crystalline and amorphous forms, depending upon the prior combustion conditions. Though there are limitations in its application as filler in thermoplastics [1,2], its usage in various industries varies depending upon its purity and particle characteristics [3,4]. Industrial by-products (e.g., saw dust, rice husks) and a recent entry in the form of silica ash – an industrial waste material – obtained by burning rice husks. Rice hulls possess an unusually high percentage of 'opaline silica'. Its annual worldwide output is more than 80 million tons, which corresponds to 3.2 million tons of silica. Rice husk on burning gives 14–20% ash which contains 80–95% silica in the crystalline form and minor amounts of metallic elements. By controlling the burning conditions like temperature and time, amorphous silica of ultrafine size and reactivity can be produced. Pretreatment of the husk with mineral acids followed by controlled ashing gives silica with high purity. Rice husk is a byproduct in rice mills and creates disposal and pollution problems. It is reported that about 30 million tons of rice husk per annum is produced in India. Therefore, an efficient utilization of RH is urgently needed. A number of papers have been published on various aspects of rice husk [5-7]. The major constituents of rice husk are cellulose, lignin and ash [7]. The chemical constituents is found, it varies from sample to sample which may be due to the different geographical conditions, type of paddy, climatic variation, soil chemistry and fertilizers used in the paddy growth (Table 1). The silicon atoms are concentrated in the protuberances and hairs

Constituents	weight
Silica (SiO_2)	94.50
Calcium oxide (CaO)	00.48
Manganese oxide (MnO)	01.09
Magnesium oxide (MgO)	00.23
Iron oxide(Fe_2O_3)	00.54
Aluminum oxide(Al_2O_3)	00.21
S,P_2O_5,K_2O,Na_2O	Traces

Table 1: A typical composition of rice husk

***Corresponding author:** Rajesh Ghosh, Department of chemical engineering, Calcutta Institute of Technology, Uluberia, Howrah, West Bengal-711316, India on the outer and inner epidermis of the husk [8]. The rice husk ash is used as a good pozzolana in cement industry [9-11] and also as a support material for metal catalysts. Preparation of different value added products like SiC, porous carbon, zeolites, cordierite, etc from rice husk ash has been reported [12-19]. It has extensive uses as filler, additive, abrasive, oil adsorbent, sweeping component, suspension agent for porcelain enamels etc. It is also used for soil treatment, water purification, glazing pottery ware, in ceramics and refractory materials, making special quality bricks [20-24].

History of Contribution/Background

Silica is not a new commodity in the plastics market. Its usage as extenders and reinforcing fillers, as pozzolanic material and as glass-microspheres for specific engineering applications are well known in the market [25-28]. Because of its high silica and lignin content, rice husk is tough, woody and abrasive in nature with low nutritive properties and resistance to weathering [29]. With growing environmental concern, open burning has been outlawed in many major rice-producing countries (Figure 1). World rice production shown in percentage wise, are obtained from the U.S. department of agriculture and show that total production is around 80 million ton per annum. The U.S. department of agriculture shows that about 27 million metric ton of milled rice was globally traded for the year 2002. Based on 20% silica content of the rice hulls, it is estimated that effectively 1.2 MT (million ton) of silica ash can be produced from commercial rice. Burning rice hulls - as a preparative step for energy production - is a useful solution, but the desirable situation would be a better economic use of the resulting ash. Silica is the predominant component of the ash with trace amount of various elements such as potassium, sodium, magnesium and calcium (<0.5%). Rice productions around the world in 1999 [1.CHINA-40%, 2.INDIA-20%, 3.INDONESIA-10%, 4. OTHERS-20%].

Silica ash composites It is understood that characteristics such as a fine particle size (<50 micron) and high reactivity (due to its irregular shape, porosity and presence of surface hydroxyl groups) mar the performance of silica ash as a filler. Also, impurities reduce its efficiency as filler in composites. The impurity level of silica is significantly affected by the production mode. With open air burning, a variety of silica ash can be produced with acceptable level of carbon impurities [30]. Amorphous white (low-carbon) silica ash is obtained at temperatures above 500°C and depends on the time and the temperature of combustion [31]. It was postulated that at high temperature (800°C), the presence of potassium led to surface melting and carbon fixation in the lattice [32]. The effect of treating the husk with dilute HCl on the formation of black particles has been recently reported by him. Sidheswaran et al. [33] prepared pure white amorphous silica by pre-treating the husk with HNO_3 followed by ashing and measured the whiteness using a color scanner. Amorphous silica of high purity, small particle size and high surface area has tremendous potential as an adsorbent and catalyst/support in fine chemical synthesis. In order to prepare amorphous silica of high purity, treatment of the husk with chemicals before and after combustion was attempted by various researchers.

Figure 1: Major rice producing states in India.

Mineral acid leaching of rice husk is reported by Chakraverty et al. [31]. They have used acids like HCl, H_2SO_4, and HNO_3 of varying concentrations. They studied the effect of various acid treatments on the removal of metallic ingredients and reported that leaching with 1N HCl is highly effective. Yalcin et al. [34] have described pre and post treatments of husk using HCl, H_2SO_4 and NaOH. Russian inventor Raphael Osterjkio-discovered that adsorptive properties of char could be improved by treatment at higher temperatures in presence of air, steam or CO_2.

Lowitz in 1785 showed-activated carbons could be successfully used for decolorization (utilized in sugar industry). Present day carbons are produced from coal based materials like lignite, pine charcoal, soft-wood charcoal, teak waste, deoiled sandal-wood dust, coconut shells, cotton seeds hulls, and rice husk. Adsorptive properties of activated carbon can be explained as the formation of a complex network of very fine pores during its manufacture. The internal surface area of these pores contributes to a very high total internal surface area in the range of 1000 m^2/gm or more.

Materials & Methods

Analysis of rice husk ash

Precipitation of Silica - Chemical Reaction Involved

$SiO_2 + 2NaOH \rightarrow Na_2SiO_3 + H_2O$

$Na_2SiO_3 + H_2SO_4 \rightarrow SiO_2 + Na_2SO_4 + H_2O$

General process for precipitation of silica and activated carbon from rice husk

A known weight of rice husk is taken and sieved through 20 mesh size. It is then washed, cleaned and heated/carbonized to around 600°C for different time intervals (figure 2). It is treated with an activating reagent such as sodium hydroxide, zinc chloride, phosphoric acid for about 1 hour at 60°C. The activated rice husk and the solution containing the activating agent are then filtered. The activated rice husk so obtained is heated at a temperature of 900°C to get activated carbon. The filtered

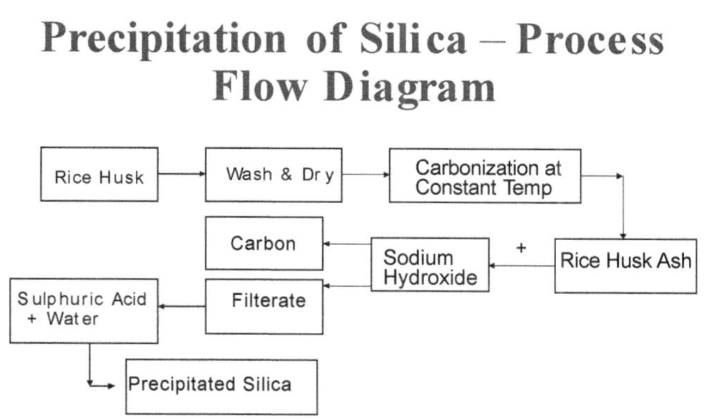

Figure 2: Precipitation of Silica – Process flow diagram.

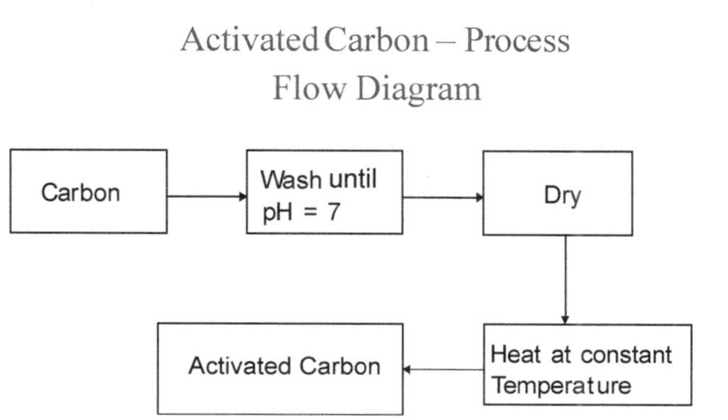

Figure 3: Activated carbon-Process flow diagram.

solution containing the activating agent is titrated with acid to precipitate silica powder.

The temperature at which rice husk is carbonized is very important since the surface area characteristic of ash depends on the temperature of formation of ash. The ash is obtained as follows rice husk is cleaned from dirt by sieving it with 20 mesh sieve. The rice husk ash so obtained produces a lot of smoke if kept directly in the furnace. In order to avoid this, the husk is first charred to black mass on a Bunsen burner. The mass is then heated & oxidized in an electric furnace at controlled temperature to obtain the ash. The ash obtained is finely ground & sieved through 150 mesh sieve. The fraction passing through this sieve is then used for experimental run. The charred husk fraction is further carbonized at 550°C, 600°C, 650°C, and 700°C for 3 hours.

The activation of the rice husk ash is done either by thermal or chemical activation. In thermal activation (eg: coal, lignite and coconut shell), raw material is carbonized to drive off the gases. A char of high carbon content is obtained. This char is treated with air at 450 to 500°C or with steam or carbon dioxide (CO_2) in the temperature range of 880 to 1100°C. In activation step the oxidizing gas reacts with carbon atoms in the interior of char to form gaseous reactant products and develop the channels and pores that contributes to high internal surface area.

In chemical activation, dehydrating agents such as zinc chloride is used to chemically decompose cellulose of the raw material (figure 3). The raw material is mixed with the chemical agent and soaked for known amount of time. It is then dried and carbonized in the absence of air. Obtained char is treated with alkali to form sodium silicate. Sodium silicate is treated with mineral acid and water to get precipitated silica. Filtrate is washed until pH is neutral.

Following are the sieving mesh sizes, composition of rice husk and the parameter employed in the procedure:

Rice Husk gradation (sieving)

Rice Husk is sieved using 20-200 mesh size sieves. The major fraction, which usually consists of 20 mesh size, is used for the experiment.

Composition of rice husk ash

SiO_2=80-90 %

Alumina=1-2.5%

Ferric oxide=0.5%

Trace elements=remaining

Carbonization of rice husk

The rice husk obtained above in step 1 is first carbonized for 2 hours at and above 700°C. If the temperature is below 500°C, the rice husk is not carbonized completely.

Effect of temperature

Below 500°C incomplete carbonization, at and above 700°C, nano-silica obtained.

Effect of time length

Minimum time duration is 2 hours

Treating with activating agent

The carbonized rice husk is then treated with activating agents such as listed below at 0.2 N for a minimum time of 2 hours at 60°C. This makes the silica soluble in the activating agent and subjects the rice husk to have more surface area during the carbonization step mentioned below.

Types of activating agents used

Sodium hydroxide, phosphoric acid and Zinc chloride

a) **Concentration:** Sodium Hydroxide - 0.2 N

b) **Length of time:** Minimum 2 hours

c) **Temperature:**

Rice Husk on being treated with activating agent - 60°C

Temperature of carbonization – 500°C, 600°C, 650°C, 700°C

Concentration of sodium hydroxide – 5%, 10%, 15%, 20% (w/v)

Temperature of activation – 800°C, 850°C, 900°C

Precipitated silica

The silica is precipitated by adding concentrated sulfuric acid to the filtered solution of Step 3 at room temperature. This yields very pure silica. The types of silica obtained depend on the carbonization temperature and the any pre-treatment before carbonization. The yield is between 95-98% of the amount of silica present in the initial rice husk.

A. Acid used: Sulfuric acid

B. Concentration used: Concentrated

C. Temperature: Room temperature

D. Types of silica obtained: Depends on carbonization temperature

E. Purity: Very pure silica can be obtained (Rice husk is burnt under controlled conditions gives amorphous silica of high purity)

F. Size: Nano-silica, powder, different grades of silica (small particle size)

G. Yield: 95-98 %

H. Surface area:

1. Nature: Amorphous powder
2. Appearance: White fluffy powder
3. Purity: > 98 %
4. Surface Area: 150 - 200 m2/gm
5. Bulk density: 120 - 200 g/liter
6. Loss on Ignition: 3.0 - 6.0 %
7. pH of 5 % slurry: 6.3 + 0.5
8. Heat loss: 4.0 - 7.0 % the properties like surface area, pH, Tap density can be tailor made for the requirement.

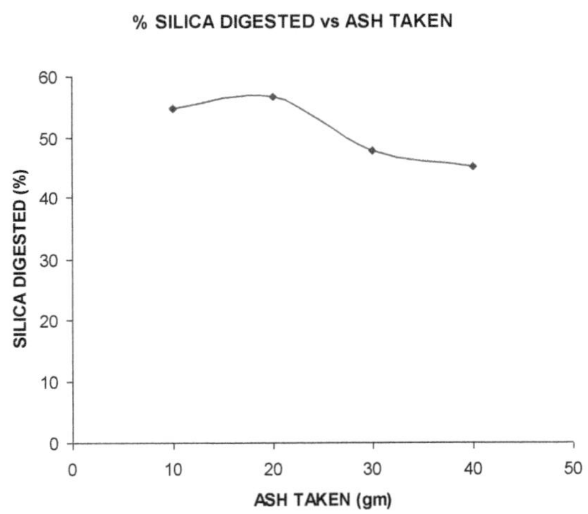

Figure 4: Digestion of rice husk ash for 1 hour in 0.5N NaOH solution.

Activated carbon

The activated rice husk which has been treated with sodium hydroxide in Step 2 and 3 is further carbonized at temperatures of 700 to 900 °C for 2 hours to obtain activated carbon of wastewater grade.

A. Temp effect: 700-900 °C

B. Duration/Time: Minimum 2 hours

C. Types/grades of carbon obtained: Mechanical grade and electronic grade

D. Surface area: High surface area

Application

Activated carbon obtained by the above procedure can be used in wastewater treatment as an adsorbent to remove organics. Elemental silica is used as a constituent of building material. In amorphous form it is used as desiccant, adsorbent, refining agent filter, catalyst compound, etc. In vitreous form, it is used in the manufacture of optical elements and glass wires. It is also used as a basic material for glass, ceramic and refractory industries. Other applications include in the manufacture of soluble silicate, silicon and its alloy, silicon carbide, silicon based chemicals and the silicones. It also finds uses in reinforcing filter, in natural and synthetic rubber and is used in adhesives to enhance bond strength. Other uses include control of porosity in battery separators, for controlling oil spillage, used in brake linings as anti-caking agent. It has also found use in cosmetics as an absorbent and thickening agent, in inks as thickening and matting agent, in tooth paste as an abrasive additive powder and in paper as filler for opacity and brightness, polishes as a mild abrasive. It can be used as a catalyst support for fine chemical synthesis.

Results and Discussion

Although there are different procedures to obtain silica from rice husk, the most important step is to first treat the rice husk with acid treatment to remove any impurities such as metals that are present in the rice husk. The acid treatment also gives a high surface area for the silica when it is precipitated. The next important aspect of producing nano-silica powder depends on the carbonization temperature and duration of carbonization. Different grades of silica with different yields can be obtained based on the carbonization temperature and duration of carbonization. The activating agent, temperature and time of heating the pre-acid washed rice husk, concentration of activating agent, all influence the yield, surface area of the precipitated nano-silica.

Sl no	Ash taken (gm)	NaOH taken (gm)	Vol of NaOH (ml)	Vol of silicate (ml)	Specific gravity	% silica in solution	Undigested ash (gm)	% silica digested
1	10	10	500	450	1.026	1.07	2.2	54.89
2	20	10	500	415	1.033	2.2	4.7	56.82
3	30	10	500	380	1.036	2.78	11.2	48
4	40	10	500	310	1.038	3.49	15.2	45.28

Table 2: Digestion of rice husk ash for 1 hour in 0.5n NaOH solution.

Sl no.	Ash taken (gm)	NaOH taken (gm)	Vol of NaOH (ml)	Vol of silicate (ml)	Specific gravity	% silica in solution	Undigested ash (gm)	% silica digested
1	10	20	500	490	1.04	1.23	1.7	63.96
2	20	20	500	450	1.052	2.33	3.6	61.28
3	30	20	500	375	1.06	3.58	6	63.24
4	40	20	500	340	1.07	5.11	8.5	68.35
5	50	20	500	310	1.077	5.66	12.6	60.96

Table 3: Digestion of rice husk ash for 1 hour in 1.0N NaOH solution.

The activated carbon grades obtained depends on the temperature and duration of carbonization after treatment with sodium hydroxide (Table 2, 3; Figures 4, 5). Also, the concentration of sodium hydroxide, time and temperature of heating produces different grades of activated carbon.

After carrying out a series of experiments and analysis it can be seen that the most optimized condition for the digestion of rice husk in sodium hydroxide solution are as follows:

- Concentration of NaOH solution: 1.5N (Tables 4, 5, Figure 6)
- Weight of ash: 120 gm/lit
- Temperature: 95 degree centigrade
- Time of digestion: 1 hour

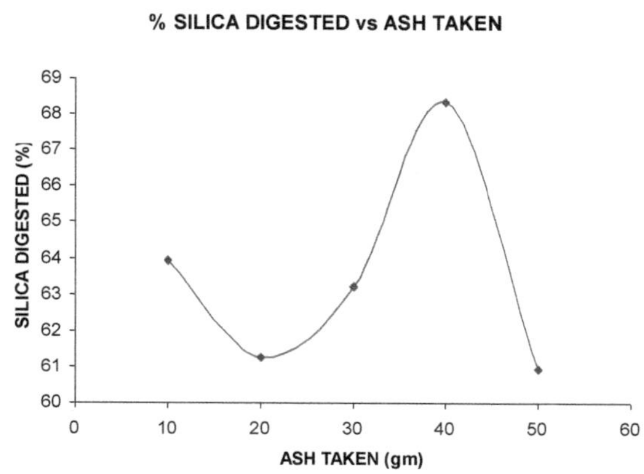

Figure 5: Digestion of rice husk ash for 1 hour in 1.0N NaOH solution.

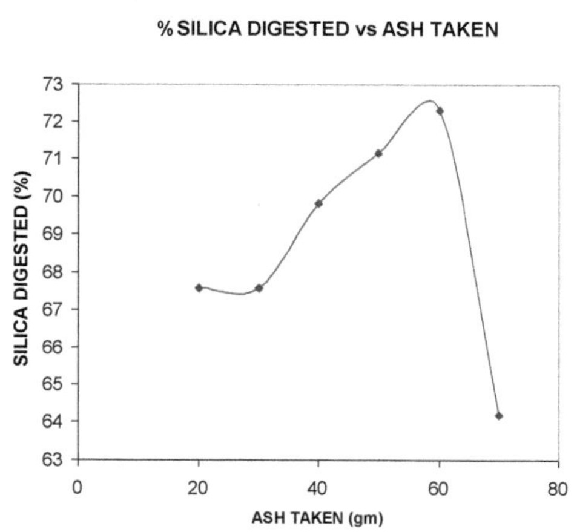

Figure 6: Digestion of rice husk ash for 1 hour in 1.5N NaOH solution.

To study the effect of temperature and time on the digestion of rice husk ash in sodium hydroxide solution (Table 6), further studies were carried out (Figure 7, 8).

Conclusion

- The procedure adopted in this project to prepare precipitated silica from rice husk ash was successful and practically very sound. This innovative idea can be used out in future in the precipitated silica industry. The manufacture of silica from rice husk ash works out to be very cheap and cost effective as the main raw material, the rice husk ash can be obtained at low cost. Activated carbon obtained as the byproduct has a good market value.
- A general procedure is outlined for silica precipitation and activated carbon from rice husk. The most important aspect is nano-silica powder is obtained (Table 7, Figure 9). The effect of different parameters like temperature of carbonization, temperature of mixture feed into the reactor, production of ash & conversion of silica has been reviewed. Precipitation of Silica carried out at 500°C, 600°C, 650°C, 700°C using 5%, 10%, 15% and 20% Sodium Hydroxide–Analysis awaiting
- Silica present in Rice Husk=30%
- Activated Carbon obtained at 850°C and 900°C–Analysis awaiting
- Precipitated silica and activated carbon was obtained from rice husk using the process outlined
- Silica in rice husk was 30%
- Analysis of yields of silica and activated carbon under progress

Sl.No.	Ash Taken (Gm)	NaOH Taken (Gm)	Vol of NaOH (Ml)	Vol of Silicate (Ml)	Specific Gravity	% Silica in Solution	Undigested Ash (Gm)	%Silica Silica Digested
1	20	30	500	425	1.077	2.44	3.8	67.58
2	30	30	500	400	1.08	3.76	5.3	67.59
3	40	30	500	355	1.09	4.97	9.4	69.84
4	50	30	500	340	1.101	6.152	9.5	71.17
5	60	30	500	295	1.105	7.64	14.1	72.32
6	70	30	500	270	1.113	7.52	13.7	64.19

Table 4: Digestion of rice husk ash for 1 hour in 1.5N NaOH solution.

Sl no	Ash taken (gm)	Naoh taken (gm)	Vol of naoh (ml)	Vol of silicate (ml)	Specific gravity	% silica in solution	Undigested ash (gm)	% silica digested
1	40	40	500	375	1.114	4.75	6.8	63.46
2	50	40	500	340	1,124	6.33	9.7	66.14
3	60	40	500	305	1.13	7.12	12.4	67.089
4	70	40	500	345	1.132	8.49	15.3	68.66
5	80	40	500	275	1.14	8.907	20.1	71.19
6	90	40	500	260	1.146	9.829	22.5	62.57

Table 5: Digestion of rice husk ash for 1 hour in 2N NaOH solution.

Scope for future work

Effect of three parameters (acid treatment, temperature and heating rate) on the formation of black particles in rice husk silica ash was studied. However, at low temperature (400°C) the oxidation of carbon in the acid treated rice husk is sluggish. Several efforts are being made to discover a suitable technology capable of producing large quantities of solar grade silicon at a cost less than $10/kg. An industrial by-product can be converted to a high value added product by using a simple inexpensive method. A powder consisted of 92% of silica was prepared after calcinations of RHA at 700°C for 5 hour. Silica from rice husk can be prepared under controlled temperature without any chemical treatment, the bulk of investigations indicate that producing silica ash composites directly or into related silica filler composites, suggest that improvement can be achieved in the composite performance by modifying the physical and chemical nature of the filler. The extent of improvement, however, is greatly influenced by the ability to control the many relevant characteristics of the filler. A systematic study from a thermodynamic point of view is expected to provide the user with a better understanding of the parameters that are significant & relevant. The husks on leaching with acetic and oxalic acids of different concentrations followed by thermal treatment resulted in products with improved properties like purity, reactivity, brightness, surface area and pore volume. The leaching with mineral acids like HCl and HNO_3 of different concentrations followed by thermal treatment and measurement of properties showed that the acid treatment is almost equally effective in improving the silica properties.

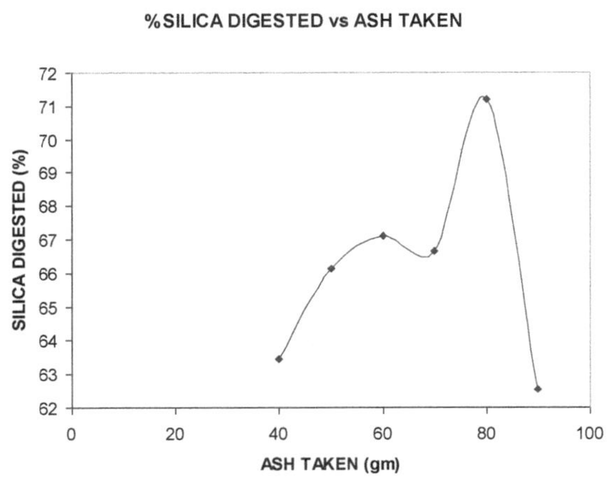

Figure 7: Digestion of rice husk ash for 1 hour in 2N NaOH solution.

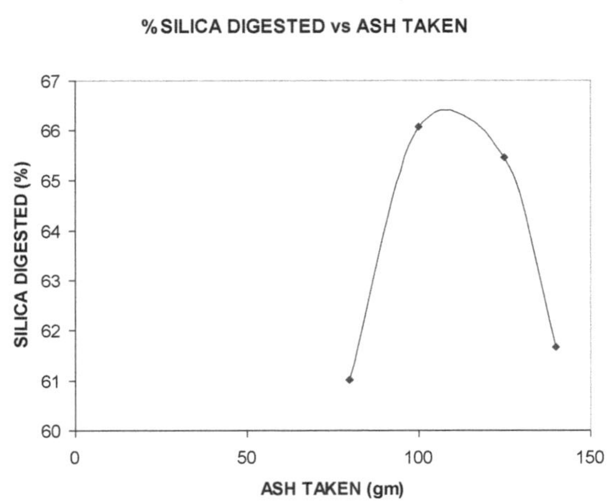

Figure 8: Digestion of rice husk ash for 1 hour in 2.5N NaOH solution.

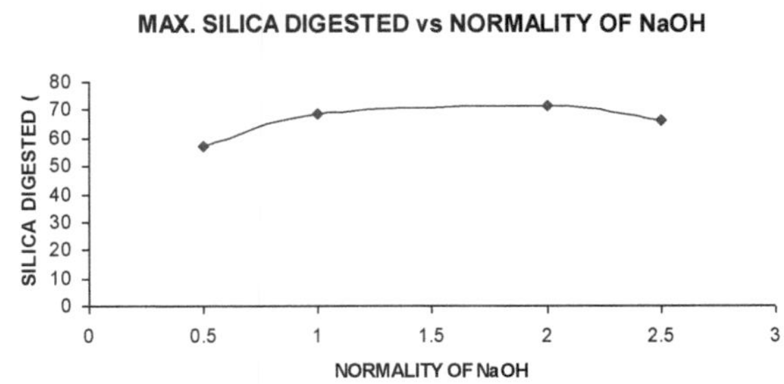

Figure 9: Comparisons of the peak values of %silica digested.

Acknowledgement

The authors thank to Dr. Ashoke Kumar Biswas, former HOD of Chemical Engineering dept. IIT Kharagpur for his constant encouragement and valuable suggestions. The author Rajesh Ghosh expresses his heartfelt gratitude and sincere thanks to Dr. Balaji Krishna Murthy, Professor of chemical engineering dept. BITS PILANI, and Hyderabad for his constant support.

References

1. Rozman HD, Lee MH, Kumar RN, Abusamah A, Ishak ZAM (2000) The Effect of Chemical Modification of Rice Husk with Glycidyl Methacrylate on the Mechanical and Physical Properties of Rice Husk-Polystyrene Composites. Journal Of Wood Chemistry And Technology 20: 93-109.

Sl no	Ash taken (gm)	Naoh taken (gm)	Vol of naoh (ml)	Vol of silicate(ml)	Specific gravity	% silica in solution	Undigested ash (gm)	% silica digested
1	80	50	500	290	1.162	8.402	14.7	61.02
2	100	50	500	250	1.184	11.16	20	66.08
3	125	50	500	185	1.2	13.64	32.3	65.472
4	140	50	500	110	1.202	14.37	57.7	61.68

Table 6: Digestion of rice husk ash for 1 hour in 2.5N NaOH solution.

Sl No.	Normality of NaOH	Ash taken	Silica digested
1	0.5	20	56.82
2	1	40	68.35
3	1.5	60	70.35
4	2	80	71.19
5	2.5	100	66.08

Table 7: Comparisons of the peak values of %silica digested.

2. Ismail H, Mega L, Khalil HPSA (2001) Effect of a silane coupling agent on the properties of white rice husk ash–polypropylene/natural rubber composites. Polymer International 50: 606-611.
3. Seymour RB (1990) Polymer Composites. Utrecht, The Netherlands 45.
4. Sae-Oui P, Rakdee C, Thanmathorn P (2002) Use of Rice Husk Ash as Filler in Natural Rubber Vulcanizates: In Comparison with Other Commercial Fillers. Journal of Applied Polymer Science 83: 2485-2493.
5. Tzong-Horng Liou (2004) Preparation and characterization of nano-structured silica from rice husk. Mat Sci Eng A 364: 313-323.
6. Prasad CC, Maiti KN, Venugopal R (2001) Ceram Intern 27: 629.
7. Govinda Rao VMH (1980) J Sci Ind Res 39: 495.
8. Park BD, Wi SG, Lee KH, Singh AP, Yoon TH, et al. (2003) Characterization of anatomical features and silica distribution in rice husk using microscopic and micro-analytical techniques. Biomass And Bioenergy 25: 319-327.
9. Nehdi M, Duquette J, Damatty El (2003) Performance of rice husk ash produced using a new technology as a mineral admixture in concrete. Cement concr Res 33: 1203-1210.
10. Stroeven P, Bui DD, Sabuni E (1999) Ash of vegetable waste used for economic production of low to high strength hydraulic binders. Fuel 78: 153-159.
11. Zhang MH, Lastra R, Malhotra VM (1996) Rice-husk ash paste and concrete: Some aspects of hydration and the microstructure of the interfacial zone between the aggregate and paste. Cement Concr Res 26: 963-977.
12. Karera A, Nargis S, Patel S, Patel M, (1986) J Sci Industr Res 45: 441.
13. Krishnarao RV, Godkhindi MM (1992) Effect of Si3N4 additions on the formation of SiC whiskers from rice husks. Ceram Intern 18: 185-191.
14. Tzong-Horng Liou (2004) Evolution of chemistry and morphology during the carbonization and combustion of rice husk. Carbon 42: 785-794.
15. Guo Y, Yu K, Wang Z, Xu H (2000) Effects of activation conditions on preparation of porous carbon from rice husk. 41: 1645-1648.
16. Wu M, Zha Q, Qiu J, Guo Y, Shang H, et al. (2004) Preparation and characterization of porous carbons from PAN-based preoxidized cloth by KOH activation. Carbon 42: 205-210.
17. Naskar MK, Chatterjee M (2004) A novel process for the synthesis of cordierite (Mg2Al4Si5O18) powders from rice husk ash and other sources of silica and their comparative study. J Eurp Ceram Soc 24: 3499-3508.
18. Dalai AK, Rao MS, Gokhale KVGK (1985) Synthesis of NaX zeolite using silica from rice husk ash. Ind Eng Chem Prod Res Dev 24: 465-468.
19. Wang HP, Lin KS, Huang YJ, Li MC, Tsaur LK (1998) Synthesis of zeolite ZSM-48 from rice husk ash. J Hazard Mater 58: 147-152.
20. Suwanpratab J, Hatthapanit K (2002) Rice-husk-ash-based silica as a filler for embedding composites in electronic devices. J Appl Polym Sci 86: 3013-3020.
21. Ismail H, Nasaruddin MN, Ishiaku US (1999) White rice husk ash filled natural rubber compounds: the effect of multifunctional additive and silane coupling agents. Polym Test 18: 287-298.
22. Ishak ZAM, Abubaker A, Ishiaku US, Hashim AS, Azahari B (1997) An investigation of the potential of rice husk ash as a filler for epoxidized natural rubber—II. Fatigue behavior. Europ Polym J 33: 73-79.
23. Ishak ZAM, Baker AA (1995) An investigation on the potential of rice husk ash as fillers for epoxidized natural rubber (ENR). Euro Polym J 31: 259-269.
24. Mbui DN, Shiundu PM, Ndonye RM, Kamau GN (2002) Adsorption and detection of some phenolic compounds by rice husk ash of Kenyan origin. J Environ Monit 4: 978-984.
25. Kamath SR, Proctor A (1998) Silica Gel From Rice Hull Ash: Preparation and Characterization. Cereal Chemistry 75: 484-487.
26. Anon, Enr (1994) (Engineering News-Record).
27. Robert B (1997) Paper 5 at Addcon Asia Seminar, Westin Plaza, Singapore 28-29.
28. Bledzki A, Kwasek A, Spychaj S (1985) Hollow Glass Microspheres Used As Fillers in Thermosetting Plastics. Kunststoffe 21: 75.
29. Juliano, Bienvenido O (1985) Rice: Chemistry And Technology. (2nd Edn), St. Paul, Minnesota, USA, AACC.
30. Liou TH, Chang FW, Lo JJ (1997) Pyrolysis Kinetics Of Acid-Leached Rice Husk. Industrial Engineering & Chemical Research.
31. Chakraverty A, Mishra P, Banerjee HD (1988) Investigation Of Combustion Of Raw And Acid-Leached Rice Husk For Production Of Pure Amorphous White Silica. J Mater Sci 23: 21-24.
32. Krishnarao RV, Subrahmanyam J, Jagadish-Kumar T (2001) Studies on the Formation of Black Particles in Rice Husk Silica Ash. Journal European Ceramic Society 21: 99-104.
33. Sidheswaran P, Bhat AN (1996) Trans Ind Ceram Soc 55: 93.
34. Yalcin N, Sevinc V (2001) Studies on silica obtained from rice husk. Ceram Int 27: 219-224.

Processing of Barium Sulphide to Barium Carbonate and Sulphur

M Masukume[1]*, J P Maree[2], S Ruto[1] and H Joubert[3]

[1]*Department of Chemical and Metallurgical Engineering, Tshwane University of Technology, Pretoria, South Africa*
[2]*Department of Environmental, Water and Earth Sciences, Tshwane University of Technology, Pretoria, South Africa*
[3]*Africa Remediation Technologies (Pty) Ltd, Cullinan, Pretoria, South Africa*

Abstract

Gauteng Province, South Africa will have a shortfall of 400 ML/d of fresh water by 2014. This shortfall can be made up either by purchasing additional expensive water from the Lesotho Highlands Scheme or by; treating acid mine water. The CSIR ABC (alkali-barium-carbonate) Process is a most cost-effective process as the sludge produced is processed back to the process raw materials and valuable by-products, such as sulphur. Coal is the main material used in processing the sludge.

The chemical processing of barium sulphide into barium carbonate and elemental sulphur has been demonstrated on a pilot scale in batch mode. The barium sulphide was reacted with carbon dioxide at ambient temperature and atmospheric pressure to produce $BaCO_3$ and H_2S. The H_2S generated, was contacted and oxidized with ferric iron solution to form elemental sulphur. Over 80% of the barium reacted during the carbonation process while all the sulphide was converted to H_2S. The reaction rate of barium carbonation was found to increase with increase in CO_2 flowrate. The pilot plant has produced $BaCO_3$ and sulphur of high quality compared with commercial products in terms of reactivity and purity, respectively. The $BaCO_3$ produced was used in acid mine drainage treatment and lowered sulphate concentration from around 2500 mg/L to 415 mg/L. The outcome of this project will assist in solving the problem of economic acid mine drainage treatment, using regenerated barium carbonate. Moreover, the project also aims to enable the generation of income through recovery of valuable by-products.

Keywords: Dissolution; Carbonation; Barium carbonate; Acid mine drainage; Sulphur

Introduction

Acid mine drainage (AMD) continues to be a serious water pollution problem in South Africa, owing to its well documented adverse effects on the environment. While there are several AMD treatment technologies that are being proposed, that include ion exchange [1] adsorption [2] freeze crystallization [3], reverse osmosis [4] and biological treatment [5-7], finding a robust and economic AMD treatment technology is still continuing. Each of these technologies has its own advantages and disadvantages which tend to limit its use.

One of the technologies with the potential of solving AMD treatment problem is the South African Council for Scientific and Industrial Research's (CSIR) Alkali-Barium Carbonate (ABC) process [8,9]. The CSIR has reported on the recovery of drinking water from Gold mine effluents via the ABC desalination process. Hence AMD could be a possible source of fresh drinking water. Moreover, the barium carbonate used in this process is recyclable by barium sulphate calcining in the presence of coal and subsequent CO_2 treatment of the BaS. This process not only has the benefit of generating income through the recovery of saleable by-products, thereby reducing operational costs of treatment but also allows the regeneration and reuse of a process chemical, barium carbonate. When the value of treated water and by-products exceeds the treatment cost, it is feasible to create enterprises that could provide economic benefits while solving environmental problems.

In this study, the processing of barium sulphide produced by the thermal reduction of barium sulphate sludge from the ABC process was carried out on pilot scale. The process was made up of three phases: (i) biological oxidation of ferrous iron, followed by, (ii) barium sulphide carbonation and finally, (iii) the absorption of hydrogen sulphide gas by ferric iron solution to form elemental sulphur. Sulphur is a key raw material for many manufacturing industries such as fertilizer, acids, steel, petroleum etc. Nengovhela et al. [10] have reported on the recovery of sulphur from gypsum on a laboratory scale.

Materials and Methods

Major reactors used and procedures followed during this study

Feeder: The feeder (250L) was made of chemically inert stainless steel with a turbulent stirrer of high speed. This is the stage where crude barium sulphide was dissolved in tap water (room temperature) by using a high powered stirrer and the remaining ash leached. Two batches were prepared to make up a volume of 500 L barium sulphide solution. After 10 min of mixing, the solution was allowed to settle for 5 min. The supernatant was transferred by pump to the barium carbonate reactor. The ash was removed from the feeder after every run.

Barium carbonate reactor: The reactor (1000L) was made of fibre glass and had maximum pressure and temperature ratings of 4 bars and 50 °C, respectively. The barium carbonation process is exothermic hence fibre glass was used to allow heat loss at a faster rate. Depending on the initial mass of BaS, the solution temperature rose to a maximum of 38 °C. The barium sulphide solution was thoroughly mixed before carbon dioxide was introduced. The solution was continuously circulated using the pump. Carbon dioxide was introduced at the bottom of the reactor from a gas cylinder at a given flowrate. Quality control checks were done on this solution as the reaction proceeded; these included pH and conductivity, sulphide and barium concentration. Samples were collected at 30 minute time intervals. As the reaction proceeded, hydrogen sulphide gas was produced and stripped off through the opening on top of the reactor. The sour gas (H_2S + CO_2 + water vapor).

***Corresponding author:** Mike Masukume, Department of Chemical and Metallurgical Engineering, Tshwane University of Technology, Pretoria, Private bag X680 Pretoria 0001

The sour gas was led by means of a PVC pipe to the ferric iron solution. The barium sulphide containing supernatant was reused to prepare a further batch of barium sulphide solution.

Hydrogen sulphide scrubber: The scrubber (1000 L) was made of fibre glass and had maximum pressure and temperature ratings of 4 bar and 50°C, respectively. The barium carbonation process is exothermic hence fibre glass was used to allow heat loss at a faster rate.

The toxic hydrogen sulphide gas was contacted with the ferric iron solution at the bottom of the reactor to ensure a longer contact time and the excess gas was recycled back to the solution by a venturi. The reaction resulted in the reduction of ferric to ferrous iron with the subsequent production of elemental sulfur. The solution pH, conductivity and ferrous iron concentration were monitored as the reaction proceeded. At the end of the reaction, the solution was transferred to the sulphur settling tank for decanting.

Biological reactor: The biological reactor was made of polyethylene. About 700 L of 10 g/L ferrous iron solution is transferred to the biological reactor for bio-oxidation using *Acidithiobacillus ferrooxidans* bacteria. The bioreactor contained a geo-textile medium as biomass support to provide enough surface area for the bacteria to grow. The bioreactor operated at ambient temperature while solution pH was around 2. Hydroponic nutrients (300 g) containing mainly nitrogen and phosphate, were introduced at the start of the preparation of each batch. Air was continuously supplied at 56.6 L/min and 206.7 kPa and was evenly distributed through a diffuser at the bottom of the reactor. Ferrous iron concentration was monitored as the oxidation process proceeded. Figure 1 shows a typical flow diagram of the carbonation process.

Analytical Tests

The standard stock solutions were prepared using deionized water. The solutions were stored in glass bottles with glass stoppers. The working solutions were prepared by dilution of standard solutions.

Ferrous iron analyses: Filtered ferrous/ferric solution (10 mL) was pipette into a 100 mL volumetric flask and the solution was made up to the mark using deionized water. 10 mL of this solution was pipette into a 250 mL conical flask. This was followed by the addition of 10 mL of 0.1N sulphuric acid, and10 mL of Zimmerman- Reinhard reagent. The solution was titrated against 0.1N potassium permanganate to a pale pink color.

pH and conductivity measurements: The solution pH and conductivity were both measured using an Orion 4 Star Combined pH and Conductivity meter.

Barium analyses: The barium sulphide containing sample (25 mL) was pipette into a 250 mL conical flask and 75 mL deionized water added. The pH of the solution was adjusted to around 12 using 1N sodium hydroxide solution. This was followed by the addition of 50mg of methylthymol blue indicator. The solution was titrated against 0.01M EDTA to a grey end point.

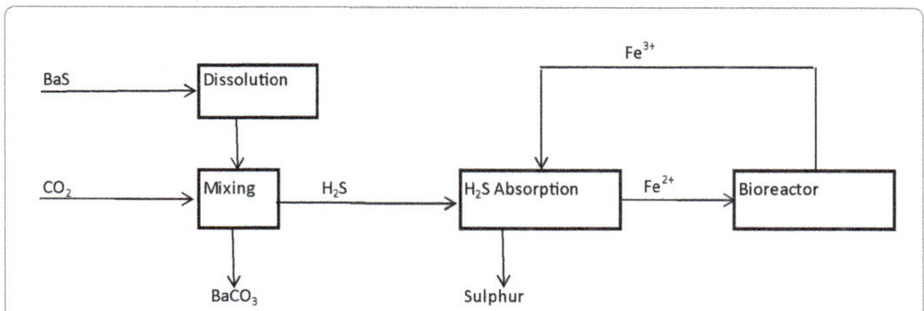

Figure 1: Process flow diagram for the production of $BaCO_3$ and elemental sulphur.

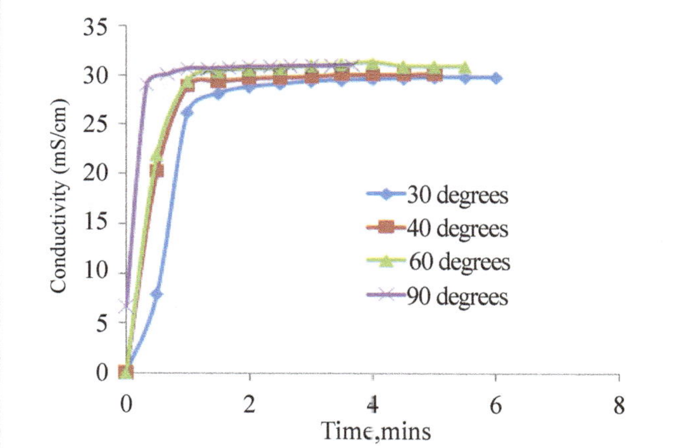

Figure 2: Effect of temperature on dissolution of BaS, Solid Load = 15kg/500 L.

Sulphide analyses: The barium sulphide containing sample (10 mL) solution was pipetted into 100 mL volumetric flask and made up to volume with water. An aliquot of this solution (10mL) was pipetted into a 250 mL conical flask. This was followed by the addition of 10 mL of 0.1N iodine solution and thereafter 5 mL of 1:1 HCl. The solution was titrated against 0.1N sodium thiosulphate to a faint yellow color, where upon, 1% starch was added and the titration continued to a colorless end point.

Sulphur recovery reactions

Dissolution of BaS and reaction with CO_2 to form $BaCO_3$ and H_2S: $BaS + H_2O + CO_2 \rightarrow BaCO_3 + H_2S$ (1)

H_2S stripped and converted to elemental sulphur via the ferric iron route: $H_2S + 2Fe^{3+} \rightarrow S + 2Fe^{2+} + 2H^+$ (2)

Results and Discussion

The processing of barium sulphide into saleable barium carbonate and elemental sulphur was technically demonstrated on pilot scale. The process involved the dissolution of barium sulphide, carbonation of barium sulphide solution and finally oxidation of the hydrogen sulphide by a ferric iron solution. Several control analyses were performed on the solutions during the different stages of the process as described above. The results of these analyses are described below.

Effect of temperature on dissolution of barium sulphide (BaS)

The effect of temperature on dissolution of BaS was monitored by measuring the changes in electrical conductivity of the BaS solution over time as shown in figure 2. While an increase in solution temperature always results in corresponding increases in dissolution rate for endothermic reactions, the results in figure 2 showed an insignificant increase possibly because BaS is readily soluble in water. Due to this small increase in solubility upon heating the barium sulphide solution, all the batches were therefore prepared using tap water at room temperature.

Leaching tests

The leaching tests, the results of which are shown in figure 3, were carried out in order to dissolve all the barium sulphide present together with process ash and to ensure that the residual ash was safe for disposal. The second run (Figure 3) showed that all the BaS was dissolved in the first run and an insignificant amount remained in the ash which posed no risk when disposed to the environment.

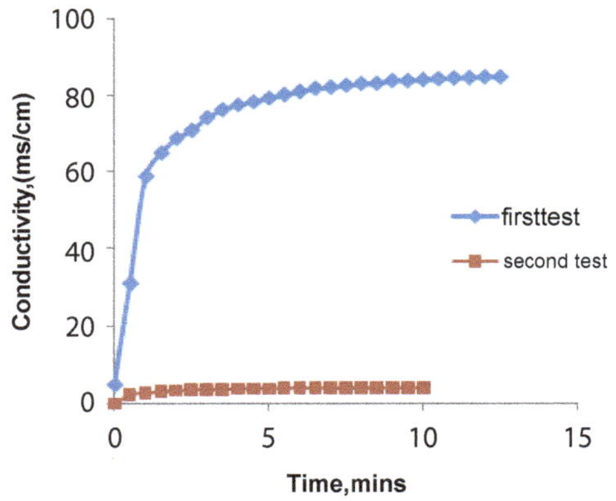

Figure 3: Leaching tests on barium sulphide. Solid load = 40kg/500 L. BaS dissolution was monitored by the electrical conductivity of the solution.

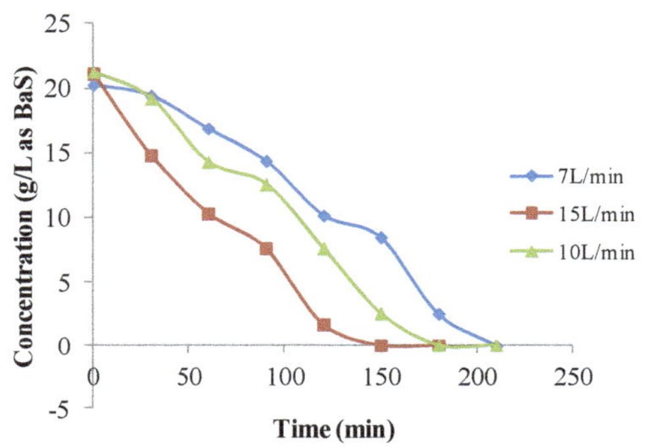

Figure 4: Effect of CO_2 flowrate on carbonation reaction rate.

Efficiency of the barium carbonation process

The barium sulphide solution was reacted with carbon dioxide at ambient temperature to form barium carbonate and hydrogen sulphide. Tables 1 and 2 show the results of chemical analyses checks where the solution pH, conductivity, sulphide and barium concentrations were monitored during barium carbonation and the oxidation of H_2S gas by the ferric iron solution.

The results (Table 1) showed that over 80 % of the dissolved barium reacted with the CO_2 whilst all the sulphide was converted to H_2S. The results (Figure 3) showed that the rate of barium carbonation was dependant on CO_2 flowrate. An increase in CO_2 flowrate from 7 to 15 L/min resulted in a corresponding increase in carbonation reaction rate (Figure 4). Hence CO_2 concentration is the major rate limiting factor in this reaction. A high flow rate of CO_2 also aided in stripping of H_2S from the barium carbonate reactor. As indicated in figure 5, the solution pH decreased as the reaction proceeded and the carbonation process was stopped when the solution pH reached 6.3 (Figure 6). Based on laboratory studies, a solution pH below 6.3 resulted in the undesired formation barium bicarbonate.

Barium carbonate reactivity

The barium carbonate produced was characterised for reactivity. Figure 7 shows reactivity of commercial and Cullinan pilot plant barium carbonate with synthetic calcium sulphate rich water over time. Results from figure 7 showed that the barium carbonate from the Cullinan pilot plant compared well with a commercial product in terms of reactivity.

Efficiency of the barium carbonate produced on AMD treatment

The $BaCO_3$ produced was used in acid mine drainage treatment and lowered sulphate concentration from around 2500 to 415 mg/L. Table 3 shows a summary of the performance of barium carbonate produced on AMD treatment. The sulphate removal efficiency of 83.4% compares well with the results reported employing the CSIR ABC desalination process that utilised commercial $BaCO_3$ as reported by Motaung et al. [9] and Mulopo [11].

Time (min)	CO_2 flowrate (L/min)	Conductivity (Sm/cm)	pH	Ba (g/L)	Ba removal (%)	Sulphide (g/L) as BaS	Sulphide removal (%)
0	7	20.3	12.7	17.6	-	20.2	-
30	7	14	12.1	14.9	15.4	19.4	4.2
60	7	11	8.3	11.5	34.6	16.9	16.7
90	7	8.3	7.7	10.1	42.3	14.4	29.2
120	7	7.2	7	7.4	57.7	10.1	50
150	7	5.5	6.5	6.8	61.5	8.5	58.3
180	7	5	6.4	5.4	69.2	2.5	87.5
210	7	3.4	6.3	3.4	80.8	0	100

Table 1: Results of analytical monitoring during the barium carbonation process.

Time (min)	CO_2 flowrate (L/min)	CO_2 used/Sulphide removed	pH	Conductivity (mS/cm)	Fe(II) (mg/L)	Fe(II) formed (eq)
0	7	-	2.3	8.8	167.6	-
30	7	3.8	2.2	8.9	167.6	-
60	7	1.9	2.2	8.7	167.6	-
90	7	1.6	2.2	8.9	446.8	2.8
120	7	1.3	2.1	10.3	2625	24.2
150	7	1.3	1.6	13.8	3071.8	28.6
180	7	1.1	1.5	16.7	3295.2	30.8
210	7	1.1	1.4	20.5	4132.9	39.1

Table 2: Results of analytical monitoring during oxidation of H_2S by ferric iron solution.

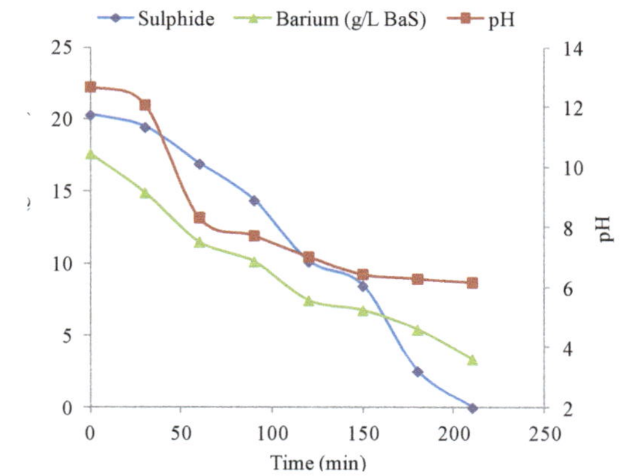

Figure 5: Changes in sulphide and barium concentrations and solution pH as the barium carbonation process proceeded.

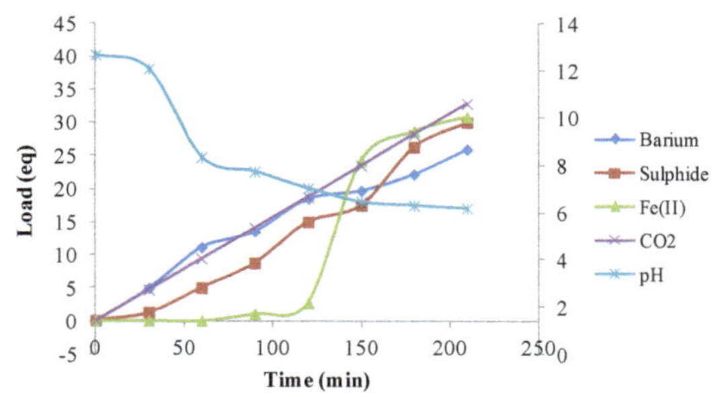

Figure 6: Changes in sulphide, barium and ferrous iron concentrations (eq) and pH over time.

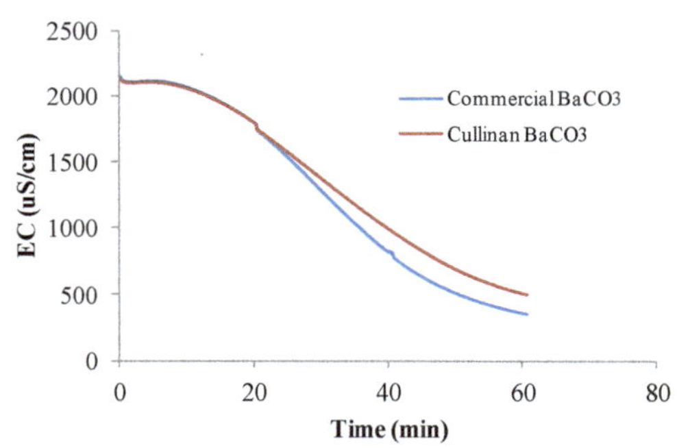

Figure 7: Reactivity of commercial and Cullinan pilot plant barium carbonate with synthetic calcium sulphate rich water over time.

Feed of $BaCO_3$ (ml/min)	**10**	**10**	**10**
Raw water			
pH	7.7	7.9	9.9
Conductivity (mS/cm)	9.4	-	9.5
Sulphate (mg/L)	2485	2500	2500
Ca (mg/L)	3360	-	3270
Mg (mg/L)	32	-	24.5
Treated water			
pH	10.5	10	10
Conductivity (mS/cm)	9	9	9
Sulphate (mg/L)	419	415	410
Ca (mg/L)	820	960	724
Mg (mg/L)	67.4	33.5	37.1
Ba (mg/L)	238.8	239.8	238.6

Table 3: Summary of the performance of barium carbonate produced in the present project on AMD treatment.

Conclusions

A process was investigated for manufacturing barium carbonate by reacting barium sulphide with carbon dioxide, on a pilot scale. The pilot plant produced $BaCO_3$ and sulphur of high quality compared with commercial products.

Acknowledgement

The authors wish to thank the Technology and Human Resources for Industrial Programme (THRIP) and Tshwane University of Technology, for providing financial support.

References

1. Feng D, Alderick C, Tan H (2000) Treatment of acid mine water by use of heavy metal precipitation and ion exchange. Journal of Minerals Engineering 13: 623-642.
2. Hendricks NR (2005) The application of high capacity ion exchange adsorbent material, synthesized from fly ash and acid mine drainage for the removal of heavy and trace metals from secondary co-disposed water. MSc dissertation, University of the Western Cape.
3. Nathoo J, Jivanji R, Lewis AE (2009) Freezing your brines off: Eutectic freeze crystallization for brine treatment. International Mine Water Conference, Pretoria, South Africa.
4. Kentish SE, Stevens GW (2001) Innovations in separations technology for recycling and reuse of liquid waste streams. Chemical Engineering J 84: 149-159.
5. Chang IS, Shin PK, Kim BH (2000) Biological treatment of acid mine drainage under sulphate reducing conditions with solid waste materials as substrate. Water Research 34: 1269-1277.
6. Maree JP, Hill E (1989) Biological removal of sulphate from industrial effluents and concomitant production of sulphur. Water Science &Technology 21: 265-276.
7. Maree JP, Greben HA, De Beer M (2004) Treatment of acid and sulphate rich effluents in an integrated biological/chemical process. Water SA 30(2): 183-189.
8. De Beer M, Maree JP, Wilsenach J, Motaung S, Bologo L, et al. (2010) Acid mine water reclamation using the ABC process. International Mine Water Association Symposium, Canada .
9. Motaung S, Maree JP, De Beer M, Bologo L, Theron D, et al. (2008) Recovery of Drinking Water and By-Products from Gold Mine Effluents. International Journal of Water Resources Development 24: 433-450.
10. Nengovhela NR, Strydom CA, Maree JP, Greben HA (2004) Chemical and biological oxidation of iron in acid mine water. Mine Water and the Environment 23: 76-80.
11. Mulopo J, Ikhu-Omeregbe D (2012) Phosphogypsum conversions to calcium carbonate and utilization for remediation of acid mine drainage. J Chem Eng Process Technol 3: 129.

4

Kinetic Modeling of the Hydrolysis of Acetic Anhydride at Higher Temperatures using Adiabatic Batch Reactor (Thermos-Flask)

Asiedu N*, Hildebrandt D and Glasser D

Centre of Material and Process Synthesis, School of Chemical and Metallurgical Engineering, University of Witwatersrand, Johannesburg, South Africa

Abstract

The kinetics of the reactions of acetic anhydride-excess water has been studied at higher temperatures up to 366K under adiabatic conditions in a thermos-flask fitted with a negative coefficient thermistor and kinetic data reported in this work. The same techniques were applied for the kinetic studies of excess acetic anhydride-water reactions at higher temperature up to 368 K. The reaction kinetics of excess acetic anhydride-water reactions which has not been determined previously was modeled using the same adiabatic analysis and the average activation energy, heat of reaction, pre-exponential factor and the order of reaction with respect to water has been determined. The studies have shown that the two processes appear to be completely different and might proceed through different reaction mechanisms pathways for the production of acetic acid.

Keywords: Kinetic modeling; Thermos flask; Hydrolysis; Esterification; Exothermic reactions; Temperature

Introduction

Classical techniques for modeling reaction kinetics have been depended on measuring or monitoring concentration-time profiles of liquid phase reactions. This method requires that at some time the researcher must have means of arresting the reaction and take measurement rapidly and continuo the process. With this approach some continuous measuring techniques are possible, pH, conductivity, Kralj [1], Fourier transfer infrared Haji et al. [2]. These methods actually require different approach when it comes to the analysis of the experimental results, and one is always at the disadvantage when reaction rates are extremely fast. Temperature-time profiles of exothermic reaction are used instead of concentration-time profiles for the determination of the kinetic parameters. This method is now used in modern laboratories and can be done with a high degree of accuracy. Most reactions have reasonable amount of heat of reaction and measuring dices such as thermistor can be used to measure temperature changes up to accuracy on 0.0001°C. It follows that in principle an adiabatic reaction covering only 10°C should be measureable to one part in 10^5, which compares very well with the accuracy of chemical titration used in isothermal experiments. Thus reactions with very small heat of reactions could be handled easily. If one is to model an exothermic reaction by estimating the kinetic parameters of a process when concentration and temperature vary simultaneously, one need to perform regression analysis directly on the differential equations describing the process [3]. Recently this regression analysis is easily performed by high speed digital computers. Previous researchers have been confined to the case where reactions occur under adiabatic conditions [4]. Bell and Clunie [5] used flow reactors and it has generally been used for fast reactions. The temperature-time profiles have been converted directly concentration-time profiles and then been analyzed by the classical methods. The method of Schmidt et al. [6] requires numerical evaluations of the first and second derivatives functions of concentration.

The current technique applied for the analysis of the results in this paper is not subject to any limitations and does not require prior knowledge of the heat of reaction. Some studies have been reported in elsewhere in the literature on the kinetic parameters of acetic anhydride hydrolysis. Eldridge and Pirect [7] obtained the pseudo-first order reaction rate constant using a batch reactor. It is well-known that the reaction rate increases with temperature increase. The rate of temperature rise can be represented as a slope of the temperature-time curve. Some studies has been reported in the past using temperature-time profiles to predict the kinetics of the hydrolysis of acetic anhydride in excess distilled water. Williams [8] reported the kinetics and stoichiometry of the reaction between hydrogen peroxide and sodium thiosulphate. An adiabatic reactor was used with a thermocouple and a strip-chart recorder to determine the kinetic parameters of the hydrolysis of acetic anhydride in dilute aqueous solutions Glasser and Williams [9]. A vessel was placed in a constant temperature bath and the heat lost from the reactor was described mathematically. A thermistor was used for the temperature readings. King and Glasser [10] studied the kinetic parameters of the hydrolysis of acetic anhydride in dilute solutions. An adiabatic reactor was used in the experiments. The walls of the reactor were kept at the same temperature as the fluid by passing a high current through the reactor walls. Temperature readings were taken by the use of thermocouples. In this work hydrolysis of acetic anhydride in excess distilled water (dilute solution) was studied and compared with hydrolysis in excess anhydride solution in adiabatic conditions. It is known that not much work has been done on the case where hydrolysis is carried out in excess acetic anhydride solution using adiabatic batch reactor. Moreover this work is also to test the current model of the hydrolysis of excess acetic anhydride under adiabatic conditions. It is claimed that the hydrolysis follows a first order kinetic model. The same method is also used to elucidate the kinetic parameters of the reaction between acetic anhydride and methanol. This process has not been studied in this manner as known to the authors.

***Corresponding author:** Nana Yaw Asiedu, Centre of Material and Process Synthesis, School of Chemical and Metallurgical Engineering, University of Witwatersrand, Private mail Bag X3, Wits 2050, Johannesburg, South Africa

The Mathematical Model of the Reacting System-Thermos-Flask

Given an adiabatic batch reactor (thermos-flask) the mathematical model is made up of a set of differential equations resulting from the mass and energy balances referred only to the reaction mixture because there is no heat transfer.

The stoichiometry of the reactions studied is given below:

$$(CH_3CO)_2O + H_2O \rightarrow 2CH_3COOH \quad \Delta H\ (298K) = -56kJ/mol \tag{1}$$

For a constant -volume batch reactor one can write:

$$-r_A = \frac{1}{V}\frac{dN_A}{dt} = -\frac{dC_A}{dt} \tag{2}$$

The energy balance equation can be established as:

Heat generated = Heat absorbed by reactor contents + Heat transferred through reactor walls (3)

$$(-\Delta H)(-rA)Vdt + Qstirrer\ dt = mC_p\ dT + UA\ (\Delta T)dt \tag{4}$$

$$\text{Given } (-r_A) = \frac{d\varepsilon}{dt} \tag{5}$$

where (ε) is the extent of reaction, we can rearrange eq 4 to give eq 6 below:

$$UA(\Delta T)dt + mC_p = (-\Delta H)\frac{d\varepsilon}{dt} \tag{6}$$

Assume that heat given by stirrer speed ($Q_{strirrer}$) is negligible. Integrating eq 6 gives

$$(T - T_o) + \frac{UA}{mC_p}\int_0^{\infty}(T - T_o)dt = \frac{(-\Delta H)}{mC_p}\Delta\varepsilon \tag{7}$$

The left hand side of eq 7 can be used to correct experimental data to adiabatic conditions. It is assumed that the reaction is independent of temperature, hence correcting the experimental data the adiabatic temperature rise parameter (ΔT_{ad}) can be obtained from the experimental result for the process. This adiabatic temperature rise is equal to the right hand side of eq 7 or we can write;

$$\Delta T_{ad} = \frac{(-\Delta H)\varepsilon}{mC_p} \tag{8}$$

From eq 8 one can easily show that the energy balance equation of an adiabatic batch reactor reduces to a linear form given by eq 9 as:

$$T = T_0 + \Delta T_{ad\varepsilon} \tag{9}$$

where (ε) is the extent of reaction with respect to the reactant of interest and its value can be computed in advance from the initial conditions (temperature and heat capacities) of the reacting species. Equation 9, also allows one to find conversion at any instant under adiabatic conditions by using only one measure of temperature. Then from the initial concentrations of the reactants the concentrations products can be monitored at any instant in the reactor.

The Experimental Set Up

The experimental set- up as shown in Figure 1 was used in all the reactions studied. The adiabatic batch reactor used in the experiments is 18/8 stainless steel thermos-flask of total volume of 500 mL equipped with a removable magnetic stirrer. The flask is provided with a negative temperature coefficient thermistor connected on-line with a data-logging system. The signal from the sensor (thermistor) is fed to a measuring and a control unit amplifier and a power interface. The acquisition units are connected to a data processor. A process control engineering support data management.

The data acquisition system called Clarity has the following part numbers: C50 Clarity Chromatography SW, single instrument, 3 x 55 Clarity Add-on instrument SW, 194 (INT9 quad channel A/D converter card). The hardware is INT9 PCI A/D 24 bit converter. The properties: Input signal range 100mV -10V, Acquisition frequency 10-100Hz, Internal A/D converter (INT9-1 to 4 channel PCI A/D converter). All physically available analog inputs and outputs as well as virtual channel are all automatically monitored and the process values are stored. The process values are transmitted in such a way that the computer screen displays profiles of voltage-time curves. Data acquisition software was used to convert the compressed data form of the history file on the hard disk into text file format. The text files are converted to excel spreadsheet and the data are then transported into MatLab 2010a for analysis.

The Thermistor Calibration

The thermistor used in the experiments was negative temperature coefficient with unknown thermistor constants. The calibrations involve the determination of the thermistor constants and establish the relationship between the thermistor's resistance and temperature. This was done by fitting both the thermistor and an electronic digital temperature measuring device in a sealed vessel and the system was slowly warmed until the voltage reaches it asymptotic state. The thermistor was connected to a computer with data acquisition software to provide data of voltage-time real-time plot. The voltage-time data was used to match the temperature-time data obtained from the electronic digital temperature device. Thermistor resistance (R_{TH}) and Temperature (T) in Kelvin was modeled using the empirical equation given developed by Considine [11]

$$R_{TH} = exp\left(\frac{B}{T} + C\right) \tag{10}$$

or

$$\ln(R_{TH}) = \frac{B}{T} + C \tag{11}$$

Where the parameters (B) and (C) are the thermistor constants. The constants were determined by fitting the "best" least square straight line plot of $\ln(R_{TH})$ against 1/T, giving the thermistor equation as:

$$\ln(R_{TH}) = \frac{1773.8}{T} + 18.178 \tag{12}$$

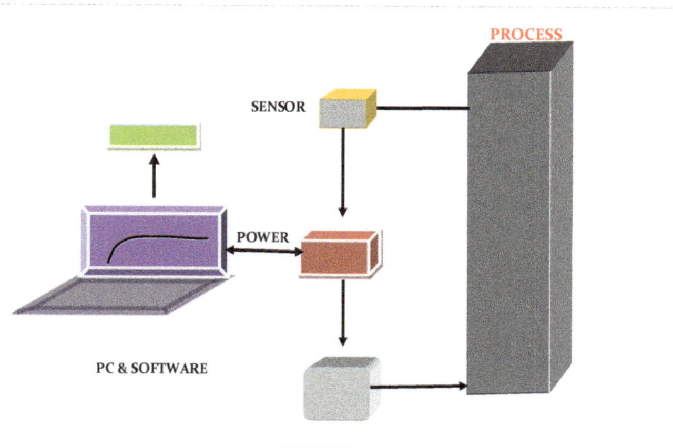

Figure 1: Block diagram of experimental set-up.

Experimental Procedures and Results

Excess water-acetic anhydride reactions-hydrolysis-A

Analytical reagent grade acetic anhydride and distilled water was used in all experiments. In all experiments 1 mol of acetic anhydride was injected into the reaction vessel followed by 10 mols of distilled water. These quantities (volumes) were used so that at least 60% of the length of the sensor (thermistor) would be submerged in the resulting initial reacting mixture. Reactants were brought to a steady-state temperature before starting the stirrer to initiate the reaction. In course of the reaction, the stirrer speed was set at 1000 rev/min, the resulting voltage-time profiles were captured and the corresponding temperature-time curve was determined using the thermistor equation derived above. Runs were carried out adiabatically at the following initial temperatures: 305K, 310K and 328K.

Experimental determination of heat transfer coefficient of the process

For a given system (reaction vessel and content) one can write:

$$\left(\frac{mC_p}{UA}\right)_{system} = \left(\frac{mC_p}{UA}\right)_{water} + \left(\frac{mC_p}{UA}\right)_{vessel} \quad (13)$$

In this experiment different amount of water (m_w) (200 g, 300 g and 400 g) was injected into the reaction flask at 343.08K, 347.77K and 347.85K and allowed to cool until the temperature reaches a steady-state temperature (T_S). The cooling process thus follows equation 14. A nonlinear least-square regression analysis was performed on all the three experimental curves and the value of UA/mCP for each cooling curve was determined.

$$T = T_s + (T_0 - T_s)\exp\left(-\frac{UA}{mC_p}t\right) \quad (14)$$

From Table 1, a plot of mass of water (m) against (mC_p/UA) is shown.

Determination of UA/mC_p for the reaction mixture

Table 2 consists of some of the physical constants of the reacting species which was used in the determination of the quantity UA/mCP.

Equation 23 was used to determine specific heat of the reacting mixture ($C_{P.mix}$) and the quantity $M_{TCP.mix.}$

$$C_{P.mix} = \frac{m_{AA}}{MT}C_{P.AA} + \frac{m_W}{MT}C_{P.W} \quad (15)$$

M_T, is the total mass of the reacting mixture and $M_{TCP.mix}$ was calculated to be 943.64 J/K. Since the reaction is not pure water we therefore determined its equivalent mass of water by dividing 943.64 J/K by the specific heat of water:

$$m_{eqv}water = \frac{(M_T C_P)_{mix}}{C_{P.water}} = 225.5g\ water \quad (16)$$

Mass of water(g)	Initial Temperature(K)(T_O)	Final Temperature(K)(T_S)	UA/mCp (min^{-1})
200	343.08	292.13	0.00215
300	347.77	293.00	0.00189
400	347.84	293.38	0.00178

Table 1: Summary of characteristics of the cooling curves.

Species	Moles	Molar mass(g/mol)	C_P (J/mol.K)
Acetic anhydride	1.0	102.9	189.7
Water	10.0	18.01	75.4

Table 2: Some constants of reacting species.

From Figure 2 the value of the quantity $(UA/mC_P)_{mix}$ was determined to be 0.00209/min. This value was plugged in eq 7 above and the correction of the experimental results to adiabatic conditions was performed. The Figures 3-5 shows the experimental and the corrected temperature profiles for all the three runs.

From eq 9 the variation of acetic anhydride concentration with temperature/time was deduces as:

$$C_{AA} = 3.64 - a\,(T-T_o) \quad (17)$$

The constant (α) in eq 17 was calculated using the initial concentration of the acetic anhydride, 3.64 mol/L and the adiabatic temperature change $(\Delta T)_{adiabatic}$ values provided in Table 3.

Variation of rate of reaction with concentration change

The average rate of a reaction ($-r_A$) which is a function of

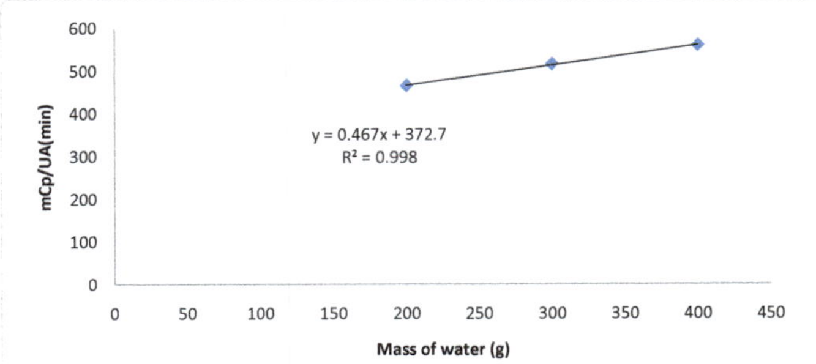

Figure 2: A plot of mass of water (m) against (mC_p/UA).

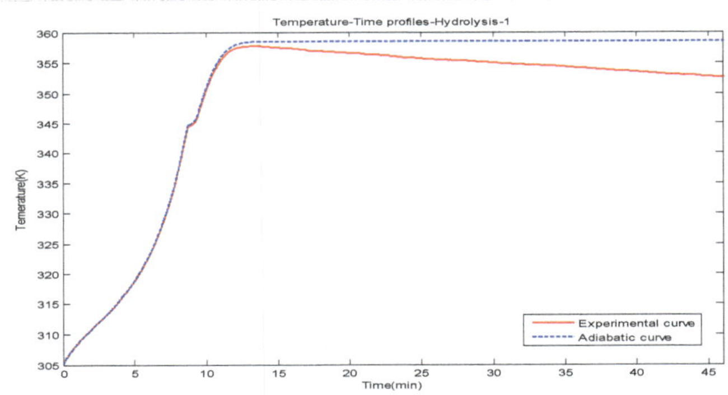

Figure 3: Temperature-time plots of acetic anhydride-excess water reaction at 305K.

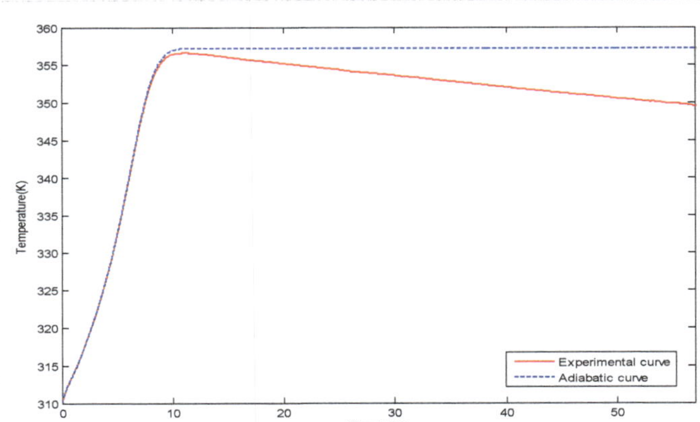

Figure 4: Temperature-time plots of acetic anhydride-excess water reaction at 310K.

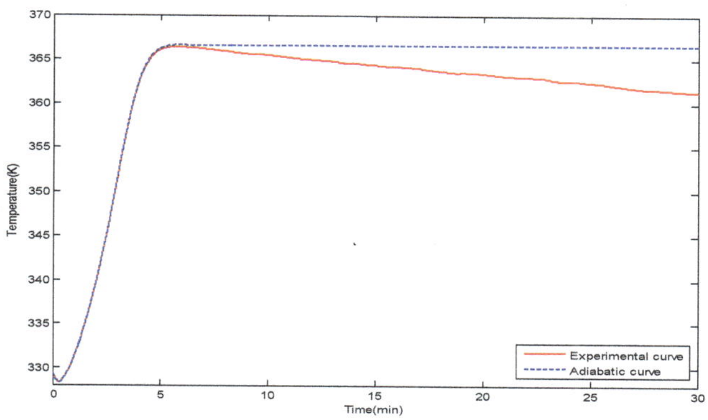

Figure 5: Temperature-time plots of acetic anhydride-excess water reaction at 328K.

Experiment	T(o)(K)	$T_{(max\ adiabatic)}$(K)	$(\Delta T)_{adiabatic}$(K)
1	305.10	358.50	52.67
2	310.10	358.70	56.40
3	328.40	366.40	38.00

Table 3: Summary of the characteristics of the above temperature-time plots of figures (3-5).

temperature and concentration and for a pseudo-first process such acetic anhydride reacting in excess water, at a particular temperature the specific rate constant value should be the same in all three cases of the processes considered (Figures 6-8).

Recall that $k(T) = \dfrac{-r_A(T)}{C_A}$ (18)

The average rate of reaction at a particular time was determined from the concentration time-time plots by determining the first derivatives at different time intervals numerically using the concentration-time data of the runs (Tables 4-6).

Kinetics and thermodynamic analysis of the experiments

The rate information obtained from the concentration-time plot enabled one to calculated specific rate constant k(T) of the processes at any time (t) and temperature (T). Arrhenius plots was thus generated from which kinetic parameters of runs was extracted (Figures 9-11).

The values obtained for the specific rate constants as a function of temperature are given in equations 27, 28 and 29 below:

$$k_1(T)=1.68*10^7\ exp\left(-\frac{6078.7}{T}\right) \quad (19)$$

$$k_2(T)=1.78*10^7\ exp\left(-\frac{6108.3}{T}\right) \quad (20)$$

$$k_3(T)=1.22*10^7\ exp\left(-\frac{6062.2}{T}\right) \quad (21)$$

The thermodynamic information extracted from the experiments was the heat of the reaction (ΔH_{rxn}). This was assumed to be independent of temperature and was determined by using equation 10. Table 7 shows ΔH_r values of the various experiments.

Experimental Procedures and Results

Excess acetic anhydride-water reactions-hydrolysis-B

Analytical reagent grade acetic anhydride and distilled water was used in all experiments. In these experiments 2.645 mols of acetic anhydride was injected into the reaction vessel followed by 0.556 mols

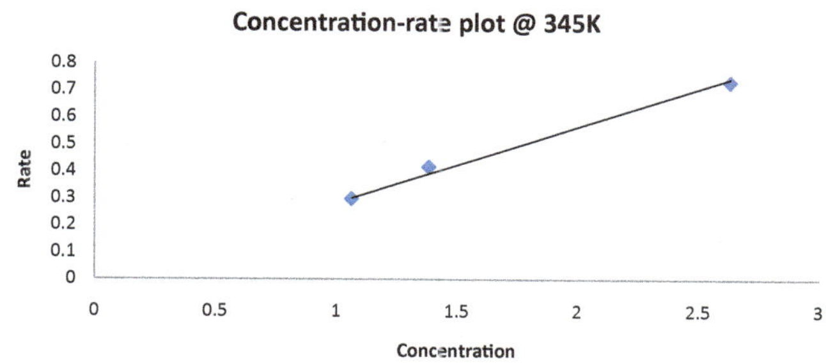

Figure 6: Concentration-rate plot at 345K.

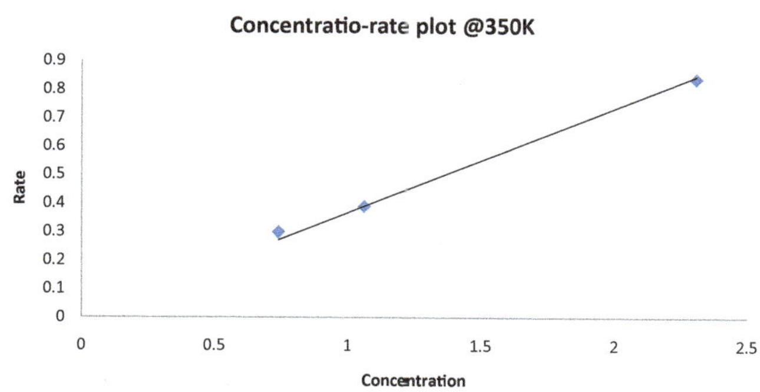

Figure 7: Concentration-rate plot @ 350K.

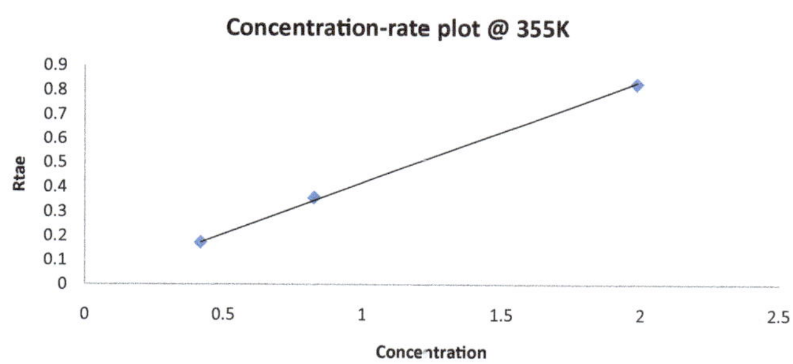

Figure 8: Concentration-rate plot @ 355K.

T=345K			
Experiment	Concentration	Rate	Rate constant (k)
1	1.0637	0.2965	0.2787
2	1.3848	0.4160	0.3004
3	2.633	0.7300	0.2772
Average (k)		**0.2854**	
Standard deviation (k)		**0.0013**	

Table 4: Variations of temperature, rate and concentration at T =345K.

T=350K			
Experiment	Concentration	Rate	Rate constant (k)
1	0.7403	0.3005	0.4059
2	1.0623	0.3913	0.3684
3	2.3109	0.8391	0.3613
Average (k)		**0.3785**	
Standard deviation (k)		**0.024**	

Table 5: Variations of temperature, rate and concentration at T=350K.

of distilled water. These quantities (volumes) were used so that at least 60% of the length of the sensor (thermistor) would be submerged in the

T=355K			
Experiment	Concentration	Rate	Rate constant (k)
1	0.4177	0.1746	0.4199
2	0.8256	0.3590	0.4348
3	1.9893	0.8283	0.4164
Average (k)		**0.4237**	
Standard deviation (k)		**0.0098**	

Table 6: Variations of temperature, rate and concentration at T=355K.

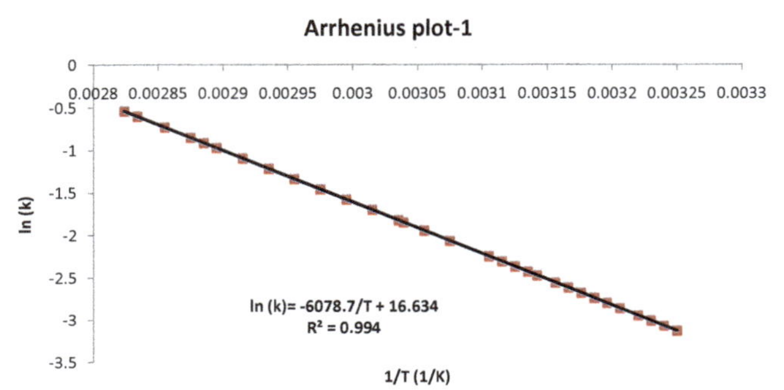

Figure 9: Arrhenius plot of experiment-1.

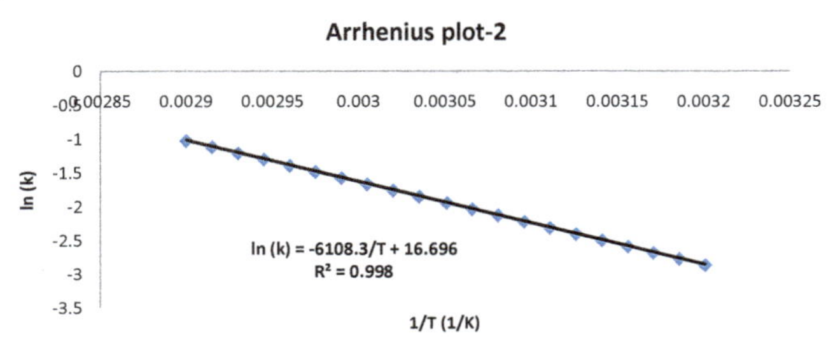

Figure 10: Arrhenius plot of experiment-2.

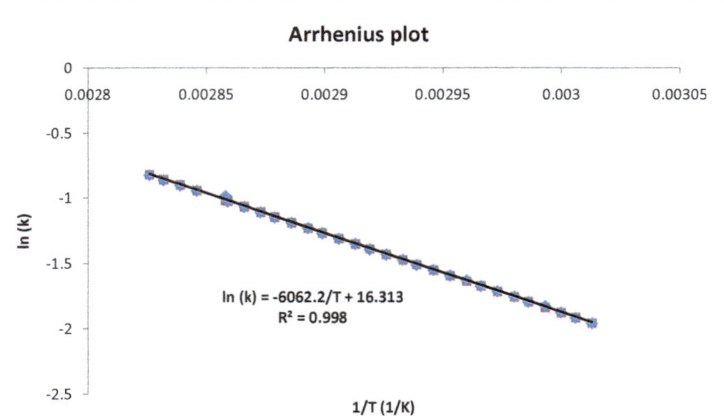

Figure 11: Arrhenius plot of experiment-3.

resulting initial reacting mixture. Reactants were brought to a steady-state temperature before starting the stirrer to initiate the reaction.

The stirrer speed was set at 1000 rev/min, the resulting voltage -time profiles were captured and the corresponding temperature-time curve was determined using the thermistor equation derived above. Runs were carried out adiabatically at the following initial temperatures: 286.8K, 293.6K and 304.5K.

Determination of UA/mC_p for the reaction mixture

The quantity UA/mC_p was determined for each of the runs using the cooling parts of the experimental curves. It is noticed from the cooling parts of the runs were similar to the curve cooling water experiments described in section above. It was an indication the cooling part of the experimental runs was real cooling process and no reaction took place during that process. The cooling parts thus follow eq 14 above. A nonlinear regression analysis was performed on all the cooling parts and the value of UA/mC_p was determined for all the experimental runs. The steady-state temperature (T_S) for the experiments was 293K. The regressed lines of the experiments cooling sections and their UA/mC_p was obtained and used for the corrections of the experimental curves, corresponding experimental and adiabatic curves are shown in Figures 12-14. The adiabatic curve was determined using equation 7 above.

Table 8 shows initial temperatures and the corresponding values of the UA/mC_p of the runs.

From eq 11 the variation of water concentration with temperature/time was deduces as:

$$C_{H20}=2.13 - a\,(T-T_o) \qquad (22)$$

The constant (α) in eq 22 was calculated using the initial

Experiment	1	2	3
ΔH_{rxn}(kJ/mol)	-53.36	-53.44	-53.51

Table 7: Thermodynamic information of the experiments.

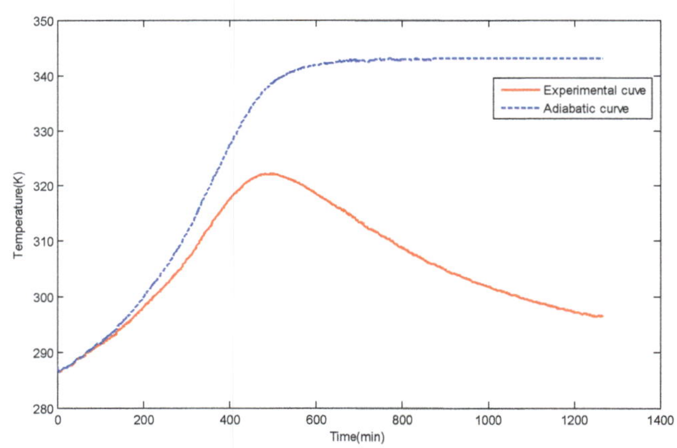

Figure 12: Experimental and Adiabatic curves of experiment-4.

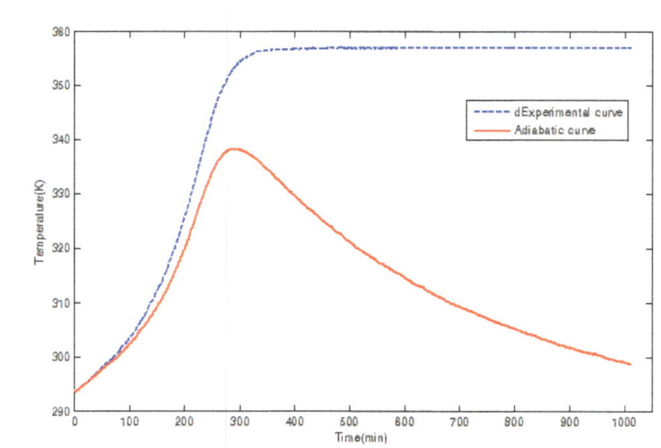

Figure 13: Experimental and Adiabatic curves of experiment-5.

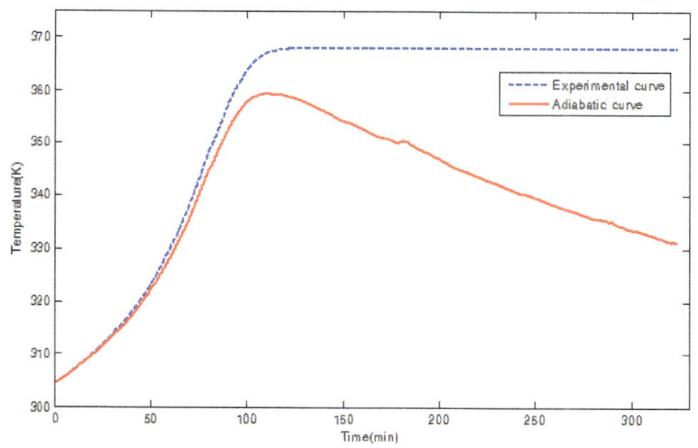

Figure 14: Experimental and Adiabatic curves of experiment-6.

Experiment	Initial Temperature(K)	UA/mC_p (min^{-1})
4	286.8	0.002030
5	293.6	0.002698
6	304.5	0.002795

Table 8: UA/mC_p of hydrolysis-B experiments.

Experiment	T(0)(K)	$T_{max(adiabatic)}$(K)	$\Delta T_{exp(adiabatic)}$(K)
4	286.8	343.2	56.4
5	293.6	357.1	63.5
6	304.5	368.3	63.8

Table 9: Summary of the characteristics of the above temperature-time plots.

T=320K			
Experiment	Concentration	Rate	Rate constant (k)
4	1.42458	0.0050506	0.0003545
5	1.55912	0.0.008286	0.0053150
6	1.79670	0.0.014893	0.0082890
Average (k)		**0.0057163**	
Standard deviation (k)		**0.0023973**	

Table 10: Variations of temperature, rate and concentration at T =320K.

concentration of the water 2.13 mol/L and the adiabatic theoretical temperature change $(\Delta T)_{adiabatic}$ (Table 9).

Variation of rate of reaction with concentration change

The tables and the profiles below show the variation of concentration, rate of reaction at specific temperatures of hydrolysis B experiments (Table 10 and Figures 15-17).

Kinetics and thermodynamic analysis of the experiments

The rate information obtained from the concentration-time was to calculated specific rate constant k(T) of the processes at any time (t) and temperature (T). Arrhenius plots was thus generated from which kinetic parameters of runs was determined. The Figures 18-20 shows the Arrhenius plots for all the three experiments.

The values obtained for the specific rate constants as a function of temperature are given in equations 23, 24 and 25 below:

$$k_1(T)=5.99*10^3\,exp\left(-\frac{4598.1}{T}\right) \quad (23)$$

$$k_2(T)=1.25*10^3\,exp\left(-\frac{4015.6}{T}\right) \quad (24)$$

$$k_3(T)=7.69*10^7\,exp\left(-\frac{4711.5}{T}\right) \quad (25)$$

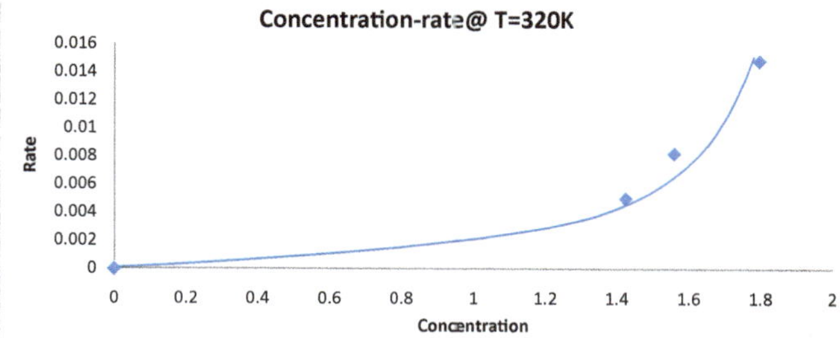

Figure 15: Concentration-rate plot at 320K.

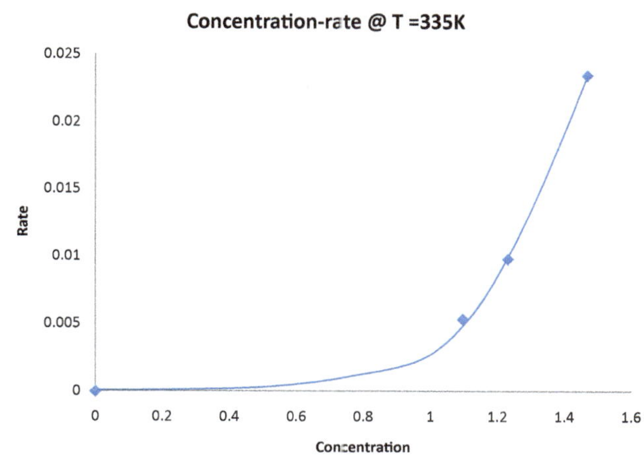

Figure 16: Concentration-rate plot @ 335K.

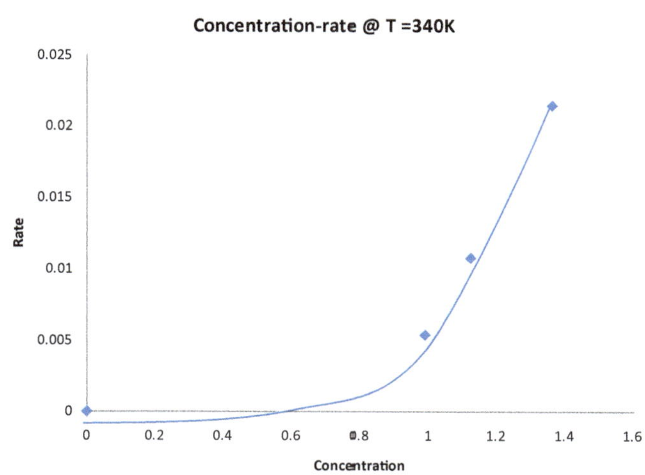

Figure 17: Concentration-rate plot at 340K.

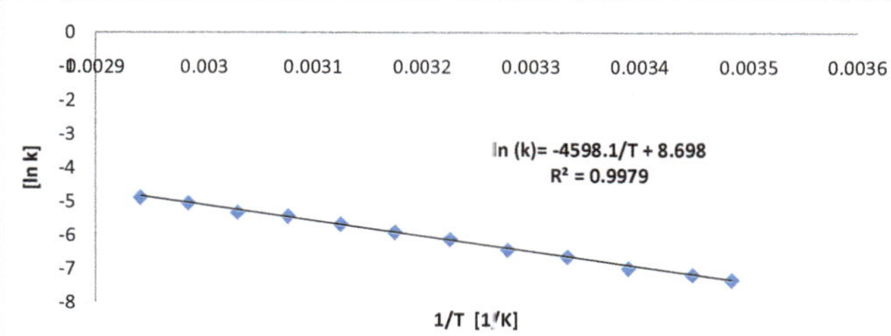

Figure 18: Arrhenius plot of experiment-4.

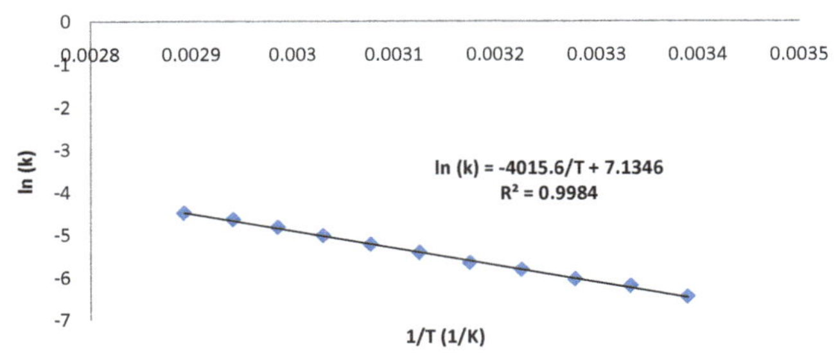

Figure 19: Arrhenius plot of experiment-5.

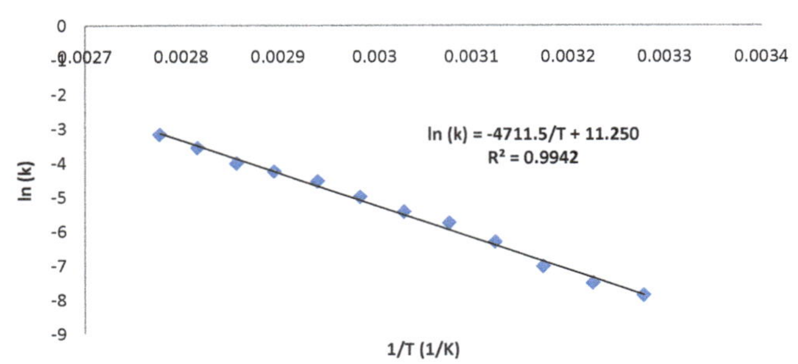

Figure 20: Arrhenius plot of experiment-6.

The thermodynamic information extracted from the experiments was the heat of the reaction (ΔH_{rxn}). This was assumed to be independent of temperature and was determined by using eq 8. Tables 11-13 shows ΔH_r values of the various experiments.

Discussion of Results

The experimental results in all the processes studied showed that a strong relationship exists between reaction rates and changes in temperature for exothermic reactions as stated in the literature [9,10]. The reactions between excess water (10 mols) and acetic anhydride (1 mol) thus hydrolysis reactions-(A) were carried out at 305K-366K. The extent of the reactions was followed by monitoring the changes in temperature in the reaction vessel (thermos-flask). The experimental and the adiabatic curves are shown in Figures 3-5. The reactions between excess acetic anhydride (2.65 mols) and water (0.55 mol) thus hydrolysis reactions-(B) were carried out at 286 K-368 K. The profiles of experimental and adiabatic curves are shown in Figures 12-14 respectively. The experimental reaction profiles (Hydrolysis process-A and Hydrolysis process-B) of this study shows that hydrolysis process-A and hydrolysis process-B are completely two different processes. One unanticipated finding was the nature of the hydrolysis-B profiles compared with the hydrolysis-A profiles. On the other hand, another interesting finding was the cooling parts of the processes. Hydrolysis-A, showed very gradual cooling process as compared with Hydrolysis-B which showed a cooling process similar to the cooling process of the experimental determination of heat transfer coefficient of the of the reaction vessel (thermos-flask). There may be several possible explanations to these observations. A possible explanation for this might be that the rate of heat loss during the cooling processes is higher for hydrolysis-B than that of hydrolysis-A under the same experimental conditions these parts thus suggests real cooling process. On the other hand, the rate of cooling of the hydrolysis-A processes, although very slow there might be some amount of hydrolysis reactions taking place. It is also interesting to note that, the processes (hydrolysis A and B) showed different reaction times for the processes to reach their maximum temperatures during their reactions. This finding was somewhat not surprising. The observed differences discussed is however very significant. The reaction order for all the hydrolysis-A experiments were first order with respect to the acetic anhydride. This is shown in the rate-concentration plots in the Figures 6-8. This result is consistent with other researchers Kralj [2] and Haji et al. [1].

T=335K			
Experiment	Concentration	Rate	Rate constant (k)
4	1.0972539	0.0053749	0.0048985
5	1.2317951	0.0098262	0.0079771
6	1.4693163	0.0235028	0.0159953
Average (k)		**0.0096236**	
Standard deviation (k)		**0.0057287**	

Table 11: Variations of temperature, rate and concentration at T =335K.

T=340K			
Experiment	Concentration	Rate	Rate constant (k)
4	0.9881278	0.0053892	0.0054539
5	1.1226940	0.01079109	0.0096116
6	1.3602177	0.0214770	0.0157889
Average (k)		**0.0125524**	
Standard deviation (k)		**0.0081873**	

Table 12: Variations of temperature, rate and concentration at T =340K.

Experiment	4	5	6
ΔH_{rxn}(kJ/mol)	-53.92	-60.71	-60.99

Table 13: Thermodynamic information of the experiments..

However, contrary to our expectations the rate -concentration plots of the hydrolysis-B was very surprising as the profiles appeared to be part of a parabola suggesting a second other process. This finding is unexpected and has not previously been reported. The rate law temperature dependence were satisfactorily correlated by means of the Arrhenius function for the hydrolysis-A. Figures 9 -11 shows the Arrhenius plots of the hydrolysis-A. From the plots the values of the activation energies and pre-exponential factors were determined as shown in the equations 19-21. Similarities of the activation energies (50.54 kJ/mol, 50.78 kJ/mol, 50.40 kJ/mol) which agrees very well with what is reported by Haji [1] and Kralj [2]. The average activation energy reported in the literature is 50.24 kJ/mol, Haji [1] which differs by +0.65% from the average value reported in this work. For the hydrolysis-B processes, Arrhenius functions for the hydrolysis-B are shown in Figures 18-20. From the plots, the values of the activation energies and pre-exponential factors were determined as shown in the equations 23-25. The activation energies (38.23 kJ/mol, 33.89 kJ/mol, 39.19 kJ/mol) are however smaller than that of the hydrolysis-A, this is not surprising since the studies has shown that the kinetics of the hydrolysis-A and hydrolysis-B appears very different processes. These activation energies of the hydrolysis-B reactions have not been reported in the literature. The heat of the reactions determined from the experiments was assumed to be independent of temperature. The values obtained for hydrolysis-A is shown in Table 7. The average value compared well with what was obtained by Glasser and Williams [9]. The values determined for the hydrolysis-B are shown in the Table 13.

Conclusions

The kinetics of the reactions acetic anhydride-excess water, excess acetic anhydride-water has been modeled as a function of temperature in an adiabatic batch reactor (thermos-flask) using a negative coefficient thermistor as a measuring device. The results has shown that the acetic anhydride-excess water process at 1: 10 mole ratio the process remains first order at higher temperatures up about 366K with the activation energy similar to that reported at 298K.

Excess acetic anhydride-water reactions which have no literature information on its kinetics were shown to follow a second order kinetics with respect to the water. The activation energy is approximately 37.1 ± 2.8 kJ/mol. The heat of reaction which was assumed to be independent of temperature for the excess acetic anhydride/water is approximately 58.5 ± 4.0 kJ/mol. The heat of mixing was assumed negligible and was not considered in the heat of reaction calculations. This studied has shown that the two processes studied appears to be completely different processes and may have completely different reaction mechanisms pathways for the acetic acid production.

Acknowledgement

Nana Asiedu would like to thank the National Research Fund (NRF) of South Africa, Directors of Centre of Material Processing and Synthesis (COMPS), Prof. David Glasser and Prof Diane Hildebrandt, University of Witwatersrand, Johannesburg, South Africa for financial support of this work.

Nomenclature

ΔH_{rxn} - heat of reaction (KJ/mol)

ΔT (ad) – adiabatic temperature change (K)

$\Delta\varepsilon$ – extent of reaction

A – preexponential factor

B,C – thermistor constants

C_A - concentration of A (mol/L)

C_P – specific heat capacity

$C_{P,AA}$ – specific heat capacity of acetic anhydride

$C_{P,mix}$ - constant heat capacity of reaction mixture (kJ/mol.K)

$C_{P,W}$ – specific heat capacity of water

E_a – activation energy

I_{total} – total current of the circuit

k_1, k_2, k_3 - specific rate constant

m – mass

M_{AA} – Mass of acetic anhydride

M_{AA} – Mass of water

M_{eqv} – Equivalent amount of water

M_T - mass of reaction mixture(Kg)

N_A – mole of A (mol)

p - probability factor

Q - heat produced by stirrer

R – molar gas constant

R_2 – known resistance in the circuit

r_A – rate of reaction of (A)

R_{TH} – Thermistor resistance

t – time

T – absolute temperature

T_{max} – maximum temperature (K)

T_o – initial temperature

T - reactor temperature (K)

T_S – steady state temperature (K)

T_s – steady-state temperature

U – heat transfer coefficient (J/m² s.K)

v – total voltage of the circuit

V -volume

V_{out} – voltage measured by thermistor

V_{TH} – voltage across thermistor resistance

Z - collission frequency

α, β - constants

References

1. Kraji AK (2007) Checking the Kinetics of Acetic Acid production by Measuring the conductivity. J Ind Eng 13: 631-636.

2. Haji S, Erkey C (2005) Kinetics of hydrolysis of acetic anhydride by in situ FTIR spectroscopy. Chemical Engineering Education 56-61.

3. King RP (1967) Estimation of parameters in systems Defined by Differential Equations. S Afr J Sci 91-96.

4. Livingstone R (1961) Technique of organic chemistry. Weissberger edn, VII, 2nd edn, Intersicience NY.

5. Bell RP, Clunie JC (1952) A Therma Method of Following Fast Reactions in Solution. Proc Roy Soc Lond A 212: 16-32.

6. Schmidt JP, Mickley HS, Grotch SL (1964) Use of adiabatic experiments for kinetic studies in fixed catalyst beds. AIChE 10: 149-154.

7. Eldridge JW, Piret EL (1950) Continuous-flow stirred-tank reactor system I. Design equations for homogeneous liquid phase reactions, Experimental data. Chem Eng Prog 46-290.

8. Williams RD (1974) Indirect Measurement of Reaction Rate. Chem Eng Educ Winter 23-30.

9. Glasser D, Williams DF (1971) The study of liquid-phase kinetics using temperature as a measured variable. Ind Eng Chem Fundam 10: 516-519.

10. King RP, Glasser D (1967) The use of Adiabatic Calorimeter for Reaction Rate Studies. S Afr Ind Chem 12-15.

11. Considine DM (1957) Process Instruments and Controls Hand book. MCgraw-Hill Book Co., Inc., New York, 2-60.

5

Adsorption of Copper Complexes on Anion Exchange Resin in Non-Ammoniacal and Ammoniacal Thiosulfate Leaching Systems

Abrar Muslim*

Department of Chemical Engineering, Engineering Faculty, Syiah Kuala University, Banda Aceh, Indonesia

Abstract

The leaching of copper using thiosulfate reagent as an alternative reagent to cyanide has been investigated since it is almost presented in gold thiosulfate leaching. This paper addressed the investigation of adsorption of copper complexes on anion exchange resin in non-ammoniacal and ammoniacal thiosulfate leaching systems.The work provided the experimental procedures for the adsorption of copper complexes on strong based anion exchange resin in non-ammoniacal and ammoniacal resin-solution systems. The result showed that increasing ammonia in the solution resulted in the diminishing total copper complexes on the resin. As the solution pH of resin-solution systems was maintained by the ammonia solution added in the system, the same effect of ammonia on the total copper complexes on resin was also shown. Overall, non-ammoniacal resin-solution system gave much better adsorption isotherm of copper complexes on resin, and the dominant species copper complexes on the strong based anion exchange resin was expected to be $Cu(S_2O_3)_3^{5-}$ based on leaching modelling.

Keywords: Adsorption; Copper; Leaching modelling; Resin

Introduction

The leaching of copper using thiosulfate reagent as alternative reagent to cyanide has been investigated since it is almost presented in gold thiosulfate leaching. In general, copper is found in the gold leaching of complex ores [1,2]. The present of copper reduces the efficiency of gold leaching and consumes additional reagent [3,4]. To improve the recovery of gold, ammonium thiosulfate and oxygen under pressure was used [5] because copper ions can be catalysts for the oxidation of thiosulfate by oxygen or other oxidant, there should only be sufficient copper present to maximize leaching of gold and minimize thiosulfate consumption, and it only improves the initial rate of gold extraction [6,7].

The chemistry of gold leaching involving thiosulfate, ammonium and copper ions are complex due to the oxidation by dissolved oxygen [8-10]. Using ammoniacal thiosulfate in the gold leaching is typically much more economical than cyanidation to lixiviate significant quantities of copper without consuming excessive reagent [11]. The leaching of gold linearly increases with the Cu(II) concentration at low concentration (less than 5 mM) while the higher Cu(II) concentration almost does not affect the gold leaching rate [12]. The oxidation of thiosulfate in the presence of copper(II) and oxygen is very complex wherein oxygen increases the rate of copper(II) reduction to copper(I) by thiosulfate [13]. Hence, limiting the presence of oxygen and ammonia in the leach solution would be worthy to investigate the adsorption of copper complexes on anion exchange resins. Anion exchange resin had been intensively used in the quantification of thiosulfate and polythionates in gold leach solution on the resins [14]. The loading of gold using strong base anion exchange resin in non-ammoniac resin-solution (NARS) systems has been studied, and polythionate concentration affects gold loaded on the resins [15]. However, there are still limited studies on the adsorption of copper complexes onto the resin at the equilibrium stage.

Therefore, the main objective of this work was to investigate the adsorption of copper complexes on strong based anion exchange resin the resin-solution systems with the present and the absent of ammonia. The specific objectives of the work were to provide the experimental procedure for the adsorption of copper complexes on strong based anion exchange resin in the resin-solution systems; to investigate the equilibrium adsorption of copper complexes in the resin-solution systems; to investigate the effect of ammonia concentration and the solution pH on the equilibrium adsorption of total copper complexes onto resin; to highlight the adsorption isotherm of total copper complexes in the resin-solution systems; and to obtain the dominant species of copper complexes in the resin-solution systems.

Experimental Methods

Apparatus and chemicals

The concentration of thiosulfate, polythionates and the total copper complexes in the experiments was analysed using the Waters 2695 HPLC separation module. For HPLC analysis, the copper must be masked with ethylenediamine [16], as copper thiosulfate peak interferes with the analysis of thiosulfate and polythionates. The HPLC separation was achieved using a Dionex IonPac AS16 ion exchange column equipped with an IonPac AG16 guard column. The mobile phase was a solution

***Corresponding author:** Abrar MUSLIM, Chemical Engineering Department, Engineering Faculty, Syiah Kuala University, Banda Aceh,IndonesiaPostCode-23111

of sodium perchlorate (0.075 –0.2 M). The species detection was via a Waters 2996 Photodiode Array Detector (UV). Empower software was used for the calculation of peak area [17]. Details about the mobile solution and the wavelengths of UV adsorption and peak retention times of the involved species are described in the literature [14]. For the ICP analysis of total copper complexes, the samples were treated with NaCN (0.0167 M).

Copper was introduced as cupric sulphate ($CuSO_4.5H_2O$). Ammonia-containing solutions were prepared using a 2 M NH_3 stock solution standardised by acid-base titration [18]. Purolite A500/2788 in the chloride form with a macrospore structure was also used. The wet form of the resin contained 53-58 % moisture, the resin capacity is 1.15 meq/mL, and the specific gravity is 1.08 g/mL [19].

Procedures

All the experiments of the adsorption of copper complexes on strong based anion exchange resin in the resin-solution systems without and with ammonia were conducted at the ambient temperature (~23 °C) and variety pH. The concentration on the resin for a particular species was calculated from the change in its solution concentration in the system. The steps for the experiments were described as follow:

1. Leaching solution 200 ml was prepared, consisting of 20 mM $(NH_4)S_2O_3$ (ammonium thiosulfate), 5 mM sodium trithionate ($Na_2S_3O_6$), 2 mM cupric sulphate ($CuSO_4.5H_2O$) and 50 mM ammonia in a 200ml volumetric flask. In dealing with ammonia loss, the ammonia solution was quickly added last before rapidly putting DI water into the flask to reach the 200 ml level. Then, it was quickly closed the flask with the cap.
2. Resin 0.5 g was prepared, and placed in a conical flask. It was then placde in a 130 ml and sealed with plastic bottle.
3. The leaching solution was quickly placed in the bottle until it was full (130 mL total volume), and tightly sealed.
4. The bottle then placed on a roller at 35 rpm. Following that, the adsorption time is started using stopwatch.
5. The solution sample 1 mL was taken for HPLC analysis at just after 2 hours adsorption time started, and the solution pH was measured using pH probe.
6. After 2 hours adsorption, the resin was decanted and washed 4 times with DI water into a large vial. The remaining H_2O was sucked using pipette.
7. $NaClO_4$ 50 mL (0.5M) was added into the vial for the dilution of copper complexes from resin, it was shaked at 150 U/minute on a Jubalo SW-20°C shaker for 30 minutes. The timer was started at the time of the 1st dilution started.
8. The solution sample 1 mL at 30 minutes was taken for HPLC analysis as the 1st dilution sample.
9. Following that, the resin was decanted and washed 4 times with DI water in the vial. The remaining H_2O was sucked.
10. $NaClO_4$ 50 mL (0.5M) was re-added into the vial for the 2nd dilution of thiosulfate from resin, and it was also shaked at 150 U/minute for 30 minutes.
11. The 2nd dilution sample was taken for HPLC analysis at 30 minutes.
12. The resin was then decanted and washed 4 times with DI water in the vial. Before placing the resin in a small glass vial, the empty vial mass was measured.
13. The resin was placed in a small glass vial, and the remaining H_2O was also sucked.
14. Finally, the resin was dried at 60°C in an oven for about 16 hours.
15. The mass of vial with resin was measured to obtain the dried resin mass; the dried resin mass = the mass of vial with resin – the empty vial mass.

The 1st to 15th steps were repeated for the resin-solution systems with the initial ammonia concentration in the range of 0-400 mM and ammonium thiosulfate concentration in the range of 20-100 mM. For the ICP analysis of total copper complexes, 2 mL NaCN (0.0167 M) was added to each sample to stabilize the copper complexes in the sample.

Results and Discussion

Effect of ammonia concentration on equilibrium adsorption of copper complexes on resin

The effect of ammonia concentration in solution on total copper complexes adsorbed on resin in the resin-solution systems with the various initial concentration of thiosulfate in solution was viewed in Figure 1. As can be clearly seen in Figure 1, the total copper complexes adsorbed on resin sharply decreased from approximately 22.322 g copper/kg dry resin for the solution without ammonia to approximately 10.7122 g copper/kg dry resin for the solution with 50 mM ammonia

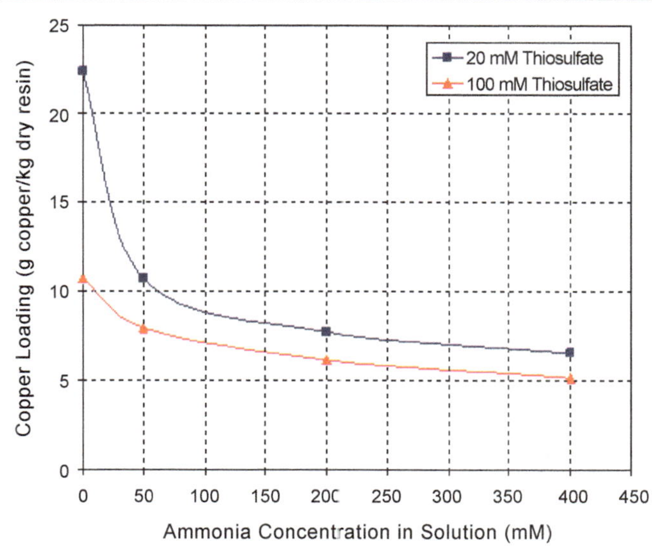

Figure 1: The copper complexes concentration on resin over initial concentration of ammonia in solution with various thiosulfate concentrations.

(see the top graph of leaching with the 20 mM initial concentration of thiosulfate). Then, as the ammonia concentration in solution increased to 200 and 400 mM, the total copper complexes adsorbed on resin gradually decreases to 7.688 and 6.552 mM, respectively.

Compared to the total copper complexes adsorbed on resin in the system with the initial concentration of thiosulfate being 20 mM, the profile of total copper complexes adsorbed on resin with the 100 mM initial concentration of thiosulfate was decreased moderately from approximately 10.646 g copper/kg dry resin for the solution without no ammonia to approximately 7.881 g copper/kg dry resin for the solution with 50 mM ammonia (see the bottom graph in Figure 1). The same shape of graphs was also shown between the one with 20 mM thiosulfate and the one with 100 mM thiosulfate. But for the one with 100 mM thiosulfate, the total copper complexes adsorbed on resin was lower proportionally, which was 6.079 and 5.097 g copper/kg dry resin for the ammonia concentration in solution of 200 and 400 mM, respectively as shown by the bottom graph in Figure 1.

Effect of pH on equilibrium adsorption of copper complexes on resin

Meanwhile, the relationship between the total copper complexes adsorbed on resin and the solution pH was sighted in Figure 2. Because the pH solution was generated by the concentration of ammonia in the solution, it was not surprising that the profile of the total copper complexes adsorbed on resin over the solution pH was almost the same as Figure 1. The more pH in the solution was applied, the less total copper complexes adsorbed on resin was obtained.

Adsorption isotherm of total copper complexes on resin

The adsorption isotherm of total copper complexes in non-ammoniacal and ammoniacal resin-solution systems was clearly shown in Figure 3. The ammonia concentration applied in the system for the adsorption isotherm of total copper on resin shown by the bottom plots of the equation Yc = -674.15x + 81630 was in the range of 50-400 with the equilibrium concentration of thiosulfate in solution being in the range of 6.932-85.841 mM. Meanwhile, when there was no ammonia applied, the system for the adsorption isotherm of total copper on resin are shown by the top plots of the equation Ya = -494.84x + 66491, and the equilibrium concentration of thiosulfate in solution was in the range of 9.016-143.073 mM. Increasing ammonia concentration in the system resulted in the lower total copper complexes adsorbed on resin.

Overall, although the slope of the bottom linear fitting line was less than the top linear fitting line, the adsorption of total copper complexes with the addition of ammonia in the system was not be lifted up to the one without the addition of ammonia because the total copper adsorbed on the top plot (see the G plot in Figure 3) of the bottom line consists 50 mM ammonia. Therefore, non-ammoniacal resin-solution (NARS) system gave much better adsorption isotherm of copper on the resin as clearly highlighted in the previous results and discussions.

The Dominant Species of Copper Complexes Based on Leaching

The leaching and adsorption of copper in non-ammoniacal and ammoniacal resin-solution systems can be a simultaneous complex process. Copper (I) in solution reacted with thiosulfate as the leaching reagent in excessive amount to form copper thiosulfate complexes in solution. In the same time, copper (I) in solution also reacted with ammonia being added to form copper amine complexes. The equilibrium reactions in the resin-solution systems for the leaching could be modelled by the following equations:

$$Cu^{+}_{(aq)} + S_2O_3{}^{2-}{}_{(aq)} <=> CuS_2O_3{}^{-}{}_{(aq)} \quad (1)$$

$$Cu^{+}_{(aq)} + 2S_2O_3{}^{2-}{}_{(aq)} <=> Cu(S_2O_3)_2{}^{3-}{}_{(aq)} \quad (2)$$

$$Cu^{+}_{(aq)} + 3S_2O_3{}^{2-}{}_{(aq)} <=> Cu(S_2O_3)_3{}^{5-}{}_{(aq)} \quad (3)$$

$$Cu^{+}_{(aq)} + NH_{3(aq)} <=> CuNH_3{}^{+}{}_{(aq)} \quad (4)$$

$$Cu^{+}_{(aq)} + 2NH_{3(aq)} <=> Cu(NH_3)_2{}^{+}{}_{(aq)} \quad (5)$$

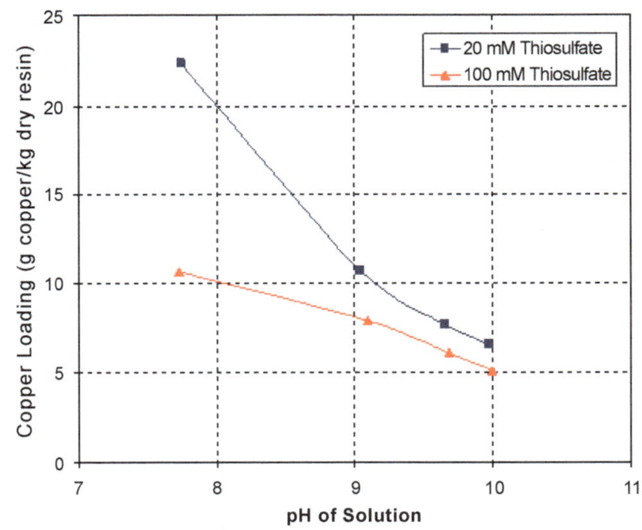

Figure 2: The copper complexes concentrations on resin over solution pH with various thiosulfate concentrations.

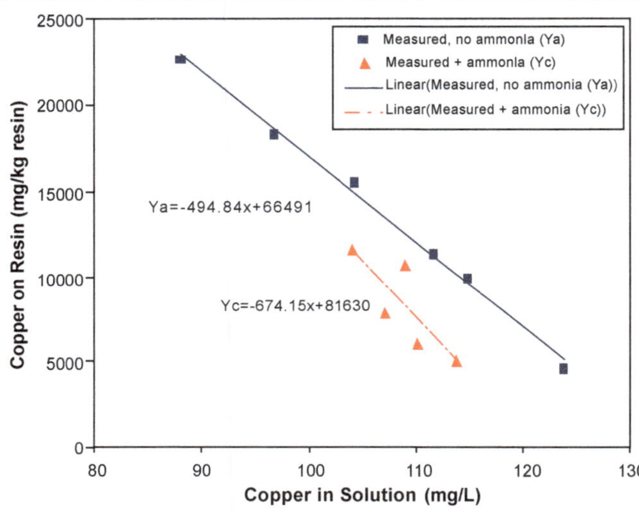

Figure 3: Adsorption isotherm of copper onto resin based on measurement in the system with and without the addition of ammonia..

$$Cu^{+}_{(aq)} + 3NH_{3(aq)} <=> Cu(NH_3)_3{}^{+}_{(aq)} \quad (6)$$

$$Cu^{+}_{(aq)} + NH_{3(aq)} + S_2O_3{}^{2-}_{(aq)} <=> CuNH_3S_2O_3{}^{-}_{(aq)} \quad (7)$$

$$Cu^{+}_{(aq)} + NH_{3(aq)} + 2S_2O_3{}^{2-}_{(aq)} <=> CuNH_3(S_2O_3)_2{}^{3-}_{(aq)} \quad (8)$$

where the stability constant for Equations (1), (2), (3), (4), (5), (6), (7) and (8) can be presented in $log\,\beta_1$, $log\,\beta_2$, $log\,\beta_3$, $log\,\beta_4$, $log\,\beta_5$, $log\,\beta_6$, $log\,\beta_7$ and $log\,\beta_8$, respectively. The value of stability constant is 9.29, 12.17, 14.48, 5.8, 10.37, 9.93, 12.67 and 14.02 respectively [20].

Due to the competitive adsorption of thiosulfate and trithionate, and trithionate is strongly loaded on resin [15], the copper thiosulfate complexes and the copper amine complexes was expected to react with trithionate more than thiosulfate on the strong based anion exchange resin. Meanwhile, non-ammoniacal resin-solution system resulted in the higher adsorption isotherm of copper on the resin as stated in the previous discussion. From these reasons and it can be expected from the value of stability constant, $Cu(S_2O_3)_3{}^{5-}$ was the dominant species in the leaching which was then adsorbed on resin to be the dominant species on resin.

However, further investigation was needed to obtain the most dominant species in the resin-solution system.

From Equations (1) to (8), there were 11 unknowns established with regards to the concentration of each species in leaching which are namely; $[Cu^+]$, $[CuS_2O_3{}^-]$, $[Cu(S_2O_3)_2{}^{3-}]$, $[Cu(S_2O_3)_3{}^{5-}]$, $[CuNH_3]$, $[Cu(NH_3)_2{}^+]$, $[Cu(NH_3)_3{}^+]$, $[CuNH_3S_2O_3{}^-]$, $[CuNH_3(S_2O_3)_2{}^{3-}]$, $[NH_3]$, and $[S_2O_3{}^{2-}]$. In order to be able to solve the 11 unknowns, at least 3 other equations were required. The required 3 equations were the 3 mass balance equations for each species, Equations (9) to (11):

$$[Cu\,(I)]_{total} = [Cu^+] + [CuS_2O_3{}^-] + [Cu(S_2O_3)_2{}^{3-}] + [Cu(S_2O_3)_3{}^{5-}] + [CuNH_3] + [Cu(NH_3)_2{}^+] + [Cu(NH_3)_3{}^+] + [CuNH_3S_2O_3{}^-] + [CuNH_3(S_2O_3)_2{}^{3-}] \quad (9)$$

$$[S_2O_3{}^{2-}]_{total} = [CuS_2O_3{}^-] + 2[Cu(S_2O_3)_2{}^{3-}] + 3[Cu(S_2O_3)_3{}^{5-}] + [CuNH_3S_2O_3{}^-] + 2[CuNH_3(S_2O_3)_2{}^{3-}] + [S_2O_3{}^{2-}] \quad (10)$$

$$[NH_3]_{total} = [CuNH_3] + 2[Cu(NH_3)_2{}^+] + 3[Cu(NH_3)_3{}^+] + [CuNH_3S_2O_3{}^-] + [CuNH_3(S_2O_3)_2{}^{3-}] + [NH_3] \quad (11)$$

To solve the 11 unknown, a numerical method utilising the multi-dimensional Newton-Raphson [21] approach in Excel software has been adopted. The column vector, x for the unknown was presented in Equation (12):

$$x = \begin{bmatrix} \ln[Cu^+] \\ \ln[CuS_2O_3{}^-] \\ \ln[Cu(S_2O_3)_2{}^{3-}] \\ \ln[Cu(S_2O_3)_3{}^{5-}] \\ \ln[CuNH_3{}^+] \\ \ln[Cu(NH_3)_2{}^+] \\ \ln[Cu(NH_3)_3{}^+] \\ \ln[CuNH_3S_2O_3{}^-] \\ \ln[CuNH_3(S_2O_3)_2{}^{3-}] \\ \ln[NH_3] \\ \ln[S_2O_3{}^{2-}] \end{bmatrix} \quad (12)$$

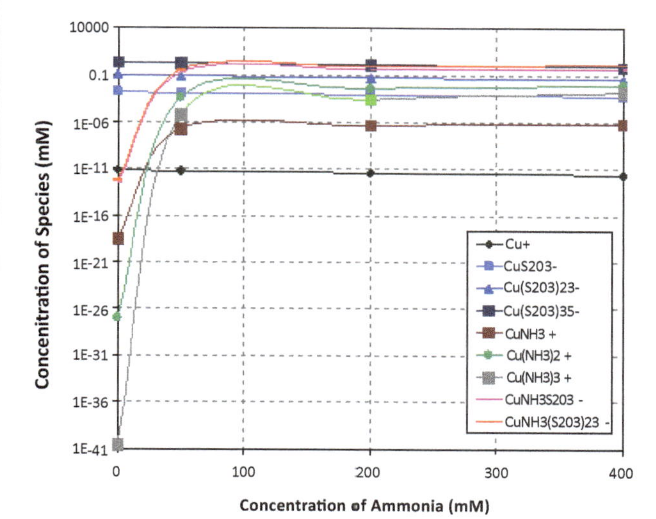

Figure 4: Concentration of copper complexes in the leaching solution over ammonia concentration based on leaching modeling.

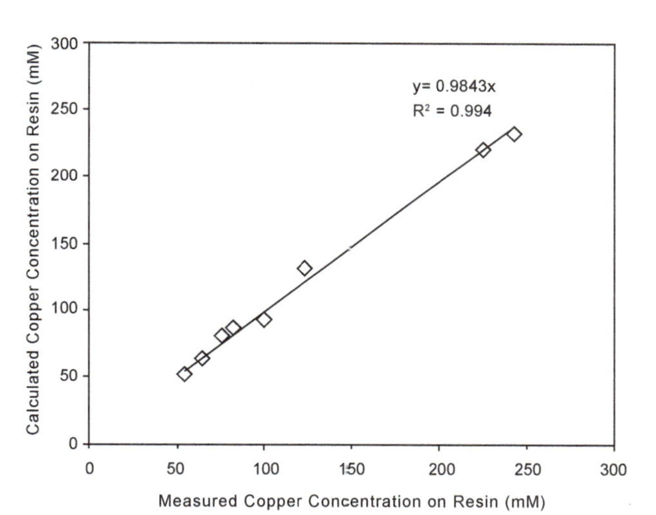

Figure 5: The relationship between the calculated copper concentration on resin and the measured copper concentration on resin.

The column vector, F for the eleven functions to solve was expressed as Equation (13), and the Jocobian matrix J was then expressed as Equations (14). As the result of leaching modelling for 100 mM thiosulfate, the concentration of copper complexes species in the solution can be predicted as shown in Figure 4. The species $Cu(S_2O_3)_3{}^{5-}$ was the dominant species in the solution of non-ammoniacal resin-solution system, which was approximately 1.899 mM. When ammonia concentration was increased to be 50 mM and 200 mM, this species was still the highest among others, which is approximately 1.515 mM and 0.944 mM, respectively (see Figure 4).

$$F = \begin{bmatrix} \ln\beta_1 + \ln\left[Cu^+\right] + \ln\left[S_2O_3^{2-}\right] - \ln\left[CuS_2O_3^-\right] \\ \ln\beta_2 + \ln\left[Cu^+\right] + 2\ln\left[S_2O_3^{2-}\right] - \ln\left[Cu(S_2O_3)_2^{3-}\right] \\ \ln\beta_3 + \ln\left[Cu^+\right] + 3\ln\left[S_2O_3^{2-}\right] - \ln\left[Cu(S_2O_3)_3^{5-}\right] \\ \ln\beta_4 + \ln\left[Cu^+\right] + \ln\left[NH_3\right] - \ln\left[CuNH_3^+\right] \\ \ln\beta_5 + \ln\left[Cu^+\right] + 2\ln\left[NH_3\right] - \ln\left[Cu(NH_3)_2^+\right] \\ \ln\beta_6 + \ln\left[Cu^+\right] + 3\ln\left[NH_3\right] - \ln\left[Cu(NH_3)_3^+\right] \\ \ln\beta_7 + \ln\left[Cu^+\right] + \ln\left[NH_3^+\right] + \ln\left[S_2O_3^{2-}\right] - \ln\left[CuNH_3S_2O_3^-\right] \\ \ln\beta_8 + \ln\left[Cu^+\right] + \ln\left[NH_3^+\right] + 2\ln\left[S_2O_3^{2-}\right] - \ln\left[CuNH_3(S_2O_3)_2^{3-}\right] \\ e^{\ln\left[Cu^+\right]} + e^{\ln\left[CuS_2O_3^-\right]} + e^{\ln\left[Cu(S_2O_3)_2^{3-}\right]} + e^{\ln\left[Cu(S_2O_3)_3^{5-}\right]} + e^{\ln\left[CuNH_3\right]} + e^{\ln\left[Cu(NH_3)_2^+\right]} + e^{\ln\left[Cu(NH_3)_3^+\right]} + e^{\ln\left[CuNH_3S_2O_3^-\right]} + e^{\ln\left[CuNH_3(S_2O_3)_2^{3-}\right]} - \left[Cu(I)\right]_{total} \\ e^{\ln\left[CuS_2O_3^-\right]} + 2e^{\ln\left[Cu(S_2O_3)_2^{3-}\right]} + 3e^{\ln\left[Cu(S_2O_3)_3^{5-}\right]} + e^{\ln\left[CuNH_3S_2O_3^-\right]} + 2e^{\ln\left[CuNH_3(S_2O_3)_2^{3-}\right]} + e^{\ln\left[S_2O_3^{2-}\right]} - \left[S_2O_3^{2-}\right]_{total} \\ e^{\ln\left[CuNH_3\right]} + 2e^{\ln\left[Cu(NH_3)_2^+\right]} + 3e^{\ln\left[Cu(NH_3)_3^+\right]} + e^{\ln\left[CuNH_3S_2O_3^-\right]} + 2e^{\ln\left[CuNH_3(S_2O_3)_2^{3-}\right]} + e^{\ln\left[NH_3\right]} - \left[NH_3\right]_{total} \end{bmatrix} \quad (13)$$

$$J = \begin{bmatrix} 1 & -1 & 0 & 0 & 0 & 0 & 0 & 0 & 0 & 0 & 1 \\ 1 & 0 & -1 & 0 & 0 & 0 & 0 & 0 & 0 & 0 & 2 \\ 1 & 0 & 0 & -1 & 0 & 0 & 0 & 0 & 0 & 0 & 3 \\ 1 & 0 & 0 & 0 & -1 & 0 & 0 & 0 & 0 & 1 & 0 \\ 1 & 0 & 0 & 0 & 0 & -1 & 0 & 0 & 0 & 2 & 0 \\ 1 & 0 & 0 & 0 & 0 & 0 & -1 & 0 & 0 & 3 & 0 \\ 1 & 0 & 0 & 0 & 0 & 0 & 0 & -1 & 0 & 1 & 1 \\ 1 & 0 & 0 & 0 & 0 & 0 & 0 & 0 & -1 & 1 & 2 \\ e^{\ln\left[Cu^+\right]} & e^{\ln\left[CuS_2O_3^-\right]} & e^{\ln\left[Cu(S_2O_3)_2^{3-}\right]} & e^{\ln\left[Cu(S_2O_3)_3^{5-}\right]} & e^{\ln\left[CuNH_3^+\right]} & e^{\ln\left[Cu(NH_3)_2^+\right]} & e^{\ln\left[Cu(NH_3)_3^+\right]} & e^{\ln\left[CuNH_3S_2O_3^-\right]} & e^{\ln\left[CuNH_3(S_2O_3)_2^{3-}\right]} & 0 & 0 \\ 0 & e^{\ln\left[CuS_2O_3^-\right]} & 2e^{\ln\left[Cu(S_2O_3)_2^{3-}\right]} & 3e^{\ln\left[Cu(S_2O_3)_3^{5-}\right]} & 0 & 0 & 0 & e^{\ln\left[CuNH_3S_2O_3^-\right]} & 2e^{\ln\left[CuNH_3(S_2O_3)_2^{3-}\right]} & 0 & e^{\ln\left[S_2O_3^{2-}\right]} \\ 0 & 0 & 0 & 0 & e^{\ln\left[CuNH_3^+\right]} & 2e^{\ln\left[Cu(NH_3)_2^+\right]} & 3e^{\ln\left[Cu(NH_3)_3^+\right]} & e^{\ln\left[CuNH_3S_2O_3^-\right]} & 2e^{\ln\left[CuNH_3(S_2O_3)_2^{3-}\right]} & e^{\ln\left[NH_3^+\right]} & 0 \end{bmatrix} \quad (14)$$

However, the dominant species in the solution was changed to be $CuNH_3(S_2O_3)_2^{3-}$ which was approximately 0.906 mM when ammonia concentration was increased to be 400 mM as can be seen in Figure 4. Meanwhile, the $Cu(S_2O_3)_3^{5-}$ concentration was approximately 0.628 mM.

The total copper complexes adsorbed the on resin can be calculated based on the initial copper concentration in leaching solution and the equilibrium copper concentration in leaching solution. The calculated total copper complexes on resin was compared to the measured total copper complexes on resin, and the relationship between the two results was illustrated in Figure 5. As obtained by Figure 5, the calculated result was in a good agreement with the measured result the correlation coefficient, R^2 being 0.994.

Conclusions

The proposed experimental procedures were considerable method for the adsorption of copper complexes on strong based anion exchange resin in non-ammoniacal and ammoniacal resin-solution systems. The effect of ammonia presence and pH was investigated. Increasing ammonia in the solution resulted in the decrease of total copper complexes on the resin. Other result showed that the solution pH of resin-solution systems was maintained by the ammonia solution added in the resin-solution system where in the higher pH was applied, the lower amount of copper complexes on resin was adsorbed. Compared to the ammoniacal resin-solution system, non-ammoniacal resin-solution (NARS) system gave much better adsorption isotherm of copper on resin. The dominant species copper complexes in solution of the NARS system was obtained to be $Cu(S_2O_3)_3^{5-}$ based on leaching modelling, and this species was expected to be the dominant species copper complexes adsorbed on resin.

Acknowledgment

The author is grateful to Curtin University of Technology and Australian Mineral Research Centre, CSIRO Mineral in Western Australia for the financial and technical support.

References

1. Bhappu RB (1990) Hydrometallurgical processing of precious metal ores. Gordon and Breach Science Publishers, New York.
2. Marsden J, House I (1992) The Chemistry of Gold Extraction, Ellis Horwood Ltd, London, UK.
3. Fagan PA, Paull B, Haddad PR, Dunne R, Kamar H (1997) Ion chromatographic analysis of cyanide in gold processing samples containing large concentrations of copper(I) and other metallo-cyanide complexes. J Chromatogr 770: 175-183.
4. Huang Q, Paull B, Haddad PR (1997) Optimisation of selectivity in the separation of metallo-cyanide complexes by ion-interaction liquid chromatography. J Chromatogr 770: 3-11.
5. Berezowsky R M, Sefton VB, Gormely LS (1978) Recovery of precious metals from metal sulfides. US Patent 4070182 Sherritt Gordon Mines Limited.
6. Zipperian D, Raghavan S, Wilson JP (1988) Gold and silver extraction by

ammoniacal thiosulfate leaching from a rhyolite ore. Hydrometallurgy 19: 361-375.

7. Abbruzzese C, Fornari P, Massidda R, Veglio F, Ubaldini S (1995) Thiosulphate leaching for gold hydrometallurgy. Hydrometallurgy 39: 265-276.

8. Weingarten S, Savoldelli D, Langhans W (1992) Enhancement or loss of the hypophagic effect of interleukin-1 upon chronic administration. Physiol Behav 52: 831-837.

9. Tozawa K, Inui Y, Umetsu Y (1981) Dissolution of gold in ammoniacal thiosulfate solution. Metallurgical Society AIME 12.

10. Li J, Miller JD, Wan RY, Levier KM (1995) The ammoniacal Thiosulphate System for Precious Metal Recovery. Littleton Co Soc for Min Metall and Explor.

11. Aylmore MG (2001) Treatment of a refractory gold–copper sulfide concentrate by copper ammoniacal thiosulfate leaching. Miner Eng 14: 615–637.

12. Jeffrey MI, Breuer RL, Choo WL (2001) A kinetic study that compares the leaching of gold in the cyanide, thiosulfate, and chloride systems. Metallurgical and Materials Transactions B 32: 979-986.

13. Breuer PL, Jeffrey MI (2003) Copper catalysed oxidation of thiosulfate by oxygen in gold leach solutions. Minerals Engineering 16: 21-30.

14. Jeffrey MI, Brunt SD (2007) The quantification of thiosulfate and polythionates in gold leach solutions and on anion exchange resins. Hydrometallurgy 89: 52-60.

15. Muslim (2009) Gold loading on ion exchange resins in non-ammoniacal resin-solution systems. Jurnal Rekayasa Kimia dan Lingkungan 7: 158-162.

16. Aksu S, Doyle FM (2002) Electrochemistry of Copper in Aqueous Ethylenediamine Solutions. J Electrochem Soc 149: B340-B347.

17. Waters (2002) Empower Software. Workgroup UPGRADE. Milford, USA.

18. Karen CT (2008) Chemistry: an introduction to general, organic and biological chemistry. Upper Saddle River, NJ, Pearson Prentice Hall.

19. Purolite (2008) Purolite® A500/2788.

20. Black SB (2006) The thermodynamic chemistry of the aqueous copper-ammonia thiosulfate system. Chemistry. Perth, Murdoch University.

21. Norman WL (2001) Applied mathematics methods for chemical engineers, Boca Raton, Florida, CRC Press LLC.

6

On Improved Mechanistic Modeling for Enzymatic Hydrolysis of Cellulose

Yang Zhang and Wen Zhou*

Department of Chemical Engineering, Michigan Technological University, Houghton, MI49931, USA

Abstract

An improved model for enzymatic hydrolysis of cellulose is developed that considers oligomer reactions with beta-glucanases, inhibition of oligomers to cellulases and enzyme decay processes during hydrolysis. Our oligomer reactions with beta-glucanases are modeled based on the enzymatic glucan chain fragmentation kinetics to describe the further fragmentation of oligomers in solution after being solubilized from the insoluble glucan chains. The inhibition effects on all cellulases by different types of cello-oligomers are then taken into account by competitive adsorption of cello-oligomers to the active site of cellulases, which is a critical factor contributing to the decrease in the rate of enzymatic hydrolysis of cellulose. As another factor affecting the kinetics of cellulose hydrolysis process, enzyme decay factor is incorporated into the model as the typical first order decay process. We consider two different processes for cellulases losing activity during hydrolysis in order to better understand the impact of enzyme decay on hydrolysis. Numerical simulation results are presented to investigate the phenomenon of hydrolysis rate slow-down commonly observed in experiments. Improvement of the predictive capability of the new model over previous one is clearly demonstrated by comparing the simulations with experimental data. By considering all the possible hydrolysis rate slow-down factors, the simulation results can agree with the experimental data very well, showing that our model is capable to fully capture the rate decrease of cellulose hydrolysis.

Keywords: Modeling; Enzymatic hydrolysis; Cellulose; Site formalism; Enzyme decay; Inhibition

Nomenclature

D_κ: Decay factor of type-κ enzymes

$I_\kappa\ (l)$: Oligomer adsorption coefficient for (κ, l) EO complexes (1/mM)

k, k': number of glucose units contained in a glucan chain

k: chain length, equal to the number of glucose units contained in a glucan chain

l_s: minimum insoluble chain length for glucan chains, =7

$L_{\kappa,\mu}$: substrate adsorption coefficient for (κ , μ) ES complexes (1/mM)

$R_S(l)$: changing rate of soluble oligomers in solution contained l glucose units

$R_{S,\sigma}(l)$: production rate of soluble oligomers contained l glucose units dissolved from glucan chains exposed on class-σ SAC surfaces (mM/min)

$T_{1/2k}$: half-life of type-κ enzymes (min)

u_κ: total concentration of type-κ enzymes (mM)

v_κ: concentration of free type-κ enzymes in solution (mM)]

$x_{S,\sigma}$: total concentration of glucose units dissolved from glucan chains exposed on class-σ SAC surfaces (mM)

$x_S\ (l)$: concentration of oligomers dissolved from glucan chains containing l glucose units

$y_{\mu,\sigma}$: concentration of free type-μ sites on glucan chains exposed on class-σ SAC surfaces (mM)

$y_S\ (l)$: concentration of free oligomers dissolved from glucan chains containing l glucose units

$z_{\kappa,\mu,\sigma}$: concentration of (κ, μ) ES complexes exposed on class-σ SAC surfaces (mM)

$z_\kappa(l)$:concentration of (κ, l) EO complexes formed by type-κ enzymes and oligomers containing l glucose units (mM)

κ: index of enzyme types

$\gamma_{\kappa,\mu}$: cutting rate coefficient for (κ , μ) ES complexes (cuts per (κ , μ) ES complex per time)

μ, μ': index of site types

σ: index of SAC classes

Introduction

The enzymatic hydrolysis of cellulose into soluble and fermentable oligomers (e.g., glucose and cellobiose) has been under intensive investigation due to the potential utilization of lignocellulosic biomass to produce sustainable biofuel and replace the non-renewable fossil transportation fuel. In order to optimize the design of reactors and the biofuel production process, it is critical to have a mechanistic model describing the hydrolysis kinetics of solid cellulosic substrates being solubilized by all kinds of cellulases in detail. Unlike non-mechanistic and semi-mechanistic models which usually include less than two substrate and/or enzyme variables and are used to fit experimental data [1], mechanistic models involve multiple substrate and enzyme variables and can provide insights on the complex chemical and physical properties of both enzyme and substrate and all the enzymatic and material transformations occurring during hydrolysis.

Over the past 10 years, many advanced mechanistic models have

***Corresponding author:** Wen Zhou, Department of Chemical Engineering, Michigan Technological University, Houghton, MI49931, USA been developed in full generality to describe the enzymatic hydrolysis process of cellulose based on more realistic representations of cellulosic substrate [2-5]. For example, Zhou et al. [2] developed a detailed mechanistic model for enzymatic hydrolysis of cellulose considering the substrate morphologies and their coupling with morphology-depended substrate hydrolysis kinetics. As described in the model, the whole process of cellulose hydrolysis could be viewed as the process of fragmentation and solubilization of glucan chains and lead to the evolution of cellulosic substrate morphology, that is, the organization of glucan chains, which, in turn, significantly influenced the cellulose hydrolysis kinetics. The concept of smallest accessible compartment (SAC) was first proposed, defined as a minimal volume by external and internal surfaces exposed to enzyme-accessible hydrated interior voids of the solid cellulosic substrate material, and used as minimal time-evolving structural unit to keep track of the cellulosic substrate morphology. In addition, a site representation formalism of enzyme hydrolytic fragmentation coupled with morphology evolution was also introduced in the model. The formalism considered all the β-(1, 4)-glycosidic bonds on glucan chains as six types of bond sites based on their locations and reactions with different kinds of cellulases that mainly act on the solid substrate. Basically, these cellulases can be categorized into two broad classes, which are endo-glucanases and exo-glucanases. Endo-glucanases usually adsorb onto the glucan chains of insoluble part of substrate and randomly cut the internal β-(1,4)-glycosidic bonds. Unlike endo-glucanases, exo-glucanases only cut the terminal β-(1,4)-glycosidic bonds at the ends of each glucan chain. Since the two ends of each glucan chain are chemically distinct from each other, exo-glucanase can be divided into two groups, which are cellobiohydrolase I (CBH I) and cellobiohydrolase II (CBH II). CBH IIs usually cut the terminal bonds from the non-reducing end of each glucan chain, while CBH Is cut the terminal bonds from the reducing end of each glucan chain. These cellulases can be produced in nature by many different cellulolytic fungi species. The most commonly used species in industry is *Trichoderma* species, especially *Trichoderma reesei* (*T. reesei*). Thus *T. reesei* EG1 (Cel7B), *T. reesei* CBH II (Cel6B) and *T. reesei* CBH I (Cel7A) are the three major cellulases that often used in model simulation work to analyze the process of enzymatic hydrolysis of cellulosic substrate. The site representation formalism was proposed to not only take into account all kinds of solid-substrate-acting cellulases but also change the view of solid cellulosic substrate from a bundle of glucan chains with different lengths into a composite of six types of bond sites, so that the total number of ordinary differential equations (ODEs) can be reduced at magnitude of two orders and solved much more efficiently.

Almost all of the mechanistic models have recognized that the enzyme accessibility of substrate, or the enzyme-accessible substrate surface area, is a critical rate-limiting factor during the process of enzymatic hydrolysis [2-4]. In order to increase the kinetics of hydrolysis and obtain more soluble oligomers, it will be helpful to increase the amount of accessible glucose units (β-(1,4)-glycosidic bonds) exposed on substrate surface by pretreatments before hydrolysis proceeding. However, most of the simulation results from the mechanistic models still could not agree with the experimental data completely, especially not be capable to capture the full extent of the hydrolysis rate slow-down, probably due to the reason that these models did not consider the complex inhibition relationships between substrate and oligomers except glucose and cellobiose, and possible enzyme decay and inactive absorption. Some of the models also did not consider the oligomer reactions which happen in liquid phase with the beta-glucanases. Unlike endo- and exo-glucanases, beta-glucanases mainly act on soluble cello-oligomers in solution which often contains no more than 7 glucose units. The most commonly used cellulolytic fungi species to created beta-glucanases in industry is *Aspergillus* species, especially *Aspergillus niger*

(*A. niger*) because *Trichoderma* species do not produce significant amount of beta-glucanases compared to other cellulases. *A. niger* beta-glucosidase (BG) is the major beta-glucanase and often used with other cellulases to increase their efficiency during enzymatic hydrolysis of cellulose, because dissolved oligomers could cause strong inhibition effects by adsorbing the free cellulase molecules. BGs can hydrolyze soluble oligomers into smaller ones mainly by releasing glucose from them and avoid the strong inhibitions between cellulases and long-chain oligomers. It is clear that the oligomer reactions with beta-glucanase will affect hydrolysis kinetics and change inhibition equilibrium. Ignoring the kinetics slow-down factors will also hinder the precise prediction of final conversion level of cellulosic substrate after a long time-scale of enzymatic hydrolysis process.

In this study, we improve the model described by Zhou et al. [2] by incorporating (1) the reactions involving beta-glucanases digesting soluble oligomer sugars in solution, (2) the comprehensive competitive inhibition effects of oligomers from glucose (G1) to cellohexaose (G6) on enzymatic hydrolysis, and (3) the enzymatic thermal deactivation and inactive adsorption. Here, we consider enzymatic thermal deactivation as the process that the hydrolytic enzymes lose the ability of binding to substrates, and consider enzymatic inactive adsorption as the process that the hydrolytic enzymes lose the ability to continue the subsequent catalytic reactions although they are able to bind to substrate. We carry out numerical analysis to investigate the impacts of aforementioned factors on the kinetics of enzymatic hydrolysis of cellulose.

Materials and Methods

Oligomer reactions with beta-glucanases

The basic part of the model we used in this study was from the mechanistic model of enzymatic hydrolysis of cellulose presented by Zhou et al. [2]. In their work, the concept of smallest accessible compartment (SAC) was first proposed to describe the geometrical construction of pretreated pure cellulosic substrate material. An SAC unit was defined as a minimal volume that is delimited by external surfaces and by internal surfaces at the internal surfaces of the voids exposed to all kinds of cellulases. SACs initially had a random distribution of sizes, and to represent the distribution, the parameter of SAC geometry class was used and labeled by "σ" with σ=1,2... MMD where MMD was the population size of SAC geometry classes. During the process of enzymatic hydrolysis, the size of each SAC would shrink due to the enzymatic ablation of glucose units from the SAC surfaces. And glucose units which were originally blocked would gradually become exposed on the SAC surfaces. In addition, Zhou et al. [2] proposed in their work an innovative surface site concentration formalism which treated each SAC as a composite of six different types of β-(1,4)-glycosidic bond sites. These site types were labeled by index "μ" and referred to as N-, O-, X-, Y-, L-, and R-sites. N-site bonds could only adsorb, and only be cut by endo-glucanases. O-site bonds could not adsorb any enzyme molecule due to the obstruction. X-site bonds could adsorb and be cut by either endo-glucanases or cellobiohydrolase IIs, which were located a distance of k_X glucose units from the non-reducing ends of glucan chains. Y-site bonds could adsorb and be cut by either endo-glucanases or cellobiohydrolase Is, which were located a distance of k_Y glucose units from the reducing ends of glucan chains.

L- and R-site bonds represented the non-reducing and reducing ends of glucan chains respectively. They were both broken bonds and could not adsorb any enzyme molecule. Then the production rate of glucose units dissolved from glucan chains of class-σ SACs was given by

$$\dot{x}_{S,\sigma} = \sum_{l=1}^{l_s-1} l \cdot R_{S,\sigma}(l) \quad (1)$$

where $R_{S,\sigma}(l)$ is the production rate of oligomers containing l glucose units ($1 \leq l < l_s = 7$) dissolved from insoluble glucan chains exposed on class-σ SAC surface caused by the enzymatic cuts on β-(1,4)-glycosidic bonds. l_s is the minimal length of insoluble glucan chains, which could be varied into any length if needed. In the mode, we considered cellohexaose (G6) as the longest cello-oligomer in solution based on the typical industrial applications.

We improved the mechanistic model to not only consider the processes that endo- or exo-glucanases cut the β-(1,4)-glycosidic bonds of insoluble glucan chains to produce oligomers, but also consider explicitly the reactions of oligomers after being released from solid substrates that involve beta-glucanases in solution. The production rates of soluble oligomers in solution are given by

$$x_s(l) = \Sigma R_{S,\sigma}(l) + R_s(l) \quad (2)$$

where $R_s(l)$ is the production rate of the oligomers containing l monomer units after being dissolved from glucan chains, which describes the reactions between oligomers and beta-glucanases. $R_S(l)$ did not exist in the model of Zhou et al. [2] where all the oligomers kept increasing in solution by being hydrolyzed from solid substrate.

Based on the study of sub-site structure of *A. niger* beta-glucosidase [6], the reaction mechanism between oligomers and beta-glucanases in solution can be considered as: an oligomer which contains l glucose units can adsorb and be hydrolyzed by an beta-glucanase molecule into a glucose unit and an oligomer containing l–1 glucose unit(s). So the expression of $R_s(l)$

$$\begin{cases} R_{Sk}(l) = \gamma_\kappa(2)\cdot z(2)_\kappa + \sum_{2}^{l_s-1} \gamma_\kappa(l)\cdot z(1) \\ R_{Sk}(l) = -\gamma_\kappa(l)\cdot z(l_\kappa) + \gamma(l+1)\cdot z(l+1)_S \quad 2 \leq l \leq l\text{-}2 \\ R_{Sk}(l_{kS} -_S 1) = -\gamma_\kappa(l_S - 1)\cdot z(l-1) \end{cases} \quad (3)$$

where $\gamma_\kappa(l)$ is the oligomer cutting rate coefficient, $z_\kappa(l)$ is the concentration of (κ, l) enzyme-oligomer (EO) complexes where the oligomers contain l glucose units adsorb the type-κ beta-glucosidase enzyme molecules. The expression of EO complexes $z_\kappa(l)$ can be written, based on the enzyme adsorption equilibrium, as

$$Z_\kappa(l) = I_\kappa(l).v_\kappa.y_S(l) \quad (4)$$

with $l \leq l < l_s = 7$. Here, $I_\kappa(l)$ is the oligomer adsorption coefficient; $y_S(l)$ is the concentration of free oligomers containing l glucose units dissolved from glucan chains, and v_κ is the concentrations of type-κ free enzyme molecules in solution which could be any type of cellulases.

Inhibition effects

Cellulase inhibition describes the process that inhibitors limit the activity of cellulase molecules. Based on the impacts on cellulase, cellulase inhibitors can be classified into reversible and irreversible inhibitors in which the oligomer products of cellulose hydrolysis (e.g. glucose and cellobiose) are reversible inhibitors [7]. Reversible inhibitors can be classified into three types, which are competitive, uncompetitive and mix (or non-competitive) inhibitors. Most of the inhibition processes between the cello-oligomer inhibitors and cellulases are competitive inhibitions [7] where inhibitors can only bind free cellulase molecules in solution, which can also be described by adsorption Equation (4), where v_κ is now the concentration of type-κ cellulose (both endo- and exo- glucanases) free enzyme molecules in solution, and $I_\kappa(l)$ is the oligomer adsorption coefficient specifically for different cellulases.

If the EO complexes are formed by oligomers and endo- (or exo-) glucanases by oligomers binding onto active sites of binding domain of cellulases, and since we assume that catalytic domain cannot act on soluble oligomers, their corresponding values of $\gamma_\kappa(l)$, oligomer cutting rate coefficient, will be 0. Similarly, we could consider the inhibition of glucose (G1) to the beta-glucanases hydrolysis, where glucose binds onto the active site of beta-glucanase and no further catalytic action can be taken. Thus for beta-glucanases, if they are adsorbed by glucose units, $\gamma_\kappa(l)$ will also be 0 in that glucose units are the finial products of oligomer reactions and cannot be further dissolved. In Equation (4), v_κ and $y_S(l)$ are related to their corresponding total concentrations u_κ and $x_S(l)$. The relationships are written as

$$u_\kappa = v_\kappa + \sum_{\mu,\sigma} z_{\kappa,\mu,\sigma} + \sum_{l} z_\kappa(l) \quad (5)$$

$$x_{s\kappa}(l) = y_s(l) + \sum_{\kappa} z(l) \quad (6)$$

$z_{\kappa,\mu,\sigma}$ is the concentration of enzyme-substrate complexes formed by type-μ bond sites and type-κ enzyme molecules on class-σ SAC surfaces. Equations (5) and (6) can be used to find the expressions of v_κ and $y_S(l)$ as functions of u_κ and $x_S(l)$ respectively, which are written as

$$v_k = \frac{u_k}{1 + \sum_{\mu,\sigma} L_{k,\mu}\cdot y_{\mu,\sigma} + \sum_{l} I_k(l)\cdot y_s(l)} \quad (7)$$

$$y_s(l) = \frac{x_s(l)}{1 + \sum_{k} I_k(l)\cdot v_k} \quad (8)$$

where $y_{\mu,\sigma}$ is the concentration of free type-μ sites on glucan chains exposed on class-σ SAC surfaces, $L_{\kappa,\mu}$ is the adsorption coefficient between type-κ cellulase molecules and type-μ sites on glucan chains. Then the more detailed expressions of $R_S(l)$ can be obtained by combining Equations (3), (4), (7) and (8).

Enzymatic thermal deactivation and inactive adsorption

Many experiments illustrated that even after alleviating the inhibition effects mostly, the rate of enzymatic hydrolysis of cellulose still decreases during hydrolysis, meaning that inhibition effect is not the only rate-limiting factor [4,8]. Many studies considered the process of cellulases losing activities as a contributing factor to the hydrolysis rate slow-down, which was called enzyme decay process and often modeled as a first order process [4,8]. Enzyme decay process naturally happen to all hydrolytic enzymes based on experimental observations [9,10]. In principle, the enzyme decay will make the enzymes lose their hydrolytic capability through two possible meanings according to the specific structure of enzymes. Most of the hydrolytic enzymes contain a carbohydrate binding domain and a catalytic domain. Typically, these two domains are away from each other and connected by a long segment of linker sequence. Some enzymes, however, only contain a catalytic core domain, such as *T. reesei* EG III (Cel12A), which bind to substrate through the substrate-binding sites in the catalytic core domain. So if the carbohydrate binding domains or the substrate-binding sites are deactivated, hydrolytic enzymes will lose the capability of binding to the cellulosic substrate chain sites. On the other hand, if the catalytic domain is deactivated, hydrolytic enzymes are still able to bind to substrate chain sites, but will not carry out the subsequent catalytic reactions.

We incorporated into the model these two different mechanisms, which are referred to as the model the enzymatic thermal deactivation and the enzymatic inactive adsorption, respectively. We defined the enzymatic thermal deactivation in the model as the process that the binding domains or sites of cellulase molecules kept losing the ability to adsorb on glucan chains during hydrolysis. By contrast, we defined enzymatic inactive adsorption in the model as the process that the catalytic domains of all cellulases kept losing the reactivity to cut glucan chains after adsorbing on the substrate. For a single cellulase molecule, the binding and catalyzing abilities could keep losing simultaneously. The decay processes of cellulase may be caused by the affection of temperature during hydrolysis, since each type of enzymes has its own optimal temperature to work efficiently for certain period of time.

The enzyme decay factor D_κ is set to $D_\kappa = (0.5)^{t/T_{1/2,\kappa}}$ in the model for type-κ enzyme molecules where $T_{1/2,\kappa}$ is the half-life for type-κ enzyme molecules, which represent the time for an enzyme lost half of its enzymatic activity. We apply the decay factor D_κ to parameters u_κ and $z_{\kappa,\mu,\sigma}$ to represent the processes of enzymatic thermal deactivation and enzymatic inactive adsorption respectively and test the impact range of enzyme decay on the cellulose hydrolysis kinetics.

Model parameters

The specific activity and adsorption equilibrium coefficients of Endo-glucanase (EG1) and Exo-glucanases including Cellobiohydrolase I and II (CBH I and II) are from the work by Zhang and lynd [3]. For beta-glucanase (BG), the values of adsorption and kinetics parameters describing the reactions with oligomers are from the literature [6,11]. In the model, the values κ=1,2,3 and 4 represent EG1 CBH II CBH I and BG. The values of inhibition parameters are from various literature sources [4,12,13] due to the fact that we considered several different inhibition effects involving 4 different types of cellulases (EG1, CBHI, CBHII and BG) and 6 types of oligomers from glucose (G1) to cellohexaose (G6). The half-life of all kinds of cellulases was set to 42.5h as reported in the literature [14] (Table 1).

Results and Discussion

Testing cellulosic substrate accessibility

In this section, we investigate the effect of enzymatic accessibility of cellulosic substrates. In the work of Zhou et al. [15], the initial value of F_a for Avicel, that is, the ratio of enzyme-accessible glucose units to the total number of glucose units in Avicel was set to 0.00620, which was originally described in the literature [1]. Not until recently, Hong et al. [9] determined that the value of F_a for Avicel should be 0.00232 from the experiments testing the adsorption of fluorescent cellulase-like molecules on the substrate surface. We test some new values of F_a and compare the new simulation results with the original ones from the work of Zhou et al. [16]. The 12 experimental data [9,10] was used in the work of Zhou et al. [16] to compare their simulation results, and will be adopted in this work.

Figure 1 clearly shows that increasing the enzymatic accessibility for substrate prior to hydrolysis could allow more enzyme molecules attack the bonds on glucan chains and thus increase the hydrolysis rate and finial conversion level of substrate. Substrate accessibility in hydrolysis is believed to govern the entire hydrolysis process and act as a critical rate-affecting factor during hydrolysis. The previous initial value 0.00620 for F_a was obtained based on the nitrogen BET measurements. In the process of such adsorption-based measurements, nitrogen was usually used to test the initial accessible surface area of substrate. However, since nitrogen molecules are much smaller than enzyme molecules, and the substrate need to be in dry conditions, the value is probably overestimated and thus not accurate. Hong et al. [9], determined the accessible surface area of cellulosic substrate by testing the adsorption of molecules containing fluorescent proteins and cellulose-binding modules (CBMs), which had similar size as cellulases could be quantitatively tested. However, the value 0.00232 may be underestimated due to the fact that only one type of cellulose-binding module was used to create cellulase-like molecules. Recently, Levin et al. [4] pointed out in their work that the initial value of F_a from the work of Hong et al. [9] might be underestimated Based on their estimation by using the Random Sequential Adsorption (RSA) simulation process, the initial accessible surface area of Avicel should be at least 3.5 m^2g^{-1}, meaning that initial value of F_a should be 0.00341 for Avicel. We believe that the initial value of F_a should be set between 0.00232 and 0.00620 so that 0.00341 is more accurate and will be used in next sections. The parameter of cellulosic substrate accessibility is critical and reflects the cellulase-accessible surface area. However, it was estimated that during the early stage of hydrolysis the substrate accessibility could either decrease or increase if using different substrate morphologies [15], remaining the impact of substrate accessibility on the hydrolysis rate unclear. Most of the experiments only tested the value of substrate accessibility at the beginning and the end of hydrolysis and did not keep track of its changing through the entire process. So in order to better understand the role of substrate accessibility, it is highly recommended to test substrate accessibility at different time points during hydrolysis. By using smaller initial values for F_a, the new hydrolysis rates are a little slower than the original ones but still could not capture the full extent of the hydrolysis rate decreases shown in the experimental data. The enormous differences between the simulation results and experimental data also indicate that only considering substrate morphology as hydrolysis rate-limiting factor is not enough to reproduce the kinetics

Parameter	Value	Units	Source
$L_{1,N}$ $L_{1,X}$ $L_{1,Y}$	3	1/mM	[2,3]
$L_{2,X}$ $L_{3,Y}$	4	1/mM	[2,3]
$L_{3,N,Glu}$ $L_{3,X,Glu}$ $L_{2,N,Glu}$ $L_{2,Y,Glu}$	0	1/mM	[2,3]
$\gamma_{1,N}$ $\gamma_{1,X}$ $\gamma_{1,Y}$	23.5	1/min	[2,3]
$\gamma_{2,X}$	10.6	1/min	[2,3]
$\gamma_{3,Y}$	6.2	1/min	[2,3]
$\gamma_{2,N}$ $\gamma_{2,Y}$ $\gamma_{3,N}$ $\gamma_{3,X}$	0	1/min	[2,3]
$I_1(1)$	0.06	1/mM	[4]
$I_2(1)$ $I_3(1)$	0.032	1/mM	[4]
$I_1(2)$ I2 (2) I3 (2)	0.13	1/mM	[12]
$I_1(3)$ I2 (3) I3 (3)	0.3	1/mM	[12,13]
$I_1(4)$ I2 (4) I3 (4)	0.37	1/mM	[12,13]
$I_1(5)$ I2 (5) I3 (5)	0.44	1/mM	[12,13]
$I_1(6)$ I2 (6) I3 (6)	0.51	1/mM	[12,13]
$I_4(1)$	0.294	1/mM	[6]
$I_4(2)$	1.136	1/mM	[6,11]
$I_4(3)$	3.846	1/mM	[6,11]
$I_4(4)$	4	1/mM	[6,11]
$I_4(5)$	2.174	1/mM	[6,11]
$I_4(6)$	1.449	1/mM	[6,11]
$\gamma_4(1)$	0	1/min	
$\gamma_4(2)$	1897	1/min	[6]
$\gamma_4(3)$	1739	1/min	[6,11]
$\gamma_4(4)$	1423	1/min	[6,11]
$\gamma_4(5)$	895.8	1/min	[6,11]
$\gamma_4(6)$	843.1	1/min	[6,11]

Table 1: Parameters used in the model.

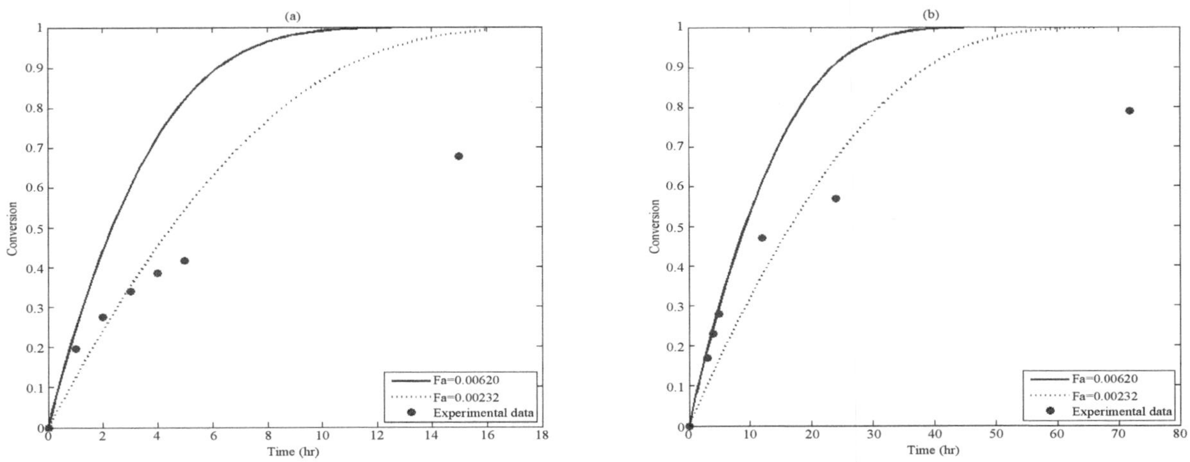

Figure 1: Simulation results using different initial values of Fa compared with experimental data. The substrate and enzyme have initial properties of (a) 20 g/L Avicel and 4.1 g/L Spezyme CP [9] and (b) 10 g/L Avicel and 0.51 g/L Spezyme CP together with 30 IU beta-glucanase/g Avicel (57.69 mg/L beta-glucanase) [10]. Commercial enzyme Spezyme CP consists EG1, CBHI and CBHII in a mass ratio of 0.17:0.24:0.13 (EG1/CBHI/CBHII) [17].

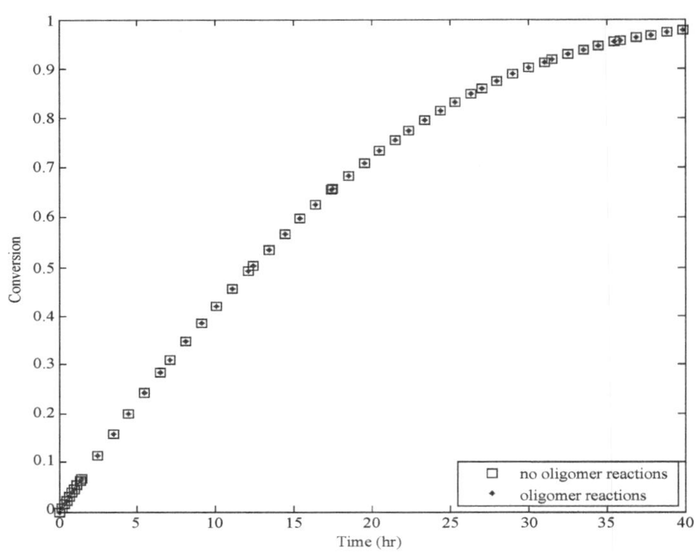

Figure 2: Simulation results of finial conversion level with and without consideration of oligomer reactions in solution. The substrate and enzyme have initial properties of 10 g/L Avicel and 0.51 g/L Spezyme CP together with 30 IU beta-glucanase/g Avicel (57.69 mg/L beta-glucanase) [10].

slow-down phenomenon during the process of enzymatic hydrolysis of cellulose. More hydrolysis rate-limiting factors, such as oligomer inhibition, should be considered when constructing a model of enzymatic hydrolysis of cellulose.

Oligomer reactions

In this section, we demonstrate the simulation results after considering the oligomer reactions in solution without inhibition effects. In the work of Zhou et al. [2], the reactions involving beta-glucanases were not concerned. Although the main rate-limiting steps of the cellulose hydrolysis process is the reactions that endo- and exo-glucanases depolymerize insoluble long glucan chains into soluble oligomers, the changes of oligomers in solution could also affect the hydrolysis kinetics in turn and thus could not be ignored.

Figure 2 shows the comparison of final conversion level of cellulosic substrate between simulations considering and not considering oligomer reactions in solution, as we excepted, there is no difference of simulation results after incorporating oligomer reactions. Because the conversion level describe the ratio of the total amount of dissolved substrate to the initial amount of substrate. According to the mass balance, although the amount of each type of oligomer will be changed after considering oligomer reaction, the total amount of soluble substrate should be the same as that considers no oligomer reaction. In Figures 3 and 4, differences are illustrated for cello-oligomers during hydrolysis, indicating that if considering oligomer inhibition effects, simulation results taking into account the oligomer reactions will be more accurate.

Inhibition effects

In this section, we demonstrate the simulation results after considering the oligomer reactions in solution and all kinds of

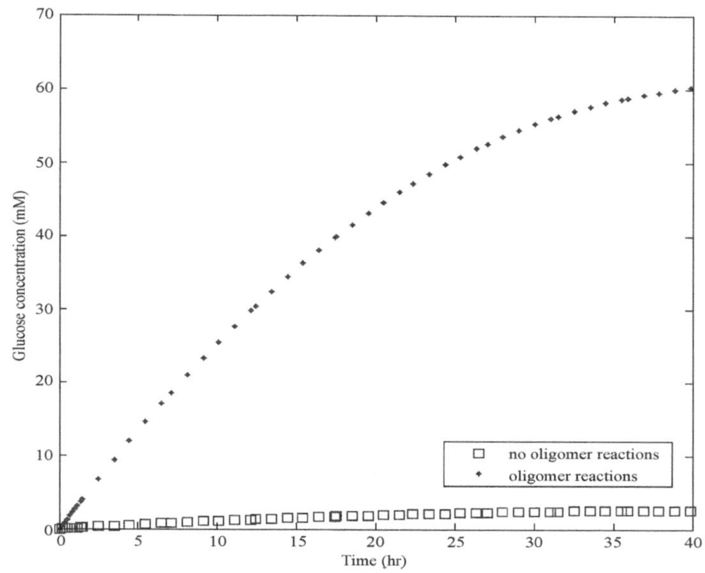

Figure 3: Simulation results of glucose concentration with and without consideration of oligomer reactions in solution. The substrate and enzyme have initial properties of 10 g/L Avicel and 0.51 g/L Spezyme CP together with 30 IU beta-glucanase/g Avicel (57.69 mg/L beta-glucanase) [10].

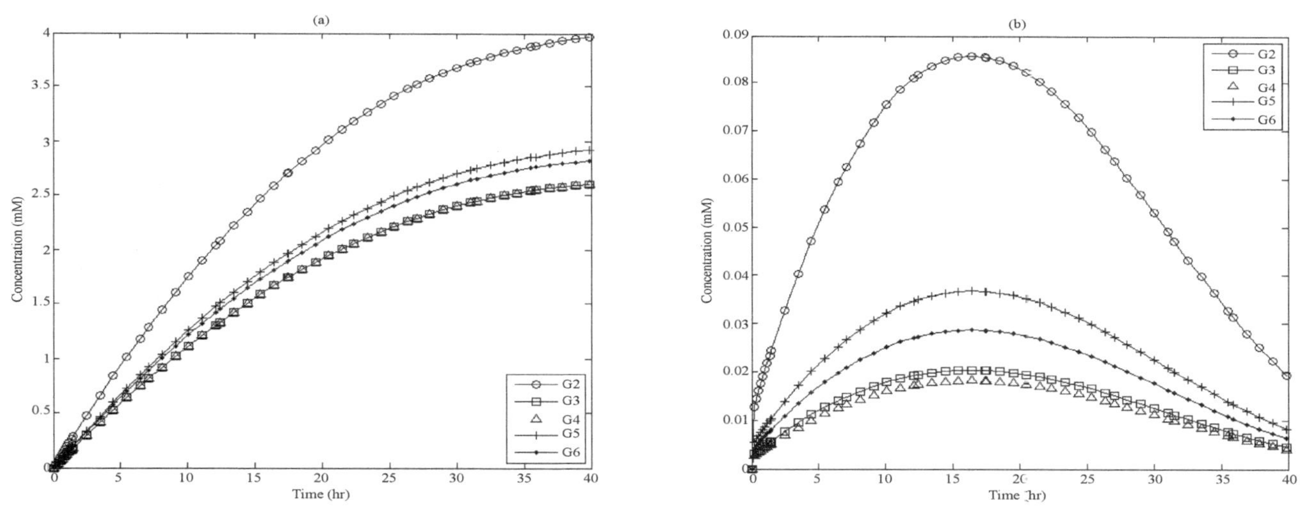

Figure 4: Simulation results of oligomer concentrations (a) without and (b) with consideration of oligomer reactions in solution. The substrate and enzyme have initial properties of 10 g/L Avicel and 0.51 g/L Spezyme CP together with 30 IU beta-glucanase/g Avicel (57.69 mg/L beta-glucanase) [10].

competitive inhibition effects between 4 types of cellulases and 6 types of cello-oligomers. As shown in Figure 5, after incorporating both the oligomer reactions and inhibitions, the simulation results can capture some extent of the decrease in the rate of cellulose hydrolysis often observed in experiments, showing that inhibition is one of the critical factors contributing to the phenomenon of hydrolysis rate slow-down.

However, the process of inhibition still remains to elucidate in more details. Although we considered all kinds of competitive inhibitions that could possibly happen in solution during enzymatic hydrolysis, it still might exist other types of inhibition relationships, such as uncompetitive inhibitions of cellobiose which can not only bind free cellulase molecules but also EO or ES complexes. Deeper understanding of the inhibitions during cellulose hydrolysis is needed to improve the kinetic models.

Enzyme decay and the slow-down of enzymatic hydrolysis kinetics

In order to reproduce the phenomenon of hydrolysis rate slow-down, we first test the enzymatic decay factor in the model. It is clear that only considering the inhibition effect during hydrolysis is not enough to investigate the rate decrease of enzymatic hydrolysis of cellulose.

Similar view was given in the work of Levin et al. [4] where they tried to fit the data by using much shorter estimated half-life values

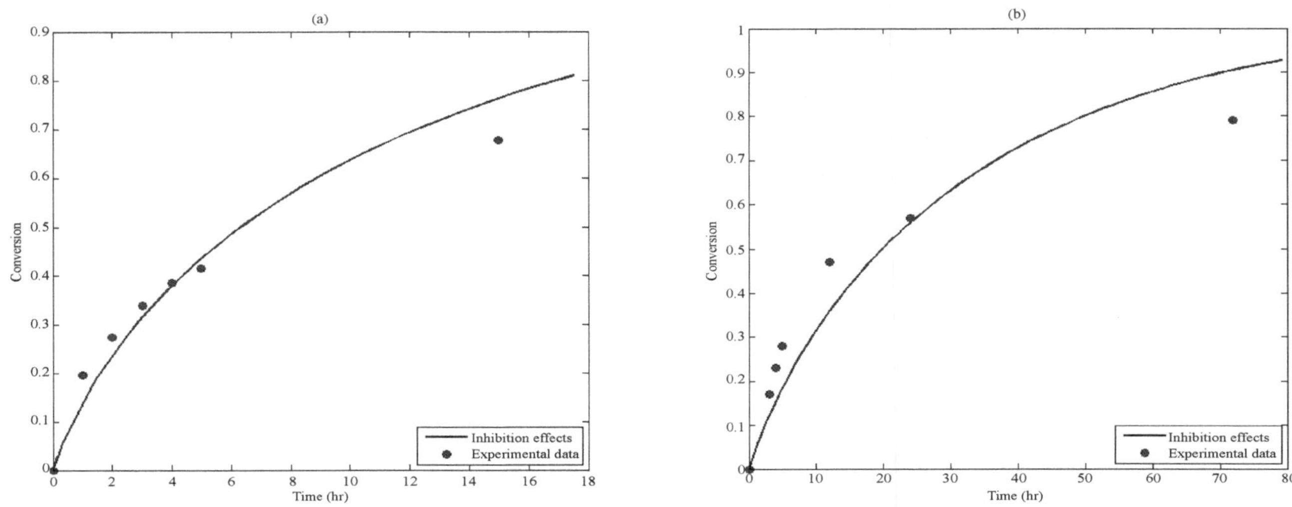

Figure 5: Simulation results considering oligomer reactions and inhibitions compared with experimental data. The substrate and enzyme have initial properties of (a) 20 g/L Avicel and 4.1 g/L Spezyme CP [9] and (b) 10 g/L Avicel and 0.51 g/L Spezyme CP together with 30 IU beta-glucanase/g Avicel (57.69 mg/L beta-glucanase) [10]. Commercial enzyme Spezyme CP consists EG1, CBHI and CBHII in a mass ratio of 0.17:0.24:0.13 (EG1/CBHI/CBHII) [17].

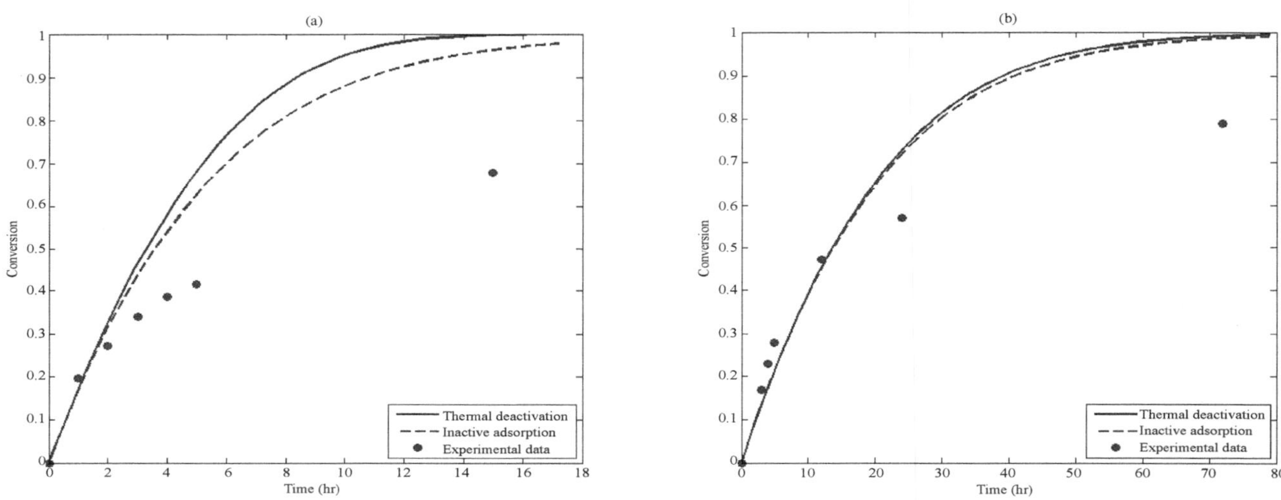

Figure 6: Simulation results considering different enzyme decay factors compared with experimental data. The substrate and enzyme have initial properties of (a) 20 g/L Avicel and 4.1 g/L Spezyme CP [9] and (b) 10 g/L Avicel and 0.51 g/L Spezyme CP together with 30 IU beta-glucanase/g Avicel (57.69 mg/L beta-glucanase) [10]. Commercial enzyme Spezyme CP consists EG1, CBHI and CBHII in a mass ratio of 0.17:0.24:0.13 (EG1/CBHI/CBHII) [17].

for cellulases. Here, we use the experimentally-determined half-life values for all the 4 types of cellulases in the model, and test the impact range between enzymatic thermal deactivation and enzymatic inactive adsorption on the hydrolysis kinetics. As shown in Figure 6 the impacts of enzymatic thermal deactivation and enzymatic inactive adsorption on the hydrolysis rate are very close to each other, probably because they are both reflecting the affection of temperature on the rate of hydrolysis. Also, both the impacts of enzymatic thermal deactivation and enzymatic inactive adsorption are weaker than the impact of the inhibition effects if comparing the results in Figures 5 and 6, indicating that inhibition is the prime factor contributing to the decrease in the rate of hydrolysis. After combining all the possible factor that cause the hydrolysis rate slow-down, the simulation results can agree with the experimental data very well in Figure 7, which indicate that considering inhibition and enzyme decay together in the model of enzymatic hydrolysis of cellulose can capture the full extent of the rate decrease phenomenon during the hydrolysis process. In order to enhance the hydrolysis rate and the conversion level of cellulosic substrate, a route is to increase the loading or improve the specific activities of cellulases. However, increasing the amount of cellulases will causes a huge waste of cellulase since only a small part of free cellulase molecules could adsorb onto the substrate and as hydrolysis proceeds all the cellulase molecules will decay and gradually lose the abilities of binding and catalyzing. Besides, since the oligomer inhibition effects impact the hydrolysis rate more than other rate-limiting factors as discussed before, highly efficient cellulases may increase the rate in the early stage of hydrolysis but will unavoidably end up with binding oligomers and could not enhance the conversion level too much. So a better approach

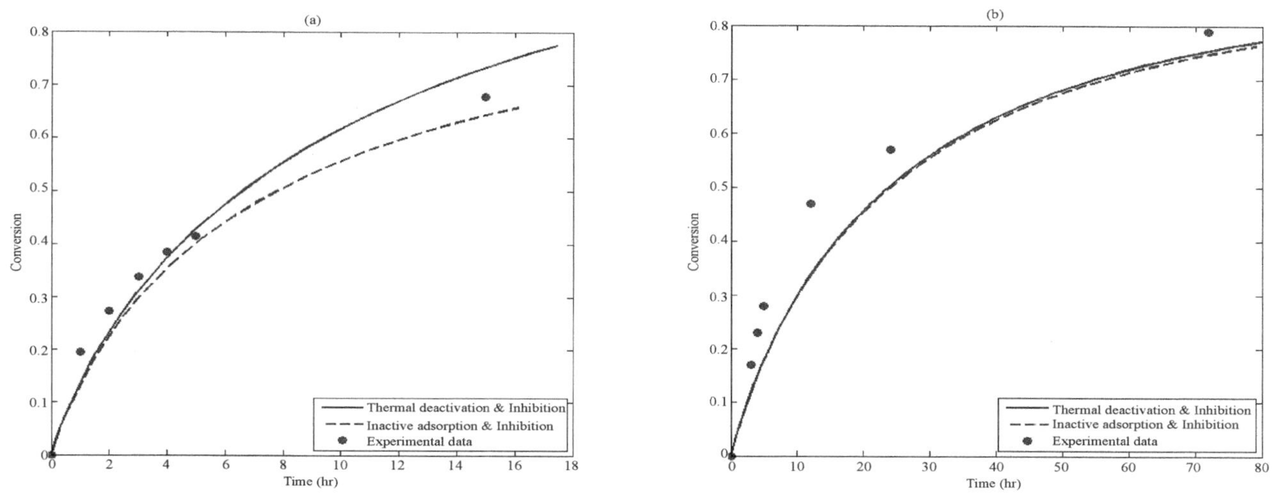

Figure 7: Simulation results considering all rate-limiting factors compared with experimental data. The substrate and enzyme have initial properties of (a) 20 g/L Avicel and 4.1 g/L Spezyme CP [9] and (b) 10 g/L Avicel and 0.51 g/L Spezyme CP together with 30 IU beta-glucanase/g Avicel (57.69 mg/L beta-glucanase) [10]. Commercial enzyme Spezyme CP consists EG1, CBHI and CBHII in a mass ratio of 0.17:0.24:0.13 (EG1/CBHI/CBHII) [17].

would be to add cellulases of small loading at different time points during hydrolysis with substrate filtered from the original solution into another reactor. This approach can theoretically reduce the impact of inhibition and keep the high efficiency of cellulases which need to be further tested by experiments.

Mechanistic modeling the entire process of enzymatic hydrolysis of cellulose is a challenging work. The differences between simulations and experiments probably come from two sources:

(1) The parameters used may not be accurate or may not be considered constants along the entire hydrolysis process (i.e. could be functions of time and operational conditions); and (2) there are some other factors that have not been taken into account, for example, lignin effect.

Conclusions

We have developed a detailed modeling framework for enzymatic hydrolysis of cellulose. The model for the first time not only consider the main hydrolysis step where long insoluble glucan chains are hydrolyzed by endo- and exo-glucanases on the surface of solid substrate, but also the reactions in solution which involve beta-glucanases and oligomers released from the conversion process of insoluble glucan chains. To investigate the phenomenon of hydrolysis rate slow-down often reported in the literature, the model considers the competitive inhibition effects of all possible cello-oligomers on cellulases and two theoretical enzyme decay processes, which are enzymatic thermal deactivation and the enzymatic inactive adsorption.

By using the model, we have presented analyses for the role of enzymatic accessibility of cellulosic substrate. As the enzyme-accessible surface area increases, both the hydrolysis rate and the finial conversion level of substrate could increase, showing that the enzymatic accessibility is a critical rate-limiting factor during the entire process of hydrolysis. After incorporating the reactions involving beta-glucanases and oligomers in solution and inhibition effects by oligomers, the simulation results can capture some extent of the decrease in the rate of cellulose hydrolysis often observed in experiments. We have tested the impacts of enzymatic thermal deactivation and the enzymatic inactive adsorption on the rate of hydrolysis and found the impacts of the two enzyme decay processes are very close to each other. We have presented detailed investigation of the phenomenon of hydrolysis rate slow-down. From the simulation results, we have found that inhibition effect of oligomer and enzyme decay are both critical factors contributing to the kinetics slow-down phenomenon during the process of enzymatic hydrolysis of cellulose.

References

1. Zhang YH, Lynd LR (2004) Toward an aggregated understanding of enzymatic hydrolysis of cellulose: noncomplexed cellulase systems. Biotechnol Bioeng 88: 797-824.

2. Zhou W, Schuttler HB, Hao Z, Xu Y (2009) Cellulose hydrolysis in evolving substrate morphologies I: A general modeling formalism. Biotechnol Bioeng 104: 261-274.

3. Zhang YH, Lynd LR (2006) A functionally based model for hydrolysis of cellulose by fungal cellulase. Biotechnol Bioeng 94: 888-898.

4. Levine SE, Fox JM, Blanch HW, Clark DS (2010) A mechanistic model of the enzymatic hydrolysis of cellulose. Biotechnol Bioeng 107: 37-51.

5. Griggs AJ, Stickel JJ, Lischeske JJ (2012) A mechanistic model for enzymatic saccharification of cellulose using continuous distribution kinetics I: depolymerization by EGI and CBHI. Biotechnol Bioeng 109: 665-675.

6. Yazaki T, Ohnishi M, Rokushika S, Okada G (1997) Subsite structure of the beta-glucosidase from Aspergillus niger, evaluated by steady-state kinetics with cello-oligosaccharides as substrates. Carbohydr Res 298: 51-57.

7. Sharma R (2012) Enzyme Inhibition and Bioapplications. ISBN: 978-953-51-0585-5, InTech.

8. Bansal P1, Hall M, Realff MJ, Lee JH, Bommarius AS (2009) Modeling cellulase kinetics on lignocellulosic substrates. Biotechnol Adv 27: 833-848.

9. Hong J, Ye X, Zhang YH (2007) Quantitative determination of cellulose accessibility to cellulase based on adsorption of a nonhydrolytic fusion protein containing CBM and GFP with its applications. Langmuir 23: 12535-12540.

10. Yang B, Willies DM, Wyman CE (2006) Changes in the enzymatic hydrolysis rate of Avicel cellulose with conversion. Biotechnol Bioeng 94: 1122-1128.

11. Chauve M, Mathis H, Huc D, Casanave D, Monot F, et al. (2010) Comparative kinetic analysis of two fungal beta-glucosidases. Biotechnol Biofuels 3: 3.

12. Tolan JS, Foody B (1999) Cellulase from submerged fermentation. Advances in Biochemical Engineering/Biotechnology 65: 41-67.

13. Leggio LL, Pickersgill RW (1999) Xylanase-oligosaccharide interactions studied by a competitive enzyme assay. Enzyme and Microbial Technology 25: 701-709.

14. Drissen RET, Maas RHW, Van Der Maarel MJEC, Kabel MA, Schols HA, et al. (2007) A generic model for glucose production from various cellulose sources by a commerical cellulase complex. Biocatal Biotransformation 25: 419-429.

15. Zhou W, Hao Z, Xu Y, Schüttler HB (2009) Cellulose hydrolysis in evolving substrate morphologies II: Numerical results and analysis. Biotechnol Bioeng 104: 275-289.

16. Zhou W, Xu Y, Schüttler HB (2010) Cellulose hydrolysis in evolving substrate morphologies III: time-scale analysis. Biotechnol Bioeng 107: 224-234.

17. Nagendran S, Hallen-Adams HE, Paper JM, Aslam N, Walton JD (2009) Reduced genomic potential for secreted plant cell-wall-degrading enzymes in the ectomycorrhizal fungus Amanita bisporigera, based on the secretome of Trichoderma reesei. Fungal Genetics and Biology 46: 427-435.

7

Isolation, Partial Purification and Characterization of Phospholipid Hydroperoxide Glutathione Peroxidase (Phgpx Enzyme from *Oryza Sativa* Seedlings

Om Prakash Verma[1]*, Pankaj Kumar Ojha[1], Shashi B Bailey[1], Pankaj Singh[1], Rajendra Dubey[2] and B.N.Mishra[3]

[1]Jacob School of Biotechnology & Bioengineering, SHIATS, Allahabad
[2]Galaxy Concepts Kolkata
[3]IET, Lucknow

Abstract

The recombinant Phospholipid hydroperoxide glutathione peroxidase isolated from *Oryza sativa* is a type of antioxidant enzyme. It is stable at pH 7-8 and temperature 27 and 62°C. The molecular weight of enzyme was found to be around 21-26 kDa by SDS PAGE. The peroxidation activity of PHGPx was inhibited by $HgCl_2$, EDTA, Potassium ferrocynate and was activated by $FeCl_2$ and $MnSO_4$ .These properties indicate that the possibilities for use of the PHGPx in the medical, clinical and cosmatic industry. This enzyme can be exploited commercially with some modification.

Keywords: PHGP, Enzyme activity, SDS Page, Antioxidant

Introduction

Glutathione peroxidase (E.C. 1.11.1.9) is an antioxidant enzyme family with peroxidase(oxidoreductase) activity whose main biological role is to protect the organism from oxidative damage.Glutathione peroxidases (GSHPx) are a group of selenium-dependent enzymes. Out of some are cytosolic GSHPxl, plasma GSHPx, phospholipid hydroperoxide PHGSHPx established the presence of glutathione peroxidase in mammalian erythrocyte. It has been characterized as a unique intracellular antioxidant enzyme that markedly reduces peroxidized phospholipids produced in cell membranes and is generally considered to be the main line of enzymatic defense against oxidative biomembrane damage in mammalian cells [5,10]. The expression of the tomato PHGPx gene *(LePHGPx)* was observed to inhibit cell death induced by Bax and oxidative stress in yeast and plants [2]. These findings suggest that plant PHGPx might play a very important role in development and stress tolerance in plants. The present investigation is carried out in view to isolate, partial purify and characterize Phospholipid hydroperoxide glutathione peroxidase (PHGPx enzyme from *Oryza sativa* seedlings.

Materials and Methods

Isolation of Glutathione Peroxidase (PHGPx enzyme)

Phospholipid Hydroperoxide Glutathion Peroxidase (PHGPx) is an antioxidant enzyme and it's expression takes place in side (intracellular) the transformed cell present inside the LB broth. So recovery of Phospholipid Hydroperoxide Glutathione Peroxidase (PHGPx) occurs after cell lysis by chemical method.

Purification of isolated of Phospholipid Hydroperoxide Glutathione Peroxidase (PHGPx)

Purification of enzyme Phospholipid Hydroperoxide Glutathione Peroxidase (PHGPx) isolated from *Oryza sativa*. The Crude enzyme was precipitated by ammonium sulfate (ammonium sulfate precipitation method). The precipitated enzyme was passed through the process of dialysis by cellular membrane filter (treated with 3% sodium bi carbonate) for overnight and centrifuged the dialyzed enzyme at 6000-10000rpm for 10 minute then transferred the supernatant to a fresh tube and pure enzyme was separated by ion-exchange chromatography by DEAE cellulose.

Ammonium sulphate precipitation: The PHGPs enzymes were also purified by ammonium sulfate saturation. The protein fraction precipitated with 85 % ammonium sulfate. The suspension was centrifuged at 10,000rpm for 15 minutes. The precipitate was collected and dissolved in 25mM NaCl, 10MmTris base buffer. Followed by its dialysis for 1-2 days to make it more purified.

Dialysis: The precipitate obtained after ammonium sulphate precipitation was dissolved in 25mM NaCl; 10Mm Tris base buffer. Then it is put inside a semi permeable membrane bag and kept immersed in Tris base buffer for dialysis at 4°C for 1-2 days. After dialysis the enzyme was taken out and centrifuged at 8000rpm for 15minutes. The supernatant is collected as the pure enzyme after dialysis.

Characterization of pure enzyme Phospholipid Hydroperoxide Glutathione Peroxidase (PHGPx):

Protein estimation by Lowry's method: The protein was estimated by [8].

Enzyme assay: The enzyme activity of isolated Phospholipid Hydroperoxide Glutathione Peroxidase (PHGPx) enzyme was measured by decreasing in absorbance at 340nm during detoxification of H_2O_2 (Hydrogenperoxide) and increase in relative activity and it was measured by continuous spectrophotometric rate determination and substrate concentration dependent assay.

Effect of temperature, pH, inhibiter and activator on enzyme activity: The enzymes was mixed with enzyme assay buffer and then

***Corresponding author:** Om Prakash Verma, Jacob School of Biotechnology & Bioengineering, Shiats, Allahabad incubated at different temperature for 30 minutes and absorbance was taken at 340 nm.

Estimation of molecular weight: The electrophoresis was conducted to determine the molecular weight of the visualized protein bands by comparing them with the molecular weight marker.

Zymography of Recombinant Phospholipid Hydroperoxide Glutathione Peroxidase (PHGPx): Zymography was done for visualization of protein bands.

Results and Discussion

Ammonium sulphate precipitation

Ammonium sulphate precipitation of recombinant Phospholipids Hydro peroxide Glutathione Peroxidase (PHGPs) enzyme was found at 85-90% saturation.

Spectrophotometric qualitative estimation of purified Recombinant enzyme (PHGPx)

Protein estimation by Lowry's method: The graph obtained from protein estimation was clearly showed that the concentration of enzyme PHGPs expressed by *E. coli* plasmid, P^{BR322}, and Pet EMBL21 was higher than standard (BSA) concentration respectively as shown in Figure 1.

Protein estimation of different fraction of purified intracellular enzyme: The graph obtained from protein estimation it was clearly showed that concentration of enzyme PHGPs higher in 50% fraction for Pet EMBL21 expressed enzyme, 50% fraction for PBR^{322} expressed enzyme and 25% fraction for *E. coli* plasmid expressed enzyme respectively as shown in Figure 2 and Table 1.

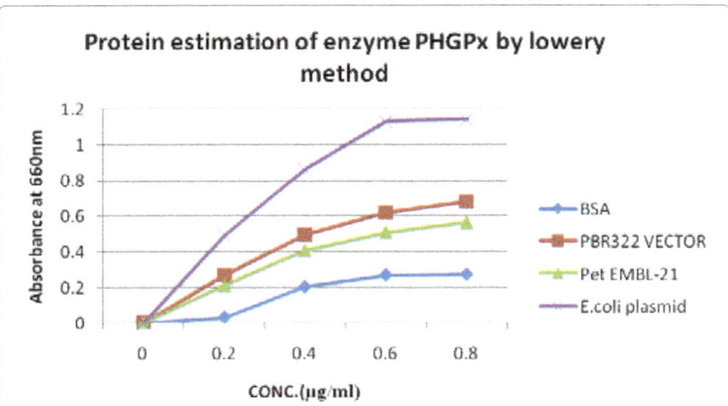

Figure 1: Protein Estimation of PHGPx enzyme by Lowry's method.

Phospholipids Hydro peroxide Glutathione Peroxidase (PHGPs)in different fraction of buffer	Absorbance at 280 nm	Absorbance at 280 nm	Absorbance at 280 nm at 280 nm
	P^{BR322}	PEMBL21	Natural *E.coli* plasmid
Crude	2.371	2.211	2.848
0% fraction of PHGPs (first elution)	1.228	0.880	1.732
25% fraction of PHGPs	1.32	1.116	1.983
50% fraction of PHGPs	1.018	1.095	1.357
75% fraction of PHGPs	0.556	0.86	0.853
100% fraction of PHGPs	0.478	0.592	0.603

Table 1: Absorbance of different fraction of enzyme.

Serial number	*E.coli* Plasmid	PBR^{322} VECTOR	Pet EMBL21
1 spot RF value	0.18	0.18	0.304
2 spot RF value	0.64	0.64	0.54
3 spot RF value	0.79	0.91	0.79
4 spot RF value	0.975	0.96	0.975

Table 2: Different RF value.

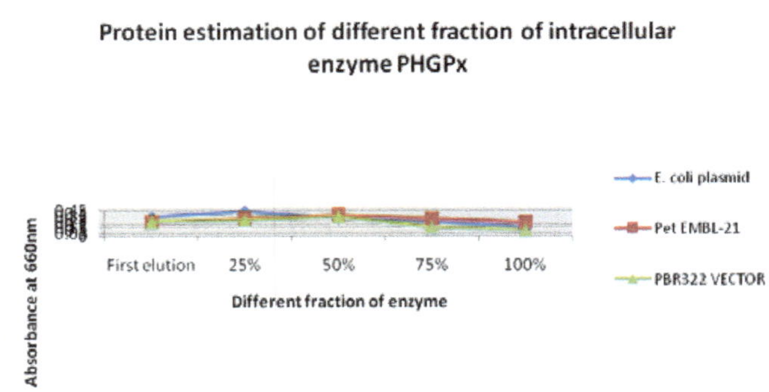

Figure 2: Protein Estimation of different fraction of PHGPs enzyme by Lowry's method.

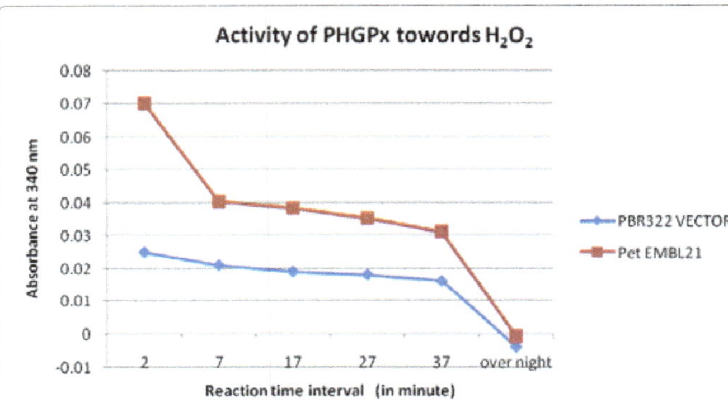

Figure 3: Activity of PHGPx towards H_2O_2 according to time.

Continuous spectrophotometric rate determination: Absorbance and activity of enzyme by Continuous spectrophotometry rate determination at different time interval as shown in Figure 3.

Characterization of enzyme Phospholipid Hydro peroxide Glutathione Peroxidase (PHGPx)

Effect of temperature on intracellular enzyme activity: It is clearly showed that all intracellular enzymes showed maximal activity at 27°C and 62°C, but at 37°C activity of enzyme was slightly decreased and as the temperature was gradually increases the activity decreases due to denaturation of the enzyme. It indicates that the enzyme was most active at an optimum temperature of 62°C (Figure 5).

Effect of temperature on extracellular enzyme activity: It is clearly showed that all intracellular enzymes showed maximal activity at 27°C and 62°C, but at 37°C activity of enzyme was slightly decreased and as the temperature was gradually increases the activity decreases due to denaturation of the enzyme. It indicates that the enzyme was most active at an optimum temperature of 27°C (Figure 6).

Effect of pH on intracellular enzyme activity: It is clearly showed that all intracellular enzymes showed maximal activity at range of pH 5-8 respectively, but as the pH was gradually increased above pH 8 and decreases below pH 5, the activity decreases due to denaturation of the enzyme. It indicates that the enzyme was most active at an optimum pH of 8 as shown in Figure 7.

Effect of pH on extracellular enzyme activity: It is clearly showed that all extracellular enzymes showed maximal activity at range of pH 5-8 respectively, but as the pH was gradually increased above pH 8 and decreases below pH 5, the activity decreases due to denaturation of the enzyme. It indicates that the enzyme was most active at an optimum pH of 8 Figure 8).

Effect of inhibitor and activator on P^BR322 and Pet EMBL21 expressed intracellular enzyme activity: It is clearly showed that NaN_3, SDS and $AgNo_3$ showed minimum inhibition on enzyme activity

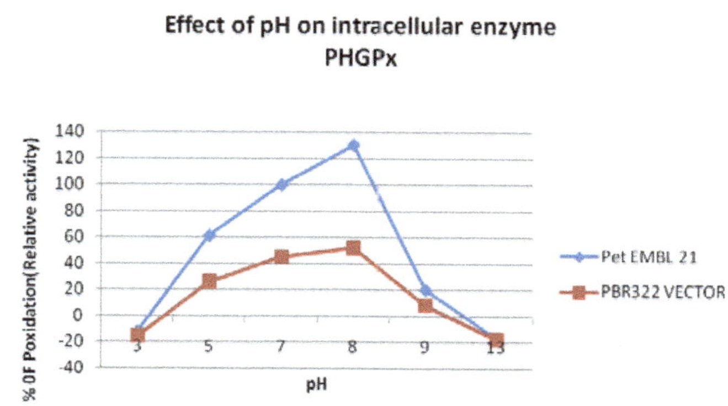

Figure 7: Effect of different pH on intracellular PHGPx enzyme.

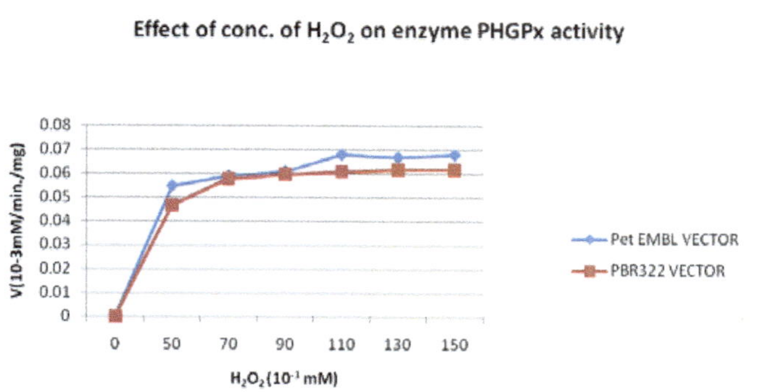

Figure 4: Effect of concentration of H_2O_2 on enzyme PHGPx activity.

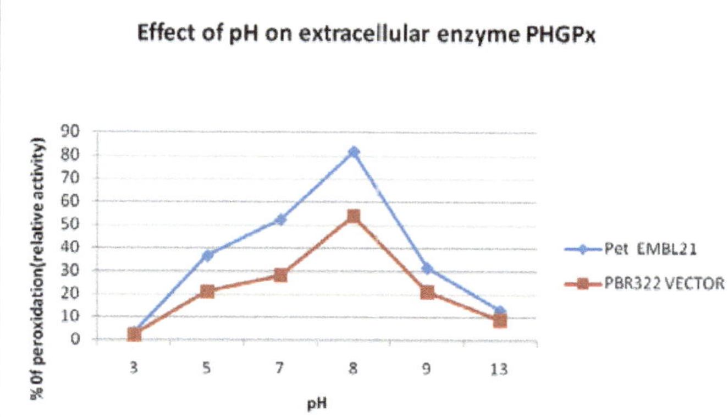

Figure 8: Effect of different pH on extra cellular PHGPx enzyme.

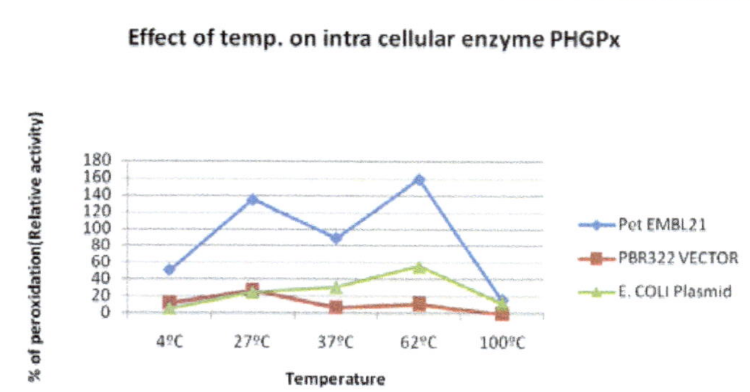

Figure 5: Effect of different temperature on intracellular PHGPx enzyme.

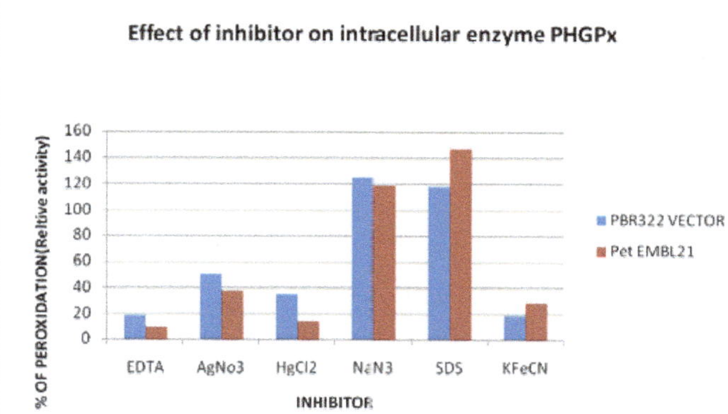

Figure 9: Effect of inhibitor on intra cellular PHGPx enzyme.

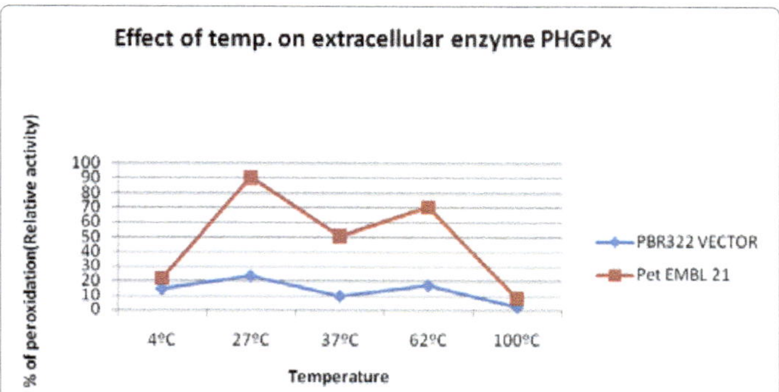

Figure 6: Effect of different temperature on extra cellular PHGPx enzyme.

and $HgCl_2$, EDTA and KFeCN completely effect the activity of P^BR322 expressed intracellular enzyme and in case of Pet EMBL21 expressed intracellular enzyme NaN_3, SDS and $AgNO_3$ showed minimum inhibition on enzyme activity and $HgCl_2$, EDTA and KFeCN completely effect on enzyme activity. So it indicates that NaN_3, SDS and $AgNo_3$ are the good inhibitor (Figure 9-12)

It is clearly showed in Figure 10 that P^BR322 expressed intracellular enzyme activity was greatly activated in the presence of $FeCl_2$, $Cacl_2$ and $FeSO_4$ and partially activated in the presence of $MnCl_2$, $MnSO_4$ and $MgSO_4$. So it indicates that that $FeCl_2$, $Cacl_2$ and $FeSO_4$ is the good activator.

In case of Pet EMBL21 expressed intracellular enzyme, enzyme activity was greatly activated in the presence of $FeCl_2$, $MnSO_4$ and $MnCl_2$ and partially activated in the presence of $Cacl_2$, $FeSO_4$ and $MgSO_4$. So it indicates that that $FeCl_2$, $MnSO_4$ and $MnCl_2$ is the good activator.

Estimation of molecular weight of recombinant PHGPs protein by Sodium dodecyl sulphate polyacrylamide gel electrophoresis (SDS-PAGE)

The protein bands was obtained on SDS PAGE was compared to the standard marker (BSA -66.2 kDa, ova albumin- 45kDa and lysozyme-14 kDa).The molecular weight of PBR32,Pet EMBL21 and *E.coli* plasmid expressed protein was about 21 - 25kDa compared with standrd marker respectively Figure 13.

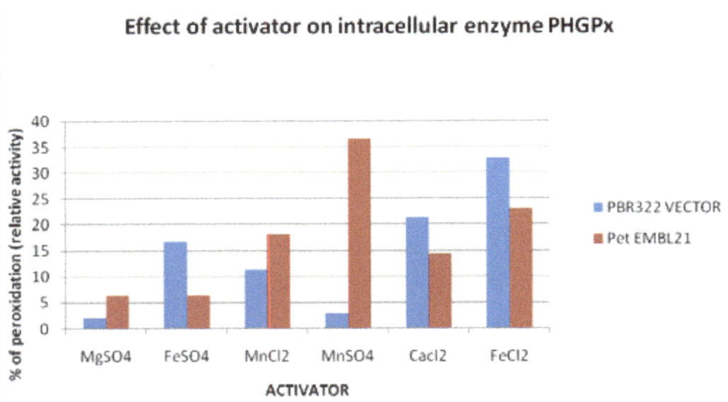

Figure 10: Effect of activator on intra cellular PHGPs enzyme.

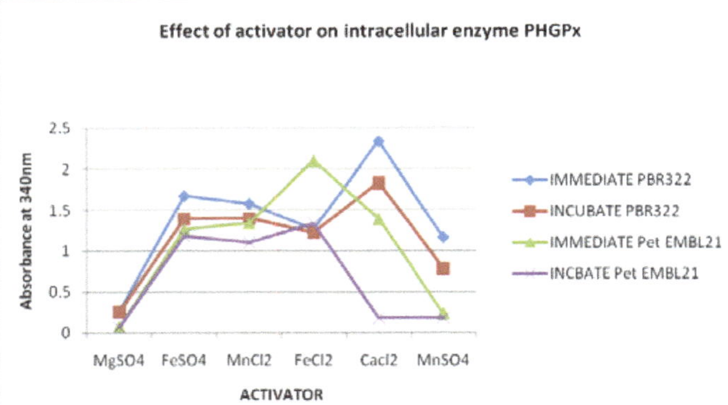

Figure 11: Effect of activator on intra cellular PHGPs enzyme.

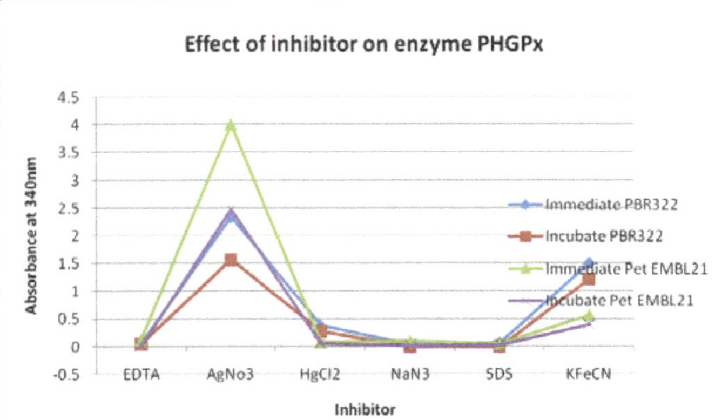

Figure 12: Effect of inhibitor on intra cellular PHGPx enzyme.

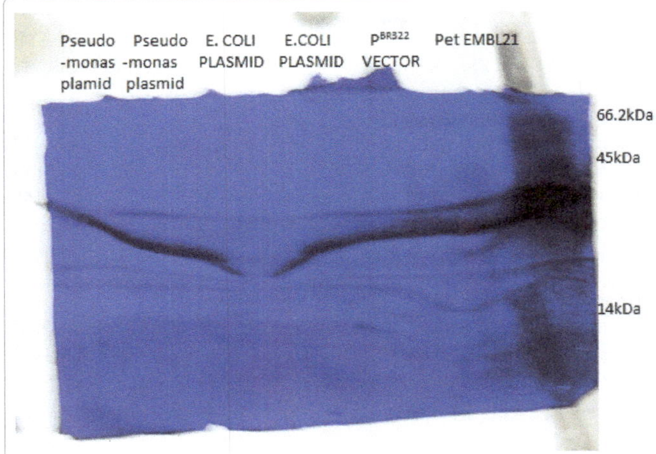

Figure 13: Protein band on polyacrylamide gel.

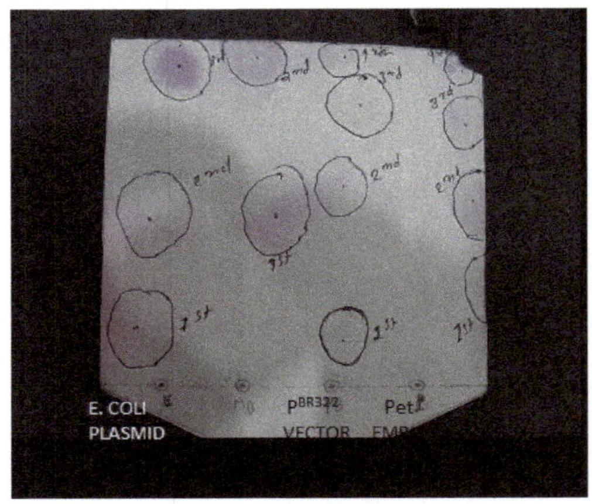

Figure 14: Paper chromatograph.

Paper chromatography of recombinant PHGPs protein

The activity of OsPHGPx towards several hydroperoxides was determined in the presence of glutathione/glutathione reductase or Trx/TrxR, which involved following NADPH oxidation by measuring changes in absorbance at 340 nm. First, GSH-dependent enzymatic activity of recombinant OsPHGPx towards H_2O_2 was measured. A linear decrease of A_m (absorbance of NADPH) following the addition of H_2O_2 was observed, whereas no obvious declines were detected when GSH, H_2O_2 or OsPHGPx was omitted, indicating that the peroxidase activity of OsPHGPx was remarkably GSH-dependent Figure 14. When the effects of pH and temperature on GSH-dependent OsPHGPx activity towards H_2O_2 were evaluated, according to [12], the pH optimum was in the range of pH 9.0-10. With a maximum of activity at approximately pH 9.3, but in contrast to our study the optimum pH was observed in the range of pH 7.0-8.0 with a maximum of activity at approximately pH 8. The dependence of enzyme activity *versus* temperature from 10 to 45°C yielded an obvious temperature optimum at approximately 27°C, with activity demonstrated in a wide temperature range of 10-35°C, in contrast to present study enzyme stable in the range 27-62°C with maximum activity approximately 27°C that is same result was observed. In the present study indicated that EDTA, $HgCl_2$ and potassium ferrocynate acts as a good inhibitor for PHGPx while $MnSO_4$ and $FeCl_2$ acts as a good activator. These results show the temperature- and pH-dependence of GSH-dependent

OsPHGPx enzymatic activity. The optimum pH range for catalysis of OsPHGPx was consistent with that of its well-known mammalian counterpart [3], while its optimum temperature coincides with that required for plant growth.

Considering the broad specificity for hydroperoxide substrates of PHGPx [9], the hydroperoxide substrate preference for recombinant OsPHGPx was investigated. Peroxidase activity towards different concentrations of H_2O_2 was assayed with a fixed GSH concentration of 15mM. As illustrated in, the plot of GSH-dependent OsPHGPx activity versus H_2O_2 concentration was consistent with normal Michaelis-Menten kinetics. From this plot, the apparent K_m and V_{max} value in case of P^{BR322} expressed enzyme and Pet EMBL21 expressed enzyme for H_2O_2 was 55 × 10^{-3} mM, 52 × 10^{-3} mM and 0.068 mM/min/mg, 0.061 mM/min./mg respectively.

The V_{max} value of OsPHGPx was at least 50% higher than those of all other known plant homologues, including citrus, tomato, and sunflower PHGPx [2,4]. The results also demonstrated that OsPHGPx was active on H_20_2 indicating a similar broadness of substrate specificity to mammalian PHGPx [9].

It is also noteworthy that the overall enzymatic activity of plant PHGPxs recorded so far with phospholipid hydroperoxide as a substrate is generally three orders of magnitude lower than those of the mammalian proteins. This low activity has made it difficult to explore the potential physiological role of PHGPx in higher plants and also raises the question of whether higher plants indeed do not maintain their PHGPx with higher activity equal to that of their mammalian counterpart. In this report, the presented results showed an unusual finding: recombinant OsPHGPx exhibits enzymatic activity against not only phospholipid hydroperoxide, but also H_2O_2 using GSH as the electron donor. It is reasonable to believe that if the cysteine at the catalytic site of OsPHGPx is replaced by a selenocysteine residue, OsPHGPx's catalytic activity may be enhanced to a higher level. Our kinetic characterization displayed obvious similarity in substrate preference between plant OsPHGPx and its mammalian counterpart: both of them could use H_2O_2, organic hydroperoxide, and phospholipid hydroperoxide as substrates, but preferred phospholipid hydroperoxide [9,11]. Interestingly, OsPHGPx showed higher catalytic efficiency and V_{max} for H_2O_2, but it displayed a lower affinity to H_2O_2. This probably means that under normal physiological conditions OsPHGPx is predominantly concerned with reduction of phospholipid hydroperoxides rather than competing with other H_2O_2 detoxifying peroxidases for H_2O_2 reduction, such as ascorbate peroxidase, which has higher affinity to H_2O_2 [6]. On the other hand, OsPHGPx may also participate in the detoxification of H_2O_2 when endogenous H_2O_2 accumulates to a very high concentration. By analogy with mammalian PHGPx whose main *in vivo* oxidative substrate has proven to be phospholipid hydroperoxides [9,11], we suppose that phospholipid hydroperoxides might be the main natural substrates of OsPHGPx.

Conclusion

Plant phospholipid hydroperoxide peroxidase, OsPHGPx, was produced in *E. coli* and purified to near homogeneity. The purified recombinant protein showed GSH-dependent peroxidase activity towards H_2O_2. In addition, the enzyme (OsPHGPx) was also found to have a glutathione-dependent hydroperoxide peroxidase activity towards H_2O_2. The results clearly demonstrated that OsPHGPx is an enzyme with broad specificity for hydroperoxide substrates. Previous works have demonstrated that the OsPHGPx gene is expressed at a relatively high level in rice flag leaves and markedly induced by a variety of oxidative stress and defense-related stimuli, strongly suggesting that the product of the gene plays a key role in defense against oxidative damage in rice [7,1]. Taken together, this evidence *in vivo* and *in vitro* has provided significant insight into the induction expression of the gene, the biochemical and biophysical properties of the protein, and the dynamics of the OsPHGPx enzyme, and will greatly help us in the future to uncover the exact physiological function of this important enzyme in plant cells.

References

1. Imai H, Nakagawa Y (2003) Biological significance of phospholipid hydroperoxide glutathione peroxidase (PHGPx, Gpx4) in mammalian cells. Free Radic Bio Med 34: 145-169.
2. Ursini F, Heim S, Kiess M, Maiorino M, Roveri A, et al. (1999) Dual function of the selenoprotein PHGPx during sperm maturation. Science 285: 1393-1396.
3. Lowry OH, Rosebrough NJ, Fano AL, Randall RJ (1951) Measurement with the Folin phenol reagent. J Boil Chem 193: 265-275.
4. Wang Z, Duan R, Yuan LJ, Liu JY (2007) Purification and Physicochemical Characterization of a Recombinant Phospholipid Hydro peroxide Glutathione Peroxidase from *Oryza sativa*. J Biochem Mol Biol 40: 412-418.
5. Flohe L, Gunzler WA (1984) Assays of glutathioneperoxidase. Methods Enzymol 105: 114-121.
6. Maiorino M, Gregolin C, Ursini F (1990) Phospholipidhydroperoxide glutathione peroxidase. Methods Enzymol 186: 443-457.
7. Chen S, Vaghchhipawala Z, Li W, Asard H, Dickman MB (2004) Tomato phospholipid hydroperoxide glutathione peroxidase inhibits cell death induced by bax and oxidative stresses in yeast and plants. Plant Physiol 135: 1630-1641.
8. Herbette S, Lenne C, Leblanc N, Julien JL, Drevet JR, et al. (2002) Two GPX-like proteins fromLycopersicon esculentum and Helianthus annuus are antioxidant enzymes with phospholipid hydroperoxide glutathione peroxidase and thioredoxin peroxidase activities. Eur J Biochem 269: 2414-2420.
9. Ursini F, Mariorino M, Brigelius-Flohé R, Aumann KD, Roveri A, et al. (1995) Diversity of glutathione peroxidase. Methods Enzymol 252: 38-53.
10. Kvaratskhelia M, Winkel C, Thorneley RNF (1997) Purification and characterization of a novel class III peroxidase isoenzyme from tea leaves. Plant Physiol 114: 1237-1245.
11. Li WJ, Feng H, Fan JH, Zhang RQ, Zhao NM, et al. (2000) Molecular cloning and expression of a phospholipid hydroperoxide glutathione peroxidase homolog in *Oryza sativa*. Biochi Biophys Act 1493, 225-230.
12. Agrawal GK, Rakwal R, Jwa NS, Agrawal VP (2002) Effects of signaling molecules, protein phosphatase inhibitors and blast pathogen (*Magnaporthe grisea)* on the mRNA level of a rice *(Oryza sativa L.)* phospholipid hydroperoxide glutathione peroxidase (OsPHGPx) gene in seedling leaves. Gene 283: 227-236.

8

Biorefinery of Energy Crop Cardoon (*Cynara cardunculus l.*) - Hydrolytic Xylose Production as Entry Point to Complex Fractionation Scheme

Anatoly A. Shatalov* and Helena Pereira

Centro de Estudos Florestais, Universidade Técnica de Lisboa, Instituto Superior de Agronomia, Tapada da Ajuda, 1349-017 Lisboa, Portugal

Abstract

Response surface methodology (RSM) was employed for statistical modeling and optimization of low temperature dilute sulfuric acid hydrolysis of hemicellulose fraction of energy agro-crop cardoon (*Cynara cardunculus L.*), as an entry point to complex biorefinery scheme. The 2^3 central composite rotatable design (CCRD) was used to assess the effect of the principal independent process variables (reaction time, temperature and acid concentration) on efficiency and selectivity of heteroxylan conversion to xylose. The second-order polynomial model was fitted to experimental data to find optimal reaction conditions of xylan-to-xylose hydrolysis by multiple regression analysis. The effect of acid concentration (linear and quadratic) was found as a more significant ($p=0.001$-0.007) for monomeric xylose recovery in solution. The maximal xylose yield of ca. 86% (18.08 g /100 g biomass) was achieved after cardoon hydrolysis at 138.5ºC in 1.28% sulfuric acid solution for 52 min, vs. 87% predicted by model. The resulting xylose-enriched substrate revealed low concentration of toxic substances (1.04% furfural, 0.33% 5-hydroxymethylfurfural, 2.03% glucose), providing required quality for subsequent xylose (bio)conversion to final products (e.g. to xylitol). The enzymatic saccharification/digestibility of insoluble residue after hemicellulose removal was improved in four times, resulting in cellulose conversion to fermentable glucose by 76% vs. 19% for unhydrolyzed cardoon.

Keywords: Xylan; Xylose; Hydrolysis; Statistical modeling; Optimization; Cardoon

Introduction

The biorefinery concept of biomass treatment is an area of much current scientific interest and research activity. The multi-step processing (fractionation) of different biomass feedstocks into a wide range of value-added products through combination of physical, chemical and biological approaches is viewed as a more potential way to guarantee the sustainability of the future bio-based economy [1,2]. Of all existing the large-scale bioconversion schemes, the particular attention has been given to biorefinery of lignocellulosic biomass (LCF, or Lignocellulosic Feedstock Biorefinery), as a more diverse and valuable source of chemicals [3,4]. Lignocellulosics (wood, straw, reeds, grasses, agro-crop and forest residues, paper and cellulosic municipal wastes, etc.) basically consist of three major chemical constituents: cellulose, hemicelluloses and lignin. Assuming that carbohydrate fraction account for up to 80% (and more) of lignocellulosic raw materials [1], the efficient access and conversion of carbohydrates to chemical bulk products (industrial intermediates) and the corresponding final products is a key factor for successful commercialization of LCF biorefinery [5]. The so-called "sugar platform" of biomass fractionation was developed based on biochemical conversion processes with focus on the fermentation of sugars extracted from biomass feedstocks to produce fuel, such as ethanol, or other building block chemicals [6].

The agro-based lignocellulosics, such as annual crops and perennial herbaceous species (grasses), represent abundant and low-cost feedstocks for LCF biorefinery. Some perennial rhizomatous herbs, i.e., switchgrass, alfalfa, miscanthus, canarygrass and giant reed, were indicated as the more potential energy crops for bioenergy production in both the US and Europe [7]. *Cynara cardunculus L.* (cardoon or artichoke thistle) is an abundant naturally growing perennial herb native to the Mediterranean region. For biomass production, cardoon can be cultivated as a perennial field crop in dry farming with annual harvest of the whole aboveground biomass (yielding up to 30 t of dry matter per ha [8] for several years [9]. As a raw material for industrial applications, the use of cardoon biomass as a solid fuel or for production of seed oil, biodiesel, paper pulp, green forage and pharmacologically active compounds was reviewed [10]. The high biomass productivity and appropriate chemical composition makes cardoon as a very attractive lignocellulosic feedstock for LCF biorefinery schemes. Hemicelluloses are the second largest (after cellulose) constituent of cardoon biomass accounting for about 25-30% of the stalk-wall material [9]. Similar to other agro-crop species, the hemicellulosic fraction of cardoon consists almost entirely of xylan heteropolysaccharide [9], representing a valuable source of xylose for subsequent (bio) chemical conversion to final products, e.g., xylitol. The careful xylan isolation and depolymerization to monomeric sugars can therefore be an important first step (entry point) in the complex scheme of cardoon fractionation (biorefinery) to individual chemical components. The most widely used and tested approach of xylan conversion to monosaccharides is based on hydrolysis reaction [11,12]. Particularly, the dilute sulfuric acid hydrolysis under moderate temperatures was proved to be a reliable and easily performed low cost method for quantitative conversion of xylan to xylose and reactivity (accessibility) improvement of cellulose and lignin in solid residue [13-17]. The yield of xylose after hydrolysis is strongly dependent on the type of raw material used and operation conditions applied. Under controlled (optimized) hydrolysis conditions the minimal cellulose

***Corresponding author:** A.A. Shatalov, Centro de Estudos Florestais, Universidade Técnica de Lisboa degradation and by-products (basically furans and acetic acid) formation takes place, providing the high efficiency and selectivity of the overall process.

Only limited published data related to acid hydrolysis of cardoon are available [18-20], basically reporting biomass (pre)treatment under drastic reaction conditions for the purpose to improve the enzymatic accessibility/reactivity of cellulose for ethanol production. The high temperatures used at these studies (up to 200-230°C) led to incomplete xylan conversion to monomeric sugars (xylose) during auto-hydrolysis process (when no mineral acid was used as catalyst), or caused substantial monosaccharide (basically xylose) decomposition (up to 4.2% of furfural) as well as significant cellulose degradation (by 10-18%) in the presence of 0.1-0.2% of sulfuric acid as catalyst, thereby providing unacceptable quality of sugar hydrolyzate for xylose utilization.

The low temperature (below 150°C) dilute sulfuric acid hydrolysis of energy crop cardoon (*Cynara cardunculus* L.) has been examined to produce a high quality xylose-rich substrate ready for (bio)chemical conversion to final products and reactive hemicellulose-free solid residue ready for subsequent fractionation within complex biorefinery scheme. The response surface methodology (RSM) was used for hydrolysis modeling and optimization with the aim to maximize efficiency and selectivity of xylan polysaccharide conversion to monomeric units within one-step reaction. The principal results of this study are reported in the present paper.

Materials and Methods

Materials

The cardoon (*Cynara cardunculus L.*) used in this study was sampled from the university experimental plantation field (Institute of Agronomy, Technical University of Lisbon). The air dried stalks were manually stripped of leaves, milled and screened to particle size of 40-60 mesh and stored in sealed plastic bags at room temperature until using for chemical analysis and hydrolysis experiments. The moisture content of prepared material was determined according to TAPPI standards.

All chemicals used were of analytical grade purity and purchased from Sigma, Ardrich and Fluka Chemical Co.

Acid hydrolysis: For optimization study, 5 g (on oven-dry basis) of cardoon sawdust were hydrolyzed with 75 ml of diluted (0.5-1.5%) sulfuric acid in the stainless steel autoclaves (ca. 100 cm^3 capacity) rotated in an oil bath at 130-150°C for 30-60 min, with replication of each experimental conditions set. After heating-up period (ca. 2min), the loaded autoclaves were held at desired temperature for pre-determined residence time. Reaction was terminated by immediate autoclave plunging into an iced-water bath. The solid residue was separated by filtration, thoroughly washed by deionized water and kept frozen for subsequent analysis and treatment. The hemicellulosic hydrolyzate, combined with washing water, was collected in the volumetric flask and analyzed on degree of monosaccharide recovery and degradation after hydrolysis.

Enzymatic hydrolysis: The enzymatic digestibility of insoluble lignocellulosic residue after acidic hydrolysis was checked by NREL standard procedure [21]. Residue (0.15 g) was mixed with commercial preparations of cellulases (60 FPU/g cellulose) and β-glucosidases (64 *p*NPGU/g cellulose) in 50mM sodium citrate buffer (pH 4.8) and incubated at 50°C for 24-72 hours under 150 rpm rotation. Sodium azide (2% solution) was added as antibiotic to prevent any microbial infection during digesting. The release of soluble sugars (basically glucose) was determined by HPLC and corrected by blank tests on substrate and enzymes. The enzymatic digestibility (saccharification) was defined as a ratio of cellulose digested (g) to cellulose added (g).

Analytical methods

Extractives were determined gravimetrically after successive Soxhlet extraction by dichloromethane, ethanol and water. Ash, silica (as SiO_2), acid-insoluble (Klason) and acid-soluble lignin were determined according to TAPPI standards. Holocellulose was isolated by modified chlorite method in the presence of sodium acetate buffer [22]. Hemicelluloses were successively extracted from holocellulose by 5% and 24% potassium hydroxide solutions and recovered by ethanol addition, giving, respectively, fractions A and B. The insoluble residue after alkaline extraction of holocellulose was accepted as α-cellulose [22].

The monosaccharide composition of holocellulose and hemicelluloses was determined by GC as alditol-acetate derivatives using the following conditions: FID, SP-2330 column, injector temperature 240°C; detector temperature 250°C; initial column temperature 230°C; final column temperature 240°C; 2°C min^{-1} rate; 2-deoxy-D-glucose as an internal standard.

The total content of reducing sugars and furans in hemicellulosic hydrolysates after the preliminary tests was determined by DNS method and UV-spectroscopy, respectively [23,24].

The concentration of xylose, glucose, acetic acid, furfural and hydroxymethylfurfural (HMF) in hydrolysates was analyzed by HPLC using Aminex HPX-87H column (Bio-Rad, Hercules, CA, USA) operating at 50°C, in combination with a cation H^+-guard column (Bio-Rad, Hercules, CA, USA) in a Waters (Waters, Milford, USA) HPLC system. A RI detector (Waters 2410) was used for monosaccharides and aliphatic acids quantification whereas a UV-Vis detector (Waters 486) set at 280 nm was used for furfural and HMF measurement. The mobile phase was 5mM sulfuric acid and the flow rate 0.4 ml min^{-1}.

Xylan conversion after hydrolysis (Y_1) was defined as a ratio of xylose content in hydrolysate to xylose content in cardoon. Hydrolysis selectivity (Y_2) was defined as a ratio of xylose to glucose in hydrolysate.

Experimental design and statistical treatment: Response Surface Methodology (RSM) was employed for statistical data treatment and optimization of hydrolysis conditions by multiple regression analysis using Statistica 6.0 (Statsoft, USA) software. The 2^3 central composite rotatable design (CCRD) with three independent variables at five different levels, six star (axial) points and five central points (total 19 runs) was adopted to find linear, quadratic and interaction effects of independent process variables on experimental responses. A second-order polynomial model was fitted to each set of experimental data to predict optimal reaction conditions by following equation:

$$Y = b_0 + \sum_{i=1}^{3} b_i X_i + \sum_{i=1}^{3} b_{ii} X_i^2 + \sum_{i<j,j=2}^{3} b_{ij} X_i X_j \qquad (1)$$

where Y is a predicted response (xylan conversion or hydrolysis selectivity), b_0 is an interception coefficient (regression coefficient at central point), b_i are the linear coefficients; b_{ii} are the quadratic coefficients, b_{ij} are the interaction coefficients, X_i and X_j are the independent variables (temperature, time and acid concentration).

The statistical significance of regression coefficients and effects was checked by analysis of variance (ANOVA).

Results and Discussion

Compositional analysis of raw material

Since biomass composition may be variable due to season, growing conditions etc., it is extremely important to be able to relate process performance to actual experimental material. The chemical composition of cardoon whole stalk (with pith) material used in this study is presented in Tables 1 and 2. Similar to other commercially important agro-crops, cardoon has less lignin but more extractives, as compared with wood species [25], what indicates the higher degree of accessibility and therefore reactivity of carbohydrate complex to chemical processing, e.g., in acid hydrolysis, under relatively mild reaction conditions. The content of cellulose (ca. 39%) and hemicelluloses (ca. 30%) in cardoon is close to that of hardwoods (38-50% and 20-28%, respectively) [25], underlining the value of this plant as an important agro-source of sugars for subsequent (bio)-chemical conversion. Generally, the found proportion between principal chemical constituents (i.e., lignin, cellulose, hemicelluloses and extractive) does not differ greatly from the Spanish cardoon reported in [19,20], displaying however some obvious deviations connected first of all with the reduced content of extractives and the elevated portion of carbohydrate polymers.

The hemicellulose fraction of cardoon stalk consists mainly (by ca. 90%) of xylan polysaccharide (Table 2), what comprises about 27% of total crop carbohydrates and gives the value of maximal (theoretical) xylose yield of 21.49% (on oven-dry material) after complete hydrolytic depolymerization of xylan.

Preliminary hydrolysis experiments

To define the current levels (settings) of the independent process variables to be used in statistical experimental design on xylose recovery, a series of experiments on cardoon hydrolysis was performed under variable conditions of reaction temperature (130, 140 and 150°C), reaction time (30 and 60 min) and sulfuric acid concentration (0.5, 1.0 and 1.5%) and the yield of the principal monosaccharides (glucose and xylose) and decomposition side-products (acetic acid, furfural and 5-hydroxymethylfurfural) was determined in hydrolyzate.

As can be seen from Table 3, even moderate reaction conditions applied for xylan hydrolysis inevitably cause the partial hydrolysis of cellulose (as a more accessible amorphous cellulose portion) and decomposition of formed monosaccharides to furans, decreasing thereby selectivity and efficiency of xylan conversion as a whole. Generally, increase in reaction temperature and acid concentration, while accelerating xylan hydrolysis to xylose, intensifies substantially the secondary degradation reactions of monomeric sugars. The process temperature of 130°C provides the most preserving reaction conditions, revealing only traces of furfural and limited cellulose degradation. But, the xylose recovery in solution is low pointing to incomplete xylan conversion. In contrast, the harsh conditions of 150°C cause substantial monosaccharide degradation (up to 4% of furfural and 6% of acetic acid) and cellulose hydrolysis (up to 3.6% of glucose); what also lowers the yield of xylose, particularly under increasing acid concentration.

The most balanced interrelation between hydrolysis efficiency and selectivity can be achieved under reaction temperature about 140°C. Within the tested range of experimental conditions, the maximal xylose recovery of ca. 81% (or 17.48% of oven-dry material) was observed after cardoon hydrolysis at 140°C for 30 min in 0.5% acid solution. The low content of furfural (0.46%) and glucose (1.72%) indicated only limited monosaccharide decomposition and cellulose hydrolytic depolymerization under these reaction conditions.

Hydrolysis modeling and optimization

The response surface methodology (RSM) was used for statistical modeling and optimization of cardoon hydrolysis reaction [26]. The 2^3 central composite rotatable design (CCRD) was employed to optimize the effect of the principal independent variables, i.e., reaction time (X_1), temperature (X_2) and acid concentration (X_3), on efficiency (Y_1) and selectivity (Y_2) of xylan conversion to xylose.

Based on results of the preliminary tests, the current settings (range and levels) of independent process variables to be used in RSM were defined (Table 4) and the RSM experimental design matrix for three coded independent variables at five levels each, with six star (axial) points and five replicates at the central point (total 19 runs) was developed according to CCRD (Table 5). The obtained experimental data were used to calculate the significant effects having the greatest impact on reaction outputs. Figure 1 illustrates the Pareto charts of significant effects estimated, respectively, for xylan conversion and hydrolysis selectivity. The graphs show the standardized linear, quadratic and interaction effects of process factors (independent variables), sorted by their absolute magnitude in relation to the statistical significance (p) level of 0.05. Apparently, the linear and quadratic effects of acid concentration, along with quadratic effect of reaction temperature and interaction effect between temperature and concentration, are the most important in determining the resultant degree of xylan conversion to xylose. The hydrolysis selectivity, in its turn, is controlled almost solely by linear effect of reaction temperature.

Component	Content (% on oven-dry matter)
Ash	5.44 ± 0.02
silica SiO_2	0.08 ± 0.01
Extractives	7.36 ± 0.14
dichloromethane	0.41 ± 0.01
ethanol	3.25 ± 0.19
water	3.70 ± 0.21
Lignin	18.19 ± 0.05
Klason (acid-insoluble)	15.46 ± 0.03
acid-soluble	2.73 ± 0.07
Holocellulose	69.87 ± 0.03
α-cellulose	39.28 ± 0.05
hemicelluloses	30.59 ± 0.08
fraction A	25.30 ± 0.06
fraction B	5.29 ± 0.09

Table 1: Chemical composition of cardoon stem-wall material.

	Holocellulose	Hemicellulose	
		Fraction A	Fraction B
Rhamnose	0.96	2.01	1.28
Arabinose	1.48	2.05	2.34
Xylose	27.07	91.98	77.43
Mannose	1.42	0.70	3.22
Galactose	1.74	2.56	4.28
Glucose	67.33	0.70	11.45

Table 2: Carbohydrate composition of holocellulose fraction of cardoon stem-wall material (%).

Temperature (°C)	Time (min)	Acid Concentration (%)	Yield (% on oven-dry matter)				
			Glc[1]	Xyl	AcA	F	HMF
130	30	0.5	0.35	3.72	Traces	Traces	Traces
		1.0	0.63	9.46	2.10	Traces	Traces
		1.5	0.98	12.89	2.60	Traces	0.23
	60	0.5	0.47	8.17	1.23	Traces	0.20
		1.0	0.93	13.35	2.64	0.09	0.27
		1.5	1.38	15.61	3.72	0.33	0.29
140	30	0.5	0.57	9.43	1.24	Traces	0.27
		1.0	1.33	15.78	2.95	0.16	0.28
		1.5	1.72	17.48	3.82	0.46	0.29
	60	0.5	1.06	13.91	2.71	0.20	0.28
		1.0	2.01	15.81	3.05	0.95	0.30
		1.5	2.59	16.12	3.93	1.78	0.33
150	30	0.5	1.20	12.36	2.15	0.20	0.28
		1.0	2.20	16.43	3.67	0.95	0.31
		1.5	2.88	15.97	4.35	2.00	0.36
	60	0.5	2.21	17.09	3.83	1.01	0.31
		1.0	2.82	12.49	4.84	3.38	0.44
		1.5	3.57	10.95	6.30	4.15	0.47

[1]Glc, Xyl, AcA, F,HMF as glucose, xylose, acetic acid, furfural and 5-hydroxymethylfurfural, respectively

Table 3: Experimental data on monosaccharide formation and decomposition under variable conditions of dilute acid hydrolysis of cardoon stalks.

Variable		Range and levels				
		-α	-1	0	+1	+α
Temperature (°C)	X_1	123.2	130	140	150	156.8
Time (min)	X_2	19.8	30	45	60	70.2
Acid concentration (%)	X_3	0.2	0.5	1.0	1.5	1.8

Table 4: Range and levels of independent process variables used in experimental design.

Run N°	Coded variables			Responses	
	X_1	X_2	X_3	Y_1	Y_2
1	-1	-1	-1	0.18	10.55
2	-1	-1	+1	0.61	13.18
3	-1	+1	-1	0.39	17.28
4	-1	+1	+1	0.74	11.30
5	+1	-1	-1	0.58	10.27
6	+1	-1	+1	0.76	5.55
7	+1	+1	-1	0.82	7.88
8	+1	+1	+1	0.52	3.07
9	-α	0	0	0.53	19.29
10	+α	0	0	0.51	3.28
11	0	-α	0	0.53	16.32
12	0	+α	0	0.83	7.73
13	0	0	-α	0.08	3.43
14	0	0	+α	0.87	6.85
15 (C)	0	0	0	0.81	9.72
16 (C)	0	0	0	0.81	9.42
17 (C)	0	0	0	0.80	9.86
18 (C)	0	0	0	0.81	9.91
19 (C)	0	0	0	0.83	9.37

Table 5: Central composite rotatable design (CCRD) applied for cardoon hydrolysis and the corresponding experimental responses on xylan conversion (Y_1) and process selectivity (Y_2) used for RSM modeling.

The significance of each estimated effect was checked by analysis of variance (ANOVA). The low p values ($p<0.01$) of the main effects influencing the xylan conversion and hydrolysis selectivity, along with results of the Student's (t) and Fisher's (F) tests (Table 6), indicate the high statistical significance of the estimated relation between variables within a 99% confidence interval.

The second-order polynomial model (Eq.1) was fitted to experimental data to determine the optimum levels (conditions) of the independent process variables by multiple regression analysis. The regression coefficients for process modeling were calculated and the statistical significance of each coefficient was estimated by p-values.

Two model equations (Eqs. 2,3) describing, respectively, xylan conversion (Y_1) and hydrolysis selectivity (Y_2) were obtained using more statistically significant regression coefficients:

$$Y_1 = -26.2065 + 0.3218X_1 - 0.001X_1^2 + 0.0644X_2 + 4.8553X_3 - 0.4642X_3^2 - 0.023X_1X_3 - 0.0093X_2X_3 \quad (2)$$

$$Y_2 = 120.6866 - 1.7215X_1 + 0.8381X_2 + 0.0041X_2^2 - 6.0414X_3^2 - 0.0081X_1X_2 - 0.1452X_2X_3 \quad (3)$$

The values of the determination coefficient R^2=0.91 and R^2=0.89, found respectively for Eqs. 2 and 3, pointed to good correlation between experimental and predicted data (model fit) and indicated the high statistical significance of the models as a whole, which can explain 89-91% of the total variability in responses within examined range of reaction conditions.

The 3D response surfaces and the corresponding contour plots illustrating the modeled effects of independent variables on reaction outputs are shown in (Figures 2-4). As can be seen, the hill-shaped surfaces of xylan conversion during hydrolysis (upper plots of Figures 2-4) reveal some maximum values (stationary points) near the center point of the experimental design, thereby allowing locating and characterizing the optimum responses under variable reaction conditions. Evidently, keeping constant acid concentration at 1% as a center point, the temperature range of 135-145°C and reaction period of 40-60 min are desirable to maximize the yield of xylose (Figure 2). In its turn, at constant temperature of 140°C as a center point, the highest efficiency of xylan hydrolysis can be achieved within the range of acid concentration 1-1.5% and reaction period 40-60 min (Figure 3). The marked desirable ranges of reaction temperature and acid concentration are particularly notable in Figure 4, where the reaction time was kept constant at 45 min as a center point.

The effect of process inputs on hydrolysis selectivity is more complex and depicted by saddle-shaped response surfaces with maximum and minimum values encountered at various combinations of independent variables (bottom plots of (Figures 2-4). As mentioned above, the reaction temperature has a vital importance for hydrolysis selectivity. Under fixed acid concentration of 1% as a center point, the

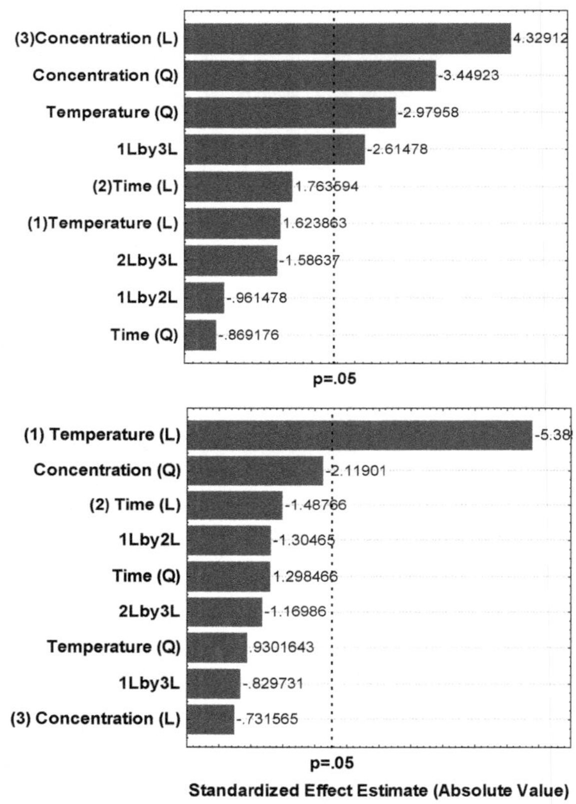

Figure 1: Pareto charts of standardized effects for xylan conversion to xylose (top) and hydrolysis selectivity (bottom).

Factor	Y_1				Y_2			
	Estimated effect	*t*-test	*F*-test	*p*	Estimated effect	*t*-test	*F*-test	*p*
(1) Temperature (L)	0.1092	1.6239	2.6369	0.1389	-7.6815	-5.3898	29.0504	0.0004
Temperature (Q)	-0.2005	-2.9796	8.8779	0.0155	1.3260	0.9302	0.8652	0.3766
(2) Time (L)	0.1186	1.7636	3.1103	0.1116	-2.1202	-1.4877	2.2131	0.1710
Time (Q)	-0.0585	-0.8692	0.7555	0.4073	1.8510	1.2985	1.6860	0.2264
(3) Concentration (L)	0.2912	4.3291	18.7413	0.0019	-1.0426	-0.7316	0.5352	0.4830
Concentration (Q)	-0.2321	-3.4492	11.8972	0.0073	-3.0207	-2.1190	4.4902	0.0631
1L by 2L	-0.0845	-0.9615	0.9244	0.3614	-2.4294	-1.3047	1.7021	0.2244
1L by 3L	-0.2298	-2.6148	6.8371	0.0281	-1.5450	-0.8297	0.6885	0.4282
2L by 3L	-0.1394	-1.5864	2.5165	0.1471	-2.1784	-1.1699	1.3686	0.2721

Table 6: ANOVA of estimated linear (L), quadratic (Q) and interaction effects for xylan conversion (Y_1) and process selectivity (Y_2).

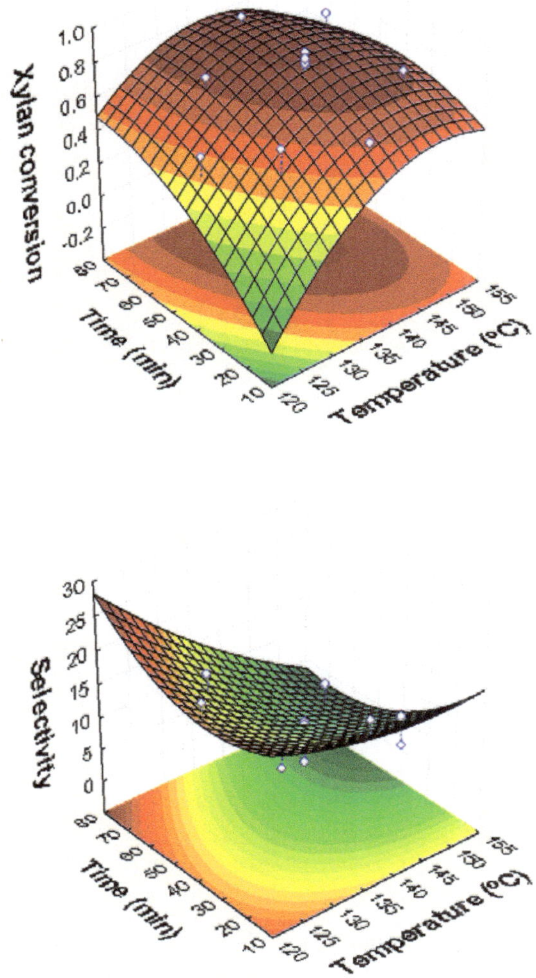

Figure 2: Response surfaces and contour plots of modeled xylan conversion to xylose (top) and hydrolysis selectivity (bottom) as a function of reaction time (*t*, min) and reaction temperature (*T*,°C) at fixed acid concentration of 1% set as a central point.

maximum estimated selectivity (ca. 20) of acid hydrolysis of cardoon hemicelluloses, i.e., the minimal hydrolytic cellulose depolymerization, can be obtained under temperatures below 120°C (Figure 2), what is however far removed from the desirable temperature range needed for maximal xylan conversion to xylose. The effect of acid concentration on selectivity is shown in Figure 3. It can be seen that at constant temperature of 140°C as a center point, the highest selectivity can be provided only by short (less than 20 min) reaction under high acid concentrations or by prolonged (over 70 min) reaction under low acid concentration, what also falls outside the desirable range of reaction conditions required for maximal xylan conversion. Hence, in support of experimental data of the preliminary tests, the statistical modeling showed the practical unfeasibility to achieve maximum xylan conversion to xylose without partial degradation of cellulose, i.e., without some loss in hydrolysis selectivity. This is obviously connected with close reactivity/accessibility of hemicelluloses and the less-ordered (amorphous) cellulose portion in the cell-wall material of plants [27]. In fact, there is a good agreement between amount of glucose released into solution during cardoon hydrolysis and the content of alkali-extractable (amorphous) cellulose portion quantified during hemicellulose analysis of cardoon stalks (Fraction B, Table 2). Based on Fig. 4, the estimated selectivity level of only ca. 6-10 (1.5-3% glucose in solution) can be expected when hydrolysis reaction is performed under conditions designed to maximize the xylose yield in solution. It is significantly better of data reported for high temperature (180-200°C) dilute sulfuric acid (0.1-0.2%) hydrolysis of cardoon [19,20], where release of 10-20% glucose was observed lowering substantially hydrolysis selectivity. The higher cellulose degradation (4.4-11.2% glucose released) was also reported for dilute acid hydrolysis of other herbaceous species [28,29].

The partial differentiation of the multivariate function described by Eq. 2 was used to find an optimal combination of the independent process variables in the stationary point of the fitted polynomial model corresponding to maximum xylan conversion to xylose during hydrolysis. The partial derivatives of Y_1 with respect to X_1, X_2 and X_3 gave the following equations:

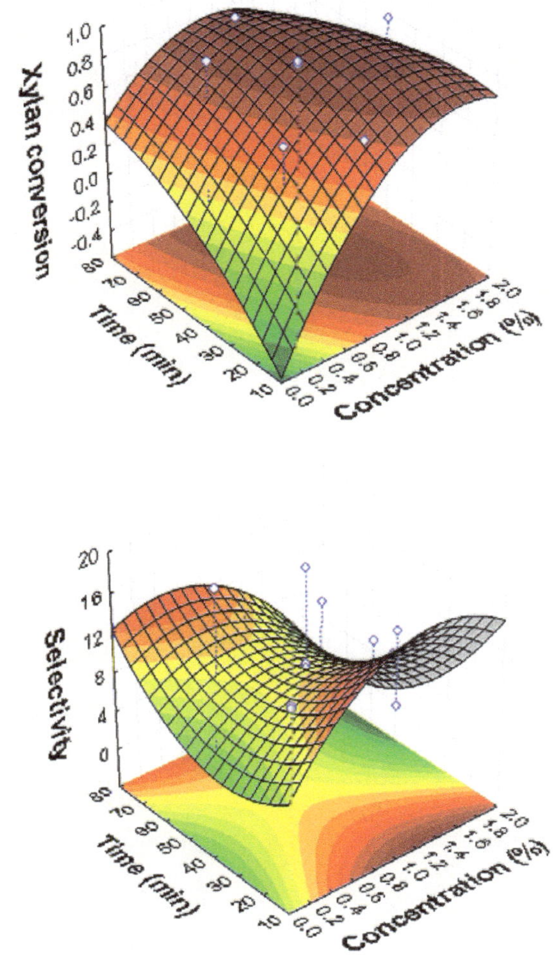

Figure 3: Response surfaces and contour plots of modeled xylan conversion to xylose (top) and hydrolysis selectivity (bottom) as a function of reaction time (*t*, min) and acid concentration (*C*, %) at fixed reaction temperature of 140°C set as a central point.

$$\partial Y_1/\partial X_1 = 0.3218 - 0.002X_1 - 0.023X_3 = 0 \quad (4)$$

$$\partial Y_1/\partial X_2 = 0.0644 - 0.0093X_3 = 0 \quad (5)$$

$$\partial Y_1/\partial X_3 = 4.8553 - 0.9284X_3 - 0.023X_1 - 0.0093X_2 = 0 \quad (6)$$

Resolution of Eqs 4-6 revealed the optimum condition set required for maximal predicted response of xylan conversion to xylose: reaction time of 51.7 min, reaction temperature of 138.5°C and sulfuric acid concentration of 1.28% with a maximum expected xylan conversion of 0.87 (87% of xylose recovery in solution).

The validity of the developed statistical model was checked by replicated (four times) control experiments performed under found optimal conditions. The obtained experimental values on xylan conversion (0.86 ± 0.01) and hydrolysis selectivity (7.85 ± 0.11; with 2.03% of glucose) showed very close agreement with calculated (predicted) data thereby confirming the high model validity and suitability as a whole. As it shown on the mass balance flow diagram (Figure 5), only limited monosaccharide degradation (furfural 1.04 g, 5-hydroxymethylfurfural 0.33 g and acetic acid 3.61 g per 100 g of dry

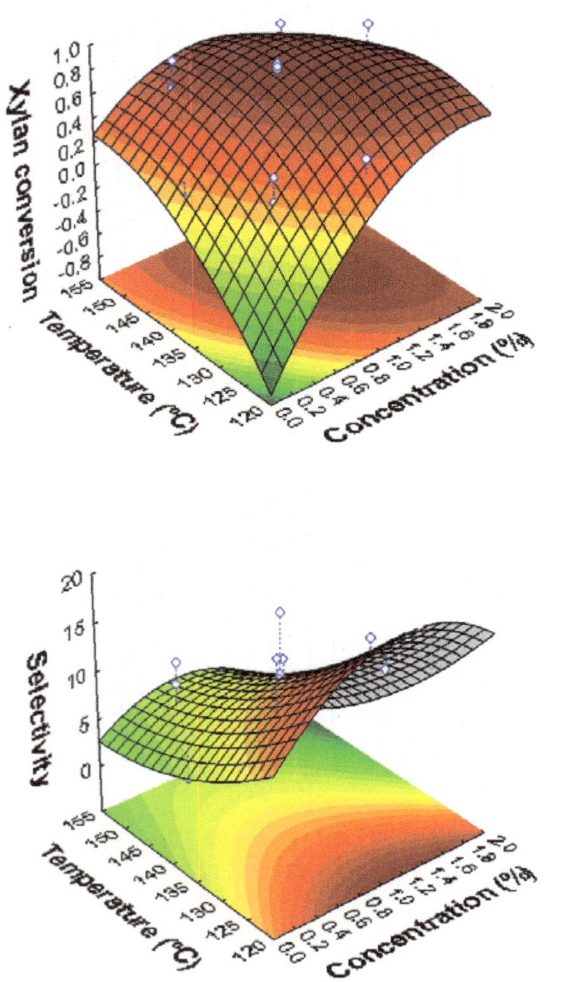

Figure 4: Response surfaces and contour plots of modeled xylan conversion to xylose (top) and hydrolysis selectivity (bottom) as a function of reaction temperature (*T*,°C) and acid concentration (*C*, %) at fixed reaction time of 45 min set as a central point.

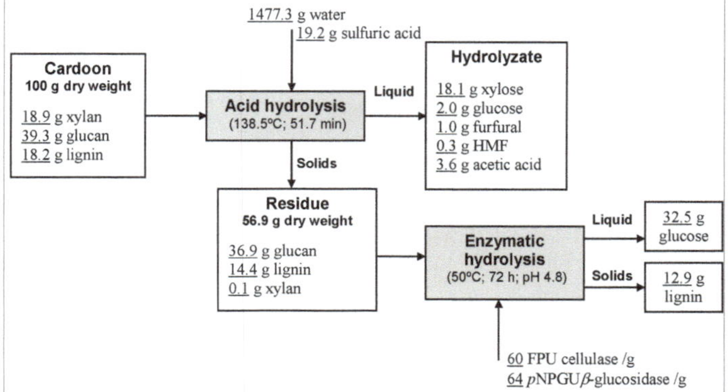

Figure 5: Mass balance flow diagram for acid hydrolysis of cardoon (performed under optimized conditions designed for total xylan conversion to monomeric xylose) and enzymatic digestibility of unhydrolyzed solid residue.

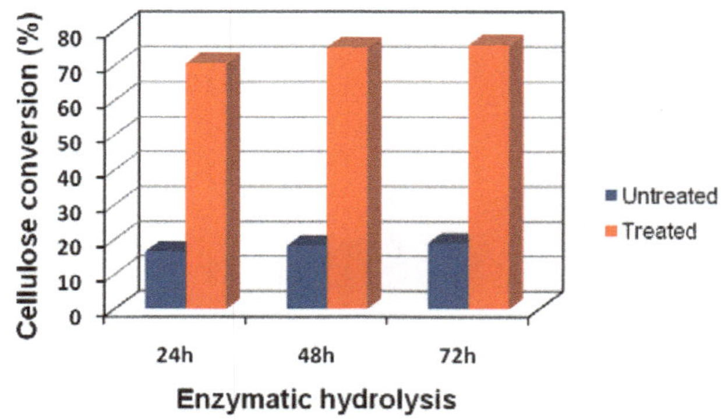

Figure 6: Enzymatic digestibility improvement of cardoon stem-wall material after dilute sulfuric acid hydrolysis performed under optimized conditions.

biomass) was observed under optimized reaction conditions providing the desired quality of sugar substrate (hydrolyzate) for subsequent (bio)chemical processing. The similar results on xylose recovery after dilute sulfuric acid hydrolysis were also reported for other perennial herbs, such as silvergrass (xylose yield of ca. 75%) [29], switchgrass, reed canarygrass and alfalfa (xylose yield of 73-86%) [28], correlating well with the data of the present study.

Enzymatic digestibility/saccharification of insoluble residue

Since hemicellulose removal is presumed to destroy the lignin-carbohydrate matrix shielding cellulose microfibrils in the cell wall material [5,11], the dilute acid hydrolysis of cardoon should therefore substantially improve enzymatic digestibility (saccharification) of cellulose portion in insoluble (unhydrolyzed) residue, providing a valuable source of fermentable sugars (glucose) for biofuel (ethanol) production.

As can be seen from Figure 6, the enzymatic digestibility of cellulose in pre-hydrolyzed biomass was substantially improved over the control (unhydrolyzed cardoon). After 72 h enzymatic hydrolysis performed under standard NREL conditions [21], the cellulose conversion to monomeric glucose of ca. 76% (32.5 g Glc / 100 g dry biomass, Figure 5) was achieved for pre-hydrolyzed biomass, in contrast to ca. 19% conversion (8.2 g Glc / 100 g dry biomass) for unhydrolyzed biomass.

Evidently, the degree of cellulose digestibility can be further improved by optimization study of enzymatic saccharification.

Conclusion

The low temperature one-step dilute sulfuric acid hydrolysis was very effective to convert the cardoon (*Cynara cardunculus* L.) heteroxylan to monomeric sugars providing a quality xylose-rich substrate for subsequent (bio)chemical processing and reactive hemicellulose-free solid residue ready for further fractionation within complex biorefinery scheme. The statistical modeling using Response Surface Methodology made it possible to identify the main factors of the multi-variable hydrolysis process affecting efficiency and selectivity of xylan conversion to xylose and to define the optimum set of reaction conditions for maximal xylose recovery in solution. Under optimum reaction conditions (138.5°C, 51.7 min and 1.28% acid concentration) the xylan conversion of 0.86 (86% of xylose recovery or 18.08 g /100 g cardoon) was achieved vs. 0.87 predicted by model, with limited cellulose degradation and furans formation (glucose 2.03 g and furfural 1.04 g per 100 g of cardoon). The insoluble residue after xylan hydrolysis can be easily digested to fermentable sugars for ethanol production, providing cellulose-to-glucose conversion of 76% vs. 19% for untreated cardoon.

Acknowledgement

The financial support of the Fundação para a Ciência e a Tecnologia (FCT, Portugal) within research contract PTDC/AGR-CFL/103840/2008 is gratefully acknowledged. The authors thank D. Neiva and Dr. L. Duarte for their technical assistance.

References

1. Kamm B, Kamm M (2007) Biorefineries – Multi Product Processes. Adv Biochem Engin Biotechnol 105: 175-204.
2. Fernando S, Adhikari S, Chandrapal C, Murali N (2006) Biorefineries:Current Status, Challenges, and Future Direction. Energy Fuels 20: 1727-1737.
3. Kamm B, Schneider BU, Hüttl RF, Grünewald H, Gusovius HJ, et al. (2006) Lignocellulosic Feedstock Biorefinery – Combination of technologies of agroforestry and a biobased substance and energy economy. Forum der Forschung 19: 53-62.
4. FitzPatrick M, Champagne P, Cunningham MF, Whitney RA (2010) A biorefinery processing perspective: Treatment of lignocellulosic materials for the production of value-added products. Biores Technol 101: 8915-8922.
5. Zhang YH (2008) Reviving the carbohydrate economy via multi-product lignocellulose biorefneries. J Ind Microbiol Biotechnol 35: 367-375.
6. Werpy T, Petersen G (2004) Top Value Added Chemicals from Biomass. US Department of Energy, Office of Scientific and Technical Information.
7. Lewandowski I, Scurlock JMO, Lindvall E, Christou M (2003) The development and current status of perennial rhizomatous grasses as energy crops in the US and Europe. Biomass and Bioenergy 25: 335-361.
8. Dalianis C, Christou M, Sooter C, Kyritsis S, Zafiris C, Sanmiotakis G (1994) Adaptation and biomass productivity of *Cynara cardunculus* in abandoned, marginal fields. Proceedings of the 7th Europe Conf. on Biomass for Energy and Industry, Ponte Press, Florence.
9. Fernández J (1998) Cardoon. In Bassam N ed, Energy plant species, James and James Sci. Pub. Ltd, London.
10. Fernández J, Curt MD, Aguado PL (2006) Industrial applications of *Cynara cardunculus* L. for energy and other uses. Ind Crops Prod 24: 222-229.
11. Sun Y, Cheng J (2002) Hydrolysis of lignocellulosic materials for ethanol production: a review. Biores Technol 83: 1-11.
12. delCampo I, Alegría I, Zazpe M, Echeverría M, Echeverría I (2006) Diluted acid hydrolysis pretreatment of agri-food wastes for bioethanol production. Ind Crops Prod 24: 214-221.
13. Martin C, Alriksson B, Sjode A, Nilvebrant NO, Jonsson LJ (2007) Dilute sulfuric acid pretreatment of agricultural and agro-industrial residues for ethanol production. Appl Biochem Biotechnol 140: 339-352.
14. Rahman SHA, Choudhury JP, Ahmad AL, Kamaruddin AH (2007) Optimization studies on acid hydrolysis of oil palm empty fruit bunch fiber for xylose production. Biores Technol 98: 554-559.
15. Jeong TS, Um BH, Kim JS, Oh KK (2010) Optimizing dilute-acid pretreatment of rapeseed straw for extraction of hemicellulose. Appl Biochem Biotechnol 161: 22-33.
16. Akpinar O, Levent O, Bostanci S, Bakir U, Yilmaz L (2011) The optimization of dilute acid hydrolysis of cotton stalk in xylose production. Appl Biochem Biotechnol 163: 313-325.
17. Castro E, Díaz MJ, Cara C, Ruiz E, Romero I, Moya M (2011) Dilute acid pretreatment of rapeseed straw for fermentable sugar generation. Biores Technol 102: 1270-1276.
18. Martinez J, Negro MJ, Sáez F, Manero J Sáez R, Martin C (1990) Effect of acid steam explosion on enzymatic hydrolysis of *O. nervosum* and *C. cardunculus*. Appl Biochem Biotechnol 24-25: 127-131.
19. Ballesteros M, Negro MJ, Manzanares P, Ballesteros I, Sáez F, Oliva JM (2007) Fractionation of *Cynara cardunculus* (Cardoon) biomass by dilute-acid pretreatment. Appl Biochem Biotechnol 140: 239-252.
20. Ballesteros I, Ballesteros M, Manzanares P, Negro MJ, Oliva JM, Sáez F (2008) Dilute sulfuric acid pretreatment of cardoon for ethanol production. Biochem Eng J 42: 84-91.
21. Selig M, Weiss N, Ji Y (2008) Enzymatic saccharification of lignocellulosic biomass. Laboratory analytical procedure.
22. Browning BL (1967) Methods of Wood Chemistry. Vol 2, John Wiley & Sons, New York.
23. Miller GL (1959) Use of dinitrosalicylic reagent for determination of reducing sugars. Anal Chem 31: 426-428.
24. Martinez A, Rodriguez ME, York SW, Preston JF, Ingram LO (2000) Use of UV absorbance To monitor furans in dilute acid hydrolysates of biomass. Biotechnol Prog 16: 637-641.
25. Atchison JE (1993) Data on non-wood plant fibres. In Kocurek MJ ed, Pulp and Paper Manufacture, Vol 3, TAPPI, CPPA.
26. Myers RH, Montgomery DC, Anderson-Cook CM (2009) Response surface methodology: Process and product optimization using designed experiments. (3rdedn) John Wiley & Sons, New York.
27. Fengel D, Wegener G (1984) Wood (Chemistry, Ultrastructure, Reactions). Walter de Gruyter Berlin, New York.
28. Dien BS, Jung H-JG, Vogel KP, Casler MD, Lamb JFS, et al. (2006) Chemical composition and response to dilute-acid pretreatment and enzymatic saccharification of alfalfa, reed canarygrass, and switchgrass. Biomass and Bioenergy 30: 880-891.
29. Guo GL, Chen WH, Chen WH, Men LC, Hwang WS (2008) Characterization of dilute acid pretreatment of silvergrass for ethanol production. Biores Technol 99: 6046-6053.

Simulation of the Effects of Turbine Exhaust Recirculation on the Composition of Flue Gas for A CO_2 Capture Unit

Ahmed Mohamed Al Menhali, Abdulla Ali. AL Marzooqi, Abdul Rahman Saleh Alameri, Abdulla Mohamed Al Ameri and Zin Eddine Dadach*

Chemical Engineering Department, Abu Dhabi Men's College, Higher Colleges of Technology, Saada street, Abu Dhabi (Abu Dhabi), UAE

Abstract

The main objective of this paper is to study the effects of Flue Gas Recirculation (FGR) ratio and Excess Air (EA) in a natural gas-fired turbine for an effective CO_2 absorption by amines. To achieve this goal, the flue gas should contain at least 10% (mol.) of CO_2. Moreover, in order to avoid technical problems related to the oxidative degradation of amines, the flue gas should also contain less than 5% (mol.) of O_2. The simulation results indicate that, for a gas turbine that limits the temperature of the exhaust gas leaving the combustor at 1100°C, an excess air (EA) of 200% and a Flue Gas Recirculation (FGR) ratio of 0.65 are needed to fit the requirements of an effective absorption process by amines. For a turbine that allows temperatures as high as 1500°C, the operating parameters (EA=100% and FGR ratio of 0.4) will be selected.

Keywords: Carbon capture; Flue gas composition; Exhaust gas recirculation ratio; Excess air; Turbine metallurgical limitation

Introduction

One of the largest EOR (Enhanced Oil Recovery) projects worldwide using anthropogenic CO_2 is the Weyburn project in Canada. The CO_2 required for this project is produced at Dakota Gasification Company's synthetic fuel plant in Beulah, North Dakota, USA. The project is expected to produce 122 million bbls of incremental oil, extending the field life by 20-25 years and increasing the oil recovery to 34% of OOIP (Original Oil In Place) [1,2]. In order to provide large volumes of CO_2 to serve EOR projects in the oil fields of the UAE, carbon dioxide need to be captured from industrial facilities such as power and desalination plants, oil refineries, gas processing facilities and/or petrochemical complexes [3]. As shown in Table 1, 76 million tons of CO_2 were emitted in Abu Dhabi and nearby (UAE) during the year 2008.

The valuable fuel gas actually used for EOR techniques in the UAE could be utilized to produce electricity in power generation plants. The captured CO_2 from these plants could be transported and stored in the large oil fields. The use of CO_2-EOR technique will also enhance the production of crude oil in the UAE.

Challenges of the absorption of CO_2 from the exhaust gas of gas turbines

Unlike Pre-combustion and Oxy fuel strategies, Post-combustion CO_2 Capture can be integrated with the existing power generation stations. At the moment, the post-combustion CO_2 capture using amine scrubbing is the most mature technology, and arguably the preferred technology, since the process of absorption has been used for decades in gas treatments [4,5]. The existing size of commercial CO_2 amine scrubbing plants is relatively small (few 100,000 metric tons/year). However, larger CO_2 capturing plants (>1000,000 metric tons/year) might be required for any future CCS (Carbon Capture and Storage) project [6].

Industry	Millions Tons of CO_2
Power generation Plants	33 (43%)
Gas processing and Refineries	25 (33%)
Metal Industries	15.8 (21%)
Petrochemical Plants	2.2 (3%)
Total	76

Table 1: Estimated amount of CO_2 by industry in Abu Dhabi and nearby for the year 2008 [3].

The absorption plants using amine mixtures are most effective at around 10-15 (mol.%) of CO_2 in the flue gas [7]. This percentage depends mainly on the composition of the combustible, the flue gas recirculation (FGR) ratio and the excess air (EA) used in the combustor. Lower CO_2 concentrations in the flue gas require a leaner solvent loading to achieve an adequate "driving force" between the solvent and the flue gas stream for high levels of CO_2 Capture. However, leaner solvent loadings require also a higher CO_2 removal from the solvent, which will increase the amount of energy utilized in the stripper and the cost of the CO_2 Capture plant.

In power-generation plants, natural gas (NG) is preferred to the other heavier fuels because the environmental problems are minimized and the total cost of a CCS plant is reduced by avoiding corrosion and other technical problems due to impurities (H_2S, SOx, NOx, HCl...) in the captured CO_2. In a simplified NGCC (Natural Gas Combined Cycle) power plant configuration (Figure 1), the exhaust gas from the combustor is expanded in a gas turbine (GT) to produce electricity. The flue gas leaving the CTG (Combustion Turbine Generator) at high temperature will generate steam in a heat recovery steam generator (HRSG) before entering the CO_2 capture unit. The steam is utilized to generate electricity from a steam turbine (ST).

In order to keep the temperatures in the gas turbine at permissible levels, the combustor has a typical overall excess air ratio in the range 3-3.5. As a result, flue gas has a percentage of CO_2 in flue gas of about

***Corresponding author:** Zin Eddine Dadach, Chemical Engineering Department, Abu Dhabi Men's College, Higher Colleges of Technology, Saada street, Abu Dhabi (Abu Dhabi), UAE

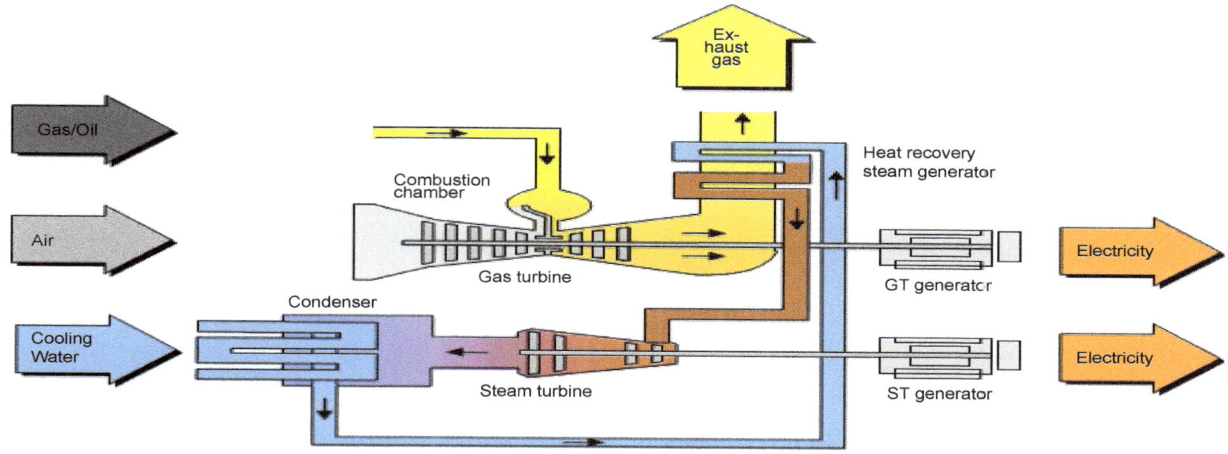

Figure 1: Schematic representation of a Natural Gas Combined Cycle [2].

3-3.5% [8]. Flue gas from natural gas-fired turbines has also relatively high levels of O_2 which can cause corrosion problems and degrade amine solvents. According to Chakravarti et al. [9], oxidative degradation of amines often occurs in CO_2 capture plants when the flue gases contain a high O_2 content such as 5%. To counter the influence of oxygen, the approach currently practiced is the use of chemical inhibitors. For example, the processes licensed by Kerr-McGee/ABB Lummus Global Inc. and by Fluor Daniel [10] use inhibited monoethanolamine solutions. However, corrosion inhibitors could also act as a catalyst towards the solvent degradation. Praxair has been granted two patents for improved oxygen tolerant absorption processes [11,12]. The key aspect of Praxair's successful approach is handling the dissolved oxygen in amine mixture through process modifications instead of the introduction of additional chemicals.

Flue Gas Recirculation (FGR) in gas turbines for an effective CO_2 absorption by amines

One possible option for increasing the concentration of CO_2 and decreasing the amount of O_2 in the flue gas from power plants using natural gas is recirculation of a part of the flue gas. The effects of Flue Gas Recirculation (FGR) on the concentration of CO_2 and O_2 in the flue gas have been investigated [7,8,13,14]. Akram et al. [7] studied the effects of recirculating part of flue gas in a 100 Kw (plus 150 Kw hot water) CHP gas turbine Turbec T100. Their preliminary results indicate that a recirculation ratio of 0.45 increased the molar percentage of CO_2 from 1.5% to 3.0% and the molar percentage of O_2 decreased from 18.2% to 16.0%. Their results show also that recycling part of the flue gas decreases NOx emissions by decreasing the flame temperature. Using a natural gas-fired power plant of 700 MW, the results presented by Bolland and Saether [8] show that, using a recirculation ratio of 0.4, the molar percentage of CO_2 increased from 3.3 % to 5.5 % and the molar percentage of O_2 decreased from 13.8% to 9.8%. To maintain high cycle efficiency, flue gas was cooled before reinjection in the combustor.

Abu Zahra [14] used Aspen plus and the Peng-Robinson equation of state, to simulate the effect of flue gas recirculation in an IGCC power plant using a low-pressure gasifier. With all the other parameters of the gas turbine constant, the results indicated that the concentration of CO_2 increased from 7.3 (% mol.) to 14.0 (% mol.) with the increase of the flue gas recirculation ratio from 0% to 45%. Moreover, the concentration of O_2 decreased from 14.2 (% mol.) to 8.0 (% mol.) with the same increase of flue gas recirculation.

Maximum value of flue gas recirculation (FGR) ratio in gas turbines

To reduce the cost of a CCS project, high Flue Gas Recirculation (FGR) ratios are desirable to achieve the maximum CO_2 concentration in the flue gas to be treated in the absorption unit. However the combustion process sets a limit due to the corresponding reduction in O_2 concentration. Due to the importance of the recirculation rate on the overall performance of the CO_2 removal process, combustion studies are required to define the maximum achievable recirculation rate. The maximum FGR ratio value will depend mainly on the combustible, the amount of excess air used and the turbine mechanical limitations. Using natural gas in a turbine of 700 MW, Bolland and Saether [8] concluded that the maximum allowable recirculation ratio is about 0.65. Higher values affected significantly the combustion process due to a low percentage of oxygen. Røkke and Hustad [13] used a 65 kW gas turbine combustor to study the effects of FGR on the combustion process. The effects of adding N_2, CO_2 and O_2 in the combustion process were investigated focusing on stability of the flame and emissions of NOx. Their results show that adding N_2 and CO_2 decreases the NOx emissions by decreasing the combustion temperature, whereas O_2 addition increases the NOx emissions. The results show also that a significant decrease of NOx is seen when flue gas is premixed with the fuel, whereas the addition into the air stream has less effect on the NOx emissions. For the stability of the flame in the combustor, the maximum values of the mass ratios O_2/CH_4, CO_2/CH_4 and N_2/CH_4 were respectively 0.79, 1.29 and 1.15. Carbon monoxide (CO) was also detected at high FGR ratios when combustion stoichiometry was approached [15].

The minimum air quantity needed to get a complete oxidation of the fuel gas is known as theoretical air [16]. It is known that the hot temperatures at the gas turbine inlet are a metallurgical limitation for each gas turbine model. These temperatures vary from 800°C to 1700°C depending on turbine blade materials and whether or not they have cooling systems [15]. As a consequence, industrial gas turbines use excess air (EA) values from 100% to 600%. The reason for using high EA in a gas turbine is to keep the turbine inlet temperature at a permissible level. A high turbine inlet gas temperature will then require a lower quantity of secondary and tertiary air to cool down the combustion gas to the specified temperature. The Modelisation results of Martinez et al. [17] show that, at a gas temperature of 1200°C at the inlet of a turbine,

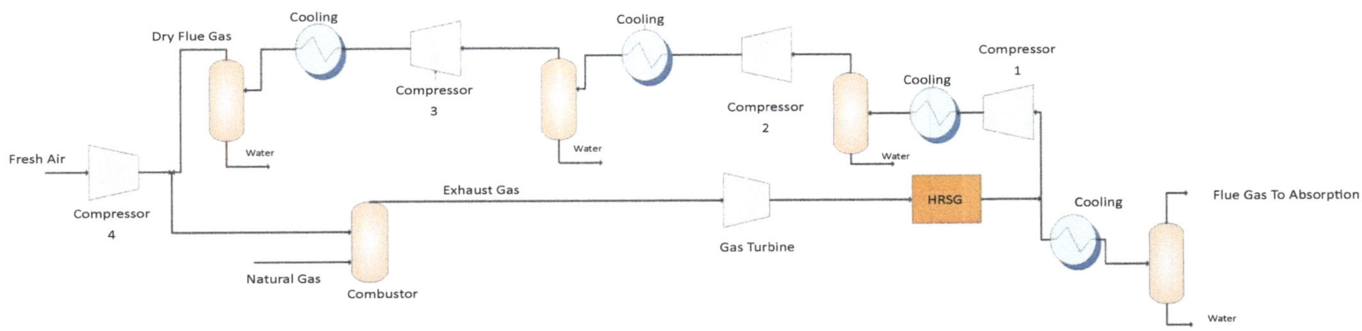

Figure 2: Proposed Combustion Turbine Generator with Flue Gas Recirculation.

Fuel Composition	Mole %
CO_2	1.0
CH_4	93.0
C_2H_6	3.0
C_3H_8	1.0
O_2	2.0
Total	100

Table 2: Natural gas composition.

the excess air is198.79%. If the temperature is 1000°C, the excess air increases to 324.51%. According to the authors, the difference between the calculations of the excess air using dry air and wet air is negligible.

Based on the literature review, the Flue Gas Recirculation (FGR) ratio is optimized in order to minimize the content of fresh secondary and tertiary air used for cooling by recycling part of the flue gas without affecting the amount of oxygen necessary for the stoichiometry of the combustion. The main objective of this study is to determine the operating conditions (flue gas recirculation ratio and amount of excess air) in a gas turbine in order to obtain at least 10% (mol.) of CO_2 and no more than 5% of O_2 in the flue gas for an effective CO_2 capture process with minimum technical problems related to the oxidative degradation of the amine mixture.

Methodology

Proposed layout of gas turbine

The Combustion Turbine Generator (CTG) of a NGCC (Natural Gas Combined Cycle) power plant producing 500 MW is under consideration in this study. It is assumed that natural gas (NG) is available in a battery limit of the plant at 3 MPa and 40°C [18]. Air is considered to enter at 101 kPa and 30°C and compress up to 3.1 MPa to mix with natural gas in the burners of the combustor. In order to increase the efficiency of the cycle, the recycled flue gas is cooled at 40°C. In the proposed recycling process, the flue gas is added to the compressed air. The pressure of the recycled gas will increase from 110 kPa to the pressure of the compressed fresh air of 3.1 MPa. The main equipment used in the proposed recycling process (Figure 2) is an air compressor, three flue gas compression stages (compression ratio for each stage=3.1) with intercooling and separators, a combustor and a gas turbine The proposed recycling process of the flue gas will have the advantage of condensing part of the water in the intermediate separators and decreasing the total energy used for the compression of recycled flue gas and the fresh air.

Estimation of the mass flowrate of the natural gas

The composition of the natural gas under consideration is presented in Table 2 [2].

Assuming that the water component of a combustion process is in a vapor state at the end of combustion, the Lower Heating Values (LHV) of the components of the natural gas will be used: LHV (CH_4)=50 MJ/kg; LHV (C_2H_6)=47.8 MJ/kg; LHV (C_3H_8)=46.35 MJ/kg. For the given composition, the average LHV of the natural gas is equal to 46.7 MJ/kg. Based on Low Heat Value (LHV), the net efficiencies of the NGCC plants with post-combustion capture are around 47.4-49.6% [19]. Therefore, to produce 500 MW with an efficiency of 48.5%, 1030.9 MW is needed from the combustion process. The needed mass flowrate of the methane, ethane and propane is 22.07 kg/s and the corresponding mass flowrate of the natural gas stream is equal to 23.54 kg/s.

Optimum value of mass flowrate of air for combustion

The stoichiometric combustion is a theoretical point in which the optimum amount of oxygen and fuel mix generates the most heat possible and maximum combustion efficiency is achieved. According to the composition of the natural gas, the following reactions are used to calculate the theoretical air for combustion:

$$CH4+(2O_2 + 7.52\ N_2) \rightarrow CO_2+2H_2O+7.52\ N \quad (1)$$

$$C_2H_6+(3.5\ O_2+13.17\ N_2) \rightarrow 2CO_2+3H_2O+3.17\ N_2 \quad (2)$$

$$C_3H_8 + (5O_2 + 18.81\ N_2) \rightarrow 3CO_2 +4H_2O +18.81\ N_2 \quad (3)$$

The AFR (Air Fuel Ratio) is calculated from the mass balance of the atoms of Carbon, Oxygen, Hydrogen and Nitrogen in equations (1) to (3). Solving the four atomic material balances, the molar stoichiometric AFR (Air Fuel Ratio) is 9.89 and the corresponding mass AFR is 17.16. As a result, 378.8 kg/s. of stoichiometric air need to be added to 23.54 kg/s. of natural gas stream for complete combustion. However, for any combustion process there is a balance sought between losing energy from using too much air, and wasting energy from running too rich and producing toxic products like carbon monoxide. The optimum combustion efficiency occurs at the optimum AFR and controlling this provides the best efficiency. This optimum value is often estimated by measuring O_2 and CO concentrations in the flue gas [20].

Excess Air (EA) is expressed as a percentage increase over the stoichiometric requirement and is defined as;

$$EA = \frac{ActualAFR - StoichiometricAFR}{Stoichiometric\ AFR} \times 100\% \quad (4)$$

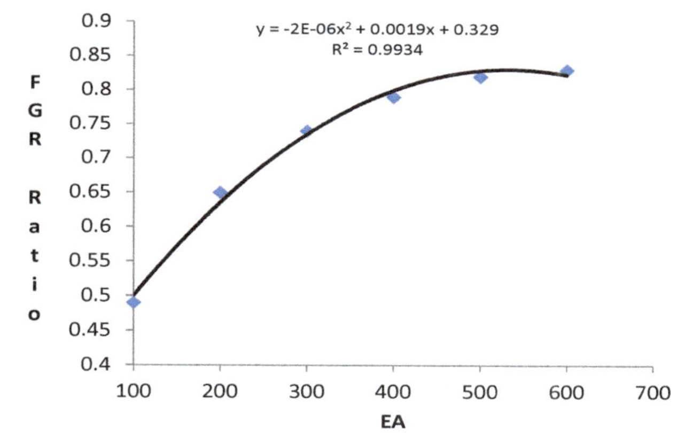

Figure 3: Relationship between excess air (EA) and the maximum value of FGR ratio.

EA (%)	Mass AFR	Total air (kg/s)	(Air+NG) (kg/s)	Maximum Flue gas recycled (kg/s)	Maximum FGR ratio
100	34.32	757.62	781.16	340.92	0.436
200	51.49	1136.43	1159.97	743.27	0.640
300	68.65	1515.23	1538.77	1122.07	0.729
400	85.82	1894.04	1917.58	1500.88	0.782
500	102.98	2272.85	2296.39	1879.69	0.818
600	120.15	2411.41	2434.95	2018.25	0.828

Table 3: Maximum values of Flue gas Recirculation (FGR) Ratio for different values of EA (Excess Air).

It is sometimes convenient to use the Excess Air Ratio (EAR) defined as:

$$EAR = \frac{Actual\ AFR}{Stoichiometric\ AFR} \qquad (5)$$

For a turbine using natural gas, no more than 10% excess air is normally added to the stoichiometric air for combustion [21,22]. Using equation (4) and for 10% excess air, the optimum value of the molar and the mass AFR are 10.88 and 18.88 respectively. The corresponding optimum flowrate of primary air for combustion is 416.70 kg/s.

Maximum values of flue gas recirculation (FGR) ratio

Since gas turbines use different values of EA (excess air), the maximum value of FGR ratio will be estimated in this investigation for excess air varying from 100% to 600%. For example, if a gas turbine use 100% excess air, the corresponding mass AFR is equal to 34.32 Therefore, the needed amount of air for combustion and cooling is equal to 757.62 kg/s. (416.70 kg/s are used as primary air and 340.92 kg/s are used for cooling). Therefore, the maximum value of the flowrate of the recycled flue gas is equal to the amount of air (340.92 kg/s) used for cooling the turbine. This value corresponds to a maximum value of FGR ratio of 0.436. Table 2 shows the corresponding values of the maximum FGR for EA (Excess Air) changing from 100% to 600%.

The maximum FGR ratio of 0.64 for 200% Excess Air (EA) is very close to the value 0.65 given in the literature for an Excess Air Ratio (EAR) of 3-3.5 (Excess Air = 200%-250%) [10]. The relationship between the maximum FGR ratio and the excess air (EA) values is shown in Figure 3.

According to Figure 3, the effects of EA (Excess Air) on the FGR (Flue gas Recirculation) ratio follow the equation:

$$(FGR)_{Ratio} = 0.329 + 0.0019(EA) - 2\times10^{-6}\times(EA)^2 \qquad 6)$$

Simulation Results

The effects of Flue Gas Recirculation (FGR) ratio and the Excess Air (EA) on the temperature of the exhaust gas leaving the combustor and on the percentage of CO_2 and O_2 in the flue gas stream to the absorption unit have been simulated using Aspen Hysys V 8.0 and the Soave-Redlich-Kwong (SRK) equation of state (Table 3).

Effects of flue gas recirculation and excess air on the temperature of exhaust gas

In a gas turbine, the blades have a metallurgical limitation and the temperatures could vary from 800°C to 1700°C depending on their materials and whether or not they have cooling systems [15]. Since the amount of excess air depends on the thermal limitation of the turbine, the first part of this investigation is to simulate the effects of excess air and flue gas recirculation ratio on the temperature of the exhaust gas leaving the combustor. The corresponding simulation results are shown in Figure 4.

For this case study, the data indicate that the temperature varied from 1666°C (EA=100% and FGR ratio=0) to 588°C (EA=600% and FGR ratio=0.75). It is shown that both the increase of excess air and flue recirculation ratio decrease the temperature of the exhaust gas. However, above an excess air of 400%, the effects of excess air on the temperature of the exhaust gas become less important.

Effects of flue gas recirculation and excess air on the amount of CO_2 in the flue gas

For an effective absorption in a CO_2 capture plant, the molar percentage of CO_2 in the flue gas should be at least equal to 10%. The simulation outputs of the effects of the flue gas recycle ratio (FGR) and the excess air (EA) on the percentage of carbon dioxide in the flue gas entering the absorption unit are shown in Figure 5.

It is clear that increasing the flue gas recirculation (FGR) ratio increases the CO_2 content in the flue gas and increasing the excess air (EA) decreases the amount of CO_2 at the exhaust of the turbine. For example, for an excess air (EA) at 100%, the percentage of CO_2 increased from 5.4% to 11% with the change of FGR from 0% to 50%. On the other hand, for a FGR of 50%, the percentage of CO_2 decreased from 11% to 5% by varying the excess air (EA) from 100% to 300%. It is also shown that for values of EA (Excess Air) ratio higher than 400%, the effects of excess air on the amount of CO_2 become very small.

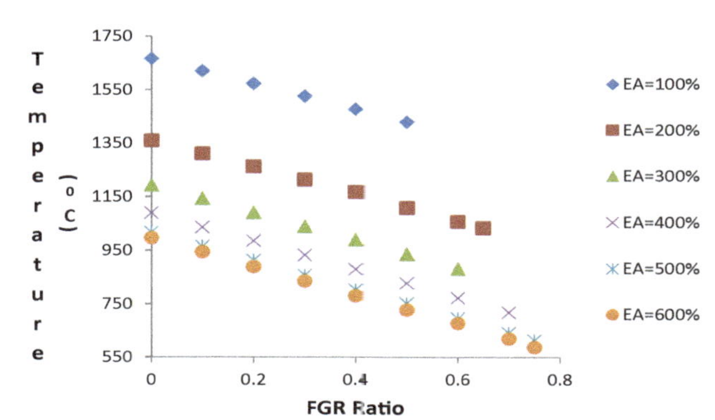

Figure 4: Effects of Excess Air and Flue Gas Recirculation on the Temperature of exhaust gas.

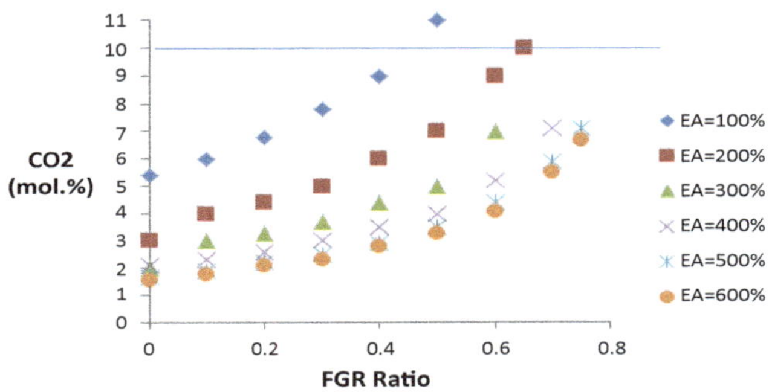

Figure 5: Effects of Flue Gas Recirculation Ratio and Excess Air on the content of CO_2 in the flue gas.

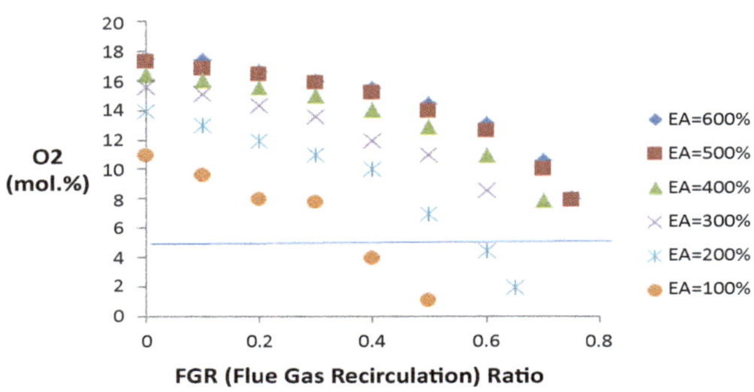

Figure 6: Effects of Flue Gas Recirculation Ration and Excess Air on the content of O_2 in the flue gas.

Effects of flue gas recirculation and excess air on the amount of O_2 in the flue gas

To reduce technical problems linked to oxidative degradation of amine in the CO_2 capture plant, the percentage of O_2 in the flue gas should be lower than 5%. In Figure 6, the effects of excess air (EA) and flue gas recirculation (FGR) ratio on the concentration of O_2 of the flue gas entering the absorption unit are shown.

The flue gas recirculation (FGR) ratio decreased the O_2 content in the flue gas and increasing the excess air (EA) increased the amount of O_2. For an excess air (EA) of 100%, the percentage of O_2 decreased from 11% to 1.1% with the change of FGR from 0% to 50%. On the other hand, for a FGR of 50%, the percentage of O_2 increased from 1.3% to 11% by increasing the excess air (EA) from 100% to 300%. After an excess air of 400%, the effects of excess air on the content of oxygen become insignificant.

Discussion and Conclusion

For an effective carbon capture by an amine mixture, the molar percentage of CO_2 in the flue gas should be at least equal to 10%. Moreover, in order to reduce technical problems due to amine oxidative degradation, the molar percentage of O_2 in the flue gas should be limited to 5%. Because the captured CO_2 will be stored through a CO_2-EOR process, it is also important to limit the content of NOx in the flue gas.

For a NGCC (Natural Gas Combined Cycle) power plant producing 500 MW, Aspen Hysys V 8.0 and the Soave-Redlich-Kwong (SRK) equation of state were utilized in order to simulate the effects of Flue Gas Recirculation (FGR) ratio and excess air (EA) on the temperature of the exhaust gas leaving the combustor and the concentrations of CO_2 and O_2 of the flue gas entering the absorption unit.

For the conditions of an effective absorption process (the molar percentage of CO_2 at least equal to 10%), the simulation results indicate that only a gas turbine with 100% excess air with at least a FGR ratio of 0.45 and a gas turbine with 200% excess air and at least a FGR of 0.65 fit the requirements (Figure 5). The corresponding exhaust gas temperatures are 1450°C and 1035°C respectively. If the metallurgical limitation of the turbine does not allow temperatures higher that 1035°C, the second operating conditions (200% EA and FGR ratio of 0.65) will be the preferred option. For a turbine allowing temperatures as high as 1450°C, the first operating parameters (100% EA and FGR ratio of 0.45) will be selected.

In order to avoid the oxidative degradation of amine solutions, the simulation data show that, only a gas turbine with 100% excess air with at least a FGR ratio of 0.4 or a gas turbine with 200% excess air and at least a FGR ratio of 0.6 are in concordance with the condition (molar percentage of $O_2 \leq 5\%$ in Figure 6). The corresponding exhaust gas temperatures are 1480°C and 1060°C respectively. If the metallurgical limitation of the turbine does not allow temperatures higher that 1060°C, 200% excess air and a FGR ratio of 0.6 should be selected. This option will also decrease the percentage of NOx in the flue gas. For a turbine allowing temperatures as high as 1480°C, the first operating parameters (EA=100% and FGR ratio of 0.4) will be selected.

In conclusion, Flue Gas Recirculation (FGR) ratio increased the content of CO_2 in the flue gas and decreased both the amount of O_2 in the flue gas and the temperature of the exhaust gas. The simulation results indicate also that for excess air values higher than 400%, the effects of excess air on the temperature of the exhaust gas and the amount of carbon dioxide and oxygen in the flue gas become small. Finally, the simulation outputs suggest that, a gas turbine that allows a maximum temperature of 1000°C, an excess air of 200% is needed. For this particular gas turbine, a FGR of 0.65 was required to fit the conditions of an effective absorption process by amines. The molar percentage of CO_2 has increased from 3% to 10 %, while the molar concentration of O_2 decreased from 14% to 2%. Moreover, the combustion temperature has also decreased from 1359°C to 1035°C. These operating conditions (EA=200% and FGR ratio=0.65) are selected because the NOx emissions will eventually be decreased. The corresponding amount of CO_2 produced is 221485 kg/hr. (1.94 Million Tons/year). Therefore, this flowrate is suitable as a feed for a CO_2 Capture unit for a future CCS project.

Acknowledgment

The authors wish to thank the Higher Colleges of Technology of the UAE for supporting this research.

References

1. The Bellona Foundation (2007) Fact sheet: CO_2 Capture and Storage.
2. Allahyari N (2013) Economic evaluation of capturing CO_2 from natural gas power plant and injecting CO_2 for enhanced oil recovery as an integrated system. University of Regina, Canada.
3. de Reus JA (2007) Carbon Capture From Industrial Emission Sources For Use in CO_2 Miscible Flood Applications in the Middle East. 2nd International CO_2 Forum, Abu Dhabi, UAE.
4. Kothandaraman A (2006) Carbon Dioxide Capture by Chemical Absorption: A Solvent Comparison Study. Massachusetts Institute of Technology.
5. Zioudas A, Dadach Z (1986) Absorption rates of CO_2 and H2S in sterically hindered amine. Chemical Engineering Science 41: 405-408.

6. Ciferno J, Litynski J, Plasynski S (2010) DOE/NETL Carbon Dioxide Capture and Storage: RD&D Roadmap. National Energy Technology Laboratory, US Department of Energy.

7. Akram M, Khandelwal B, Blakey S, Wilson CW (2013) Preliminary calculations on post combustion carbon capture from gas turbines with flue gas recycle.

8. Bolland O, Sæther S (1992) New concepts for natural gas fired power plants which simplify the recovery of carbon dioxide. Energy Convers Mgmt 33: 467-475.

9. Chakravarti S, Gupta A, Hunek B (2001) Advanced Technology for the Capture of Carbon Dioxide from Flue Gases. First National Conference on Carbon Sequestration.

10. Sander MT, Mariz CL (1992) The Fluor Daniel Econamine FG Process: Past Experience and Present Day Focus. Energy Conversion Management 33: 341-348.

11. U. S. Patent 6,146,603. System for Recovering Carbon Dioxide from a Lean Feed.

12. U. S. Patent 6,174,506 B1. Carbon Dioxide Recovery from an Oxygen containing mixture.

13. Røkke PE, Hustad JE (2005) Exhaust Gas Recirculation in Gas Turbines for Reduction of CO2 Emissions; Combustion Testing with Focus on Stability and Emissions. Int J Thermodynamics 8: 167-173.

14. Abu Zahra MRM (2009) Carbon Dioxide Capture from Flue Gas- Development and Evaluation of Existing

15. and Novel Process Concepts. Delft University of Technology, Netherlands.

16. Costa M, Ruao M, Calvalho MG (1996) On the influence of flue gas recirculation on pollutants emissions from a small scale laboratory furnace. Archivum Combustionis 16: 77.86.

17. Seume J (2003) Stationäre Gasturbinen. Springer 327-514.

18. Martínez FR, Martínez AAR, Velázquez MT, Diez PQ, Eslava GT, et al. (2011) Evaluation of the Gas Turbine Inlet Temperature with Relation to the Excess Air. Energy and Power Engineering 3: 517-524.

19. National Energy Technology Laboratory (NETL) (2011) Cost and performance baseline of fossil energy plants: Natural gas combined cycle at elevation. U.S. Department of Energy (U.S. DOE) 3.

20. Davison J (2007) Performance and costs of power plants with capture and storage of CO2. Energy 32: 1163-1176.

21. Yuki Y, Murata A (2010) Optimum Combustion Control by TDLS200 Tunable Diode Laser Gas Analyzer. Yokogawa Technical Report English Edition 53: 19-24.

22. Piper JE (1999) Operations and Maintenance Manual for Energy Management. Library of Congress Catalogues 93.

Aspen Plus® Simulation of Saponification of Ethyl Acetate in the Presence of Sodium Hydroxide in a Plug Flow Reactor

Wijayarathne UPL and Wasalathilake KC*

Department of Chemical and Process Engineering, University of Moratuwa, Sri Lanka

Abstract

This work presents the modelling and simulation of saponification of ethyl acetate in the presence of sodium hydroxide in a plug flow reactor using Aspen Plus simulation software. Plug flow reactors are widely used in the industry due to the non-mixing property. The use of plug flow reactors becomes significant when there is a need for continuous large scale reaction or fast reaction. Plug flow reactors have a high volumetric unit conversion as the occurrence for side reactions is minimum. In this research Aspen Plus V8.0 has been successfully used to simulate the plug flow reactor. In order to simulate the process as accurately as possible HYSYS Peng-Robinson EOS package was used as the property method. The results obtained from the simulation were verified by the experiment carried out in the EDIBON plug flow reactor module. The correlation coefficient (r^2) was 0.98 and it proved that simulation results satisfactorily fit for the experimental model. The developed model can be used as a guide for understanding the reaction kinetics of a plug flow reactor.

Keywords: Aspen Plus; Modelling; plug flow reactor; Simulation

Nomenclature

T - Absolute Temperature; C - Concentration; x - Fractional Conversion; k_0 - Frequency Factor; λ_m - Molar Conductivity of an ion; k - Rate Constant; r - Rate of Reaction; V - Reactor Volume; $T_{reference}$ - Reference Temperature; κ_T - Solution Conductivity; $\kappa_{Treference}$ - Solution Conductivity at $T_{reference}$; θ - Temperature Coefficient; R - Universal Gas Constant; F - Volumetric Flow Rate

Introduction

Plug flow reactors (PFRs) also known as Continuous Tubular Reactors (CTRs) play a key role in chemical manufacturing facilities where a transformation of chemical substances takes place. In a plug flow reactor, feed is introduced from one end of a cylindrical tube and flows continuously through the length of the reactor as a series of plugs. The product is then collected from the other end of the reactor. In modelling it is assumed that Inside a PFR, the fluids flow consistently and there is no radial variation in velocity, concentration, and temperature or reaction rate [1]. The main advantages of PFRs are that they have a high volumetric unit conversion and the capability of running for longer periods without maintenance. For isothermal reactions greater than zeroth order, the PFR volume will usually be lower than the Continuously Stirred Tank Reactor (CSTR) volume for the same conversion and reaction conditions (temperature, flow rate, etc.) [1].

For the industry, PFR can be configured as a single long tube or a bunch of short tubes depending on the application. The diameter of the cylindrical reactor varies from a few centimeters to several meters. Construction cost, pumping cost, residence time and heat transfer requirements are the main factors which affects the choice of diameter. PFRs are widely used in the industry for gas and liquid phase systems. Gasoline Production, Oil cracking, oxidation of sulfur dioxide to sulfur trioxide, synthesis of ammonia, polymer manufacturing are some the industries where PFRs are being used broadly.

Though they are a number of applications, the controlling of PFR has become a challenging task when it comes to chemical manufacturing. Time varying characteristics of the process have made the operation of the reactor difficult and it needs extensive monitoring. For overcome such issues, modelling and simulation of reactors has become useful when it comes to process optimization and to investigate the effect of various operating parameters towards the process.

Practice of process simulation has evolved over last two decades in the field of engineering. When it comes to process engineering, there is a wide range of process simulation packages which are specifically developed for particular process areas namely, Aspen Plus, Chemcad and Petro-SIM. All these simulators follow the sequential modular approach where the user is able develop a process by combining standard modules in a logical manner. Aspen Plus is one of the industry's leading process simulation software which has the ability to design and simulate a chemical process comprehensively.

Aspen Plus has been used by investigators to simulate and optimize various kinds of process plants and unit operations. Sohail Rasool Lone et al. [2] developed a model to simulate ethyl acetate reactive distillation column using a RADFRAC model Harjeet Nath et al. [3] developed a model to simulate the abatement of fluorine inside a fluidized bed reactor using RCSTR and RGIBBS models. Rajul Nayak et al. [4] developed a model to simulate coal gasification using RYIELD, RGIBBS and RSTOIC models.

The primary objective of this study is to develop a model for a Plug Flow Reactor and to simulate the saponification of ethyl acetate in the presence of sodium hydroxide by varying the reactor temperature and reactant flow rates using Aspen Plus. In order to validate the applicability of the developed model, simulation results were compared with the experimental results collected using EDIBON PFR module

***Corresponding author:** Wasalathilake KC, Department of Chemical and Process Engineering, University of Moratuwa, Sri Lanka

[5]. During the laboratory experiment, fractional conversions of the reaction are determined using the conductivity values of the reactants and products under different temperatures.

Theory

Saponification

Saponification is the hydrolysis of a carboxylic acid ester in a basic medium (equation 1). Saponification is essentially a soap-manufacturing process that takes place by reacting the fatty acid with alkali, especially triglycerides producing glycerol and carboxylate salts (soap).

$$NaOH + CH_3COOR \rightarrow CH_3COO^-Na^+ + ROH \quad (1)$$

In this experiment, saponification of ethyl acetate in the presence of sodium hydroxide is used to demonstrate the saponification reaction. The saponification of ethyl acetate reaction is as follows.

$$NaOH + CH_3COOC_2H_5 \rightarrow CH_3COO^- Na^+ + C_2H_5OH \quad (2)$$

When it comes to chemical kinetics, this reaction is a prominent second order reaction identified in literature [6,7].

In a PFR, reactant fluids are pumped through a tubular section where the reaction is carried out. At the inlet of the PFR, reactant concentration is very high; hence the rate of reaction is also high. But, when it passes through the reactor, reactants get consumed and products are formed. Therefore rate of reaction drops through the PFR with respect to reactor length.

Conductivity theory

Conductivity of a solution is given by, $\kappa=\lambda_m C$ (3) [8]. Where, λ_m is the molar conductivity of an ion and C is the ion concentration in the solution. Feed stream for the PFR consists with equal volumetric flow rates of $NaOH_{(aq)}$ and $CH_3COOC_2H_{5(aq)}$. Since the conductivity value for $CH_3COOC_2H_{5(aq)}$ is negligible with reference to $NaOH_{(aq)}$, conductivity of the feed solution can be estimated as,

$$\kappa_{feed} = [\lambda_{Na+} \times C_{Na+}] + [\lambda_{OH^-} \times C_{OH^-}] \quad (4)\ [8]$$

Similarly for the product stream, since it consists with both the products $CH_3COO^-Na^+ + C_2H_5OH$ as well as the unreacted, conductivity NaoH of the product stream could be expressed as,

$$\kappa_{product} = [\lambda_{Na+} \times C_{Na+}] + [\lambda_{OH^-} \times C_{OH^-}] + [\lambda CH_3COO^- \times C_{CH3COO^-}] \quad (5)\ [8]$$

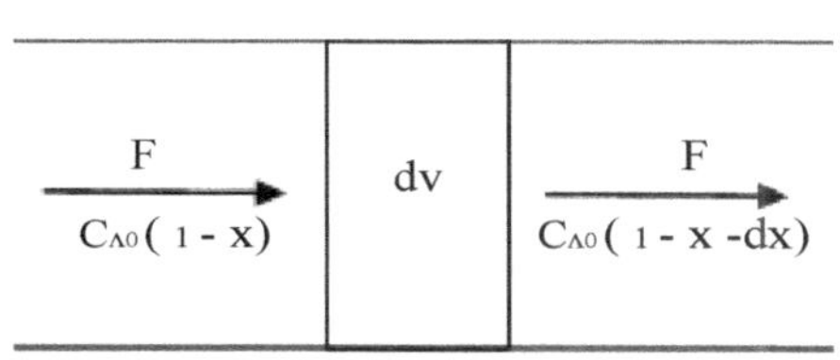

Figure 1: Elemental material balance for a PFR.

	A	B	C	D
t=0	C_{A_0}	C_{A_0}	0	0
t=t	$\frac{C_{A_0}(1-x_A)}{2}$	$\frac{C_{A_0}(1-x_A)}{2}$	$\frac{C_{A_0} x_A}{2}$	$\frac{C_{A_0} x_A}{2}$

Table 1: Concentration of streams with time.

Furthermore, online conductivity meter displays the conductivity value for the solution as at 25°C. Therefore, it should be corrected to the operating temperature using the expression,

$$\kappa_{Treference} = \frac{100}{100 + \theta(T - T_{reference})} \times \kappa_T \quad (6)\ [9]$$

Hydrolysis reaction of ethyl acetate with sodium hydroxide is in the form of A+B→C+D. Moreover, the concentration of both the NaOH and $CH_3COOC_2H_5$ are similar, the concentration of components at initially and any time t, can be expressed as follows (Table 1).

By replacing the concentration terms in above equations (4) and (5),

$$\kappa_{feed} = [\lambda_{Na+} \times C_{A0}] + [\lambda_{OH^-} \times C_{A0}] \text{ and}$$

$$\kappa_{product} = \left[\lambda_{Na^+} \times C_{A_0}\right] + \left[\lambda_{OH^-} \times \frac{C_{A_0}(1-x_A)}{2}\right] + \left[\lambda_{CH_3COO^-} \times \frac{C_{A_0} x_A}{2}\right]$$

Fractional conversion could be determined by solving for x.

PFR theory

Since the saponification of ethyl acetate with presence of sodium hydroxide is a second order reaction [6], rate of the reaction can be expressed as follows.

$$r = kC_{NaOH}C_{CH3COOC2H5} \quad (7)$$

According to the Arrhenius equation

$$k = k_0\, exp\left(\frac{-E}{RT}\right) \quad (8)$$

Tubular reactors falls under the category of flow reactors where the key assumption is that the fluid is perfectly mixed in the radial direction but not in the axial direction (zero back mix). Each plug of differential volume is considered as a batch reactor where the volume tends to be zero.

The PFR model is well applied for many fluids. Ideal PFR consist with a constant residence time which is same as the space time if other conditions such as pressure and temperature are same throughout the reactor [10-12].

Plug flow reactor design equation (Figure 1)

General assumptions made for mathematically model an ideal plug flow reactor,

- There is no mixing in the axial direction (direction of flow)
- Complete mixing in the radial direction
- Uniform velocity profile across the radius

Applying material balance for component A (NaOH) over small elemental volume dv under the steady state operation results,

$$C_{A0}(1-x)F - (-r_A dv) = C_{A0}(1-x-dx)F + 0$$

$$-r_A dv = C_{A0} F(dx)$$

$$\frac{V}{F} = CA0\int_0^x \frac{dx}{-r_A} \quad (9)$$

Residence time is same as the space time and given by

$$\tau = \frac{V}{F} \quad (10)$$

Therefore, $\tau = CA0\int_0^x \frac{dx}{kC_{A0}^2(1-x)^2}$ (11)

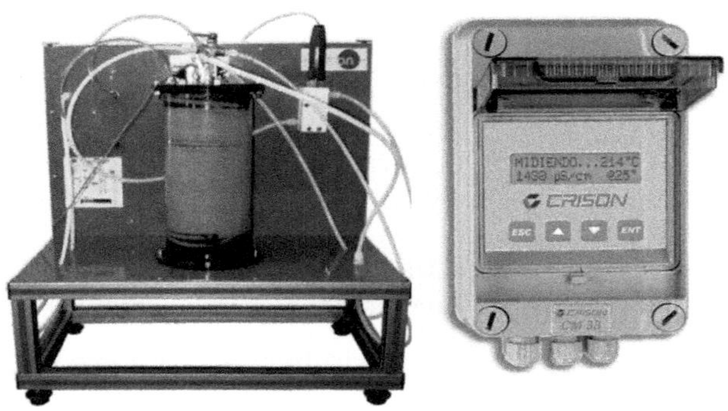

Figure 2: PFR and Conductivity meter.

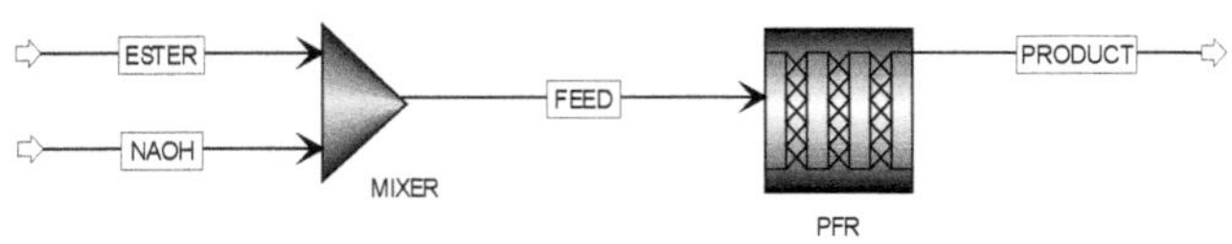

Figure 3: Aspen Flow Sheet.

Methodology

Materials

All chemicals used for this work are of analytical reagent grade. The stock solutions of NaOH (~0.1 M) and ethyl acetate (~0.1 M) were prepared using distilled water. Ethyl acetate from Sigma-Aldrich Corporation, Spruce St., St. Louis, Missouri United States with 99% assay (density: 0.902 g cm^{-3}) was directly used as stock solution. 99.8% assay NaOH pellets from Merck Specialties Private Limited, Mumbai, India was used and the stock solution of NaOH was standardized against freshly prepared standard hydro chloric acid.

Experimental setup

Reactor module: Plug Flow Reactor in EDIBON Computer Controlled Chemical Reactors Training System was used to carry out the hydrolysis reaction. The experimental setup was mainly equipped with following components.

- Two computer controlled feed pumps coupled with flow meters to measure the flow of liquids in the range of 0.7-7 l/h and 0.54-5.4 l/h
- Thermostatic bath of 9 liters capacity that regulates the temperature between ambient temperature and 70°C with a level switch and a pump which are computer controlled
- An online feed pre-heater prior to introduce the reactants to the tubular section where the reaction is carried out
- Tubular section is having a diameter of 5.8 mm and a length of 16.5 m
- Type "J" temperature sensor to get the temperature of the reactor in a continuous way
- An online conductivity meter (CRISON CM38)
- Three Pyrex-glass tanks of 10 liters capacity, two of them for the reagents and the other one for the accumulation of product

Conductivity monitoring instrument

CRISON CM38, an online conductivity and temperature meter available with EDIBON PFR module was used to measure the conductivity and the temperature of the product stream. According to the product specifications, this online device is capable of measuring the conductivity of a solution in the range of 0.01 µS/cm to 199.9 mS/cm with error percentage ≤ 1% and also it is capable of measuring the temperature of a solution in the range of -20°C to 150°C with error ≤ 0.5°C. Reference temperature for the measuring device is 25°C and all the conductivity values display as at reference temperature [13].

Method: 10 liter of 0.1 M NaOH and 0.1 M $CH_3COOC_2H_5$ (ethyl acetate) solutions were prepared and stored in feed vessels. Plug Flow Reactor module was switched on and the water bath temperature were set to 30°C. Circulation pump was switched on in order to circulate the hot water to the shell side of the PFR. Flow rate of 0.1 M NaOH was set to 2.4 l/h and allowed the system to reach steady state. Once the conductivity value was stable, it was recorded from the online conductivity meter. Water bath temperature was gradually increased to 35°C, 40°C and 45°C while maintaining the flow rate of NaOH solution at 2.4 l/h. Conductivity values corresponding to these temperatures were recorded for latter calculation requirements. Then the PFR was thoroughly washed using distilled water while allowing the water bath to cool down.

Once the temperature of the PFR reached to 30°C, flow rates of NaOH and ethyl acetate was set as 2.4 l/h and both the feed pumps were switched on simultaneously to feed the reactants to PFR.

Hydrolysis reaction was carried out inside the PFR according to the reaction (2).

System was allowed to reach steady state operation and the conductivity of the product stream was recorded using the online conductivity meter. Water bath temperature was again gradually increased to 35°C, 40°C and 45°C while maintaining the flow rates of the reactants at 2.4 l/h. Conductivity values of the product stream corresponding to these temperatures were also recorded.

Fractional conversion for the reaction (2) was calculated using the conductivity values recorded from the product stream at different temperatures. These results were compared with the simulation results from Aspen Plus V8.0 (Figure 2).

Aspen simulation: PFR was modeled in Aspen Plus according to the dimensions and the operating conditions of the EDIBON reactor module. Then it was compared with the experimental results. HYSPR was used as the property method for the Aspen model.

There are various physical property methods available in Aspen Plus V8.0 which can be used for the estimation of physical properties. HYSPR (HYSYS Peng-Robinson EOS package) property model was used in this study [14].

Two inbuilt models of a Mixer and a PFR were used to model the system (Figure 3). In the experimental setup two reactants were mixed together before introducing to the PFR. Therefore a model of a Mixer was used in the aspen simulation to represent this operation. Two feed streams, Ethyl acetate and Sodium hydroxide were defined as ESTER and NAOH respectively. Concentration of the feed solutions were set as 0.1 mol/dm^3 where, the solvent was defined as water since both the reactants were in aqueous medium and the flow rate for each feed was set as 2.4 l/h. Temperature and pressure for feed streams were set as 30°C and 1 atm respectively.

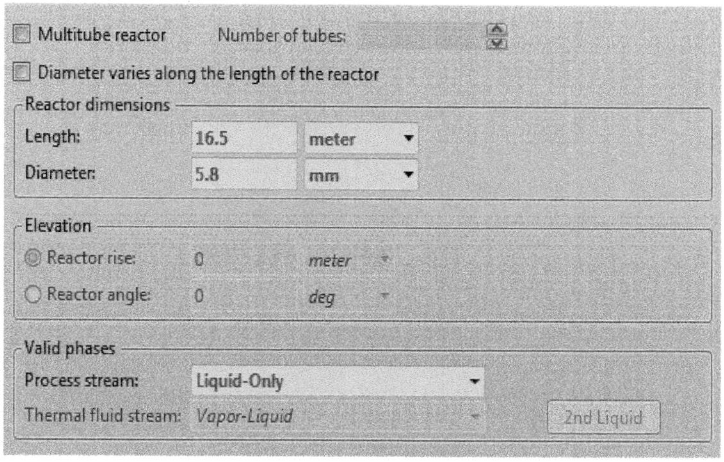

Figure 4: PFR Configuration.

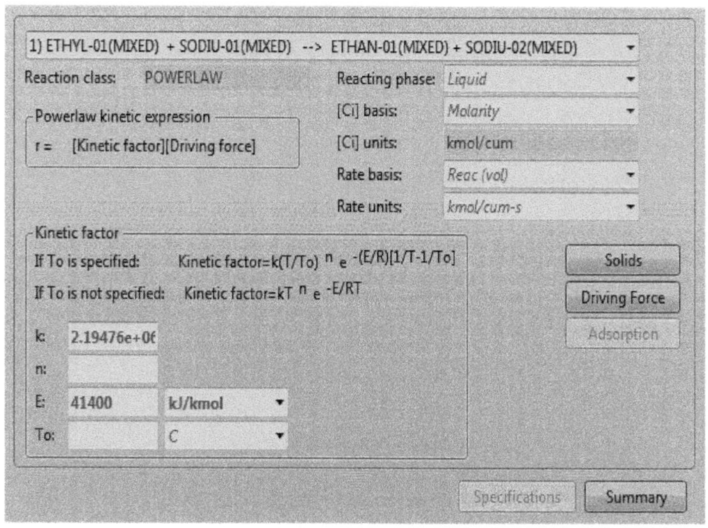

Figure 5: Reaction kinetics data.

Reactor Temperature (°C)	30	35	40	45
Experimental results	0.73	0.76	0.80	0.83
Simulated results	0.72	0.77	0.81	0.84

Table 2: Fractional conversion at different reactor temperatures.

PFR model in the aspen simulation represent the tubular reactor in the experimental module and the parameters were set accordingly. Length of the reactor was 16.5 m and the diameter of the tube was 5.8 mm. Phase of operation for the experimental model was only liquid and therefore valid phase the PFR model in the Aspen was selected as liquid- only (Figure 4). Operating condition for the PFR was set as 30°C and 1atm.

After setting up the physical parameters, reaction kinetic parameters were introduced to the Aspen model. According to the literature [8], the saponification reaction follows the power law (equation 7) and hence it was selected as the reaction class in the Aspen model. Driving force was also set as 'exponent' of one for both the reactants since the reaction rate is proportional to the concentration of each of the reactant. Activation energy (E_a) and frequency factor (k_0) were found to be 41400 kJ/kmol and 2194760 respectively [8] and those values were fed to the Aspen model (Figure 5). The results that were obtained upon the simulation of the model are given in Table 3.

Results and Discussion

Conductivity vs. Temperature

Conductivity values of both NaOH stream and product stream were adjusted to the operating temperature of the reactor according to equation (6) and were plotted against the reactor temperature (Figure 6). The conductivity values of the solution show linear dependence on the reactor temperature.

Comparison of simulated results and experimental results

Table 2 shows the calculated results for the fractional conversion of the saponification reaction at different reactor temperatures for both the experiment and the simulation. It is clear from the results that in both the cases, the fractional conversion increases as the reactor temperature increases.

	Caustic	Ester	Product at 30°C	Product at 35°C	Product at 40°C	Product at 45°C
H_2O kmol/h	1.7055E-01	1.6950E-01	3.4005E-01	3.4005E-01	3.4005E-01	3.4005E-01
$CH_3COOC_2H_5$ kmol/h	-	2.4000E-04	6.6052E-05	5.4319E-05	4.5229E-05	3.7255E-05
NaOH kmol/h	2.4000E-04	-	6.6052E-05	5.4319E-05	4.5229E-05	3.7255E-05
$CH_3COO^-Na^+$ kmol/h	-	-	1.7395E-04	1.8568E-04	1.9477E-04	2.0275E-04
C_2H_5OH kmol/h	-	-	1.7395E-04	1.8568E-04	1.9477E-04	2.0275E-04
Total Flow kmol/h	1.7079E-01	1.6974E-01	3.4053E-01	3.4053E-01	3.4053E-01	3.4053E-01
Total Flow kg/h	3.08	3.07	6.16	6.16	6.16	6.16
Total Flow l/min	0.04	0.04	0.08	0.08	0.08	0.08

Table 3: Results summary.

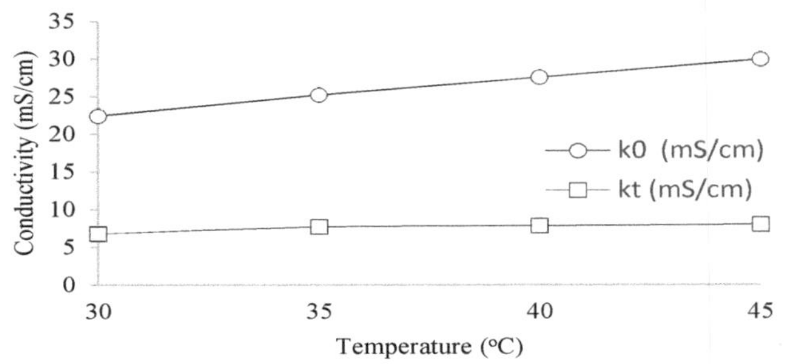

Figure 6: Solution Conductivity vs Temperature.

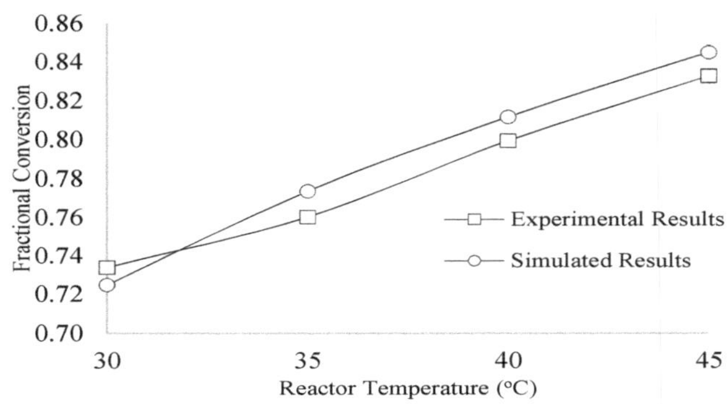

Figure 7: Fractional conversion vs reaction temperature.

The experimental values obtained for fractional conversion were correlated with the results obtained from the simulation (Figure 7). The correlation coefficient (r^2) was 0.98 and it proved that simulation results show a good fit for the experimental model.

Effect of temperature on the reaction

Figure 8-11 show the variation of molar fractions of ethyl acetate and sodium acetate at the reactor temperatures of 30°C, 35°C, 40°C and 45°C respectively. It is clear from these figures that the molar fractions of reactants decrease with the increment of reactor temperature. This is due to the fact that the reaction rate increases with the reactor temperature and hence, the molar fractions of products increase (equations 7 and 8).

Figure 13 shows the curve for molar flow of sodium acetate obtained under different reactor temperatures (From 25°C to 200°C). It can be seen that the sodium acetate molar flow rate increases with the reactor temperature and it reaches a maximum value of 0.24 mol/h when the reactor temperature is150°C.

It is also evident that the molar fractions of products increase with increment of length of the plug flow reactor. This is because the residence time of reactants increases with the length of the reactor, where it allows sufficient time for the reaction to take place (equation 11).

Effect of reactant flow rates on the reaction

A sensitivity analysis was done in Aspen Plus model to investigate the effect of reactant flow rates towards the reaction.

Figure 12 shows how the mole fractions of sodium acetate vary for different flow rates of NaOH and ethyl acetate. It shows that in each case the highest mole fraction of sodium acetate can be achieved when equal flow rates of NaOH and ethyl acetate are pumped to the plug flow reactor. According to the stoichiometry of the reaction (equation 2) it is clear that equal amounts of reactants are needed to obtain the highest yield.

When the flow rate of ethyl acetate exceeds the flow rate of NaOH as there is no more NaOH left in the reactor to react with ethyl acetate, the excess ethyl acetate remains in the product stream. Therefore though same amount of sodium acetate is produced by the reaction,

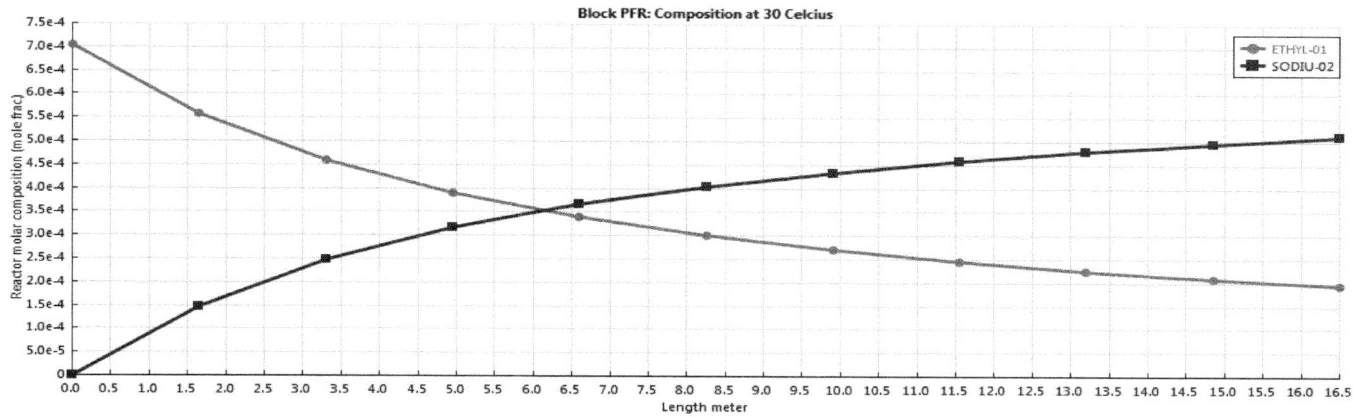

Figure 8: Ethyl acetate mole fraction vs PFR length at 30 °C.

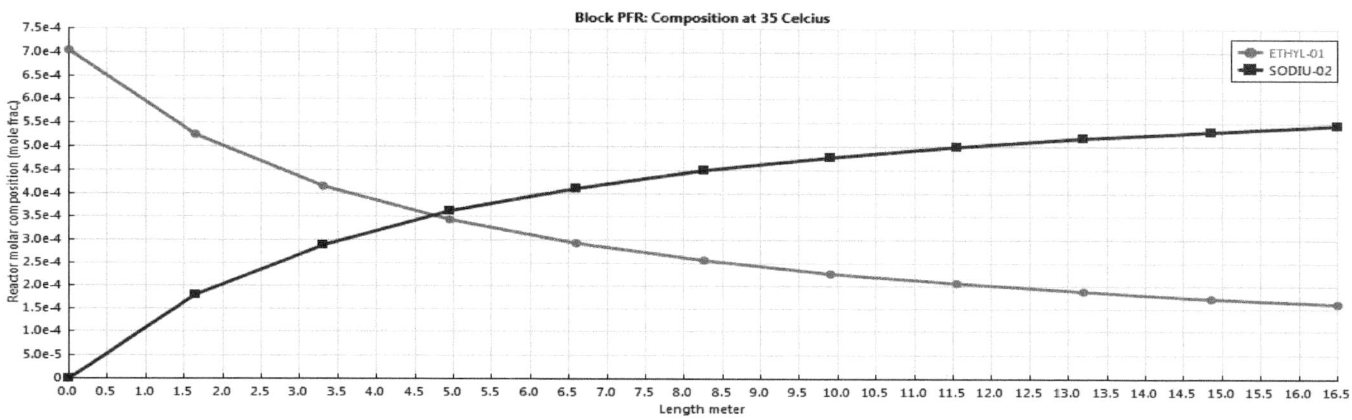

Figure 9: Ethyl acetate mole fraction vs PFR length at 35°C.

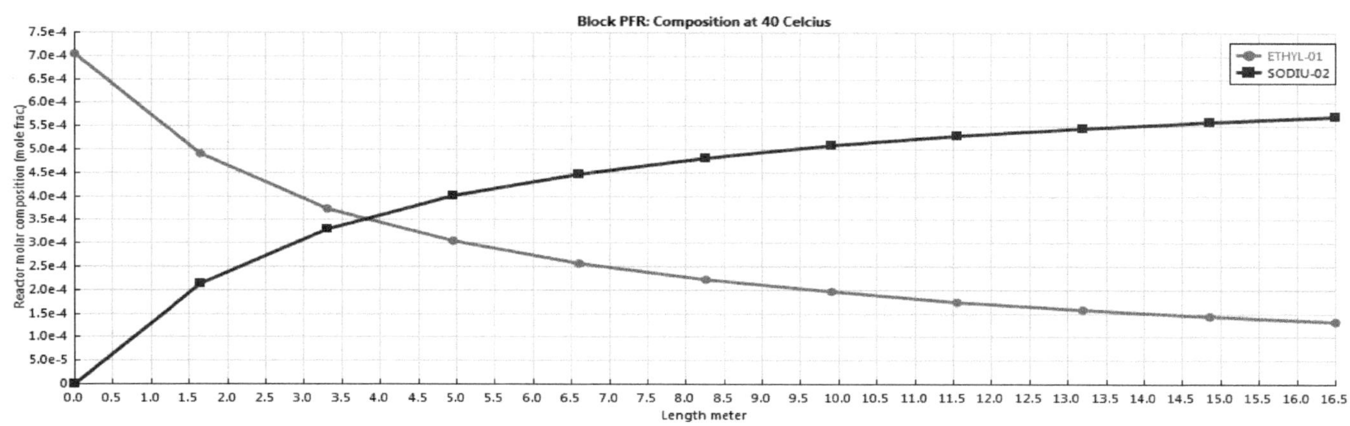

Figure 10: Ethyl acetate mole fraction vs PFR length at 40°C.

mole fraction of sodium acetate is decreased because of the presence of unreacted ethyl acetate in the product stream.

Volume comparison of CSTR and PFR

Volume required for a PFR is relatively less when compared to the volume required for a Continuous Stirred Tank Reactor (CSTR) to achieve the same conversion for a reaction while maintaining the other parameters as constant [15]. This theory is well applied for the saponification reaction of ethyl acetate with sodium hydroxide. During this study it has also proven with using Aspen simulation that the volume of CSTR to achieve the same conversion is about 3.5 times that of a PFR (Figure 14). Therefore, it is economically feasible to use a PFR

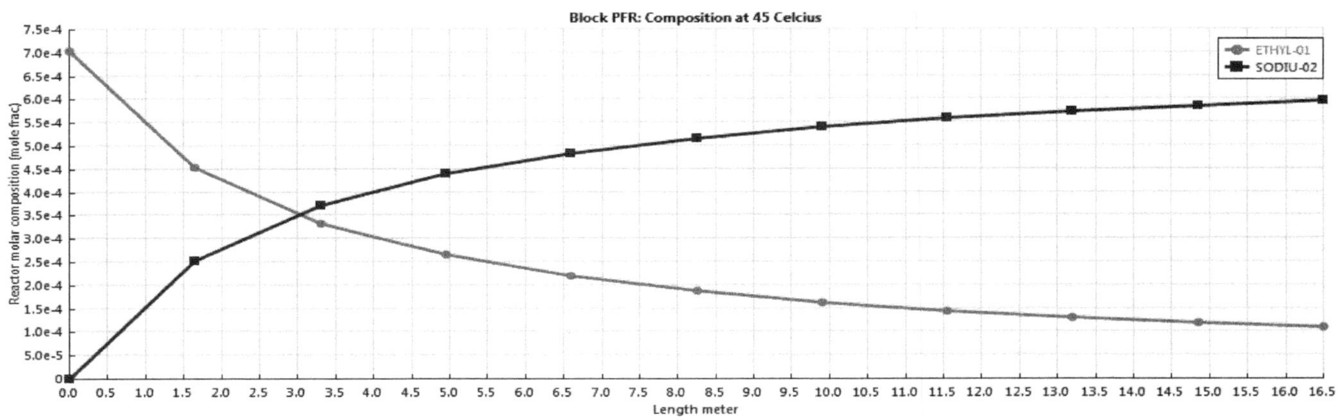

Figure 11: Ethyl acetate mole fraction vs PFR length at 45°C.

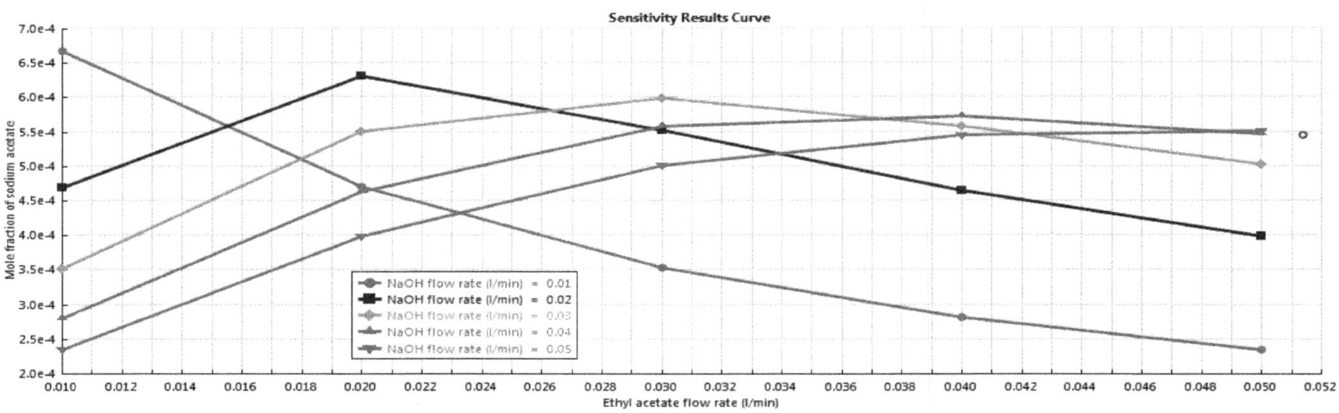

Figure 12: Variation of ethyl acetate mole fraction with acetate and NaOH flow rates.

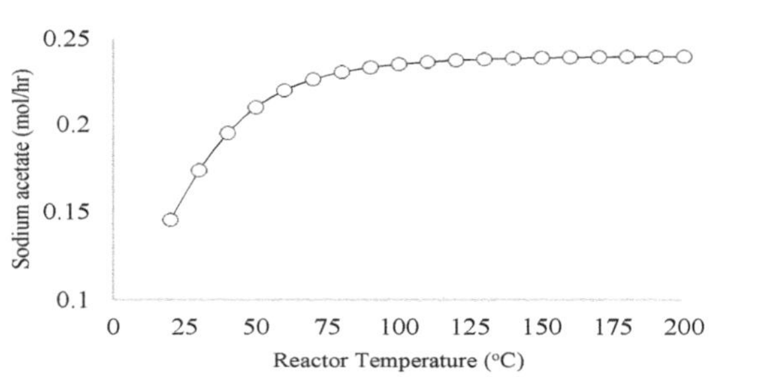

Figure 13: Variation of Sodium acetate flow rate with reactor temperature.

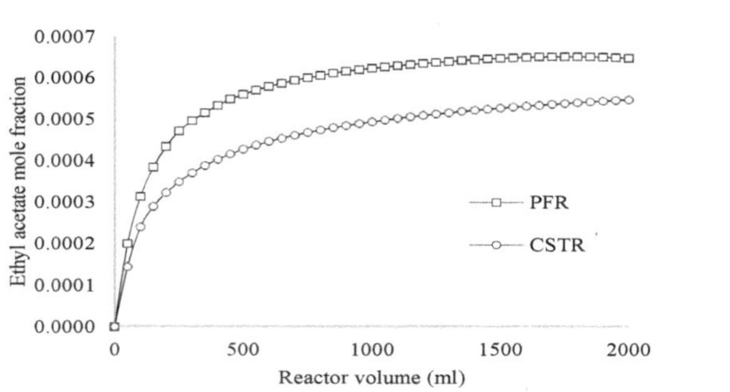

Figure 14: Ethyl acetate mole fraction vs reactor volume.

to carry out this saponification reaction in industrial scale even when producing with large quantities.

Conclusion

In this work, a simulation model was developed for the reaction of saponification of ethyl acetate in the presence of sodium hydroxide inside a plug flow reactor using Aspen Plus. The model was validated by the experimental results obtained from EDIBON Plug Flow Reactor module. The experimental results showed a good fit to the Aspen Plus model.

Aspen Plus model proved that the rate of reaction increases as the reactor temperature and the length of the plug flow reactor increase. The highest yield of the reaction was obtained at a reactor temperature of 150°C.

These results improve the understanding of reaction kinetics with reference to the saponification reaction and would be very useful in the design of plug flow reactors.

Acknowledgement

The authors would like to acknowledge Dr. Ismail F. M., Dr. Gunasekera M. Y. and Mr. Rathnayake H.H.M.P. for providing the space and all facilities needed in the research study at the Department of Chemical and Process Engineering, University of Moratuwa, Sri Lanka.

References

1. Scott FH (2006) Elements of Chemical Reaction Engineering. Prentice Hall International.

2. Lone ASR, Akhlaq S (2012) Modeling and Simulation of Ethyl Acetate Reactive Distillation Column Using Aspen Plus. International Journal of Scientific & Engineering Research.

3. Nath H, Sahoo A (2013) Abatement of Fluorine inside a Fluidized Bed Reactor Using Aspen Plus Simulation. International Conference on Frontiers in Chemical Engineering.

4. Nayak R, Mewada RK (2011) Simulation of Coal Gasification Process using ASPEN PLUS. Institute of Technology, Nirma University, Ahmedabad.

5. Edibon Technical Teaching Equipment (2013) Computer Controlled Chemical Reactors Training System with SCADA and PID Control. Edibon International, Madrid.

6. Walker J (1906) A Method for Determining Velocities of Saponification. Royal Society of London.

7. Shoemaker D (2003) Experiments in Physical Chemistry. McGraw-Hill.

8. Atkins P, de Paula J, Depaula J, Atkins PW (2006) Atkins' Physical Chemistry. 8th Edn. New York, Oxford University Press.

9. Radiometer Analytical (2004) Conductivity Theory and Practice. France.

10. Denbigh KG, Turner JCR (1971) Chemical Reactor Theory, 2nd edn, Cambridge University, Chemical Reaction Engineering, John Wiley and Sons.

11. Smith JM (1981) Chemical Engineering Kinetics, 3rd edn, McGraw-Hill, New York.

12. CRISON Instruments (1997) Controladores CM38-CM38P. Manual del usuario. SA.

13. Aspen Technology (2012) Inc. Aspen Engineering V8.0. Massachusetts: Aspen Technology, Inc.

14. Dane K (2007) Chemical and Biological Reaction Engineering. Spring.

Mass Transfer Study of Heavy Metal Removal by Ion Exchange in Batch Conical Air Spouting Bed

Yousef NS[1], Darwish S[2], Zewail TM[2] and El-Tawail YA[2]

[1]Petrochemical Department, Pharos University, Egypt
[2]Chemical Engineering Department, Faculty of Engineering, Alexandria University, Egypt

Abstract

The performance of a batch conical air spouting vessel for conducting ion exchange reactions involving heavy metal removal has been investigated. The performance was measured in terms of volumetric mass transfer coefficient (K). The effect of various parameters such as type of heavy metal ion (Ni^{2+}, Pb^{2+}), superficial air velocity and initial heavy metal concentration has been investigated. It has been found that volumetric mass transfer coefficient increases as air superficial velocity increases. Initial Ni^{+2} concentrations have two opposing effects on the volumetric mass transfer coefficient, whereas initial Pb^{+2} concentrations have negligible effect. Dimensional analysis of the present mass transfer data led to the following correlations:

For Ni^{+2} $J_D=49\ (Re*Fr)^{-0.215}$

For Pb^{+2} $J_D=117.71\ (Re*Fr)^{-0.24}$

The importance of these mass transfer correlations in the design and operation of gas spouting vessels used for conducting diffusion-controlled reactions was highlighted.

Keywords: Ion exchange; Zeolites; Heavy metals; Kinetics; Conical gas spouting bed; Mass transfer

Introduction

Water plays an important role in human beings, natural environment, and social development, but the subsequences of water use are municipal wastewaters and industrial wastewaters. Therefore, how to treat wastewaters and make them reusable is not only an important task but also an urgent problem to be solved [1]. Industrial wastewaters such as those produced by metal plating facilities, mining operations, metallurgical engineering, battery manufacturing processes, the production of paints and pigments, electroplating, nuclear power plants, and ceramic and glass industries contain heavy metals such as copper, zinc, cadmium, lead, nickel, and chromium.

These heavy metals are not biodegradable and their presence in streams and lakes leads to bioaccumulation in living organisms, thereby causing health problems in animals, plants, and human beings [2]. In the case of liquid effluents, there are many methods available for the removal of these heavy metals such as chemical precipitation, coagulation, solvent extraction, membrane processes, adsorption, and ion exchange [3]. Among the heavy metals removal processes, ion exchange process is very effective to remove various heavy metals and the ion exchange resin can be easily recovered and reused by regeneration operation [4].

Ion exchange is the process through which ions in solution are transferred to a solid matrix which, in turn releases ions of a different type but of the same polarity [5]. Ion exchange is a physical separation process in which the ions exchanged are not chemically altered [6]. The main advantages of ion exchange are recovery of metal value, selectivity, less sludge volume produced and the meeting of strict discharge specifications. Ion-exchange resins are variety of different types of exchange materials, which are distinguished into natural or synthetic resins.

Furthermore, it can be as well categorized on the basis of functional groups such as cationic exchange resins, anion exchange resins, and chelating exchange resin. Among the materials used in ion exchange processes, synthetic resins are commonly preferred as they are effective and inexpensive [7]. Ion exchange is a mass transfer process. There are two main rate-determining steps which are considered in most of the ion exchange reactions [8]. Diffusion of ions inside the material defined as intra- particle diffusion or diffusion of ions through the liquid film surrounding the particle defined as film diffusion.

The rate of ion exchange is determined by both or the slower of these two processes [9]. Industrially ion exchange reactions are conducted in either fixed bed or fluidized bed columns. In spite of the good hydrodynamic flow characteristics of spouted bed column, no attempt has been made before to investigate its performance in conducting ion exchange reactions. However a single study in literature has studied the performance of batch air sparged vessel in conducting ion exchange reactions [10].

In that study the flow pattern resembles fluidization of resin particle in solution by the aid of air sparging. An alternative flow behavior is spouted bed. Spouted beds are gas-particle contactors in which the gas is introduced through a single nozzle at the center of a conical or flat base [11]. Unlike the fluidized bed the high-velocity spout of gas penetrates the bed and carries the particles upward. The other sections of the particle move downwards at slow speed so that a fairly uniform circulation of particles is obtained as illustrated [12].

***Corresponding author:** NS Yousef, Petrochemical Department, Pharos University, Egypt

This hydrodynamic behavior could achieve more intense heat- and mass-transfer conditions. The spouted bed technique has been applied in many industrial processes, such as drying of granular materials, blending of polymer chips, coating of tablets, and granulation of fertilizers and other materials [13]. Conical spouted beds have the properties of conventional spouted beds (cylindrical conical base), while also allowing for strong gas–solid contact since they operate stably in a wide range of gas flow rates [14].

The main objective of the present work is to investigate the performance of a batch conical air spouted bed in the removal of heavy metal ions by strong cation exchange resin. To this end the following parameters have been investigated: Type of heavy metal ion: Ni^{+2} and Pb^{+2}, initial concentration of Ni^{+2} and Pb^{+2} and different air superficial velocity. In addition the present mass transfer data were correlated in terms of dimensionless correlation, which may be useful in the design and operation of such ion exchange conductors.

Materials and Methods

Materials

Fresh strong cation exchange resins, macro-porous AMBERJET 1200 Na a product of ROHM and HAAS were used for each run. (Table 1) shows the specifications of the used ion exchange resin. Nickel Sulfate, Lead Nitrate and Hydrochloric acid, were all AR-grade chemicals. The spouted bed vessel consisted of a plexi-glass cylindrical column of 16 cm in diameter and 26 cm height.

The cylindrical column was supported by a truncated cone of 10 cm height and an inclined angle of 37° 46". The truncated cone was provided with an inlet section of 2.5 cm diameter. Ion exchange resins were supported in the vessel by a sintered gla metal ion: Ni^{+2} and Pb^{+2}, initial concentration ss (G2) gas distributor placed at the end of inlet section. Air at different flow rates was introduced from the inlet section through the sintered glass to the spouted bed vessel thought a nozzle of 1 cm diameter. The flow rate of inlet air was regulated by a ball valve and measured by a calibrated rotameter. A non-return valve was fixed in the inlet nozzle to prevent back flow. A ball valve was fixed in a side pipe for drainage.

Method

Solutions of nickel and lead ions of different concentrations were prepared to be used throughout all experiments. Fresh resins were prepared before runs by immersion in 6% HCl, stirring for 15 minutes and washing with distilled water many times until pH 5.5 was reached, and finally resins were dried by air. Two liters of freshly prepared solution of heavy metal ion (Ni^{2+} or Pb^{2+}) of known initial concentration were placed in contact with 40 gm of dried resin in the spouted bed.

Physical form	Amber spherical beads
Matrix	Styrene divinylbenzene copolymer
Functional group	Sulfonate
Ionic form as shipped	Na+
Total exchange capacity	≥ 2.00 eq/l (Na+ form)
Moisture holding capacity	43 to 47% (Na+ form)
Shipping weight	850 g/l
Specific gravity	1.26 to 1.30 (Na+ form)
Uniformity coefficient	≤ 1.2
Harmonic mean size	620 ± 50µm
Fines content	< 0.300 mm : 0.1% max
Maximum reversible swelling	$Na^+ \rightarrow H^+$: 10

Table 1: Properties of macro-porous AMBERJET 1200 Na resin.

Air flow rate was adjusted by a ball valve as required. Kinetics of the ion exchange reaction was followed by withdrawing 5ml sample from the vessel every 3 mins; the samples were analyzed for the remaining heavy metal ion using atomic absorption spectrophotometer (AA 6650 - Shimadzu) made in Japan.

All experiments were carried out at 25 ± 2°C. Density and viscosity of the solutions used were measured using a density bottle and an Ostwald viscometer respectively. The diffusivities of both Ni^{+2} and Pb^{+2} ions in solution have been calculated from Stokes – Einstein correlation [15]

$$\frac{D\mu}{T} = \text{Constant.}$$

The physical properties of nickel and lead ions solutions respectively measured at 25 ± 2°C are shown in (Tables 2 and 3) respectively.

Results and Discussion

The Pseudo-first order kinetic model

The Pseudo-first order kinetic equation for solid-liquid ion exchange system is based on the following differential material balance on the present batch reactor with respect to the ion to be removed

$$-V_s \frac{dC}{dt} = kAC \tag{1}$$

This upon integration yields:

At t =0, $C=C_o$

And at t=t, $C_t=C$

$$\ln\frac{C0}{C} = \frac{kA}{V_s}t \tag{2}$$

Where: C_o=initial concentrations of heavy metal ions; C=concentrations of heavy metal ion at time t; V_s=volume of the solution; t=time; k=first order rate constant; A=surface area of resin. A typical plot of ln (C_o/C) versus time at different gas velocities for ion exchange of Ni^{+2} and Pb^{+2} respectively. It is well obvious that the present data fit well the pseudo first-order kinetic model.

It also indicates that the diffusion controlled nature of the ion exchange reaction, since as the gas velocity increases the rate of the reaction increases. From the slope of that plot (kA/V_s), volumetric mass transfer coefficient (kA) or K can be determined under different conditions. The effect of superficial air velocity on the volumetric mass transfer coefficient at different concentrations for Ni^{+2} and Pb^{+2} respectively. It is observed that the volumetric mass transfer coefficient K increases with increasing superficial air velocities. The volumetric mass transfer coefficient can be related to air superficial velocity according to the following equations: For Ni^{+2}

$$K\alpha\, V_g^{\;0.36} \tag{3}$$

Concentration (ppm)	ρ (g/cm³)	µ (gm/cm.s)	D (cm²/s)	Sc= µ/(ρ*D)
800	1.004	9.1109×10⁻³	2.8865×10⁻⁵	314.3765
1000	1.008	9.1658×10⁻³	2.8692×10⁻⁵	316.9171
1250	1.012	9.2563×10⁻³	2.8412×10⁻⁵	321.9283

Table 2: Physical properties of Ni^{2+} solutions used at 25 ± 2 °C.

Concentration (ppm)	ρ (g/cm³)	µ (gm/cm.s)	D (cm²/s)	Sc= µ/(ρ*D)
800	1.009	9.[illegible]820×10⁻³	1.9811×10⁻⁵	459.5098
1000	1.010	9.2305×10⁻³	1.9707×10⁻⁵	463.5752
1250	1.014	9.3177×10⁻³	1.9522×10⁻⁵	470.7417

Table 3: Physical properties of Pb^{2+} solutions used at 25 ± 2 °C.

For Pb^{+2}

$$K\alpha V_g^{0.28} \tag{4}$$

To explain the increase of K with the increase of superficial air velocity, it should be mentioned that with the increase of air velocity, the flow pattern within the vessel changed from stable spouting as indicated into jet spouting. The increase in K during stable spouting regime can be explained in terms of that as air velocity increases, both particles - circulation rate and particle cross rate at the bottom increase [16]. Where particle circulation rate expresses the total mass flow of particle at a given longitudinal position in the spout and particle cross flow expresses the solid flow from the annular zone into the spout zone through the interface. Thus a vigorous mixing can be achieved by increasing air velocity, which results in good mass transfer rate.

The increase in K during jet spouting regime can be explained by the fact that the upraising air stream induces radial as well as axial momentum which causes decrease in the thickness of diffusion boundary layer surrounding each particle [17].

It is worth mentioning that the power of V_g (0.28-0.36) obtained in the present work is less than the value obtained from theoretical Levich expression (0.5) [18], for diffusion controlled reaction. The deviation of the present result from the theoretical expression may be ascribed to the presence of mixed controlled reaction, i.e. the ion exchange of both Ni^{+2} and Pb^{+2} is controlled by both film diffusion and intra-particle diffusion. The lower power of V_g in case of Pb^{+2} compared to Ni^{+2} may be ascribed to that in case of Pb^{+2}, the contribution of intra-particle diffusion in the overall rate is of greater extent than in case of Ni^{+2}.

It would be of interest to compare the present results with the result obtained with other authors. (Table 4) shows the velocity exponent obtained by different authors who studied the effect of gas sparging on the rate of mass transfer [10,19]. It is obvious that the powers of V_g obtained by other authors are in fair agreement with the values obtained in the present study.

It may be concluded that the effect of spouting in mass transfer rate is equivalent and may be superior to gas sparging, as, it is well known that the total pressure drop across a spouted bed is always less than the pressure drop which would arise if the same material were fluidized [13]. The effect of initial metal ions concentration expressed as Sc on the volumetric mass transfer coefficient for Ni^{+2} and Pb^{+2} respectively. It is well remarkable that Sc has two opposite effects on K: i) As Sc increases from 314.38 to 316.92, K increases; which can be explained in terms of that the increase in metal ion concentration increases the driving force for mass transfer and thus enhances film diffusion [8]. ii) As Sc increases from 316.92 to 321.93, K decreases; which may be attributed to that at high metal ion concentration, the solution viscosity increases, and diffusivity of Ni^{+2} decreases according to Stock's Einstein law [15].

In addition the activity of Ni^{+2} decreases because of the inter ionic attraction which becomes more pronounced at high electrolyte concentration [20]. It is well observed that the initial Pb^{+2} concentration expressed as Sc has a negligible effect on K, the slight dependence of K on initial Pb^{+2} concentration may be ascribed to that the intra-particle diffusion is the predominant rate controlling step in case of Pb^{+2} ion exchange.

Author	Experimental method	n
Bohm et al.	Mass transfer	0.250
Noseir et al.	Mass transfer	0.242
Sharma et al.	Mass transfer	0.290
Fair et al.	Mass transfer	0.220
Zewail	Ion exchange of Ni^{+2}	0.58
Zewail	Ion exchange of Cd^{+2}	0.25

Table 4: Effect of the superficial gas velocity on the mass transfer ($K\alpha V_g^n$).

It is well known that at high concentrations intra-particle diffusion step predominates [8]. Comparison between % removals of Ni^{+2} and Pb^{+2} at the same superficial air velocity and initial ion concentration. It is well clear that for the same air velocity and initial concentration the % removal of Pb^{+2} is higher than the % removal of Ni^{+2} in other words the affinity of Amberjet resin (1200 Na) for Pb^{+2} is higher than of Ni^{+2}as. It has been found by Pehlivan et al. [21] that the selectivity of SO3H group of the strong cation resin increases with increasing atomic number, valance, and degree of ionization of the exchanged metals, thus selectivity increased in the series: $Pb^{+2} > Cd^{+2} > Cu^{+2} > Zn^{+2} > Ni^{+2}$

Mass transfer data correlation

In view of the complex hydrodynamic conditions of the present work, dimensional analysis was used to correlate the present data. For gas spouting the first order rate constant can be related to the other variables by the dimensionless equation:

$$J_D = a_1 (Re * Fr)^{\alpha} \tag{5}$$

Where : J_D is the mass transfer factor $(K/A_n*V_g)*(\mu/\rho D)^{0.66}$, Re is the Reynolds number $(\rho\, d_b\, V_g/\mu)$, Fr: Froude number $(V_g^2/g\, d_b)$. a_1 and α are constants show that the present mass transfer data fit the following overall correlations: For Ni^{+2}: $314.38 < Sc < 321.93$ and $230.63 < Re*Fr < 50210.24$

$$J_D = 49 (Re*Fr)^{-0.215} \tag{6}$$

With the average deviation=15.5 % For Pb^{+2}: $459.51 < Sc < 470.74$ and $229.54 < Re*Fr < 50052.67$

$$J_D = 117.71(Re*Fr)^{-0.2404} \tag{7}$$

With the average deviation=6.88 %.

Conclusion

The performance of a batch conical air spouted bed for the removal of heavy metal ions by ion exchange has been investigated. The kinetics of ion exchange has been studied under different conditions such as heavy metal type, initial concentration of heavy metal and air superficial velocities. The rate of ion exchange was expressed by the volumetric mass transfer coefficient (K).

The following conclusions have been withdrawn:

1. As the air velocity increases the volumetric mass transfer coefficient increases for both type of heavy metals ions

2. The initial concentration of nickel ions (expressed as Sc) has two opposing effects on the volumetric mass transfer, they are: As Sc increases from 314.38 to 316.92, K increases; which can be explained in terms of that the increase in metal ion concentration increases the driving force for mass transfer and thus enhances film diffusion. As Sc increases from 316.92 to 321.93, K decreases; which may be attributed to that at high metal ion concentration, the solution viscosity increases, and diffusivity of Ni^{+2} decreases according to Stock's Einstein law , and the activity of Ni^{+2} decreases because of the inter-ionic attraction which becomes more pronounced at high electrolyte concentration. While the initial concentration of lead has a negligible effect on volumetric mass transfer coefficient.

3. The present mass transfer data have been correlated in terms of J_D, Re and Fr to allow for the design and operation of industrial scale conical air spouted bed used to conduct diffusion controlled reactions or mixed controlled reactions.

4. Further studies should be conducted in future to study the effect of column geometry such as cone angle, gas inlet diameter and operating parameters such as particle diameter and stagnant bed height on the ion exchange kinetics.

Nomenclature

Symbols

Symbol	Description	Unit
A	Surface area of Resin	cm^2
a_1	Constant	-
C	Concentration of heavy metal at time (t)	ppm
C_o	Initial concentration of heavy metal	ppm
D	Diffusion Coefficient	cm^2/s
d_b	Bubble diameter	cm
g	Acceleration Gravity Time	cm/s^2
k	First order rate constant	cm/s
K	Volumetric mass transfer coefficient Mass transfer factor	cm^3/s
t	Time	s
T	Temperature	°C
V_g	superficial air velocity	cm/s
V_s	Volume of solution	L

Greek symbols

Symbol	Description	Unit
α	Constant	-
ρ	Solution Density	gm/cm^3
μ	Solution density	gm/cm.s

Dimensionless group

Symbol	Description	Unit
F_r	Froude Number	-
J_D	Mass transfer factor	-
R_e	Reynolds Number	-
S_c	Schmidt Number	-

References

1. Hsien L, Yu-Chung K, Jia-Ming C (2007) Equilibrium and kinetics of heavy metal ion exchange. J Chin Inst Chem Eng 38: 71-84.
2. Mustafa K (2012) Removal of Pb (II) from water by natural zeolitic tuff: Kinetics and thermodynamics. J Hazard Mater 199-200: 383–389.
3. Francisco JA, Manuel A, Luis JL (2004) Chromium (III) recovery from waste acid solution by ion exchange processing using Amberlite IR-120 resin: batch and continuous ion exchange. Chemosphere 57: 789-793.
4. Abo-Farha SA, Abdel-Aal AY, Ashour IA, Garamon SE (2009) Removal of some heavy metal cations by synthetic resin purolite C100. J Hazard Mater 169: 190-194.
5. Metcalf, Eddy (1991) Wastewater Engineering: Treatment, Disposal, and Reuse. McGraw-Hill, NewYork 740-741.
6. Treybal RE(1968) Mass Transfer Operation. McGraw-Hill, New York.
7. Bilge A, Sevil V (2009) Kinetics and equilibrium studies for the removal of nickel and zinc from aqueous solutions by ion exchange resins. J Hazard Mater 167: 482-488.
8. Zagorodni A (2007) Ion Exchange Materials, Properties and Applications. 1st edn 377-379.
9. Liberti L, Helfferich FG (1983) In Mass Transfer and Kinetics of Ion Exchange.
10. Zewail TM (2010) The Role of Mass Transfer in the Removal of Heavy Metal Ions by Ion Exchange Using Batch Gas Sparged Vessel 24: 437-443.
11. Duarte CR, Olazar M, Murata VV, Barrozo MAS (2009) Numerical simulation and experimental study of fluid–particle flows in a spouted bed. Powder Technol 188: 195-205.
12. Jacob M (2009) Modeling and application. Powder Technol 189: 332-342.
13. Drew Thomas B (1974) Advanced in chemical engineering. 1-312.
14. Olazar M, Lopez G, Altzibar H, Barona A, Bilbao J (2009) One-dimensional modeling of conical spouted beds. Chem Eng Process 1264-1269.
15. Cussler EL (2009) Diffusion Mass Transfer in Fluid Systems 126.
16. San José MJ, Olazar M, Alvarez AS, Izquierdo MA, Bilbao J (1998) Solid cross flow into the spout and particle trajectories in conical spouted beds. 53: 3561-3570.
17. Nosier SA, Sallam SA (2000) Removal of lead ions from wastewater by cementation on a gas-sparged zinc cylinder. 18: 93-101.
18. Makhloufi L, Saidani B, Hammache H (2000) Removal of lead ions from acidic aqueous solutions by cementation on iron. 34: 2517-2524.
19. Taha AA, Ahmed AM, Bayoumy AA (2003) Effect of Gas Sparging on the Rate of Mass Transfer at Horizontal Cylinders. Portugaliae Electrochimica Acta 21: 295-313.
20. Elshazly AH, Konsowa AH (2003) Removal of nickel ions from wastewater using a cation-exchange resin in a batch-stirred tank reactor. Desalination 158: 189-193.
21. Pehlivan E, Altun T (2006) The study of various parameters affecting the ion exchange of Cu^{+2}, Zn^{+2}, Ni^{+2}, Cd^{+2}, and Pb^{+2} from aqueous solution on Dowex 50W synthetic resin. 134: 149-156.

Evaluation of Uranium Removal from Aqueous Solution using Orange Peels in the Fixed Bed System

Mohamed Ahmed Mahmoud[1,2*]

[1]Nuclear Material Authority, Cairo, Egypt
[2]Chemical Engineering Department, Jazan University, Saudi Arabia

Abstract

In the present work, removal of uranium (VI) using orange peels studied in a fixed-bed system. The parameters of bed heights, initial concentrations, and flow rates were studied, and comparison between experimental results and Thomas, Yoon-Nelson, and Adams-Bohart models was carried out. Thomas and Yoon-Nelson models are good agreement with experimental data than the Adams-Bohart model.

Keywords: Uranium; Orange peels; Adsorption; Kinetic; Column

Nomenclature

Ct: Effluent concentration (mg L^{-1}),

C_0: Influent (Initial) concentration (mg L^{-1}),

F: Linear flow rate (L min^{-1}),

Cab: Adam-Bohart constant (L mg^{-1} min^{-1}),

Kit: Thomas rate constant (L min^{-1} mg^{-1}),

k_{YN}: Yoon-Nelson constant (L min^{-1}),

N_0: Saturation concentration (mg L^{-1}),

Q: Flow rate (ml/min),

q_e: Adsorption capacity at equilibrium, (mg of U (VI) /g adsorbate),

t: time (min),

τ: the time required for 50% adsorbate breakthrough (min),

x: Amount of adsorbent in the column (g),

Z: Bed depth of column (cm).

Introduction

Uranium is the important element in nuclear industries. So that the removal preconcentration and recovery of uranium are the main purpose of nuclear wastewater treatment. Various methods are carried out for the removal of uranium ions from nuclear wastewaters. Chemical precipitation, membrane processes, ion exchange, solvent extraction, photocatalysis and adsorption. Adsorption process is an effective method in the removal of low concentrations of uranium ion from a large volume of solution. Various sorbent materials such as activated carbon [1-3], synthetic [4-8] and natural zeolite [9,10], hematite [11], biotit [12], natural sepiolite [13], akaganeite [14], Orange peels [15], palm-shell [16], and biomasses [17] have been studied.

The Purpose of this work was to examine the removal of uranium ion by orange peels using a fixed bed column from aqueous solution. The parameters of flow rate, initial ion concentrations and bed height were studied. The evaluation of adsorption performance was carried out using Thomas, Adams-Bohart and Yoon-Nelson models.

Materials and Methods

Materials and reagents

Orange peels were prepared according to the procedures reported in my previous work [15]. 1000 mg/L of uranium solutions were prepared by dissolving appropriate amounts of $UO_2(NO_3)_2{\cdot}6H_2O$, Aldrich, USA, in distilled water. For experiments the required concentration were prepared by dilution. The concentrations of U (VI) in solution were determined spectrophotometrically employing Shimadzu UV-VIS-1601 spectrophotometer using arsenazo (III) as complexing reagent [18].

Experimental procedure

Fixed bed system studies were carried out using a glass column of 2 cm internal diameter and 50 cm height. Orange peels were packed in the column with a layer of glass wool at the bottom. Bed heights of 3, 6 and 9 cm were used. Three flow rates (1, 2 and 3 ml/min) were pumped in up flow mode using a peristaltic pump with initial ion concentrations of 50, 75 and 100 mg/L. The effluent samples were collected at regular intervals and analyzed for the residual U (VI) concentration. Column studies were terminated when the column reached exhaustion. In all experiments, the temperature was adjusted to 28°C. The column was run at a predetermined pH of 4.0 as established in my previous study [15]. Schematic of experimental set-up is shown in Figure 1.

Evaluation of break through curves in fixed-bed column

Thomas [19], Adams-Bohart [20] and Yoon-Nelson [21] models were used for kinetic studies.

Thomas model: Thomas model is one of the most widely used models in column performance studies. Thomas model is given in linear form by the following expression:

$$Ln\left(\frac{C_0}{C_t}-1\right)=\frac{K_{Th}q_eX}{Q}-K_{Th}C_0t$$

The parameters of Thomas model (k_{Th} and q_e) can be determined from a plot of $Ln\left(\frac{C_0}{C_t}-1\right)$ against time (t) at a given flow rate.

***Corresponding author:** Mohamed Ahmed Mahmoud, Chemical Engineering Department, Faculty of Engineering, Jazan University, Jazan, Saudi Arabia, P.O. Box 114

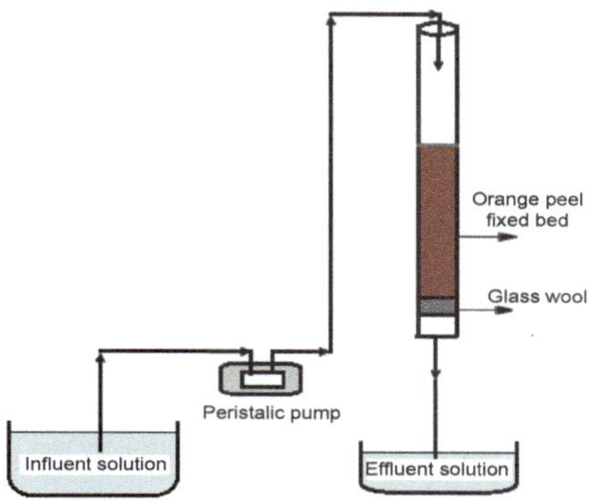

Figure 1: Schematic diagram of experimental system.

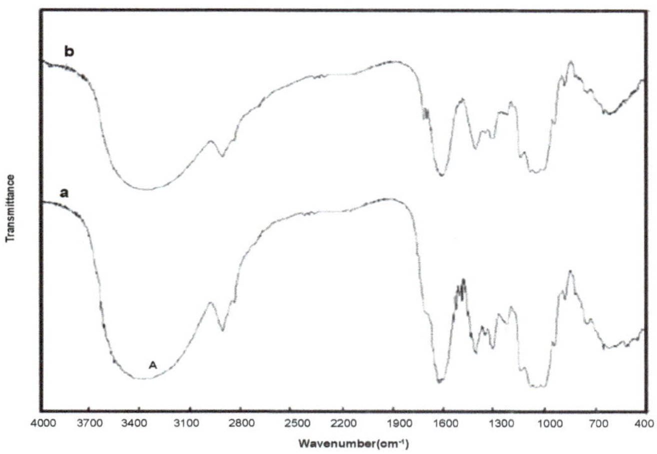

Figure 2: FTIR spectrum of OP and U (VI)-loaded OP.

Adams-bohart model: Adams-Bohart model is used for the description of the initial part of the breakthrough curve. The linear form of Adam-Bohart model is given by the following expression:

$$Ln\left(\frac{C_t}{C_0}\right) = k_{AB}C_0t - k_{AB}N_0\frac{Z}{F}$$

The parameters k_{AB} and N_0 were determined from the intercept and slope of linear plot of *ln* (C_t/C_o) against time (t), respectively.

Yoon and nelson model: Yoon and Nelson model assumes that the decreasing rate of removal of adsorbate ion is proportional to the adsorbate breakthrough on the adsorbent material. The linear form is expressed by the following expression: $Ln\left(\frac{C_t}{C_0-Ct}\right) = k_{YN}\tau - k_{YN}t$

The parameters k_{YN} and τ were determined from the intercept and slope of linear plot of *ln* [$C_t/(C_o-C_t)$] against time (t).

Results and Discussion

Characterization of adsorbent

The FTIR spectrum of OP before and after adsorption (Figure 2) displays a number of absorption peaks, indicating the complex nature of the adsorbent material. The absorption wavelengths of each peak and the corresponding functional groups are presented in Table 1.

From Table 1, after adsorption, shifts occurred in the wave numbers 3367, 2822, 1629 and 1425 cm^{-1} indicating an interaction of these functional groups with adsorbed U (VI). The appearance of wave number 1734 cm^{-1} in the U (VI) loaded spectra may indicate the interaction of this group with U (VI) ion.

Column adsorption

Effect of flow rate: The column adsorption study was carried out at different influent flow rates of 1, 2 and 3 L min^{-1} using initial U (VI) concentration of 50 mg L^{-1}, and bed height of 9 cm. Figure 3 shows that the faster breakthrough curve occurred at higher flow rate of 3 L min^{-1}. The faster breakthrough curve was attributed to faster movement of the adsorption zone along the bed and lower residence time of the influent in the column, thus reducing the contact time between U (VI) and the orange peels.

Effect of bed height: The effects of bed heights of 3, 6 and 9 cm were studied at influent concentration of 50 mg L^{-1} and 1 L min^{-1} flow rate. Figure 4 shows that the breakthrough time decreased with the

Wave number(cm^{-1})		Assignments
unloaded adsorbents	**U(VI) loaded adsorbents**	
3422	**3367**	**OH group**
2831	2822	C-H stretching
-	1734	C=O stretching
1655	1629	-COO- Carboxylic group
1549	-	-COO- Carboxylic group
1486	1425	C=O stretching
1019	1019	C-OH stretching

Table 1: FTIR data of OP and U (VI)-loaded OP.

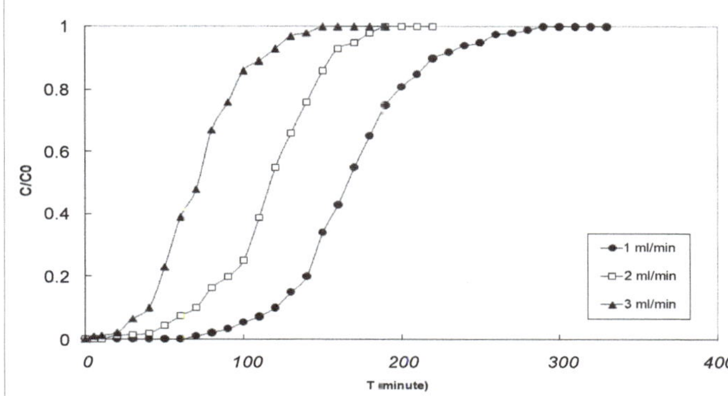

Figure 3: Breakthrough curves for U(VI) adsorption by OP at different flow rates.

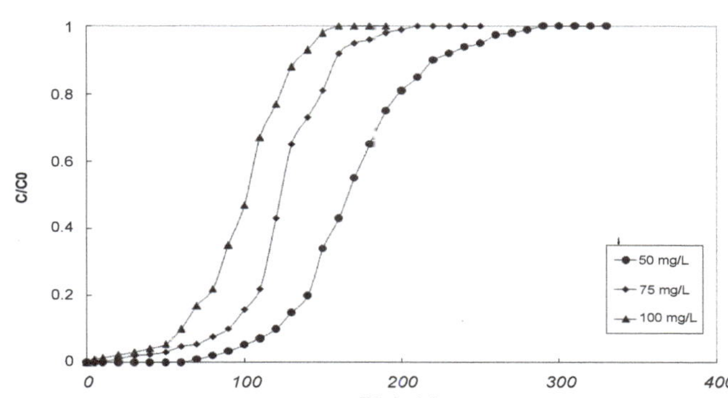

Figure 4: Breakthrough curves for U(VI) adsorption by OP at different initial U(VI) concentrations.

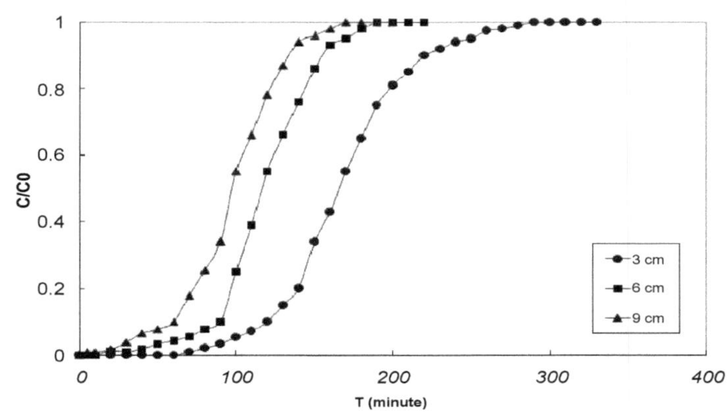

Figure 5: Breakthrough curves for U(VI) adsorption by OP at different bed heights.

Model type	Flow rate (ml/min)	Bed height (cm)	$C_{0\ (mg/L)}$	$q_{e,max\ (mg/g)}$	$k_{Th\ (mL/min.mg)}$	R^2
Thomas model	1	6	100	3.854	0. 410	1
	2	6	100	5.122	0. 511	0.99
	3	6	100	5.231	0. 630	0.99
	1	2	100	6.21	0. 352	0.99
	1	4	100	5.5	0. 381	0.98
	1	6	75	2.034	0. 650	0.97
	1	6	50	1.087	1.042	0.99
Yoon-Nelson model	Flow rate (ml/min)	Bed height (cm)	$C_{0\ (mg/L)}$	τ (min)	$k_{YN\ (1/min)}$	R^2
	1	6	100	173.4	0.041	1
	2	6	100	115.26	0.05	0.99
	3	6	100	74.55	0.063	0.99
	1	2	100	90.8	0.056	0.98
	1	4	100	118.41	0.053	0.99
	1	6	50	113.07	0.052	0.99
	1	6	75	121.46	0.049	0.97
Adams-Bohart model	Flow rate (ml/min)	Bed height (cm)	$C_{0\ (mg/L)}$	$N_{0\ (mg/L)}$	$k_{AB\ (L/min.mg)}$	R^2
	1	6	100	3.991	0. 200	0.84
	2	6	100	3.75	0.31	0.91
	3	6	100	2.88	0.781	0.85
	1	2	100	2.95	0.72	0.91
	1	4	100	3.11	0.426	0.93
	1	6	50	1.08	0.71	0.89
	1	6	75	2.022	0.4	0.91

Table 2: Thomas, Yoon-Nelson and Adams-Bohart model models parameters using linear regression analysis for U (VI) adsorption under various operating conditions.

bed height increasing. As the bed height increased, the U (VI) had more time to contact with orange peel that resulted in higher removal efficiency of U (VI) in the column.

Effect of initial concentration: The variations effect of initial U (VI) concentration from 50 to 100 mg/L at constant bed depth of 9 cm and flow rate of 1 mL/min in the breakthrough curves are shown in Figure 5. The higher the inlet concentration, the steeper is the slope of breakthrough curve. This is due to the increases of driving force and decreases in the adsorption zone length [22].

Evaluation of adsorption column models parameters

Models parameters were evaluated by linear regression analysis for U (VI) adsorption in different bed heights, initial concentrations, and flow rates. Linear regression results and values of R^2 of kinetic models are presented in Table 2. From Figures 6-8, it was observed that both Thomas and Yoon-Nelson models are appropriate models to describe fixed-bed system. But in the case of Adams-Bohart model, low correlation coefficient (R^2=0.925) is observed, which indicate that Adams-Bohart model is not as appropriate a predictor for the breakthrough curve.

Conclusions

From this study, adsorption rate strongly depends on flow rate, influent concentration, and bed height. The fixed-bed adsorption system was found to perform better with lower U (VI) inlet concentration, lower feed flow rate and higher bed height. Comparison between Thomas, Yoon-Nelson, and Adams-Bohart kinetic models with experimental data was performed, and model parameters were determined by linear regression analysis for U (VI) adsorption under various operating conditions. The column experimental data were fitted well with Thomas and Yoon-Nelson models, but the Adams-Bohart model predicted poor performance of fixed-bed column.

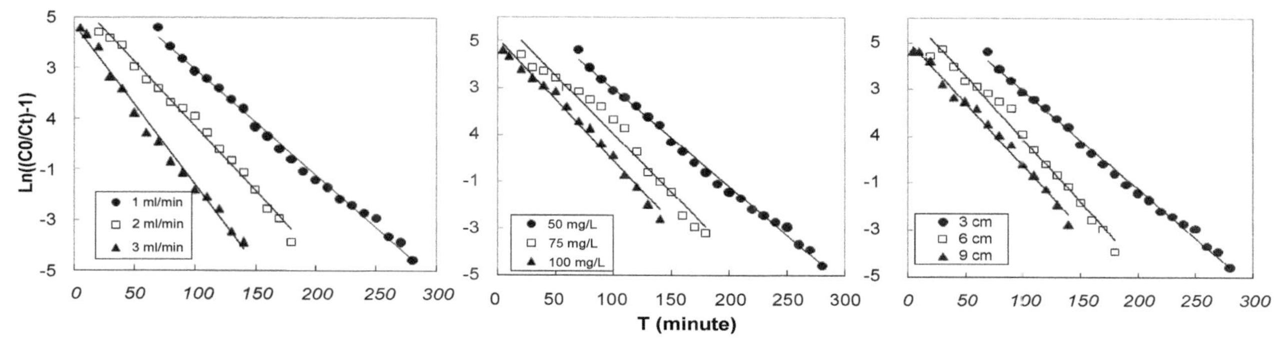

Figure 6: Linear plot of Thomas model with experimental data at different flow rates, initial U (VI) concentrations and bed heights.

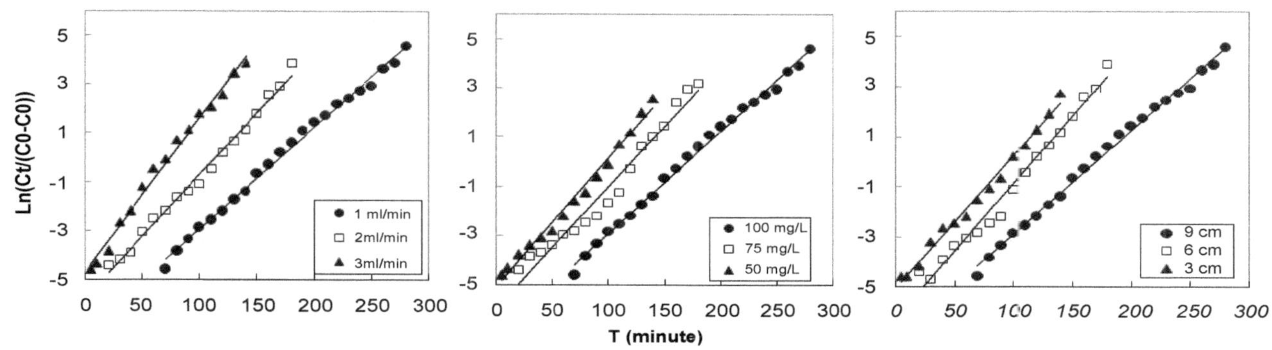

Figure 7: Linear plot of Yoon-Nelson model with experimental data at different flow rates, initial U (VI) concentrations and bed heights.

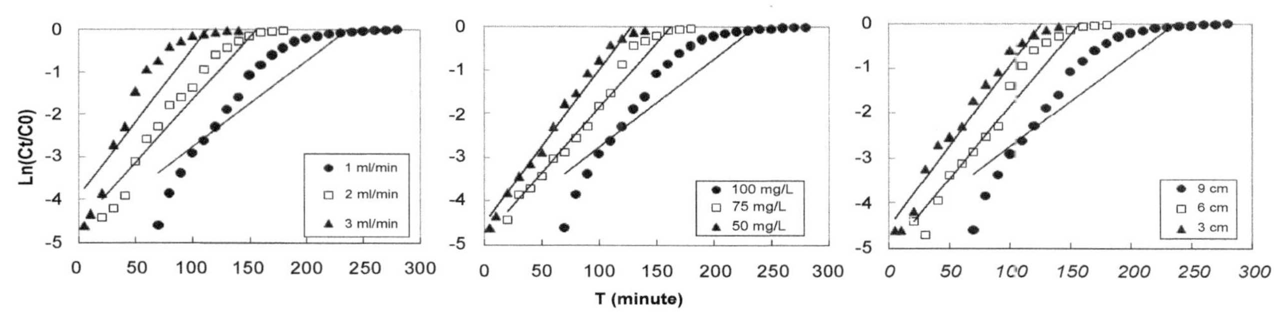

Figure 8: Linear plot of Adams-Bohart model with experimental data at different flow rates, initial U (VI) concentrations and bed heights.

References

1. Yakout SM, Metwally SS, El-Zakla T (2013) Uranium sorption onto activated carbon prepared from rice straw: Competition with humic acids. Applied Surface Science 280: 745-750.
2. Chen S, Hong J, Yang H, Yang J (2013) Adsorption of uranium (VI) from aqueous solution using a novel graphene oxide-activated carbon felt composite. J Environ Radioact 126: 253-258.
3. Mellah A, Chegrouche S, Barkat M (2006) The removal of uranium (VI) from aqueous solutions onto activated carbon: kinetic and thermodynamic investigations. J Colloid Interface Sci 296: 434-441.
4. Ilaiyaraja P, Deb AK, Sivasubramanian K, Ponraju D, Venkatraman B (2013) Adsorption of uranium from aqueous solution by PAMAM dendron functionalized styrene divinylbenzene. J Hazard Mater 250-251: 155-66.
5. Xinghui W, Guiru Z, Feng G (2013) Removal of uranium (VI) ion from aqueous solution by SBA-15. Annals of Nuclear Energy 56: 151-157.
6. Anirudhan TS, Rijith S (2012) Synthesis and characterization of carboxyl terminated poly(methacrylic acid) grafted chitosan/bentonite composite and its application for the recovery of uranium(VI) from aqueous media. J Environ Radioact 106: 8-19.
7. Sun X, Huang X, Liao XP, Shi B (2010) Adsorptive recovery of UO2(2+) from aqueous solutions using collagen-tannin resin. J Hazard Mater 179: 295-302.
8. Akkaya R (2013) Uranium and thorium adsorption from aqueous solution using a novel polyhydroxyethylmethacrylate-pumice composite. J Environ Radioact 120: 58-63.
9. Han R, Zou W, Wang Y, Zhu L (2007) Removal of uranium (VI) from aqueous solutions by manganese oxide coated zeolite: discussion of adsorption isotherms and pH effect. J Environ Radioact 93: 127-143.
10. Weihua Z, Lei Z, Runping H (2009) Removal of Uranium (VI) by Fixed Bed Ion-exchange Column Using Natural Zeolite Coated with Manganese Oxide. Chinese J Chemical Engineering 17: 585-593.
11. Shuibo X, Chun Z, Xinghuo Z, Jing Y, Xiaojian Z, et al. (2009) Removal of uranium (VI) from aqueous solution by adsorption of hematite. J Environ Radioact 100: 162-166.
12. Seung YL, Min HB, Yong JL, Young BL (2009) Adsorption of U(VI) ions on biotite from aqueous solutions. Applied Clay Science 46: 255-259.
13. Donat R (2009) The removal of uranium (VI) from aqueous solutions onto natural sepiolite. J Chemical Thermodynamics 41: 829-835.

14. Yusan SD, Akyil S (2008) Sorption of uranium(VI) from aqueous solutions by akaganeite. J Hazard Mater 160: 388-395.

15. Mohamed AM (2013) Removal of Uranium (VI) from Aqueous Solution using Low Cost and Eco-Friendly Adsorbents. J Chemical Engineering & Process Technology 4: 1-4.

16. Kushwaha S, Sudhakar PP (2013) Sorption of uranium from aqueous solutions using palm-shell-based adsorbents: a kinetic and equilibrium study. J Environ Radioact 126: 115-124.

17. Ahmed SH, El Sheikh EM, Morsy AM (2014) Potentiality of uranium biosorption from nitric acid solutions using shrimp shells. J Environ Radioact 134: 120-127.

18. Onishi H (1989) Photometric determination of trace metals. Wiley, NY, USA.

19. Thomas HC (1944) Heterogeneous ion exchange in a flowing system. J American Chemical Society 66: 1664-1666.

20. Bohart G, Adams EQ (1920) Some aspects of the behaviour of charcoal with respect to chlorine. J American Chemical Society 42: 523-544.

21. Yoon YH, Nelson JH (1984) Application of gas adsorption kinetics-II. A theoretical model for respirator cartridge service life and its practical applications. Am Ind Hyg Assoc J 45: 517-524.

22. Goel J, Kadirvelu K, Rajagopal C, Kumar Garg V (2005) Removal of lead (II) by adsorption using treated granular activated carbon: batch and column studies. J Hazard Mater 125: 211-220.

Effect of pH Solution on the Optical Properties of Cationic Dyes in Dye/ Maghnia Montmorillonite Suspensions

Elaziouti A[1]*, N. Laouedj[2] and Bekka Ahmed[1]

[1]*DR. Moulay Tahar University, Saida, Algeria*
[2]*LCPCE Laboratory, Faculty of sciences, Department of industrial Chemistry, University of the Science and the technology of Oran (USTO M.B), BP 1505 El M'naouar 31000 Oran, Algeria*

Abstract

The interaction of three cationic dyes (Methylene blue, MB; Neutral red, NR; and Malachite green oxalates, MG) with montmorillonite clay from Maghnia deposit in aqueous suspensions was investigated using visible absorption spectroscopy. The stability of the optical properties of dyes in aqueous solutions and in dye/clay suspension was assessed in terms of the absorption bands maxima as a function of pH. The optimal pH for favorable adsorption of the dyes, i.e. ≥ 90% has been achieved in aqueous solutions at 6 for NR and at 7 for both MB and MG respectively. The adsorption of MB leads to the methacromatic effect. MB agglomerates occurred predominantly in the clay with high charge density, even with low charge density. The redistribution of MB molecules over the surface of clay at higher acidic medium causing a reduction of the dye aggregation. However, the interactions of NR and MG with clay differ in certain aspects from those of MB. The adsorption of NR is accompanied by a red shift of the main band at low pH, whereas, the conjugate form of NR, remains stable at neutral and higher pH. A blue shift of the main band is observed at pH 6, which is consistent with externally and internally adsorbed dye cations. For MG, a slight stability of the main bands in clay suspension at acidic and neutral pH, while, the band attributed to MG cations shifs to longer wavelengths at pH 9. To explain the relationship between the layer charge density and the adsorption, so- colled model 2-pKa of surface complexation (SCM) was developed.

Keywords: Clay; Adsorption; Dyes; Methachromasy; 2-pKamodel

Introduction

Clay minerals, the most available, low cost and chemically active surface minerals world, are of great interest, practically because of their extensive uses in many scientific areas such as soil science (stoppage for fertilisers [1], photostabilizers of pesticides [2], industrial technology (heterogeneous catalysis [3], colloidal science (decontamination of water [4], pharmaceutical and paint industries [5] and photochemistry (guide for photo-redox reactions [6]. This clay is a dioctahedral smectite with isomorphic substitutions of Al^{+3} by Mg^{+2} with structural formula unit:

$$(Si_8)^{IV}(Al_{4-x}Mg_x)^{VI}O_{20}(OH)_4 \text{ where } x=0.5\text{-}0.9$$

However for Maghnia montmorillonite, part of the negative charge around 15-50% is due to tetrahedral substitutions of Si^{+4} by Al^{+3}. This clay has an appreciable content of iron. The interest of clay minerals for wastewater treatment in recent decade owing to their catalytic properties and high capacity in the removal pollutants too stable and resistant for conventional methods such as the redox and the exchanging resins of ions [7], coagulation/flocculation [8], membrane separation [9], adsorption [10], the biological methods [11], and more recently the advanced processes of oxidation [12]. The adsorption of cationic dyes is performed via an ionic exchange mechanism. This means that the main dye-clay interaction is a coulomb-type interaction. However, other interaction, such as hydrophobic effect between guest molecules on the clay surface [13] and /or H-bonding [14] can contribute to the adsorption of organic dye on clay surface, favouring the dye agglomeration. Dye agglomeration was found to be sensitive to many parameters, such as the dye load, surface proprieties of the clay, exchangeable cations, the age of dye-clay suspension and pH solution [15].

It is commonly accepted that in clay-aqueous systems the potential of the layer surface charge is determined by the activity of ions (e.g. H^+ or pH). A convenient index of the tendency of a surface to become either positively or negatively charged as a function of pH is the value of the pH required to give zero net charge (pH_{ZPC}) [16]. pH_{ZPC} is a critical value for determining quantitatively (sign and magnitude) the net charge (positive or negative) carried on the clay mineral surface during adsorption of reactive dyes.

In this work, the adsorption of three cationic dyes (Methylene blue, Neutral red and Malachite green oxalates) in aqueous solution with clay from a Maghnia deposit was investigated by the UV-Vis absorption spectroscopy. The effect of the optical properties of dyes in aqueous solutions and in dyes/ clay suspension was assessed in terms of the main absorption bands and the removal rate as a function of pH. To explain the relationship between the layer charge density and the adsorption, so-called Models of Surface Complexation (SCM) was developed using the 2-pKmodel.

Materials and Methods

A thiazin group cation, Methylene blue (MB) (MW= 319.5 $g.mol^{-1}$, λ_{max}=665 nm, ε =95000cm^{-1} $mole^{-1}dm^3$), a phenazin group cation, Neutral red dye, NR (MW=288.8 $g.mol^{-1}$, λ max=520 nm, ε =25000 cm^{-1} $mole^{-1}dm^3$) and triphenylemethane dye, Malachite green oxalate

***Corresponding author:** Elaziouti A, LCPCE Laboratory, Faculty of sciences, Department of industrial Chemistry, University of the Science and the technology of Oran (USTO M.B), BP 1505 El M'naouar 31000 Oran, Algeria

(MW= 507. 51 g.mol^{-1}, λ_{max}=617 nm, ε =42000cm^{-1} mole^{-1}dm^{3}) from Across product for microbiological analysis and was used as model dye without any further purification. Molecular structures of of dyes were shown in Figure 1.

Aqueous dye solution stock was prepared by dissolving accurately weighed neat dye in distilled water to the concentration of 1g/L. Experimental solutions were obtained by successive dilutions. Natural clay was supplied by the Ceramics Society of Maghnia (west of Algeria). The sodium form of clay with a particle size<0.2 µm was prepared in order to obtain good clay colloidal dispersions, as follows: clay was saturated with Na^+ by repeated exchange with NaCl (1M) aqueous solution. The <0.2 particle size fraction was separated by centrifugation and it was washed several times with distilled water until free Cl^- was not detected in the suspension ($AgNO_3$ test). The <0.2 fraction was collected by centrifugation and freeze dried at 60°C for 24 h and is crushed. The Na-saturated clay material resulting from different treatments will be indicated by clay-Na.

X-ray fluorescence was performed on a spectrometer of mark (Oxford). The sample is powder tiny room and prepared (pearl borated) is subjected to a source of X-radiation of fluorescence characteristic of its chemical composition.

Adsorption experiments were carried out as a single-stage batch test using a magnetically stirrer. A suspension containing 0.03 g of adsorbent sample was mixed with a 100 ml aqueous solution of dye at a known initial concentration and initial pH in a flask at constant temperature of 298 K. The initial pH values of the solutions were previous adjusted with 0.1M HCl or NaOH using digital pH-meter (WTN: WISSENSCHAFLLICH TECHNISECHE WERKSTÄTTEN; weilehein Allemagne pH-330) pH-meter. Aliquot of the solution were withdrawn at a predetermined time intervals and were centrifuged (EBA-Hetlich) at 3500 rpm for 15 min to remove any adsorbent particles. Dye concentrations in the supernatant solutions were subsequently estimated by measuring absorbance at maximum wavelengths of dyes (UV – 2401 (PC) SHIMADZU – corporation spectrometer). The adsorption tests were continued until the equilibrium concentration was reached. The experiments were done by varying the initial pH from 3 to 13. The data obtained from the adsorption experiments were then used to calculate the removal rate of adsorbed dye A (%) Equation (1):

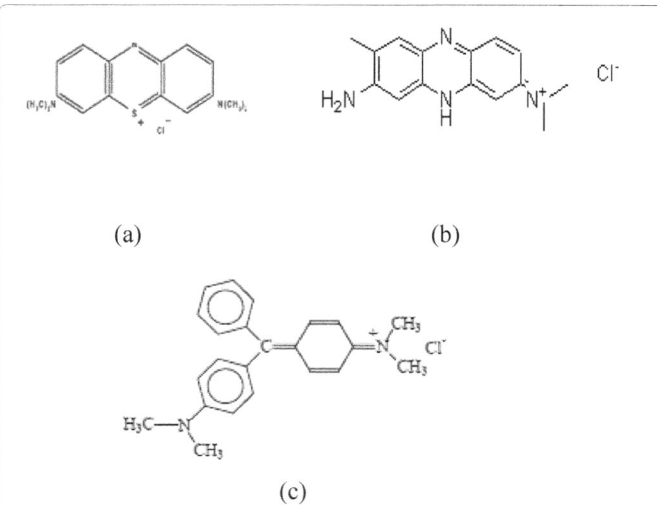

Figure 1: Molecular structures of dyes (a) Methylene blue (MB) (b) Neutral red (NR) and (c) Malachite green oxalates (MG).

Element	Percentage by weight/unit / %
SiO_2	60.75
Al_2O_3	24.15
Fe_2O_3	2.51
CaO	1.56
MgO	2.54
Na_2O	3.63
K_2O	0.65
SiO_2/Al_2O_3	2.52
SiO_2+ Al_2O_3	84.90
LIO	5.21

LOI: Loss on ignition

Table1: XRF analysis of the clay-Na sample.

$$A(\%) = (\frac{C_0 - C_e}{C_0})100 \quad (1)$$

where C_0 and C_e are the initial and equilibrium solution concentrations (mg/L) respectively.

Results and Discussion

XRF analysis of the clay-Na

Table 1 displays the results of chemical composition in the Na-clay sample. The major elements of the clay-Na, expressed in terms of oxide, are primarily of silica, and alumina with appreciable iron content. The CaO and Na_2O (K_2O) reflect the contents of interlayer cationic in clay. The other oxides present in traces amounts such. In the perfect montmorillonite structure, the total theoretical content of SiO_2 and Al_2O_3 is 92% [17] and the content ratio of SiO_2 to Al_2O_3 is 2.6. In the Maghnia clay, the total content of SiO_2 and Al_2O_3 is 84.90% and the corresponding ratio of SiO_2 to Al_2O_3 is 2.52, so this type of clay belongs to the Na-montmorillonite category. The cation capacity of exchange and total specific surface of the clay-Na sample were estimated by the methylene blue (MB) method [18] are of 101.72 meq /100gr of clay-Na and 673.93 m^2/g respectively.

Qualitative study: The evolution of maxima absorption bands of dyes in aqueous solution and in dyes/clay- Na suspensions at various pH.

Figure 2 illustrates the evolution of absorption bands maxima of dyes in aqueous solution and adsorbed onto clay-Na at various pH values. The location of the absorption bans maxima of dyes are collected in Table 2.

The visible spectra of MB in aqueous solution show that MB exhibits a main band at 668 nm assigned to the absorption of monomers and associated with a shoulder attributed to the 0-1 vibronic transition of monomers MB^+ at about 605 nm. The MB monomers have nearly constant absorption bands and absorbance intensities over the pH values in the range of pH between 3 and 13, while the shoulder is most sensitive to the pH solution. The absorbance of the vibronic shoulder at about 605 nm increases and shifts slightly to 624, 621 and 610 nm at pH 7, 11 and 13 respectively, whereas no change is observed in 605 nm-shoulder at pH 3, 5 and 9 respectively. The increase and the bathochromic shift in 605 nm-shoulder are attributed to the partially self-association of MB monomers as cationic dimmers in face-to-face arrangement to minimize their hydrophobic interaction with water [19].

In contrast, Neutral red, in solution at various pH values, exhibits two forms: NR^+H and NR, which are in equilibrium with each other. NR, which has absorption bands in the range of 451-454 nm under neutral and alkali conditions (pH≥ 6), whereas NR^+H, is found to lie between 518 and 521.5 nm under acidic media (pH<6). On the other hand, for MG, the pH has no effect on the absorption bands (at around 617 nm is attributed to the absorption of monomer cations, MG^+ and near 425 nm is assigned to the protonated cations, $MG^{2+}H$) of MG dye, thus affirming their high stability over the pH range of 3-9. At high alkaline medium (pH>9) the colorless carbinol base is formed. Such an effect was observed for the similar malachite green [20], methyl green [21], and crystal violet [22].

The behavior of the bands maxima of MB adsorbed onto clay-Na with the pH is characterized by a hyposochromic shift of the main absorption of MB^+ with respect to those in aqueous solutions. The maxima appear at 585, 588, 584, 567 and 566 nm at pH 5, 7, 9, 11 and 13 respectively. The shifts exceed 50 nm, which is consistent with that observed with MB [23]. In the most case, this change consists of a substitution of the main absorption bands, which correspond to the absorption of monomers, MB^+, by new absorption bands placed at shorter wavelengths, the so-called methachromatic effect. The methachromasy observed in several clay/dye systems been attributed to different processes:

(i) An intermolecular interaction between the electronic π-system of the dye with the electron lone-pairs of the oxygen atoms at the clay surface,

(ii) The self-association of dye molecules when they are adsorbed on the clay surface.

The π- interaction is proposed for clays with partial tetrahedral substitution of Si^{+4} by Al^{+3}. Since Clay-Na has this type of substitution, methachromatic effect has to be related to the π- interaction [24]. Such profile was reported for the analogous Methylene blue [25].

However, the main band of MB^+ at pH 3, gradually shifts from 648 nm (in aqueous solution) to longer wavelengths, almost reaching 658 nm (in clay-Na suspension) and the relative intensity of vibronic shoulder (around 605 nm) increases. The observed shift can be attributed to the change in the environmental polarity/acidity of negatively-charged clay surface with respect to water surrounding, suggesting that the adsorption process is instantaneous. Shift to higher wavelengths of ≈ 10 nm are always observed for dye monomers adsorbed at the clay surface.

As seen in Figure 2, the interactions of NR and MG with clay-Na differ in certain aspects from those of MB. The adsorption of NR^+H is accompanied by a slight red shift of the main absorption bands to longer wavelengths under pH range of 3-5, whereas the NR (conjugate base) form has a nearly constant absorption bands over the pH range of 8-12. At pH 6, the band maximum shifts from 518 nm (in aqueous solution) to shorter wavelengths, almost reaching 460 nm (in clay-Na suspension). These results are consistent with externally and internally adsorbed monomer cations (pH<6), monomer cation-conjugate base equilibrium at pH=6 and shifting the monomer cation-conjugate base equilibrium to the conjugate base form (pH>6) respectively.

On the other hand, for MG, the results (given in Table 2) show a slight stability of the main bands of both forms (MG^+ and $MG^{2+}H$) in clay-Na suspension under pH values in the range of 3–7, while, the absorption band attributed to MG^+ shifted to longer wavelengths at pH

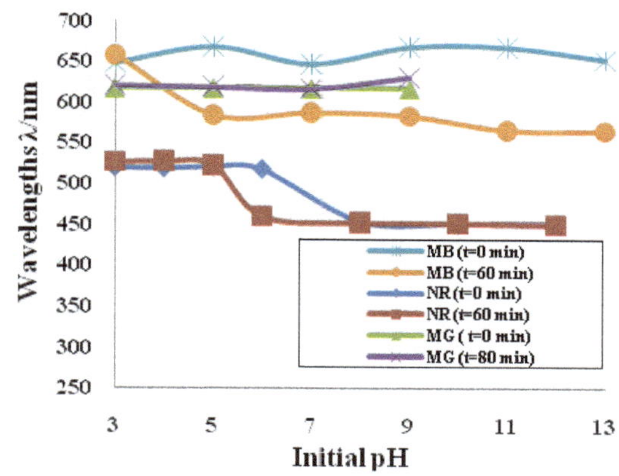

Figure 2: Evolution the main absorption bands of dyes in aqueous solution and in dyes/clay-Na suspensions as a function of pH for [MB]= [NR]= [MG]=50mg/L, [Clay-Na]=0.3g/L, T=298K and contact time=60 min (MB and NR) and 80min (MG).

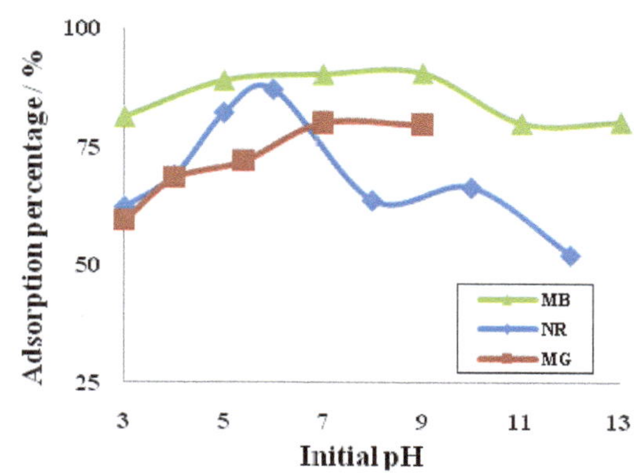

Figure 3: Effect of the pH on the absorption bands maxima of MB in aqueous solution and in dyes/ clay-Na suspensions for [MB]= [NR]= [MG]=50mg/L, [Clay-Na]=0.3 g/L, T=298K and contact time of 60 min (MB and NR) and 80min (MG).

9, almost reaching 631 nm with respect to that in aqueous solution.

The shift did not exceed 10 nm over the pH range of 3- 5 and 3-7 for NR and MG respectively, but is significantly higher for NR at pH 6 and MG at pH 9, being 14 and 58 nm respectively. Such an effect was observed for the similar Malachite green [26].

Shift to higher wavelengths of ≈10 nm are always observed for dye monomers adsorbed at the clay surface. The observed shift can be attributed to the change in the environmental polarity/acidity of negatively-charged clay surface with respect to water surrounding, suggesting that the adsorption process is instantaneous [27].

In addition, this change is accompanied by the appearance of new band at 596 nm placed in shorter wavelengths assigned to an influence on π-orbital. Such influence might be caused by an adjacent platelet, indication flocculation or by the formation of dimers on the surface of the clay-Na [28].

Quantitative study: The evolution adsorption rate of dyes in dyes/clay-Na suspensions at various pH.

The effects of initial pH solution on the adsorption rate of BM and NR onto clay-Na were investigated over a range of pH values from of 3 to 13, while for GM; the experiments were conducted over a range

MB /aqueous solution			MB⁺/clay-Na		NR /aqueous solution		NR /clay-Na		MG/aqueous solution		MG/ clay-Na	
pH	MB^+	$(MB^+)_2$	MB^+	$(MB^+)_n$	NR^+H	NR	NR^+H	NR	MG^+	$MG^{2+}H$	MG^+	$MG^{2+}H$
3	648	-	658	*608	520.5	-	527	-	617	424	621	426
4	-	-	-	-	519.5	-	527.5	-	-	-	-	-
5/ 5.4	668	624	-	585	521.5	-	523.5	-	617	425	619	-
6.5/ 7	647	-	-	588	518	-	-	460	617	424	617	426
8/ 9	668	-	-	584	-	454	-	452.5	617	423	631 / 596	-
10		-	-	-	-	451.5	-	452	Col.	Col.	Col.	Col.
11	668	621		567	-		-		Col.	Col.	Col.	Col.
12	-	-	-	-	-	451	-	450.5	Col.	Col.	Col.	Col.
13	653	610	-	566	-		-		Col.	Col.	Col.	Col.

MB^+, $(MB^+)_2$ and $(MB^+)_n$: monomer, dimer and agglomerates forms of Methylene blue dye.
NR^+H and NR: monomer and conjugate base form of Neutral red dye. Col.: colorless.
MG^+ and $MG^{2+}H$: monomer and protonated cations forms of Malachite green oxalate dye.

Table 2: Evolution the main absorption bands of dyes in aqueous solution and in dyes/clay-Na suspensions as a function of pH for [MB]= [NR]= [MG]=50mg/L, [Clay-Na]=0.3g/L, T=298K and contact time=60 min (MB and NR) and 80min (MG).

of pH values from 3 to 10 for avoiding dye degradation. As shown in Figure 3 & Table 2, the pH significantly affected the adsorption percentage of all the three dyes. The dye adsorption rate increased from 81.47% to 90. 81% for MB and from 62.52% to 87.19% for NR as the pH was increased from 3 to 7 for MB and from 3 to 6.5 for NR respectively, and then decreases to 80. 46 and 52.39 % at pH =13 and 12 for MB and NR respectively. However, for MG, the adsorption raised from 59.84 to 80.40% when the pH value was increased from 3 to 7 then it remains constant beyond pH=7. The maxima adsorptions of dyes of 90.81, 87.19 and 80.40 % for MB, NR and MG were achieved at pH 7, 6 and 7 respectively.

The adsorption of these charged dye (thiazin, MB; phenazin, NR; and triphenylemethane, MG) groups onto the adsorbent surface is primarily influenced by the surface charge of the adsorbent, which is in turn influenced by the solution. For clay minerals the potential determining ions are H^+ and OH^- and complex ions formed by bonding with H^+ and OH^-. The broken Si-O bonds and Al–OH bonds along the surfaces of the clay crystals result in hydrolysis, is an amphoteric surface hydroxyl group that can associate (Equation 2) and dissociate (Equation 3) proton.

To explain the relationship between the layer charge density and the adsorption, so-called the Surface Complexation Model of (SCM) was developed [29]. In the 2-pKa approach we assume two reactions for surface protonation.

At low pH the reaction might be:

$$Si\text{-}OH + H^+ \leftrightarrow Si\text{-}OH_2^+ \qquad (2)$$

Lower adsorption rate of dyes at low pH is probably due to the presence of excess H^+ ions competing with the cations groups on the dyes for the adsorption sites.

However, the high removal rate observed for NR at pH 6 (low pH) is attributed to the formation of surface hydrogen bonds between the surface of the hydrogen bonds of the hydroxyl group on the clay-Na surface and the nitrogen atoms of neutral red cations [30]. At high pH the reaction is:

$$Si\text{-}OH + OH^- \leftrightarrow Si\text{-}O^- + H_2O \qquad (3)$$

In our studies the adsorption maximum for both MB and MG dyes are achieved at pH 7 respectively. So at neutral and high pH, the solution in contact with the basal oxygen surface of the tetrahedral sheet will contain excess hydroxyls. The surface will then exhibits a cation exchange capacity (Equation.4):

$$Si\text{-}O^- + Dye^+ \leftrightarrow (Si\text{-}O^-, D^+) \ (4)$$

The adsorption rate of MB dye decreases with increase in pH (above pH 9), then becomes constant up to pH 13. Such decrease is assigned to the presence of dimers $(MB^+)_2$ species in the clay-Na environment competing with the monomers (MB^+) on the dyes for the active sites.

On the other hand, at neutral and higher pH (>10), most NR dye occurs in the form of undissociated molecules. Indeed, the presence of the neutral form of NR molecules in the clay-Na atmosphere increases the local concentration favoring the aggregation followed by precipitation of the dye.

Conclusion

The effect of the optical properties of dyes (MB, NR and MG) in aqueous solutions and adsorbed onto Maghnia clay was assessed in terms of the absorption bands maxima and the adsorption rate as a function of pH solutions. The adsorption of MB dye onto clay surface leads to the methacromatic effect in the absorption spectrum of the dye. Formation of MB dimers and higher agglomerates reflected sensitively the potential of the layer surface charge density. MB agglomerates occurred mainly in the clay with high charge density, even at low charge density.

The adsorption of NR onto clay-Na is accompanied by a red shift of the main absorption at low pH, whereas, that of NR, has a constant absorption bands at high pH. A blue shift of the main band is observed at pH 6, consistent with externally and internally adsorbed monomer cations and shifting the monomer cation-conjugate base equilibrium to the conjugate base form respectively.

For MG, a slight stability of the main bands in clay-Na suspension at acidic and neutral pH, while, the band attributed to MG^+ shifs to longer wavelengths at pH 9, almost reaching 631 nm. The observed shift can be attributed to the change in the environmental polarity/acidity of negatively-charged clay surface with respect to water surrounding, suggesting that the adsorption process is instantaneous.

The optimal pH for favorable adsorption of the dyes, i.e. ≥ 90% has been achieved in aqueous solutions at 6 for Neutral red and at 7 for both Methylene blue and Malachite green respectively. The relationship between the layer charge density and the adsorption is explained by the 2-pKa surface complexation model (SCM).

Acknowledgment

The authors thank U.S.T.MB University and the University DR T.M. for materially supporting this research.

References

1. Theng BKG (1974) The Chemistry of Clay-organic Reactions. Adam Higle 343.
2. Margulies L, Rozen H, Cohen E (1985) Energy transfer at the surface of clays and protection of pesticides from photodegradation. Nature 315: 658-659.
3. Ortego JD, Kowalska M, Cocke DL (1991) Interactions of montmorillonite with organic compounds adsorptive and catalytic properties. Chemosphere 22: 769-798.
4. Srinivasan KR, Fogler HS (1990) Use Of Inorgano-Organo-Clays In The Removal Of Priority Pollutants From Industrial Wastewaters: Structural Aspects. Clays Clay minerals 38: 277-286.
5. Murray HH (1991) Overview - Clay minerals application. Applied Clay Science 5: 379-395.
6. Rabek JF (1990) Photochemistry and Photophysics. (1stedn), CRC Press, Florida.
7. Dusart O, Serpaud B (1991) La tribune de l'eau 44: 15-22.
8. Linsheng Z, Dobias B (1992) Water Treatment 7: 221-232.
9. Ciardelli G, Corsi L, Marucci M (2001) Membrane separation for wastewater reuse in the textile industry. Resources, Conservation and Recycling 31: 189-197.
10. Paprowicz J, Sodczyk S (1988) Application of biologically activated sorptive columns for textile waste water treatment. Environmental Technology 9: 271-278.
11. Milano JC, Loste-Berdot P, Vernet JL (1995) Photooxydation du Vert de Malachite en Milieu Aqueux en Presence de Peroxyde D'Hydrogene : Cinetique et Mecanisme Photooxidation of Malachite Green in Aqueous Medium in the presence of Hydrogen Peroxide: Kinetic and Mechanism. Environmental Technology 16: 329-341
12. Yener J, Kopac T, Dogu G, Dogu T (2005) Adsorption of Basic Yellow 28 from aqueous solutions with clinoptilolite and amberlite. J Colloid Interface Science 294: 255-264
13. Schoonheydt RA (1995) Clay Mineral Surfaces. In: Mineral Surfaces. Edited by Vaughan DJ, Pattrick RAD. Chapman & Hall, London.
14. Meifang H, Beiyan M, Fangbai L, Hongfu W, Yijie M (2003) Chemical composition of bentonites from various parts of china and differences among samples. Geology-Geochemistry30: 70-73.
15. Bujdak J, Komadel P (1997) Interaction of Methylene Blue with Reduced Charge Montmorillonite. J Phys Chem B: 101: 9065-9068.
16. Allen SJ, Koumanova B (2005) Decolourisation of water/wastewater using adsorption. Journal of the University of Chemical Technology and Metallurgy 40: 175-179.
17. Meifang H, Beiyan M, Fangbai L, Hongfu W, Yijie M (2003) Chemical composition of betonies from various parts of caina and differences among ore samples. Geology-Geochemistry 31: 70-73
18. Pavan PC, Crepaldi EL, Valim JB (2000) Sorption of Anionic Surfactants on Layered Double Hydroxides. Journal of Colloid and Interface Science 229: 346-352.
19. Pavan P C, Crepaldi E L, Valim JB (2000) Journal of Colloid and Interface Science 229: 346-352.
20. Rytwo G, Nir S, Crespin M, Margulies L (2000) Adsorption and interaction of methyl green with montmorillonite and sepiolite. Journal of Colloid and Interface Science 222: 12-19.
21. Zolinger H (1991) Color Chemistry. (2ndedn) VCH New York.
22. Bujdák J, Janek M, Madejová J, Komadel P, Komadel P (2001) Methylene Blue Interactions With Reduced-Charge Smectites. Clays and Clay Minerals 49: 244-254.
23. Coine APP, Neumann MG, Gessner F (1998) Time-Dependent Spectrophotometric Study of the Interaction of Basic Dyes with Clays: III. Mixed Dye Aggregates on SWy-1 and Laponite. Journal of Colloid and Interface Science 198: 106-112.
24. Lopez AF, Arbeloa T, Martinez VM, Banuelos PJ, Lopez AI (2008) Characterization of Surface Properties of Clay Particles in Aqueous Suspensions by Electronic Spectroscopy of Adsorbed Dyes. Clay Sci 12 Supplement: 31-37.
25. Elaziouti A, Laouedj N (2010) Comparison Study on the Removal of Cationic Dyes from Aqueous Suspension of Maghnia Montmorillonite. Journal of the Korean Chemical Society 54: 3.
26. Yariv S, Ghosh DK, Hepler LG (1991) Metachromasy in clay-mineral systems : adsorption of cationic dyes crystal violet and ethyl violet by kaolinite from aqueous and organic solutions. Journal of the Chemical Society Faraday Transactions 87: 1201-1207.
27. Yates DE, Levine S, Healy TW (1974) Site-binding model of the electrical double layer at the oxide/water interface. Journal of the Chemical Society Faraday Transactions 1: 1807-1818.
28. Davis A, James RO, Leckie JO (1978) Surface ionization and complexation at the oxide/water interface: I. Computation of electrical double layer properties in simple electrolytes. Journal of Colloid and Interface Science 63: 480-499.
29. Davis A, James RO, Leckie J (1978) Journal of Colloid Interface Sci 36: 480-499.
30. Allen SJ, Koumanova B (2005) Decolourisation of water/wastewater using adsorption. J of the University of Chemical Technology and metallurgy 40: 175-195.

14

Simulation and Study of the Effect of Pressure Oscillations on Linear Combustion Instability in a Double Base Solid Rocket Motor

Ali Seifollahzadeh and Ali Aminian*

Department of Chemistry and Chemical Engineering, MAUT, Tehran, Iran

Abstract

The combustion instability remained as a problem during solid rocket motors progress. In this study, combustion instability of a double base solid rocket motor was estimated wherein acoustical and erosive burning was considered. The ballistic parameters of the solid rocket motor were presented and the results were compared with simulations, which show a good agreement. The result shows that specific impulse of the solid propellant rocket motor was obtained 211 s and the maximum pressure of the combustion chamber reaches 13.8 MPa. Combustion instability analysis indicates that the motor became unstable after 0.5 s.

Keywords: Combustion instability; Double base solid propellant; Internal ballistic; Erosive burning; Acoustic modes

Introduction

Combustion instability is referred to the pressure oscillations duo to unsteady burning rate in the combustion chambers. Rocket motor combustion instability is caused by pressure fluctuations and acoustic resonances in the combustion chamber, which may reduce engine performance, induce structural vibration, increase wall heat transfer, and possibly lead to catastrophic failure in combustion chamber.

Current remedy for reducing instabilities occurring in solid rocket motors is either by changing propellant composition, or improving the design of the combustion chambers. Non-steady burning and combustion instabilities appearing in solid propellants have been investigated extensively both theoretically and experimentally.

Due to considerable cost, time and difficulties encountered during motor firing tests, cold flow experiments and numerical simulations can be used to study the type of flow field in solid rocket motors [1,2]. The comparison between the results of such simulations and experimental findings help to better understanding flow pattern in combustion chambers and acquire useful information on combustion instability driven processes.

One important elements of the combustion instability are obstacles and corners in the combustion chamber cavity, which are leads to the vortex generation. Flandro and Jacobs [3] was first made acoustic mode excitation by vortex shedding in solid propellant rocket motors. The effect of nozzle cavity on suppression of vortex shedding was clearly stated by Fabignon et al. [4].

The acoustic energy lost as a result of energy transfer from acoustic pressure oscillations to the vortical gas waves produces axial gas motion known as flow turning. Flandro [5] has shown that for cylindrical grain covering 80% of the total chamber length the flow turning growth rates were -1, -0.25, +0.25 for first, second, and third axial modes, respectively. Also, the inclusion of flow turning correction term in growth rate calculations make better agreement between exact values and predicted ones.

Another important feature of the combustion instability is the effects of Combustion stability additives like zirconium carbide (ZrC), aluminum oxide ($A1_2O_3$), and zirconium ortho-silicate ($ZrSiO_4$) on suppressing combustion instability in solid propellant rocket motors. The additives appear to have effects on changing the response behavior of the propellant. In addition, high specific impulse in a solid propellant motor is generally achieved by metals. Metal combustion increases the temperature and the pressure inside the chamber [6].

The experiments show that propellants with stabilizing additives have no instability under different pulses and mild operating conditions. Higher pressure solid rocket motors are more susceptible to pulsed instability at lower pulse amplitudes. The thermal and momentum relaxation times of particles play an important role in dictating the two-phase flow interactions with oscillatory motor internal flows [7]. Small particles with a wide range of particle distribution have higher relaxation time and usually exert greater influence on the dispersion of acoustic energies. In general, the effect of energy exchange tends to be more profound for low-frequency acoustic oscillations [8].

Aluminum combustion occurs close to the burning surface, implying that this have a little impact on the oscillatory behavior. On the other hand, turbulence has a little influence on the acoustic wave oscillation. The turbulent-flow-induced particle dispersion phenomena have considerable effect on damping out the turbulent region by the turbulence-induced eddy viscosity [9].

In order to study the effects of grain configurations on combustion instability, a series of small test motor firings with various grain configurations was used [10]. Their tests showed that combination of STAR-circular perforation configuration offers a great stability margin during entire motor firing.

Since, in a solid rocket motor, pressure is spanned between 5-25 MPa and temperature varies between 2300-3600 K, real measurements in a practical motors are impossible, using only research burners can obtain the desire parameters with changing the operating conditions.

In order to predict the stability characteristics of the solid propellant

***Corresponding author:** Ali Aminian, Department of Chemistry and Chemical Engineering, MAUT, Tehran, Iran rocket motor combustion, it is require defining acoustic boundary conditions at the propellant surface, namely how the instantaneous propellant burning rate responds to pressure or cross flow velocity fluctuations.

The pressure perturbation generates perturbations in the burning rate, since the burning rate of the solid propellants are pressure dependent, and consequently there are perturbations in gas flow from burning surface. Whether or not the acoustic wave is amplified upon reflection depends on the nature of the gas flow/pressure connection. To include this effect it is required to solve the unsteady gas flow equations. An effective model combines the effects of the unsteady flow, the transient combustion process, and the boundary conditions.

The objective of this work is to utilize a simulation capability that can be used to further investigate the flow field generated through the entire combustion chamber and analyzing the combustion instability of the concerned solid rocket motor. To further proceed, two major models have been implemented: 1) the chamber pressure was assumed to be uniform in the entire chamber and the time evolution of pressure history was calculated. Other ballistic characteristics of the solid rocket motor were also obtained including thrust and specific impulse. Furthermore, combustion instability analysis was performed during the burning out of the propellant, 2) transient combustion for studying chamber flow dynamics.

Dynamic Combustion Models

This work uses the Lax-Wandrroff technique to solve for the 1-D flow field in a double base solid propellant rocket motor. In this study, non-aluminized double base solid propellant have been the primary interest, the following simplifying assumptions were made: (1) the solid propellant combustion process can be described in a single spatial dimension, (2) the unburned propellant is homogeneous, incompressible, and inert, (3) the solid is converted into gas by a global pyrolysis reaction which occurs at an infinitesimally thin interface, and (4) the gas-phase flame is quasi-steady and remains anchored to the interface.

Therefore, the 1-D inviscid flow equation with source terms to incorporate the mass injection from the burning propellant was used for modeling the transient combustion of the solid propellant rocket motor [11]:

$$\frac{\partial U}{\partial t}+\frac{\partial F}{\partial x}=S \quad (1)$$

Where

$$U=[\rho,\rho u,\rho e_t A]$$

$$F=\left[\rho uA,(\rho u^2+p)A,(\rho e_t+p)Au\right]$$

$$S=\left[\rho_p r_b P,\ p\frac{\partial A}{\partial x}-\rho_p r_b Pu,\ \rho_p r_b P(h_c+\frac{1}{2}u_f^2)\right]$$

It was assumed that the flow in the combustion chamber is Quasi-1D unsteady inviscid flow, in which ρ, u, p, A, ρ_p, r_b, u_f, h_c and P are gas density, velocity, pressure, port area, propellant density, mass burning rate, injection velocity of combustion products, heat of combustion, and perimeter of solid propellant, respectively. Total energy in the combustion chamber and injection velocity of combustion products are defined as

$$e_t=\frac{p}{(\gamma-1)\rho}+\frac{u^2}{2} \quad (2)$$

$$u_f=\frac{\rho_p r_b}{\rho} \quad (3)$$

$$\dot{m}_g=S_b\rho_p r \quad (4)$$

For 1-D simulations, numerical calculations were performed to obtain the axial pressure, velocity, temperature, and other internal flow parameters using Eq 1-3. Mass injection rate is calculated by the product of the burning rate (from burning rate model), burning area, and solid propellant density [see Eq. (4)]. The boundary conditions are u = 0, since the head end of the motor is inert and initial p is the medium pressure. Because of numerical stability considerations, the time step is evaluated using the Courant-Friedrichs-Lewy (CFL) number and spatial grid size *h* defined as below,

$$dt=\frac{CFL\times h}{max(a)} \quad (5)$$

The CFL number must be less than unity and *a* is the speed of sound.

In order to guarantee stability of the solution, another term named as artificial viscosity is added to numerical solution of partial differential equations (1).

Major ingredients of the propellant are 55.25% NC, 33.84% NG. The global pressure-dependent burning rate model was used to predict the pressure variations on the burning rate for quasi-steady state simulation as [12]

$$r_b=0.01372(P)^{0.2276} \quad (6)$$

Where P in Mpa and the unit of r_b is m/s.

The erosive burning contribution to the over-all burning rate can be written as follows:

$$r=r_b+r_e \quad (7)$$

Since two burning rates are additive, the pressure-dependent burning rate (r_b) calculated from mean chamber pressure, while erosive burning term r_e depends on the gas flux, flame and surface temperatures. Erosive burning rate is written as [13,14]

$$r_e=\alpha\, G_g^{0.8}exp\left(-\beta r_b\rho_p/G_g\right)/L^{0.2}$$

$$\alpha=\frac{0.028c_{p_g}\mu_g^{0.2}pr_g^{-0.67}}{\rho_p C_{p_s}}\left(\frac{T_c-T_s}{T_s-T_0}\right) \quad (8)$$

Where β is an empirical constant about 53 [13], c_{p_g} and c_{p_s} are specific heats of gas and solid propellant respectively, μ_g gas viscosity, p_g Prantle number for the gas phase, T_s surface temperature, T_0 initial grain temperature, and G_g is mass flux of the combustion gases leaving the burning surface.

The double-base solid propellant density, adiabatic flame temperature, grain length, chamber radius, web thickness, nozzle throat

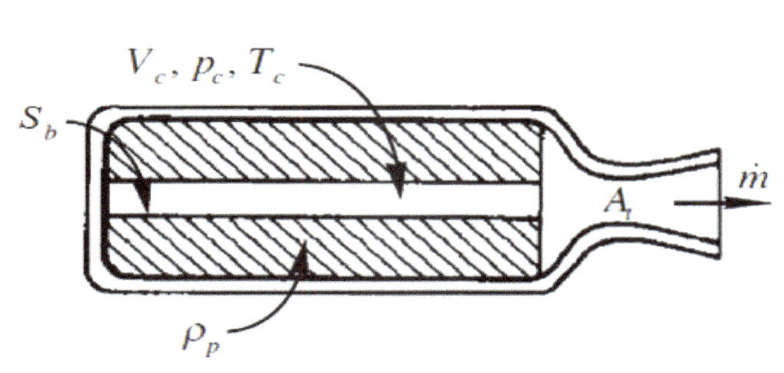

Figure 1: Schematic of the solid propellant rocket motor.

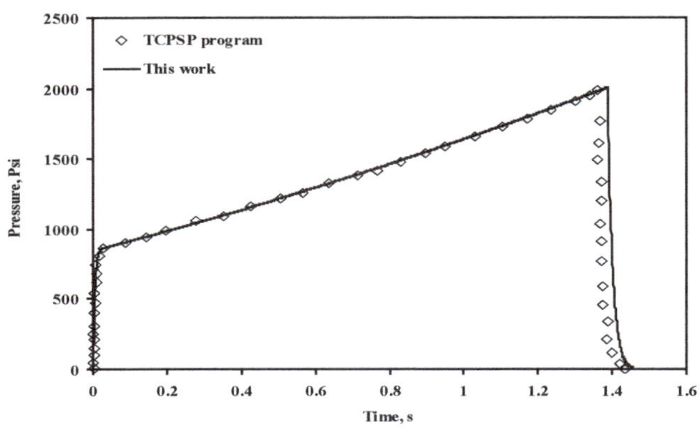

Figure 2: Comparison between the calculated pressures of this work and TCPSP program [12].

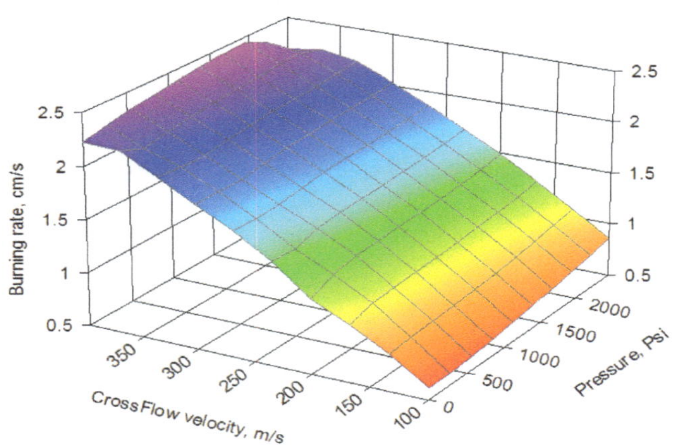

Figure 3: 3-D plot of the effect of pressure and cross flow velocity on the burning rate.

diameter, and outer diameter of the nozzle are 1600 kg/m³, 2351 K, 0.52 m, 120 mm, 22 mm, 24.4 mm, 79.5 mm, respectively. Figure 1 shows the schematic representation of the solid rocket motor.

The thrust for a rocket motor is

$F(t) = P_c(t) A_t C_F(t)$ (9)

Where the thrust coefficient, C_F, can be expressed in terms of chamber pressure and exit pressure [12]

$$C_F(t) = \eta_F \sqrt{2\frac{\gamma^2}{\gamma-1}\left(\frac{2}{\gamma+1}\right)^{(\gamma+1)/(\gamma-1)}\left[1-\left(\frac{P_e}{P_c}\right)^{(\gamma-1)/\gamma}\right]} + \frac{P_e - P_{atm}}{P_c}\left(\frac{A_e}{A_t}\right) \quad (10)$$

The total impulse of the solid rocket motor is obtained by integration of thrust over burning time

$$I = \int_0^t F(t)\,dt \quad (11)$$

For uniform spatial but time varying pressure distribution simulations along the chamber axis, the mass balance equation is used to calculate the chamber pressure. The chamber pressure is assumed to be uniform throughout the rocket chamber; therefore, the burning rate is uniform as well. With the above assumptions described above, the chamber pressure is solved from mass balance equation:

$$\frac{\partial}{\partial t}\int_{V_c} \rho dV + \oint_{S_c} \rho u . dS = 0 \quad (12)$$

The second term in eq 12 is mass of the gas in every instantaneous time in the combustion chamber. We can obtain the pressure-time relation from eq. 12 as follows

$$P_c(t) = \left[\left(P_{atm}^{1-n} - \frac{a(\rho-\rho_p)s_b}{C_D A_t}\right)\exp\left(\frac{(n-1)RT_c C_D A_t t}{V_c}\right) + \frac{a\rho_p S_b}{C_D A_t}\right]^{1/1-n} \quad (13)$$

Where P_c, T_c, V_c, S_b, A_t, C_D, n, and t are chamber pressure, temperature, volume burning area, throat area, nozzle discharge coefficient, pressure exponent, and time, respectively.

Results and Discussion

The solid rocket motor uses non-metalized double base propellant. Figure 2 shows the comparison between the calculated pressures and simulated by TCPSP program [12] for the quasi-steady state model. As can be seen from Figure 2, there is a good agreement between the calculated pressures and simulated by TCPSP program for quasi-steady state model. Solid rocket motor has no pressure spike addressing erosive burning normal to the burning surface can be neglected in this motor. The characteristic length of the solid rocket motor L^* $(=V/A_t)$ start with a value of approximately 115 cm.

As shown in Figure 2, there is no induce bulk-mode or L^* instability in the solid rocket motor under consideration. Figure 3 shows the 3-D plot of the effect of pressure and cross flow velocity of combustion gases released from burning surface on the burning rate. The burning rate increases almost linearly with cross flow velocity mainly due to enhanced heat transfer by turbulence in the fizz zone [11].

Figure 4 shows the thrust vs. time for double base solid rocket motor.

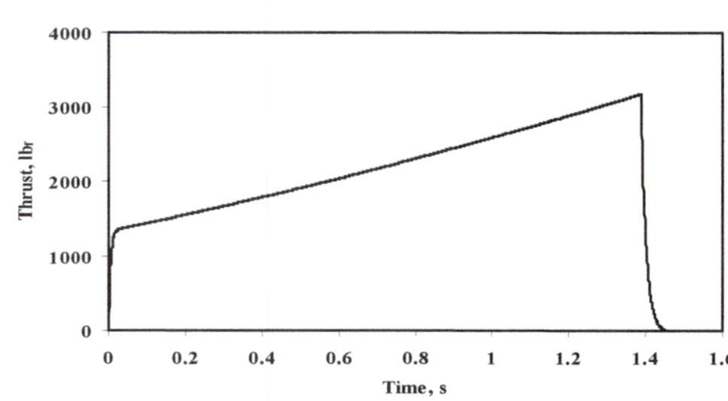

Figure 4: Calculated thrust of the solid rocket motor.

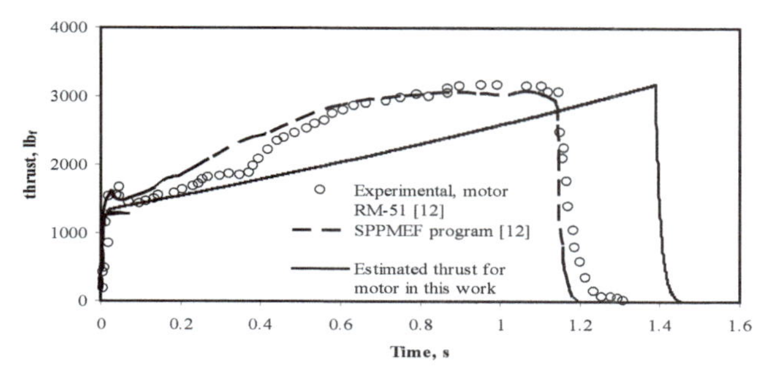

Figure 5: Comparison between the thrust vs. time for the solid rocket motor of this work and the thrust obtained for RM-51 using SPPMEF program [12].

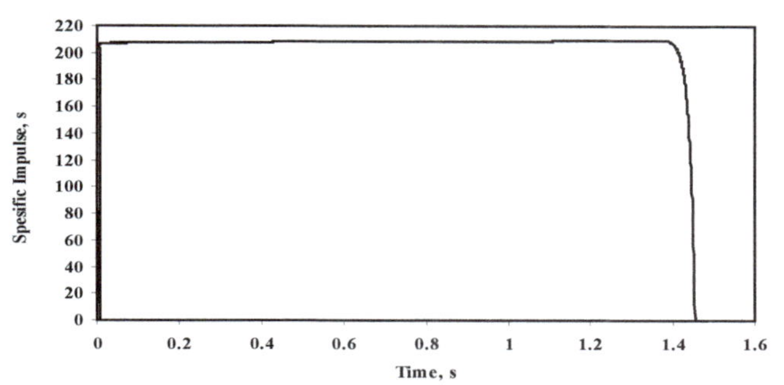

Figure 6: Variation of specific impulse vs. time for the circular perforation grain.

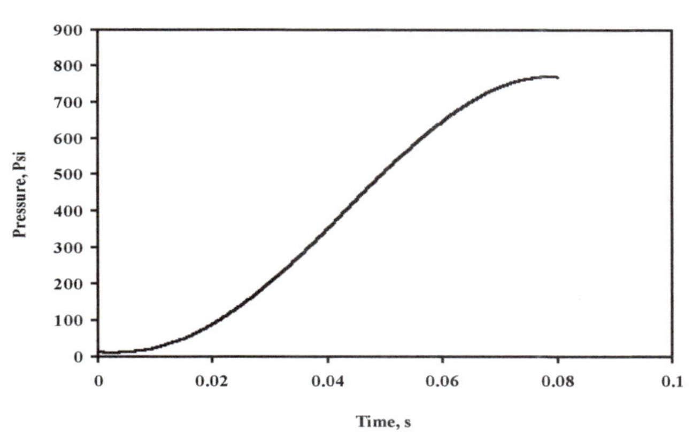

Figure 7: Transient combustion of the solid rocket motor.

As can be seen from Figure 4, the thrust of the solid rocket motor with circular perforation grain was progressive, reaches its maximum during the end of the burning time.

Comparison between the calculated thrust of the solid rocket motor in this work and the thrust obtained from solid rocket motor RM51 wit experimental thrust data for RM51 is shown in Figure 5. As in Figure 5, there is a good agreement between the thrust of this work and the calculated thrust in Ref [12].

A parameter that is used to identify the energy efficiency of propellant combustion named as Specific impulse. This parameter represents the performance of a rocket motor during its operating flight. The plot of specific impulse versus time is depicted in Figure 6.

The specific impulse of the double base solid rocket motor was obtained 211 s. currently, the performance of the solid rocket motors is improved by adding to the propellant ingredients some metal powders and/or ammonium percholorate [6].

The result of the transient combustion simulation for the pressure can be seen in Figure 7. It is interesting to note that after ignition transients have stabilized, the solid rocket motor pressure follows the trend of the motor's total burning surface area [15]. As mentioned before, the optimum grain configuration comprises a star follows a circular perforation, because erosive burning is expected to be most significant at the beginning of the burn. There is a balance between the enhanced burning rate due to erosive burning and increased burning area due to star shape of the grain at the head end.

In order to analyze the stability of the solid rocket motor, linear contributions of combustion process, flow turning, and nozzle damping were considered. Since pressure amplitudes are function of mean chamber pressure, overall growth constant have the general form

$\alpha_T = \alpha_C + \alpha_F + \alpha_N$

The amplitude of the oscillations increases with time when $\alpha_T > 0$ and therefore is unstable. If $\alpha_T < 0$, the amplitude decreases with time and the oscillations are stable. Assuming linear combustion instability and one dimensional flow in the combustor, it is possible to write the growth rates of combustion α_C, flow turning α_F, and nozzle α_N as follows

$$\alpha_c = \frac{\bar{a}}{2}\bar{M}_b . R_b^{(r)} \left[\frac{\iint \psi_n^2 dS_b}{\int_0^L \psi_n^2 S_c dz}\right] \quad (14)$$

$$\alpha_F = \frac{\bar{M}_b}{2E_1^2} . \iint_{S_b} [\cos(2k_1 z)] dS_b \quad (15)$$

$$\alpha_{Nozzle} \equiv \alpha_N = -\frac{1}{2E_1^2}\left[\left(\gamma\frac{\hat{u}^{(r)}}{\hat{\eta}_1}\psi_1 + \bar{u}\psi_1^2\right)S_c\right]_{z=L} \approx -\frac{\bar{a}}{2}\left(\frac{\gamma+1}{2}\right)\bar{M}_n\left[\frac{\iint \psi_n^2 dS_n}{\int_0^L \psi_n^2 S_c dz}\right] \quad (16)$$

Where $E_1^2 = \int_0^L \psi_1^2 S_c dz = S_c \frac{L}{2}$ and $\psi_l^2 = \cos^2(k_l L)_{z=L} = 1$, $k_l = l\pi / L$, $\bar{M}_n$ =Mach number flow at the nozzle entrance, $\bar{M}_b$= the Mach number of flow at propellant surface, S_n = Nozzle inlet area, S_b= burning surface area, S_c= cross sectional area, $R_b^{(r)}$= real part of propellant response function, and $\bar{a}$ is the speed of sound.

The common response function that was used is Denison and

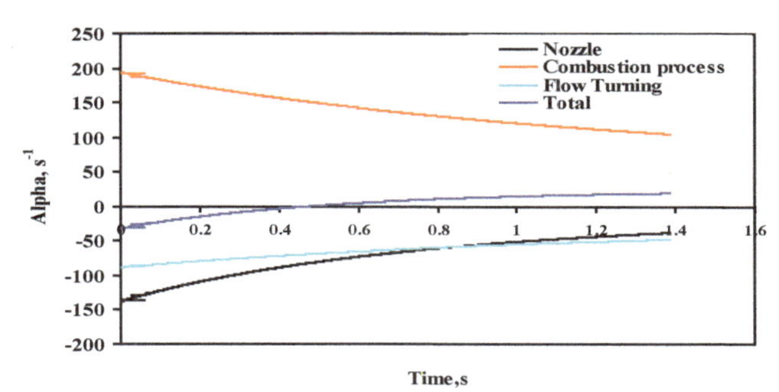

Figure 8: Predicted stability boundary for the solid propellant rocket motor.

Baum model based on flame model with zero heat flux absorption in depth of solid:

$$R_b = \frac{nAB + n_s(\lambda - 1)}{\lambda + \frac{A}{\lambda} - (1+A) + AB - \frac{Q_r A(\lambda - 1)}{\lambda(\beta + \lambda - 1)}} \quad (17)$$

For double base propellants Arrhenius surface pyrolysis law (which carries with it the condition n_s=0) can be applied.

With the assumption of negligible radiant flux, it follows that $Q_r \rightarrow 0$, response function of the double base propellant under consideration can be obtained with λ that satisfies the equation

$$\lambda(\lambda-1) = i\Omega \quad (18)$$

$$\Omega = \frac{k_p \rho_s \omega}{\bar{m}^2 C} \quad (19)$$

Where $\bar{m}$ is the average mass flux of combustion gases. The solution of the eq. (17) with real part λ_r and imaginary part λ_i are

$$\lambda_r = \frac{1}{2}\left\{1 + \left[\frac{1}{\sqrt{2}}\right]\left[\left(1+16\Omega^2\right)^{1/2} + 1\right]^{1/2}\right\} \quad (20)$$

$$\lambda_i = \left[\frac{1}{2\sqrt{2}}\right]\left[\left(1+16\Omega^2\right)^{1/2} - 1\right]^{1/2} \quad (21)$$

Assuming linear gas dynamic flow, the stability parameters of the system including combustion process, flow turning, and nozzle were obtained from eqs. 14-16. Figure 8 shows the growth rates of combustion, flow turning, nozzle, and overall growth rate of the system. The contributions of velocity coupling and structural damping were not considered. As shown in Figure 8, because the sum of stability elements is partly negative-partly positive during the burning, this motor became unstable after 0.5 s.

In order to remedy this problem the utilized capabilities are particle damping, adding nozzle cavity, changing the grain configuration, chemical modification of solid propellant which change its response to the pressure variation at various frequencies of the chamber, and active control strategies.

Conclusion

The linear combustion instability for a double base solid propellant has been investigated. The solid rocket motor uses non-metalized double base propellant. The combination of the Lax-Wendroff scheme and the artificial viscosity scheme were implemented to obtain the transient combustion of the solid rocket propellant. Also, the numerical analysis on the characteristics of the interior ballistics has been conducted. The pressure, thrust, specific impulse, and cross flow velocity were obtained for the solid propellant rocket motor. There was a good agreement between the calculated pressure and thrust with simulated ones. Linear contributions of combustion, flow turning, and nozzle were accounted for combustion instability analysis. As a result, the combustion instability was observed after 0.5 s in the solid rocket motor. The applicability of the current work is directed to help motor designers for preventing or eliminating such instability problems in solid rocket motors.

References

1. Avalon G, Ugurtas B, Grisch F, Bresson A (2000) Numerical computations and visualization tests of the flow inside a cold gas simulation with characterization of a parietal vortex shedding. AIAA Paper 2000-3387.
2. Blomshield FS (2001) Historical perspective of combustion instability in motors: case studies. AIAA Paper 2001-3875.
3. Flandro GA, Jacobs HR (1973) Vortex generated sound in cavities. AIAA Paper 1973-1014.
4. Fabignon Y, Dupays J, Avalon G, Vuillot F, Lupoglazoff N, et al. (2003) Instabilities and pressure oscillations in solid rocket motors. Aerospace & Technol 7: 191-200.
5. Flandro GA (1995) On flow turning. AIAA J.
6. Kubota N (2002) Propellants and explosives: Thermodynamical aspects of combustion. Wiley-VCH.
7. Cai W, Ma F, Yang V (2003) Two-Phase Vorticoacoustic Flow Interactions in Solid-Propellant Rocket Motors. J Propulsion Power 1: 19-28.
8. Thomas HD, Flandro GA, Flanagan SN (1995) Effects of Vorticity on Particle Damping. AIAA Paper 1995-2736.
9. Temkin S (2000) Attenuation and dispersion of sound in dilute suspensions of spherical particles J Acoust Soc Am 108: 126-146.
10. Golafshani M, Farshchi M, Ghassemi H (2002) Effects of Grain Geometry on Pulse-Triggered Combustion Instability in Rocket Motors. J propulsion power 18: 123-130.
11. Tseng IS, Yang V (1994) Combustion of a Double-Base Homogeneous Propellant in a Rocket Motor. Combustion & flame 96: 325-342.
12. Terzic J, Zecevic B, Serdarevic-kadic S, Catovic A, Baskarad M (2012) Research of influence of nozzle geometry on internal ballistics performances of solid propellant rocket motors using numerical simulations. Proceedings of 15th Seminar "New Trends in Research of Energetic Materials", Part II, Czech Republic.
13. Terzic J, Zecevic B, Baskarad M, Catovic A, Serdarevic-kadic S (2010) Prediction of internal ballistic parameters of solid propellant rocket motors. Proceedings of VIIIth international armament conference, Scientific aspects of armament and safety technology, Pultusk, Poland.
14. Lenoir JM, Robillard G (1957) A Mathematical Model to Predict Effects of Erosive Burning in Solid Propellant Rockets. Proceedings of the 6th International Symposium on Combustion, Reinhold, New York 6: 663-667.
15. Willcox MA, Brewster MQ, Tang KC, Stewart DS, Kuznetsov I, et al. (2007) Solid Rocket Motor Internal Ballistics Simulation Using Three-Dimensional Grain Burnback. J Propulsion & Power 23: 575-584.

Investigation of Oxygen Adsorption Property of Partially Reduced Tio_2 (Tio_2-X)

YoonKuen Chae[1], Shinsuke Mori[2], Masaaki Suzuki[2] and JinWon Park[1]*

[1]*Department of Chemical and Biomolecular Engineering, University of Yonsei, 262 Seongsanno, Seodaemun-gu, Seoul 120-749, Korea*
[2]*Department of Chemical Engineering, Tokyo Institute of Technology, 2-12-1 O-okayama, Meguro-ku, Tokyo, 152-855*

Abstract

The partial reduction of TiO_2 was investigated by hydrogen plasma treatment and it was performed to examine the oxygen adsorption property of partially reduced TiO_2 (TiO_{2-x}). In this experiment, TiO_{2-x} was analyzed by X-ray diffraction (XRD) and oxygen adsorption of partially reduced TiO_2 was measured by gas chromatography (GC). Consequently, it was shown that hydrogen plasma treatment can partially reduce TiO_2 by plasma effects and TiO_{2-x} can adsorb oxygen. The H_2O formation and the oxygen adsorption were found to be dependent on the heating temperature. The oxygen adsorption property was attributed to the formation of Ti_2O_3.

Keywords: Hydrogen plasma; Reduction; TiO_2; Oxygen adsorption; Ti_2O_3

Introduction

Plasma treatment was applied to the reduction of metallic oxide [1,2]. The reduction of metallic oxide has been examined using low pressure plasma containing hydrogen via radio-frequency, microwave, photochemical or electrochemical methods and heating [3-5]. Titanium dioxide (TiO_2), a representative metallic oxide, is especially attractive for environmental applications because of its chemical stability, non-toxicity and other advantageous properties. As a result, it has become by far the best-characterized oxide material. It is used in heterogeneous catalysis such as a photocatalyst in solar cells for the production of hydrogen and electric energy and as a gas sensor [6-11]. In addition, it is expected to remove a fetor through oxygen adsorption by the partially reduced TiO_2 in this study.

In heterogeneous catalysis, interaction of TiO_2 surface with oxygen (O_2) is necessary. The chemistry of oxygen is an important, but often overlooked, component of many chemical and photochemical processes that take place on TiO_2-based materials. Oxygen is widely used in photochemical oxidation studies on TiO_2 and is usually considered to act primarily as a scavanger of the photoexcited electrons, thus preventing negative charge accumulation on the surface of catalyst. In some solution-phase photochemical studies, the resulting superoxide radical anion (O_2^-) species are envisioned to participate in homogeneous-phase reactions with water molecules to form hydrogen peroxyl species that may play a role in organic oxidation, either in solution or on the surface. Only a few groups have proposed that surface-bound O_2 or O_2^- species may be directly involved in photooxidation of coadsorbed species. The step of O_2^- formation upon chemisorption of oxygen requires the presence of surface vacancy sites. No oxygen adsorption would occur on a stoichiometric surface. However, the exact nature of adsorbed O_2 on these TiO_2 surfaces is not well understood. Some studies have been carried out for the reduction of TiO_2 in oxygen vacancy sites obtained by high-temperature annealing in ultrahigh vacuum (UHV) [12]. TiO_2 decomposes at high temperatures and low oxygen pressures or at elevated temperatures in the presence of reducing gases, thereby forming suboxides TiO_x with $x < 2$. Molecular oxygen is typically used by UHV researchers to restore the stoichiometry of ionsputtered or vacuum-reduced TiO_2 surfaces.

This study aims to further examine the partial reduction of TiO_2 by hydrogen plasma treatment and to investigate the oxygen adsorption property of partially reduced.

Experimental

Figure 1 illustrates the experimental set-up for the partial reduction of TiO_2 and the oxygen adsorption. The plasma reactor was made of a quartz tube set in an electric furnace. A sample of anatase TiO_2 (JA-1, 0.18μm, Tayca) particles was set in the quartz boat and placed inside the plasma reactor. RF discharge plasma (13.56 MHz) was used to generate glow plasma discharge between copper electrodes attached to the outer surface of the quartz tube. The distance between the electrodes was defined as the electrode gap (50 mm). The reactor temperature, which

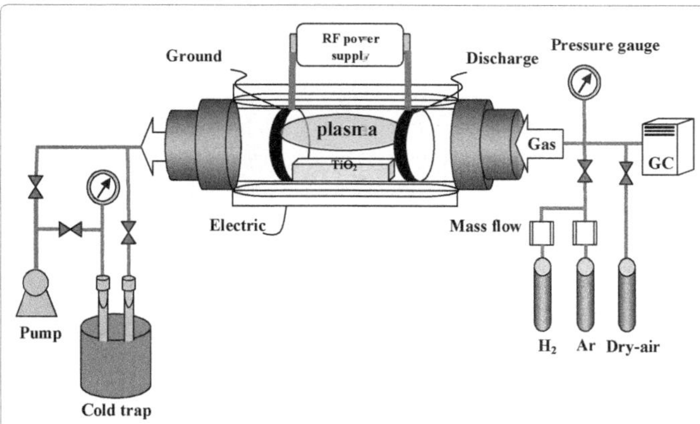

Figure 1: Experimental set-up for the partial reduction of TiO_2 and the oxygen adsorption.

***Corresponding author:** JinWon Park, Department of Chemical and Biomolecular Engineering, University of Yonsei, 262 Seongsanno, Seodaemun-gu, Seoul 120-749, Korea was monitored by thermocouples attached to the inside of the furnace, was changed by controlling the electric power of the furnace.

Table 1 shows the experimental conditions for the partial reduction of TiO_2. All experiments of this set were repeated twice to assess reproducibility. Heating temperature was an experimental variable in this experiment. The flow rate, treatment time, and reactor pressure were optimized through repeat experiments. The oxygen adsorption rate and the H_2O formation rate were derived from experimental results. On the first part of the experiment (section 3.1), pretreatment was carried out with argon (flow rate : 40 sccm) and a heating temperature of 500°C for 30 min. This was carried out in order to remove H_2O absorbed on the TiO_2 surface and to quantify the amount of H_2O formed. On the second part of the experiment (section 3.2), reduction was carried out introducing argon and hydrogen (flow rate: 20 sccm, respectively) into the reactor with a fixed pressure at 300 Pa. The RF discharge power delivered to the plasma was 90 W. The heating temperature ranged between 500°C to 1000°C for 40 min. The flow rates of Ar and H_2 were controlled with a mass flow controller (Model 3660, KOFLOC). The reactor pressure was measured at the upstream and downstream points of the reactor using pressure gauges, and the average pressure value was defined as the system pressure. In these experiments, the amount of H_2O produced and the formation rate of H_2O was measured. Water produced by reduction was trapped by a liquid-nitrogen cold trap set downstream of the reactor. The accumulated H_2O was measured every 40 min, and the average H_2O formation rate over a 40-min period was calculated.

After the reduction treatment, it was performed to investigate the oxygen adsorption property of partially reduced TiO_2. After evacuating the reactor with a rotary pump, dry-air (0.1-atm) gas was introduced in a batch type reactor and the oxygen adsorption was carried out at room temperature. The oxygen was adsorbed on the oxygen vacancy site of TiO_2 formed by reduction. The oxygen adsorbed was measured using a gas chromatograph (GC-8A, SHIMADZU, Japan). The presence of TiO_2 and other titanium oxides (e.g., Ti_2O_3) were observed with X-ray diffraction (XRD) and their intensities were analyzed quantitatively.

Results and Discussion

Partial reduction of TiO_2 by hydrogen plasma treatment

Hydrogen (H_2) can be dissociated to produce monatomic hydrogen and ionized species using a number of technologies. Monatomic and ionized species of hydrogen, as reducing agents, can be used for the reduction of metals [13,14]. The TiO_2 reduction mechanism investigated in this experiment presumes that monatomic hydrogen (H), ionized species (H_2^+, H^+) and electrons (e^-) produced by the hydrogen plasma treatment has an influence on the reduction of TiO_2. Partially reduced TiO_2 (TiO_{2-x}) and H_2O were produced by reacting diatomic hydrogen (H_2) and monatomic oxygen (O) from TiO_2. The H_2O formation rate was calculated from H_2O produced by reduction and the TiO_2 reduction rate was calculated by measuring the H_2O formation rate trapped by liquid-nitrogen. The reaction equation of the partially reduced TiO_2 is as follows:

Heating temp. (°C)	500 ~ 1000
TiO_2 sample (g)	0.8
Ar : H_2 gas(sccm)	20 : 20
Treatment time (min)	40
Reactor pressure (Pa)	300
Plasma input power (W)	90

Table 1: Experimental conditions for the partial reduction of TiO_2.

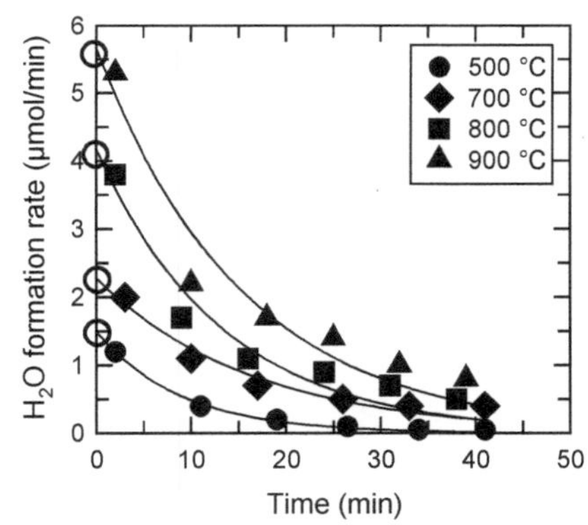

Figure 2: A plot of the H_2O formation rate as a function of the reaction time used to determine the initial H_2O formation rate. Open circles represent the initial H_2O formation rate. At reaction time zero, the initial H_2O formation rate was estimated by a first-order rate equation, $R=R_0 \exp(-kt)$.

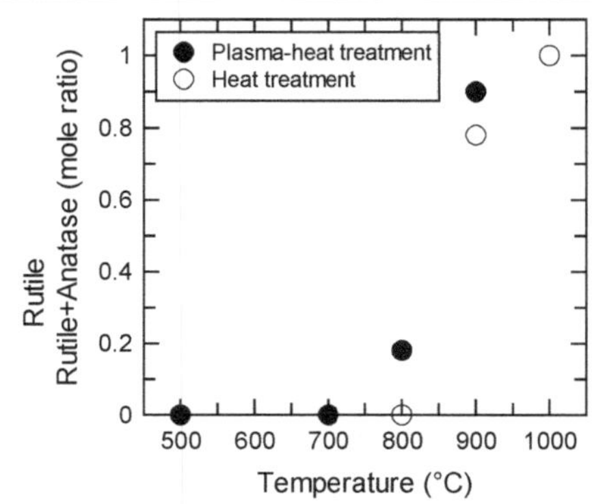

Figure 3: Rutile mole ratio of partially reduced TiO_2 as a function of the heating temperature between 500°C and 1000°C.

$$TiO_2 + \chi H_2 \rightarrow TiO_{2-x} + \chi H_2O \quad (1)$$

Figure 2 shows a plot of H_2O formation rate as a function of the reaction time which was used to determine the initial H_2O formation rate. This was estimated by varying the heating temperature from 500°C to 900°C. The H_2O formation was measured until no more H_2O was produced and it took approximately 40 min. At zero reaction time, the initial H_2O formation rate was estimated by a plot following the first-order rate equation. It is given as:

$$R=R_0 \exp(-kt) \quad (2)$$

where R is the H_2O formation rate (μmol min^{-1}), R_0 is the initial H_2O formation rate (μmol min^{-1}), t is the time (min), k is the rate constant (min^{-1}). R_0 can be determined by least-squares regression analysis of R versus t.

Figure 3 shows the rutile mole ratio of partially reduced TiO_2 as a function of the heating temperature between 500°C and 1000°C. At 800°C, the rutile mole ratio for plasma-heat treatment was higher than that for heat treatment. Phase transformation anatase to rutile was observed at 800°C for plasma-heat treatment and it was observed at 900°C for heat treatment. It has been reported that phase transformation anatase to rutile occurs at temperatures greater than 600°C [15]. This suggests that the effect of plasma treatment on the phase transformation is due to a mere temperature increase by heat generation of thermal reaction in plasma discharge

Figure 4 shows the H_2O formation rate as a function of the heating temperature between 500°C and 1000°C. This figure shows that the H_2O formation rate by plasma-heat treatment at 500°C was almost equal to that by heat treatment at 1000°C. In Figure 3, the rutile peak for heat treatment was observed at 1000°C while, no peak was observed at 500°C for plasma-heat treatment. From results of Figures 3 and 4, it is indicated that the plasma-heat treatment has almost the same as the H_2O formation rate comparing that of heat treatment at lower temperature (500°C) and, it has no rutile peak. These results suggest that the plasma-heat treatment at low temperature is almost equal to that by heat treatment at high temperature due to plasma effects such as ionized species (H_2^+, H^+) and electrons (e^-). This indicates that the increase of H_2O formed by the plasma-heat treatment was not only caused by thermal effects but also caused by plasma effects. In addition, the TiO_2 sample by heat treatment and plasma-heat treatment respectively revealed the color change of surface layer. Figure 5 shows the photograph of partially reduced TiO_2 sample by heat treatment and plasma-heat treatment. The TiO_2 sample by plasma-heat treatment revealed a dark navy surface layer after about 10 min of exposure to the plasma and the TiO_2 layer below the dark navy surface layer revealed a discoloration of gray to white. The TiO_2 sample by heat treatment revealed only a gray layer. The dark navy surface layer was observed only in TiO_2 samples treated by plasma. It is reported that the color of TiO_2 treated by H_2 plasma was changed white to navy blue due to oxygen vacancy [20]. This indicates that the color of TiO_2 layer by H_2 plasma treatment showed dark navy due to oxygen vacancy.

Oxygen adsorption properties of partially reduced TiO_2

Some recent research related to an isolated O vacancy has been dealt with the adsorption properties of oxygen [17,18]. The oxygen removal corresponds to a reduction of the surface. The oxygen adsorption is the reverse to the formation of defects on the TiO_2 surface. For defective TiO_2, a vacancy is created by removal of one bridging oxygen atom. Bridging oxygen atoms are more easily removed than other surface atoms because they are outward the surface plane and the least coordinated ones. Hence, Ti^{4+} surface atoms near the vacancy are formally reduced into Ti^{3+} (sixfold coordinated Ti_6 on the perfect surface becoming fivefold coordinated Ti_{5v} on the defective one). In this section, we investigate the oxygen adsorption property of partially reduced TiO_2 by hydrogen plasma treatment. This is carried out with

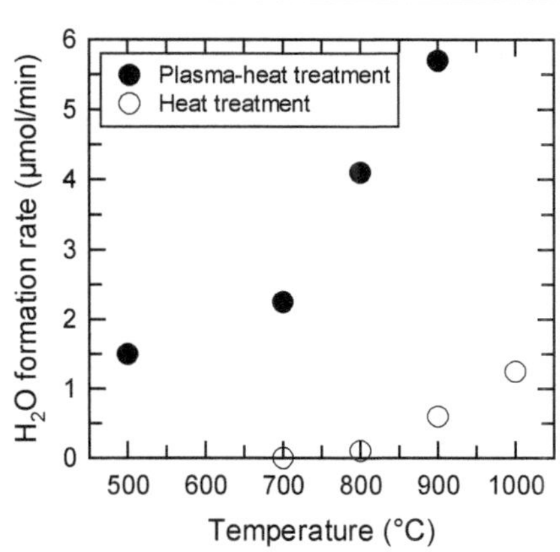

Figure 4: The H_2O formation rate as a function of the heating temperature between 500°C and 1000°C.

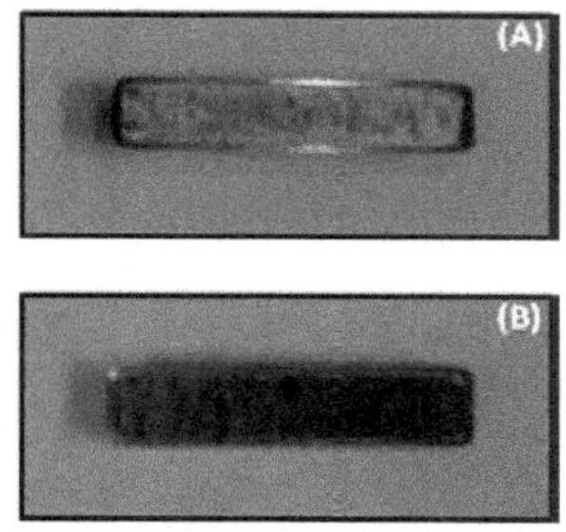

Figure 5: Photograph image of partially reduced TiO_2 sample by heat treatment (A) and plasma-heat treatment (B).

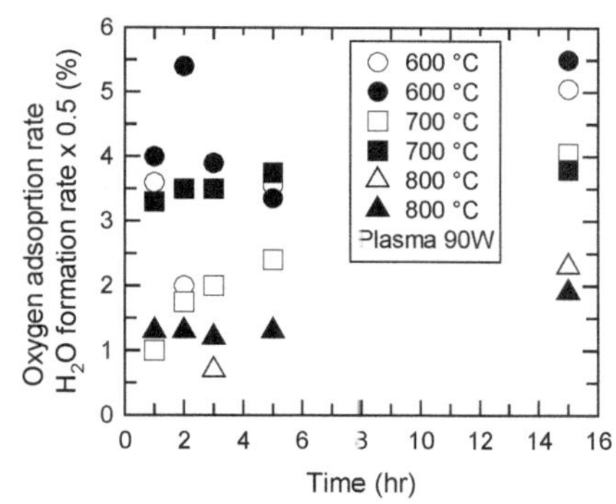

Figure 6: Ratio of the oxygen adsorption rate to the H_2O formation rate as a function of adsorption time. Plasma input power is 90 W at 600, 700, and 800°C.

plasma input power of 90 W at 600, 700, and 800°C and is repeated twice to assess reproducibility.

Figure 6 shows the ratio of the oxygen adsorption rate to the H_2O formation rate as a function of the adsorption time. As shown in Figure 6, an irregular pattern is observed during the initial oxygen adsorption which become stable as the adsorption time goes by. This result shows that oxygen adsorption to H_2O formation occurs more effectively at lower temperature. Table 2 shows the result data of the H_2O formation amount and the O_2 adsorption amount generated by plasma treatment at 600, 700, and 800°C. As can be seen in Table 2, with increase of the heating temperature increases, the amount of H_2O formation also increases and the amount of oxygen adsorption decreases. The amount of oxygen adsorption was small comparing with the amount of H_2O formed. It is due to the decrease of the specific surface by the increase of heating temperature. From results of Figure 6 and Table 2, it is indicated that the amount of oxygen adsorption is not proportional to the amount of H_2O formed by reduction and is dependent on the heating temperature

Figure 7 shows a plot of the relationship between the amount of H_2O formation and the Ti_2O_3 intensity. Figure 8 shows (included as supplementary data) X-ray Diffraction pattern of partially reduced TiO_2 by heat treatment and plasma-heat treatment at 700°C. The Ti_2O_3 peak was obtained from XRD analysis of partially reduced TiO_2 by plasma treatment. As shown in Figure 7, the Ti_2O_3 intensity increased with the increase of heating temperature. The Ti_2O_3 intensity was dependent on the heating temperature and was proportional to the amount of H_2O formed by reduction. In this study, the formation of H_2O by reduction implies that the formation of oxygen vacancy takes place in bulk TiO_2. In general, oxygen vacancy takes place by the lost of O ions that bridges Ti ions coordinated. The general type of surface vacancy of TiO_2 is formed by missing O ion that bridges two sixfold-coordinated Ti ions.

Temperature (°C)	○600	●600	□700	■700	△800	▲800
H_2O formation amount (μmol)	153.2	160.8	203.2	212	232	274.5
O_2 adsorption amount (μmol)	3.8	4.4	3.6	4	2.6	2.5

Table 2: The H_2O formation amount and the O_2 adsorption amount (Plasma input power is 90 W at 600, 700, and 800°C).

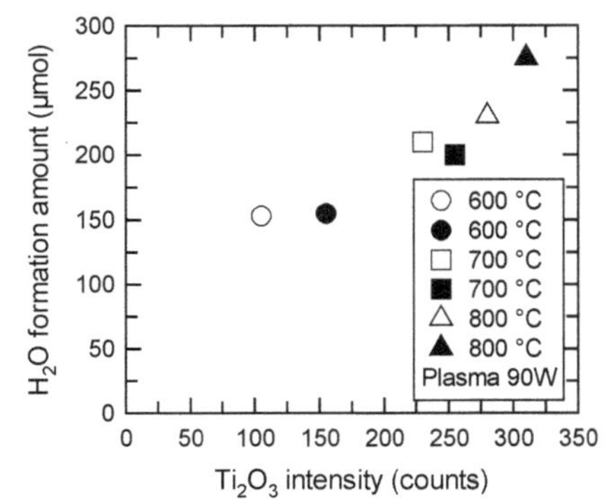

Figure 7: The relationship between H_2O formation amount and Ti_2O_3 intensity. Plasma input power is 90 W at 600, 700, and 800°C.

When a bridging O ion is removed by reduction, the two neighboring Ti ions become fivefold coordinated. For this vacancy, the two Ti ions are very poorly connected from each other, which are somewhat similar to the state in bulk Ti_2O_3 [19,20]. This is in accordance with reports that oxygen adsorption requires oxygen vacancy sites [21]. The amount of oxygen adsorption of partially reduced TiO_2 decreased with the increase of Ti_2O_3 intensity. It is indicated that the decrease of the oxygen adsorption amount is due to the formation of Ti_2O_3. It is deduced that the formation of Ti_2O_3 has an effect on oxygen adsorption because two fivefold-coordinated Ti ions is similar to the state in bulk Ti_2O_3. Unfortunately, the detail reason has not been clearly understood yet. The relation of Ti_2O_3 and oxygen adsorption is needed more research.

Conclusions

TiO_2 was partially reduced by hydrogen plasma treatment and the partially reduced TiO_2 showed oxygen adsorption property. The color of TiO_2 reduced by H_2 plasma treatment was changed white to dark navy due to oxygen vacancy. Oxygen adsorption of the partially reduced TiO_2 was occurred with oxygen vacancy. As the heating temperature increases, the H_2O formation amount and the Ti_2O_3 intensity increased and the oxygen adsorption amount decreased. From the relationship of H_2 formation and Ti_2O_3 intensity, the oxygen adsorption was dependent on the formation of Ti_2O_3. The reduction mechanism is perhaps not simple because of the relation between H_2O formation and oxygen adsorption. Further research is necessary to elucidate the reduction mechanism of TiO_2.

Acknowledgement

The authors are grateful to 'The 21st Century COE Program (Chemistry Field), Tokyo Institute of Technology'

References

1. Kim SS, Lee H, Na BK, Hyung KS (2004) Plasma-assisted reduction of supported metal catalyst using atmospheric dielectric-barrier discharge. Catal Today 89: 193-200.
2. Belmonte T, Thiébaut JM, Michel H, Cardoso RP, Maliska A (2002) Reduction of metallic oxides by late Ar–H_2–N_2 postdischarges. II. Applications to iron oxides. J Vac Sci and Technol A 20: 1353.
3. Aarthi T, Madras G (2008) Photocatalytic reduction of metals in presence of combustion synthesized nano-TiO_2. Catal Commun 9: 630-634.
4. Kase M, Moriya I, Yotsumoto T, Taoda H (2004) Japanese Patent No. 2004-137087.
5. Taoda H, Watawabe E, Nonami T, Fukuya M (2002) Japanese Patent No. 3288265.
6. Yoong LS, Chong FK, Dutta BK (2009) Development of copper-doped TiO_2 photocatalyst for hydrogen production under visible light. Energy 34: 1652-1661.
7. Changlin Yu, Jimmy Yu, Wanqin Z, Kai Y (2010) WO_3 Coupled P-TiO_2 Photocatalysts with Mesoporous Structure. Catalysis Letters 140: 172-183.
8. Bozorgtabar M, Rahimipour M, Salehi M (2010) Effect of thermal spray processes on anatase-rutile phase transformation in nano-structured TiO_2 photo-catalyst coatings. Surface Engineering 26: 422-427.
9. Lefèvre G, Fédoroff M, Čerović LJ (2008) Particles adhesion onto massive substrates: experimental measurement and deposition processes. Surface Engineering 24: 272-275.
10. Moon J, Park JA, Lee SJ, Zyung T, Kim ID (2010) Pd-doped TiO_2 nanofiber networks for gas sensor applications. Sensors and Actuators B: Chemical 149: 301-305.
11. Park H, Kim WR, Jeong HT, Lee JJ, Kim HG, et al. (2011) Fabrication of dye-sensitized solar cells by transplanting highly ordered TiO_2 nanotube arrays. Solar Energy Materials and Solar Cells 95: 184-189
12. Michael AH, William SE, Craig LP, Charles HF (1999) Interaction of Molecular Oxygen with the Vacuum-Annealed TiO_2(110) Surface: Molecular and Dissociative Channels. J Phys Chem B 103: 5328-5337.
13. Izumi A, Ueno T, Miyazaki Y, Oizumi H, Nishiyama I (2008) Reduction of oxide layer on various metal surfaces by atomic hydrogen treatment. Thin Solid Films 516: 853-855.
14. Belousov VM, Vasylyev MA, Lyashenko LV, Vilkova NY, Nieuwenhuys BE (2003) The low-temperature reduction of Pd-doped transition metal oxide surfaces with hydrogen. Chem Eng J 91: 143-150.
15. Diebold U (2003) The surface science of titanium dioxide. Surf Sci Rep 48: 53.
16. Chae YK, Mori S, Suzuki M (2009) Visible-light photocatalytic activity of anatase TiO_2 treated with argon plasma. Thin Solid Film 517: 4260-4263
17. Ménétrey M, Markovits A, Minot C (2007) Adsorption of chlorine and oxygen atoms on clean and defective rutile–TiO_2 (1 1 0) and MgO (1 0 0) surfaces. J Mol Struct (THEOCHEM) 808: 71-79.
18. Bredow T, Pacchioni G (2002) Electronic structure of an isolated oxygen vacancy at the TiO_2(1 1 0) surface. Chem Phys Lett 355: 417-423.
19. Göpel W, Rocker G, Freierabend R (1983) Intrinsic defects of TiO_2(110): Interaction with chemisorbed O_2, H_2, CO, and CO_2. Phys Rev B 28: 3427.
20. McKAY JM, Henrich VE (1984) Preparation and properties of the Ti_2O_3(047) surface. Surf Sci 137: 463-472
21. Uner D, Tapan NA, Özen İ, Üner M (2003) Oxygen adsorption on Pt/TiO_2 catalysts. Appl Catal A Gen 251: 225-234.

Tomato Dehydration in a Hybrid-Solar Dryer

Reyes A*, Mahn A, Huenulaf P and González T

Department of Chemical Engineering, University of Santiago of Chile, Av. L. B. O"Higgins 3363, Santiago, Chile

Abstract

Tomato pieces were dehydrated in a hybrid solar dryer provided with a 3 m^2 solar panel and electric resistances. At the outlet of the tray dryer 80 or 90% of the air was recycled and the air temperature was adjusted 50 or 60°C. At the outlet of the solar panel the air temperature raised between 5 and 18°C above the ambient temperature.

Temperature and sample size significantly affected critical moisture content. The color parameters of dehydrated tomato indicate a notorious redness. Rehydration was achieved in less than 50 minutes.

Drying kinetics was adequately adjusted with three empirical models. Sorption isotherms acjusted soundly well with the Guggenheim-Anderson-de Boer and Peleg models. The solar energy input resulted in 6.6-12.5% energy saving.

Keywords: Drying; Solar energy; Hybrid solar dryer; Tomato; Water activity

Nomenclature

a: Activity

a*: Colour parameter

Ap: Solar panel area m^2

b: Parameter s/m

b*: Colour parameter

c: Dimensionless parameter

C: Peleg parameters

E: Colour parameter

K: Page parameter

I: Solar radiation W/m^2

L*: Colour parameter

Mo: Initial moisture content on dry basis Kg/kg

Mw: Moisture content on dry basis Kg/kg

n: Page parameter

N: Number of experimental data

Qe: Electric energy input Joule

Qsp: Absorbed solar energy Joule

R: Air recycle %

RH: Relative air humidity

RMSE: Root mean squared error

T: Drying air temperature °C, K

t: Time:minutes

v: Air velocity, inside the chamber m/s

X: Average moisture content of tomato particles water kg/kg wb

Xo: Initial moisture content of tomato particles water kg/kg wb

Xeq: Equilibrium moisture content of tomato water kg/kg wb

Greek Letters

a: Tilt of the solar panel (a=11°)

Δ: Difference between two values

Sub-indexes

amb: ambient

c: critical

o: initial

sp: solar panel outlet

eq: equilibrium

w: water

Introduction

Drying is one of the most energy-intensive processes in the food industry [1-3]. For this reason, the partial (or total) replacement of traditional fuels by renewable sources of energy in drying processes will result in considerable energy savings. Additionally, solar energy is an attractive non-polluting alternative [1,4,5]. However, the daily and seasonal fluctuations in the irradiation level are some drawbacks that make it necessary to use additional energy sources that permit the operation during the low irradiation periods.

Chile is located in the southwest hemisphere, at the west from Greenwich meridian, between 17° 30' and 90° south latitude, with marked irradiation differences along the country. In this sense, the Metropolitan region, where the capital Santiago is found, exhibits an irradiation level similar to that of the central zone of Chile. In addition, energy consumption projections in Chile indicate a serious energy

***Corresponding author:** Reyes A, Department of Chemical Engineering, University of Santiago of Chile, Av. L. B. O"Higgins 3363, Santiago, Chile deficit for the next decade, and therefore solar energy constitutes an interesting alternative.

The drying process involves simultaneous: (i) heat transfer from the surrounding to the surface of the food particle being dried followed by heat conduction within the material; and (ii) mass transfer from inside the food to its surface, followed by external transport of moisture to the surroundings. The heat is transferred to the surface of food particle by conduction and convection from adjacent air at temperature above that of the material being dried. The drying air absorbs moisture from the solid only if its relative humidity is below saturation [6].

In indirect solar drying, the foodstuff is kept inside a drying chamber, in which the pre-heated air from the flat plate collectors enters. In order to allow using the solar dryer in low or null irradiation periods, hybrid solar dryers that consider the use of additional energy sources (biomass, hydrocarbons or electricity) have been developed [7-10].

Flat plate collectors consist of: (1) a dark flat-plate absorber of solar energy, (2) a transparent cover that allows solar energy to pass through, (3) a space through which drying air flows, and (4) a heat insulation lower base [2,11,12]. The use of roughness elements or porous matrixes acting as absorber has been used as an alternative to improve heat transfer.

Several experimental works related with fruit and vegetables dehydration in hybrid solar dryers that use electric or biomass supplementary energy have been reported [6,9,13-25].

There are phenomenological models based on the second Fick's law, which use different approaches, to describe the drying process [26]. Besides, there are several empirical models which are useful when the solid has a heterogeneous composition (skin, seeds, pomace, etc.), geometry is difficult to characterize and/or changes dramatically during the drying process. The main similarity of most empirical models is the exponential term. Among these models are found the Page model [27], given by Eq. 1, the modified Page model [28], given by Eq.2, and the Henderson and Pabis model [29], given by Eq. 3.

$$\frac{X(t)}{X_0} = exp(-k \cdot t^n) \tag{1}$$

$$\frac{X(t)}{X_0} = exp(-k \cdot t)^n \tag{2}$$

$$\frac{X(t)}{X_0} = aexp(-k \cdot t) \tag{3}$$

Water activity (i.e. the ratio of the vapor pressure in a food material to that for pure water at the same temperature) is related with the equilibrium moisture content through a water sorption isotherm, at constant temperature. The water sorption isotherms of most foods are nonlinear, exhibiting a sigmoid shape. Isotherms are obtained by exposing a small sample of food to different constant relative humidity atmospheres. Once reached equilibrium, the water content of the sample is determined gravimetrically. In many cases, the process of sorbing or desorbing water may change the solid matrix and consequently affect the sorption behavior, thus triggering the hysteresis phenomena [30]. One of the most relevant equations used to represent sorption data is the Guggenheim-Anderson-de Boer (GAB), given by Eq. 4, which is reported to be the best for fitting sorption isotherm data for the majority of food products up to a_w levels of approximately 0.9 [31].

$$M_w = \frac{M_o \cdot C \cdot K \cdot a_w}{(1-K \cdot a_w)(1-K \cdot a_w + C \cdot K \cdot a_w)} \tag{4}$$

Another one, the Peleg model (1993) [32], has four parameters and is given in Equation. 5:

$$M_w = C_1 \cdot a_w^{C_3} + C_2 \cdot a_w^{C_4} \tag{5}$$

where C_1, C_2, C_3 and C_4 are constants.

Tomato (*Lycopersicom esculentum*) is the world's most commercially produced vegetable [33]. It is a rich source of minerals, vitamins, organic acid, and dietary fiber [34], and therefore its consumption has been intensified in different presentations: fresh, juice, puree, canned, and dressings. Tomato has a high content of lycopene, ascorbic acid and flavonoids, which offer important healthy effects [33-35].

Currently more than 20 000 tomato cultivars are known worldwide, which differ in size, color, shape, flavor and chemical composition.

Considering that tomato production is markedly seasonal, a preservation alternative is dehydration, in order to increase its availability. Dried tomatoes have become a highly attractive product and their demand has increased significantly during the last years [35-37].

The purpose of the present work was to analyze the drying of tomatoes (plum tomato) using a hybrid solar dryer, the determination of water activity and some quality parameters.

Material and Methods

Equipment

The hybrid solar dryer (HSD) consists of a 3 m^2 (3 m length and 1 m width) solar panel composed by a glass sheet cover (5 mm thickness) and a black wavy zinc plate. Below this zinc plate a thermal insulating material was inserted (50 mm thickness). The air passes through the free space (30 to 50 mm height) between the glass and the zinc plate until reaching the mixing point with the recycled air. This air mixture enters the electric heating system composed by 5 kW electric resistances where the air temperature is adjusted to the desired level (50 or 60°C). After that, the air enters the drying chamber (0.5 m*0.5m *1.2 m), where it distributes to pass over 10 perforated plate trays made of stainless steel (0.45m*0.5m), located in two sections of 5 trays each one.

The air is sucked through a 3 kW blower and directed to the electric resistances zone, although a 10-20 % fraction is removed from the system through valve V_2.

The HSD is provided with a data acquisition system OPTO 22 (model SNAP-PAC-R1), which allows registering air temperature at the outlet of the solar panel (Ts), temperature and relative humidity in the drying chamber, just before the inlet of the drying chamber (T_1 and RH_1) [Hanna Instruments, HI 8666, RH/°C s] and temperature and relative humidity in the drying chamber immediately before the outlet (T_2 and RH_2). Air flow rate was determined by the pressure drop registered through two Pitot tubes located at the outlet of the centrifugal fan and the outlet of valve V_2. The solar radiation was determined by a Pyranometer (LI-200SA) located at one side of the solar panel. This information was stored in a PC. The solar panel was located on the roof of a four-floor building corresponding to Unit Operation Laboratory of the Department of Chemical Engineering. The drying chamber and the control system were installed in the 4th floor, just below the solar panel. Air passage valves opening was controlled from a PC.

Materials

Drying runs were performed using plum tomatoes, purchased at the local market, characterized by an oval shape, intense red color. The tomatoes were manually chosen, having average dimensions of 4.0 ± 0.3 cm diameter and 9.0 ± 0.3 cm length, measured with a Vernier calipter. After that, tomatoes were thoroughly rinsed. Then, tomatoes were cut longitudinally in quarters or eighths. This was necessary because the tomatoes skin represents a barrier to moisture removal, and a higher exposure of the tomatoes flesh improves water diffusion, thus shortening the drying process. Additionally, substrate size must take into account the expectations of consumers.

Experimental designs

Selection of the experimental factors considered initial tomato size (1/4 and 1/8) because it affects the diffusion path of moisture; air flow recycle (80 and 90%) since this implies minimizing energy supply to heat the drying air, and drying air temperature (50 and 60°C). The temperature levels were chosen based on the fact that temperatures lower than 50°C result in an excessively slow drying process, and temperatures above 60°C may impair the product quality. A factorial experimental design, with the above-mentioned factors in two levels (2^3), was used to analyze the drying process. Table 1 shows the experimental design matrix. Runs were carried out in a random order.

Statistical analyses

The experimental data obtained from the execution of the experimental design (Table 1) were analyzed through ANOVA (at 95% confidence), and the statistical effect of the experimental factors on moisture content were calculated at different points during the drying process, and their significance was determined. A statistical effect represents the magnitude of the change in the response variable when a factor passes from the low level to the high level. The analyses were performed with Statgraphics Centurion.

Run	Air Recycle (%)	Temperature [°C]	Fraction pieces
T1	90	50	1/8
T2	80	60	1/8
T3	80	50	1/4
T4	90	60	1/4
T5	90	60	1/8
T6	90	50	1/4
T7	80	50	1/8
T8	80	60	1/4

Table 1: Experimental design matrix.

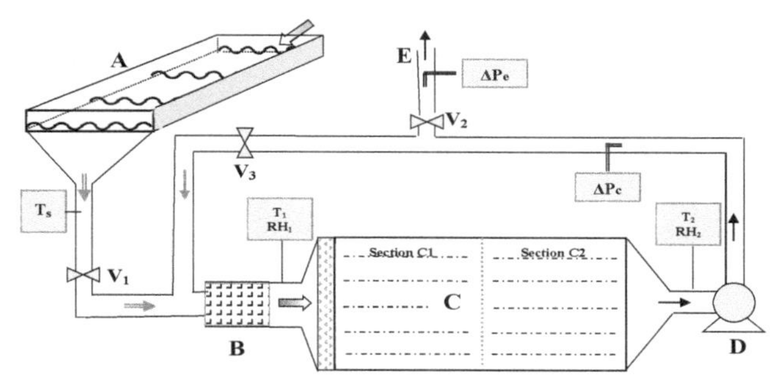

Figure 1: Hybrid-solar dryer. A. Solar panel; B. Electrical heater; C. Chamber drying; D. Centrifugal fan; E. Exit air. V_i. Valves.

Drying procedure

About 6.5 kg of tomato pieces were put in the dryer, distributing 0.650 kg in each tray (0.48 m * 0.58 m). Additionally 0.050 kg was loaded in two wire baskets, which were located in section 1 and 2, respectively. To obtain the drying curves, samples of the solids were removed from two sectors of the dryer, weighed on a digital balance with a sensitivity of 0.01 g (Boeco B3L62, Boecktel Co., Germany) and then put back in the dryer, and removed again every 30 minutes. The moisture content for each point of the drying curve was determined by calculating the difference in weight between the wet state and the dry weight after drying exhaustively at 85°C.

Water activity measurement

Relative moisture content, necessary to obtain the sorption isotherms, was determined using Thermoconstanter Humidat TH-2 of Novasina. First, the equipment was calibrated using standardized salt solutions, as suggested by the manufacturer. The desorption isotherms were determined using dehydrated tomato samples in a laboratory tunnel dryer at 70°C and air flow rate of 2 m/s. During the dehydration process, samples were withdrawn at different drying times, and consequently showing different moisture content. Moisture content was determined in a vacuum oven until constant weight according to AOAC 920.151 (AOAC, 1990). The mean of three measurements (within ± 0.01 g) was reported. Approximately 5.0 g of partially dehydrated samples were placed into Petri dishes and then into Novasina Instrument. Isotherms were determined at 15, 25 and 40°C.

Colour procedure

The colour of the samples was measured with a colorimeter (Minolta Chromo Meter CR 200b) using the L^* a^* b^* colour scale, which is an international standard for color measurements, adopted by the Commission Internationale d'Eclairage (CIE). The instrument was calibrated with white and yellow plates. Surface measurements were made under a CIE D_{65} illuminant which simulates diffused daylight. Colour measurements of tomatoes were made on a Petri dish over a white plate and replicated ten times after mixing the dried samples, and the average values of L^*, a^* and b^* were reported. Mixing of samples was done primarily to account for colour variation among dehydrated tomatoes particles.

Colour variation between fresh (denoted by the subscript 'o') and dehydrated tomatoes, was used to describe the total colour change ΔE after drying by means of Eq. 6 (http://www.cie.co.at/).

$$\Delta E = \sqrt{\left(L_0^* - L^*\right)^2 + \left(a_0^* - a^*\right)^2 + \left(b_0^* - b^*\right)^2} \qquad (6)$$

Rehydration tests

The water adsorption capacity of the dried particles was determined in a water thermostatic bath at 35 °C. 10 g of dried material were put in a wire basket and were immersed in a water bath. The basket was removed from the bath at different time intervals, the surface water was removed with an adsorbent paper and the total mass was determined. Then the basket was immersed in the bath again. The first five measurements were taken every two minutes, and after that, when the variation in mass was too low, a sample was taken every 5 minutes until 60 minutes. All measurements were made in triplicate (Figure 1).

Results and Discussion

Data acquisition

Ambient temperature, air temperature in the outlet of the solar

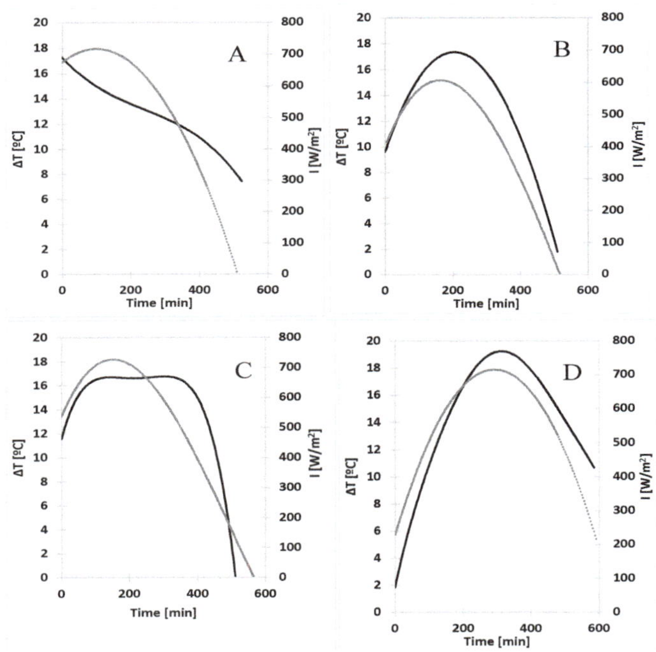

Figure 2: Solar radiation and ΔT (Solar panel temperature - Ambient temperature). A) Run T1, B) Run T2, C) Run T4 and D) Run T8. (-----Solar Radiation, _____ ΔT).

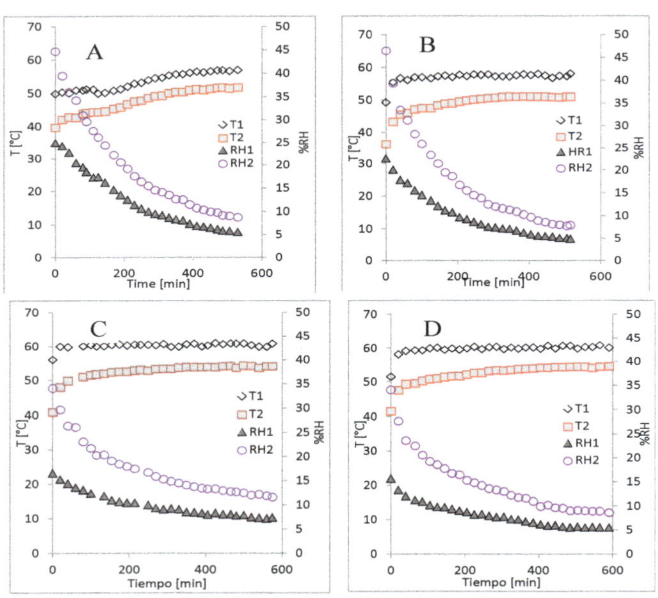

Figure 3: Temperature and relative humidity at the inlet and outlet of the drying chamber. (A) Run T1, (B) Run T2, (C) Run T4, and (D) Run T8.

panel (T_{sp}), and solar radiation were registered through the data acquisition system. This is shown in Figure 2 for runs T1, T2, T4, and T8. The ambient temperature fluctuated between 20°C and 31°C, depending on the day and time of the measurement. The solar panel allowed increasing the air temperature (T_{sp}) up to 18°C above the ambient temperature, depending on the air flow rate and mainly of the solar radiation. This temperature increase allows reducing electric energy consumption. The difference to achieve the pre-defined inlet air temperature (T_1) was obtained from the electric resistances.

Figure 2 shows that during the complete drying period, at least during 360 minutes the radiation level remained above 200 W/m², with maximum values higher than 600 W/m² for a period of 100-200 minutes period.

Air temperature (T_1 and T_2) and relative air humidity (RH_1 and RH_2) at the inlet and outlet of the drying chamber were also registered, which are presented for some runs in Figure 3.

Figure 3 shows that the air temperature at the inlet of the drying chamber kept at the pre-fixed value (50 or 60°C) with variations of ± 2°C. An exception was observed in run T1 (Figure 3A), where form the 200th minute on wards the temperature raised up to 57°C, despite the electric resistances were turned off, due to the specially high solar radiation during that day.

Figure 3 also shows that in the first 120-200 minutes the air temperature diminishes when passing through the drying chamber in about 12°C. By the end of the drying process this difference decreases to 6 °C, due to the lower remaining moisture content of the solids. The air that enters the drying chamber comes from the solar panel (fresh air) (10 or 20% of total air flow) and from the outlet of the dryer itself (recycled air) (80-90 % total air flow). For this reason the humidity of the air that enters the drying chamber varies during the drying process: ~30% humidity at the beginning of the process, and diminishing until ~10% by the end of the process. On the other hand, the relative humidity of the air at the outlet of the dryer, at the beginning of the process is in the order of 60 ± 5%, diminishing to 20% by the end of the process, as a consequence of the lower moisture content removed in the final drying period.

Experimental drying curves

Figure 4 shows the reduced drying curves (dimensionless moisture content v/s time), executed in duplicate, corresponding to runs carried out at operating conditions given in Table 1. The drying experiences were stopped when X/X0 reached 0.1 or after 600 minutes of drying because after that time the effect of solar radiation was negligible, independently from the final moisture content of the final product. In Figure 4, X_o represents the average initial moisture content (X_o=15.667 dry base). Overall the results showed deviations lower than 2 %. The drying curves exhibited the expected behavior: a linear decrease of moisture content in the initial period, until reaching the critical moisture content, where the non-bound moisture was removed; and after that the bound moisture was removed following an exponential trend in terms of time. Figure 4 shows that drying air temperature

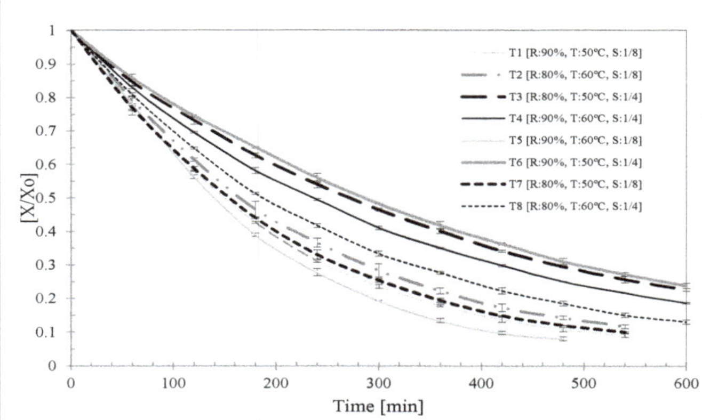

Figure 4: Drying kinetics of tomatoes (R: recycle, T: temperature, S: size).

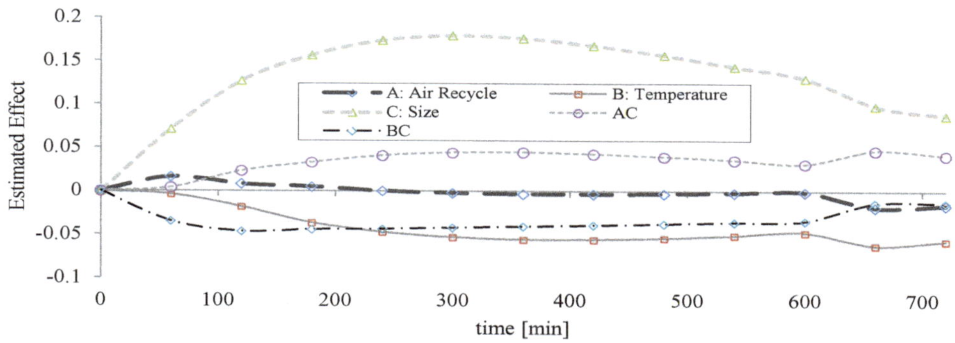

Figure 5: Evolution of significant effects on moisture content, at 95% confidence level.

	Page			Modified Page			Henderson and Pabis		
	k	n	R^2	k	n	R^2	k	n	R^2
T1	0.0033	1.1579	0.9993	0.0046	1.0089	0.9961	0.0047	1.0121	0.9965
T2	0.0028	1.1086	0.9974	0.0039	1.0667	0.9988	0.0042	1.0051	0.9989
T3	0.0035	1.1715	0.9986	0.0026	0.9726	0.9986	0.0025	0.9926	0.9988
T4	0.0047	1.1414	0.9962	0.0029	0.9860	0.9989	0.0029	0.9903	0.9992
T5	0.0014	1.2211	0.9986	0.0062	0.9184	0.9940	0.0053	1.0273	0.9954
T6	0.0017	1.1299	0.9987	0.0024	0.9897	0.9989	0.0024	0.9973	0.9991
T7	0.0023	1.1795	0.9996	0.0050	0.9221	0.9991	0.0045	1.0026	0.9991
T8	0.0056	1.1039	0.9976	0.0036	0.9756	0.9965	0.0036	0.9966	0.9992

Table 2: Page, modified Page and Henderson and Pabis models parameters with the corresponding R^2 values.

Run	X_c/X_o Block 1	X_c/X_o Block 2
T1	0.45	0.43
T2	0.43	0.48
T3	0.50	0.53
T4	0.59	0.55
T5	0.40	0.41
T6	0.54	0.51
T7	0.41	0.40
T8	0.59	0.60

Table 3: Values of X_c/X_o for tomatoes.

and sample size exhibited the expected behavior: higher temperatures resulted in faster drying kinetics and when diminishing the sample size, faster drying rates was observed, accounting for the reduction of the diffusional path. Air recycle had no significant effect probably due to its low variation imposed.

Figure 5 shows the evolution in time of the effects of the operational variables on moisture content during the drying process. From the 120th minute onwards a clear effect of sample size was observed, while air temperature begins to be statistically significant from the 200th minute onwards, agreeing with the end of the first drying stage. Air recycle had no statistical effect, but its interaction with sample size does.

Adjustment of experimental drying kinetics

Given that tomato particles showed an irregular shape, drying curves were adjusted only with the empirical models of Page [27], given by Eq. 1, the modified Page [28], given by Eq.2, and Henderson and Pabis [29], given by Eq. 3. This adjustment was performed by the Solver Tool from Microsoft Excel 2003 software, minimizing the root mean square error between the experimental values of moisture content of the tomatoes particles versus time and those calculated for the models. As shown in Table 2, the resulting R^2 values for three models were higher than 0.99, being slightly better the adjustments of Henderson and Pabis model.

Critical moisture content

In the plots of dX/dt versus X (data not shown), it was not possible to clearly visualize the change from the constant rate period to the falling rate period. For this reason, the critical moisture content (shown in Table 3) was obtained from the slope change in moisture content versus time plots. The procedure consisted in adjusting a straight line to the three initial points of each drying curve. Then, the line was extrapolated to the complete drying curve. The experimental points were compared with the corresponding points of the straight line, and the standard deviation was calculated. When standard deviation exceeded 0.035, we considered that at that point the straight line did not represent the constant rate period. Accordingly, this point corresponds to the critical moisture content.

Drying time decreased proportionally to X_c, since higher values of X_c result in shorter constant drying rate periods (highest drying rate). For this reason, and considering as unique criterion the minimization of the drying period, it is recommendable selecting operating conditions that reduce the critical moisture content. However, in the final decision

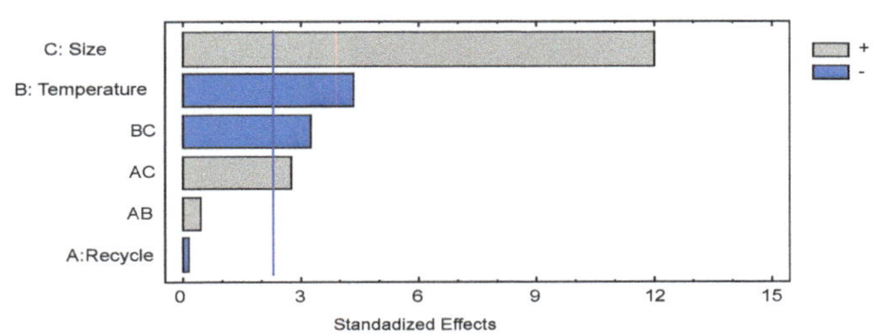

Figure 6: Pareto chart.

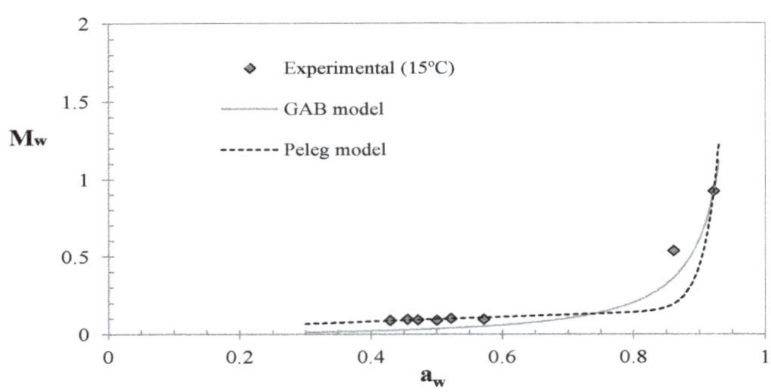

Figure 7: Desorption isotherm of tomatoes at 15°C.

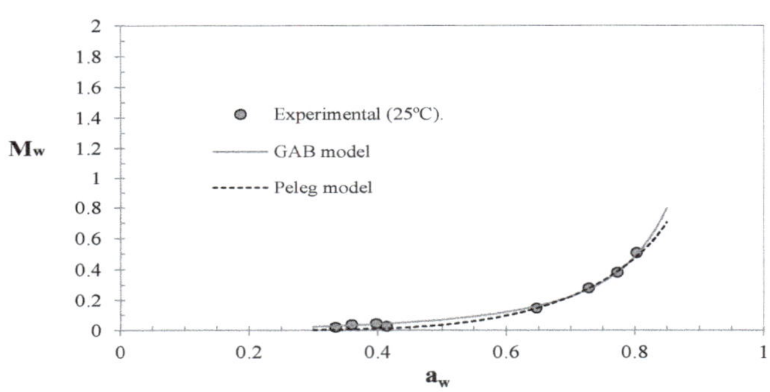

Figure 8: Desorption isotherm of tomatoes at 25 °C.

regarding the most adequate drying parameters, the effect of these parameters on product quality should also be considered.

The Pareto chart (Figure 6) shows that X_c diminished with temperature increase, and it increased with sample size, as well as the binary interactions. Air recycle did not affect X_c.

Water sorption isotherms

The desorption isotherms of tomato samples with different moisture content were determined at 15, 25 ad 40°C. Then the isotherms were adjusted with the Guggenheim-Anderson-Boer (GAB) (Eq. 4) and the Peleg model (Eq. 5). Both models offered a good adjustment quality, with RMSE lower than 0.25. Figures 7, 8 and 9 compare predictions from both models with the experimental data at 15, 25 and 40°C. Table 4 shows the parameters of both models for each temperature in the desorption process.

Color

The color parameters (a^*, b^*, ΔE, and L^*) of dehydrated tomatoes are given in Table 4. The statistical analysis of these parameters indicate that only a* is affected by the operational variables, with the air temperature showing the highest effect. The increase of a* confirms the intensification of redness in all runs (Table 5), with respect to fresh tomatoes. This effect increases directly with temperature. Redness variations was also reported by Kerkhofs et al. [37], and it was attributed to a combination of non-enzymatic browning, Maillard reaction and lycopene degradation.

Rehydration of the dried tomato particles

On checking the rate of rehydration of dried tomato particles, all experiments showed that this rate was high during the first 3 minutes of the rehydration procedure, and then it was followed by a period of asymptotic increase during the next 15 minutes, with not significant increases from this point onwards, as shown in Figure 10. The final water content after rehydration was 0.75 ± 0.05 for all runs. The time required to reach this moisture content was shorter for the thinnest slices due to the shorter diffusional path of water.

Adjustment of experimental drying kinetics

Drying curves were adjusted only with the empirical models of Page [27], given by Eq. 1, the modified Page [28], given by Eq.2, and Henderson and Pabis [29], given by Eq. 3. This adjustment was

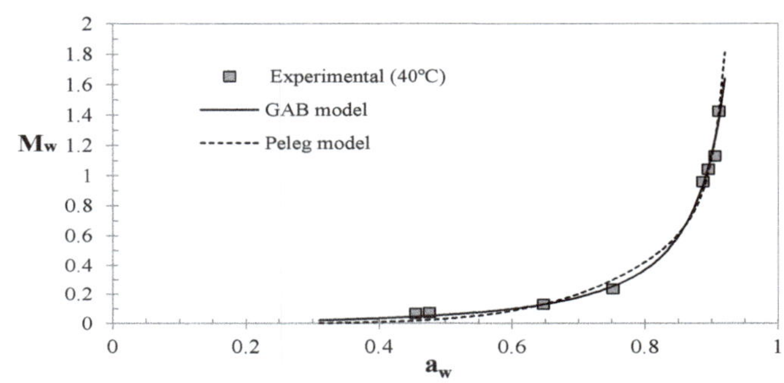

Figure 9: Desorption isotherm of tomatoes at 40 °C.

Model	Parameters	15°C	25°C	40°C
GAB	Mo	0.053	0.178	0.118
	C	0.466	0.225	0.298
	K	1.028	1.020	1.023
	RMSE	0.24	0.08	0.17
Peleg	C_1	0.166	1.616	1.468
	C_2	20.154	28.907	52.607
	C_3	0.758	5.541	5.576
	C_4	40.275	39.704	48.931
	RMSE	0.20	0.11	0.20

Table 4: Peleg and GAB model parameters for isotherms of plum tomato.

Run	ΔE	ΔL	Δa*	Δb*
T1	19.66	11.07	2.77	15.90
T2	13.461	9.39	2.11	9.41
T3	16.681	9.79	7.11	11.49
T4	14.447	8.89	5.31	10.07
T5	18.678	10.85	5.06	14.34
T6	15.300	9.61	6.15	10.20
T7	15.973	9.64	4.73	11.83
T8	20.787	12.78	1.05	16.36

Table 5: Color parameters.

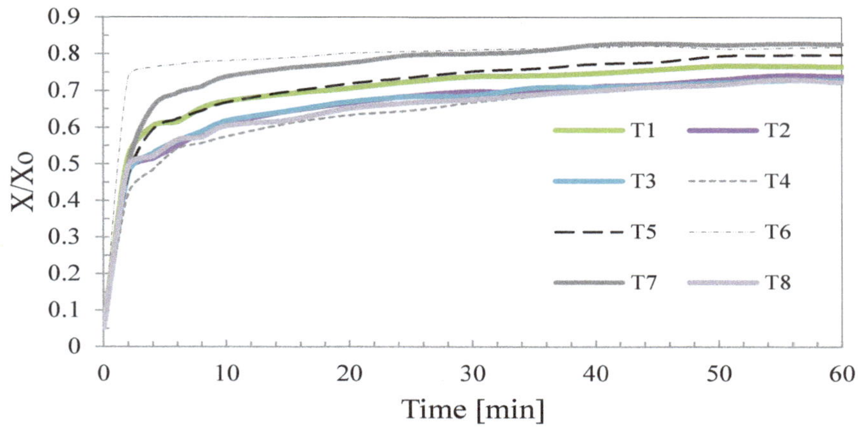

Figure 10: Rehydration profiles of tomato.

	Page			Modified Page			Henderson and Pabis		
	k	n	R^2	k	n	R^2	k	n	R^2
T1	0.0033	1.1579	0.9993	0.0046	1.0089	0.9961	0.0047	1.0121	0.9965
T2	0.0028	1.1086	0.9974	0.0039	1.0667	0.9988	0.0042	1.0051	0.9989
T3	0.0035	1.1715	0.9986	0.0026	0.9726	0.9986	0.0025	0.9926	0.9988
T4	0.0047	1.1414	0.9962	0.0029	0.9860	0.9989	0.0029	0.9903	0.9992
T5	0.0014	1.2211	0.9986	0.0062	0.9184	0.9940	0.0053	1.0273	0.9954
T6	0.0017	1.1299	0.9987	0.0024	0.9897	0.9989	0.0024	0.9973	0.9991
T7	0.0023	1.1795	0.9996	0.0050	0.9221	0.9991	0.0045	1.0026	0.9991
T8	0.0056	1.1039	0.9976	0.0036	0.9756	0.9965	0.0036	0.9966	0.9992

Table 6: Page, modified Page and Henderson and Pabis models parameters with the corresponding R^2 values.

Run	Average [Qsp/Qe]	Standard deviation
T1	0.100	0.010
T2	0.076	0.016
T3	0.082	0.034
T4	0.066	0.025
T5	0.079	0.010
T6	0.125	0.014
T7	0.120	0.034
T8	0.111	0.040

Table 7: Absorbed solar energy / electric energy input [Q_{sp}/Q_e].

performed by the Solver Tool from Microsoft Excel 2003 software, minimizing the root mean square error between the experimental values of moisture content of the tomatoes particles versus time and those calculated for the models. As shown in Table 6, the resulting R^2 values for three models were higher than 0.99, being slightly better the adjustments of Henderson and Pabis model.

Energy efficiency

The energy efficiency, defined as the ratio between absorbed solar energy (Q_{sp}), calculated with Eqn 7, and electric energy input (Q_e), fluctuated between 0.066 and 0.125, depending on the solar radiation level and also on the operating conditions (Table 7).

$$Q_{sp} = A_{col} \cdot cos(\alpha) \cdot \int_0^t I(t) \cdot dt \qquad (7)$$

Here, Q_{sp}, A_{col} and I are the heat provided by solar radiation [J/s], collector surface [m^2] and radiation [W/m^2], respectively. The alpha angle (α) corresponds to the tilt of the accumulator-panel (α=11°).

Even though these efficiency values are not very high, they result in energy saving and also in a reduction of CO_2 emission. These values could probably be improved by increasing the tomatoes load in the HSD, since in this work only 40% maximum capacity (15 kg) of the dryer was used.

Conclusions

The solar panel allowed increasing the air temperature (T_{sp}) up to 18°C above the ambient temperature, depending on the air flow rate and mainly of the solar radiation.

It was possible to obtain the drying curves at different operation conditions. The curves exhibited a good adjustment to the empirical models of Page, modified Page, and Henderson and Pabis, resulting in R^2 values higher than 0.99 for the three models, being slightly better the adjustments of Henderson and Pabis model.

The critical moisture content (Xc) was obtained indirectly from the drying curves. The statistical analysis showed that temperature and size had significant effects on Xc.

The sorption isotherms of rehydration were determined and adjusted with the Guggenheim-Anderson-Boer (GAB) and the Peleg models. Both models exhibited a good adjustment quality, with RMSE lower than 0.25.

The color parameters of dehydrated tomatoes indicate a notorious redness, in all runs. The final degree of water re-hydration was 80 ± 2% for all runs, in less than 20 minutes.

Finally, the solar energy input resulted in 6.6-12.5% energy saving. These values could even be improved by increasing the tomatoes load in the HSD.

Acknowledgment

The authors wish to thank the financial support of project FONDECYT N° 1110101.

References

1. Farkas I (2004) Solar-drying of materials of biological origin, in Dehydration of Products of Biological Origin, Ed. By Arun Mujumdar, Science Pub Inc. 318-368.
2. Kadam D, Samuel DVK (2006) Convective flat-plate solar heat collector for cauliflower drying. Biosystems Engineering 93: 189-198.
3. Belessiotis V, Delyannis E (2011) Solar drying. Solar Energy 85: 1665-1691.
4. Santos BM, Queiroz MR, Borges TPF (2005) A solar collector design procedure for crop drying. Brazilian Journal of Chemical Engineering 22: 277-284.
5. Basunia MA, Abe T (2001) Thin layer solar drying characteristics of rough rice under natural convection. Journal of Food Engineering 47: 295-301.
6. Reyes A, Mahn A, Cubillos F, Huenulaf P (2013) Mushroom dehydration in a hybrid solar-dryer. Energy Conversion Management 70: 31-39.
7. Imre L, Palaniappan C (1996) Development of solar Drying. Drying Technology 14: 1381-1388.
8. Dufour P (2006) Control Engineering in Drying Technologies: Review and Trends. Drying Technology 24: 889-904.
9. Boughali S, Benmoussa H, Bouchekima B, Mennouche D, Bouguettaia H, et al. (2009) Crop drying by indirect active hybrid-solar-electrical dryer in the eastern Algerian Septentrional Sahrta. Solar Energy 83: 2223-2232.
10. El-Sebaii AA, Shalaby SM (2013) Experimental investigation of an indirect-mode forced convection solar dryer for drying thymus and mint. Energy Conversion and Management 74: 109-116.
11. Badescu V (2006) Optimum fin geometry in flat plate solar collector systems. Energy Conversion and Management 47: 2397-2413.
12. Karim MA, Hawlader MNA (2004) Development of solar air collectors for drying applications. Energy Conversi6n and Management 45: 329-344.
13. Tsamparlis M (1990) Solar drying for real applications. Drying Technology 8: 261-285.
14. Bhattacharya SC, Ruangrungchaikul T, Pham HL (2000) Design and Performance of a Hybrid Solar/Biomass Energy Powered for fruits and Vegetables. Proceedings of the World Renewable Energy Congress VI1161-1164.
15. Abdullah K (2009) Dissemination of Hybrid ICDC Solar Drying Systems, Renewable Energy & Sustainable Development in Indonesia: Past Experience-Future Challenges, Jakarta 19-20.
16. Palaniappan C (2009) Perspectives of solar food processing in India. International Solar Food Processing Conference.
17. Pangavhane DR, Sawhney RL (2002) Review of research and development work on solar dryers for grape drying. Energy Conversion and Management 43: 45-61.
18. Bennamoun L, Belhamri A (2003) Design and simulation of a solar dryer for agriculture products. Journal of Food Engineering 59: 259-266.
19. Triebe K (2009) Practical Application of Solar Tunnel dryers. International Solar Food Processing Conference.
20. Joshi CB, Gewali MB, Bhandari RC (2004) Performance of Solar Drying Systems: A Case Study of Nepal. IE(I) Journal-ID 85: 53-57.
21. Moradi M, Zomorodian A (2009) Thin layer solar drying of cuminum cyminum grains by means of solar cabinet dryer. American-Eurasian J Agric Enviromen Sci 5: 409-413.
22. Hassanain AA (2010) Unglazed Transpired Solar Dryers for Medicinal Plants. Drying Technology 28: 240-248.
23. Simate IN (2001) Simulation of the mixed-mode natural-convection solar drying of maize. Drying Technology 19: 1137-1155.
24. Vlachos NA, Karapantsios TD, Balouktsis AI, Chassapis D (2002) Design and testing of a new solar tray dryer. Drying Technology 20: 1243-1271.
25. Yaldyz O, Ertekyn C (2001) Thin layer solar drying of some vegetables. Drying Technology 19: 583-597.
26. Crank J (1975) The Mathematics of Diffusion, Oxford University Press: London.
27. Doymaz I (2004) Convective air drying characteristics of thin layer carrots. J Food Eng 61: 359-364.
28. Wang Z, Sun J, Liao X, Chen F, Zhao G, et al. (2007) Mathematical modeling on hot air drying of thin layer apple pomace. Food Res Int 4: 39-46.
29. Brooks MS, Abou El-Hana NH, Ghatly AE (2008) Effects of Tomato Geometries and Air Temperature on the Drying Behavior of Plum Tomato. American Journal of Applied Sciences 5: 1369-1375.
30. Araya-Farias M, Ratti C (2009) Dehydration of Foods: General Concepts. In Advances in Food Dehydration, Ed. by Cristina Ratti. CRC Press, Taylor & Francis Group, Boca Raton, FL.
31. Prothon F, Ahrné L (2004) Application of the Guggenheim, Anderson and de Boer model to correlate water activity and moisture content during osmotic dehydration of apples. Journal of Food Engineering 61: 467-470.
32. Peleg M (1993) Assessment of a semi-empirical four parameter general model for sigmoid moisture sorption isotherms. Journal of Food Process Engineering 16: 21-37.
33. Gaware TJ, Sutar N, Thorat BN (2010) Drying of Tomato Using Different Methods: Comparison of Dehydration and Rehydration Kinetics. Drying Technology 28: 651-658.
34. Doymaz I (2007) Air-drying characteristics of tomatoes. Journal of Food Engineering 78: 1291-1297.
35. Hossain MA, Amer BMA, Gottschalk K (2008) Hybrid Solar Dryer for Quality Dried Tomato. Drying Technology 26: 1591-1601.
36. Kostoglou M, Chrysafis M, Andritsos N (2013) Modelling Tomato Dehydration in a Tunnel Dryer Using Geothermal Energy. Drying Technology 31: 5-15.
37. Kerkhofs NS, Lister CE, Savage GP (2005) Change in colour and antioxidant content of tomato cultivars following forced-air drying. Plant Foods Hum Nutr 60: 117-121.

Characterization of Tataouine's Raw and Activated Clay

Hattab A*, Bagane M and Chlendi M

Research Unit: Applied Thermodynamics, National Engineering School of Gabes (ENIG)-Omar Ibn Elkhatab Street, Gabes, Tunisia

Abstract

This study deals with the mineralogical and thermal analysis of red clay taken from the south of Tunisia (Tataouine area) which will be used in industrial and environmental fields. In that purpose the study had recourse to several quantitative and qualitative research instruments: chemical analysis, mineralogical study and thermal analysis. The analyses were carried out by different methods such as: (i) X-ray diffraction (XRD), (ii) infrared spectroscopy (IR), (iii) thermogravimetric analysis (TGA), (iv) differential thermal analysis (DTA) and (v) cation exchange capacity (CEC). The data collected from these techniques show that illite and kaolinite are the major clay phases. The accessory minerals detected in powdered rock is calcite.

The raw clay has undergone an acid activation using the sulfuric acid. The acid activation of south of Tunisia clay increase the specific surface area by adsorption of methylene blue from 117 m^2/g to 163 m^2/g. The experimental investigation is performed in accordance with the methodology of the factorial design of experiments. The different factors or parameters used in this methodology are: the temperature (T), the contact time (t), the liquid/solid ratio (R), and the acid concentration (C). The optimal conditions, obtained by the statistically designed experiments, for the best results of methylene blue adsorption were: 60 °C, a contact time of 1.5 h, a liquid/solid ratio of 7 and acid concentration of 20%. These parameters are applied in the preparation of the sample used in characterization study. The effect of the reaction conditions on the solid structure is investigated by comparison with the raw clay sample.

Keywords: Clay; Characterization; Activation; XRD; IR; TGA; DTA; CEC

Introduction

In recent years, many laboratories around the world were interested by the study of clays. This interest is justified by their abundance in nature (covering about 42% of the earth's crust), the importance of the specific surface areas developed, the presence of electric charges on the surface and above the exchangeability of interlayer cations. The latter one, also known as compensating cations, is the main element responsible for the hydration, swelling and plasticity.

Clay's activation is a process of improving the adsorption properties of the clay by subjecting it to thermal or chemical treatment. In chemical activation, is generally used sodium carbonate or strong acids such as sulfuric acid. Acid activation was the subject of many studies [1-4], treating either, the various reaction mechanisms, or the effect of attack on the physicochemical characteristics of clay. The supreme goal of the clay activation, as by the sulphuric acid attack [5,6], it is the increase in porosity by peripheral dissolution of clay, in order to increase the number of the active sites: Indeed, the structural cations of clay (Al^{3+}, Mg^{2+}, Fe^{3+}) are replaced during activation by protons H^+. It leads to a high capacity of adsorption.

Adsorption is a reversible and interfacial physicochemical phenomenon causing the accumulation of solute's molecules in the solid-liquid interface (or solid-gas). Often the adsorption of organic molecules is characterized by soils in the laboratory using the technique called "batch". This technique consists to agitate suspensions of adsorbents in aqueous solutions containing the adsorbate, in a closed container, until to reach adsorption equilibrium.

The MB adsorption method is currently used to measure the specific surface area of biosorbent particles in aqueous suspension. This method has been widely adopted for solids of variable nature such as iron oxides [7], clays [8], activated carbon [9,10], zeolites and silica [11]. The method has also been used to assess average pore size and pore size distribution in charcoals, silica, and alumina [12].

The work presented in this manuscript, aims to apply an acid activation on our natural clay, to characterize the raw and activated clay and to compare by emphasizing the results and shows the effect of activation.

Experimental

Material

In the present work, red clay was selected from site of 'Tataouine', located in the south of Tunisia was used as raw material. It has a BET surface area equal to 44.87 m^2/g and 117 m^2/g by adsorption of methylene blue. The received material contains many impurities. This content of impurities was decreased or removed completely by applying a purification protocol. This was achieved by dispersing the crude lumps in distilled water. The mixture was stirred vigorously and sieved several times through a different mesh sieves with the addition of more water. Then dispersed particles were collected, after allowing coarser mineral particles to settle. Finally, water was siphoned off, and it was dried in air at around 60°C.

Acid-activation

Acid-activation was carried out with sulphuric acid in a jacketed glass reactor equipped with a reflux condenser, a thermometer and a stirrer. At the end of each experiment, the solid content was immediately filtered, washed with water until the washing water was neutral and dried.

***Corresponding author:** Hattab A, Research Unit: Applied Thermodynamics, National Engineering School of Gabes (ENIG)-Omar Ibn Elkhatab Street, Zrig 6029 Gabes, Tunisia

All samples were prepared according to central composite orthogonal design (CCOD) [1]. Temperature (53.8-96.2°C), contact time (0.57-6.93 h), liquid/solid mass ratio (3.38-7.62) and acid concentration (13.8-56.2%) are chosen as numerical procedure variables for optimization for methylene blue adsorption.

The table 1 shows the variation interval of activation's factors (T, t, R et C).

ΔZ_j means the half extent of its range of variation: $\Delta Z_j = \frac{Z_j^{max} - Z_j^{min}}{2}$

The table 2 presents the matrix of acid-activation experimental design and the specific surface area (SMB) by adsorption of methylene blue.

The optimal conditions, obtained by the statistically designed experiments, for the best results of methylene blue adsorption were: a temperature of 60°C, a contact time of 1.5 h, a liquid/solid ratio of 7 and acid concentration of 20%. These parameters are applied in the preparation of the sample used in the characterization study.

Characterization

The chemical analyses of the raw material and activated clay were carried out by X-ray fluorescence. The specific surface area was calculated by adsorption of methylene blue method. The value

j	Variables	Z_j^{min}	Z_j^{max}	ΔZ_j
1	T (°C)	60	90	15
2	t (h)	1.5	6	2.25
3	R (g/g)	4	6	1.5
4	C (% mass.)	20	50	15

Table 1: Study variables and their variation ranges.

Essai (i)	T (°C)	t (h)	R (g/g)	C (%)	S_{MB} (m²/g)
1	60	1.5	4	20	146.67
2	90	1.5	4	20	146.43
3	60	6	4	20	155.04
4	90	6	4	20	138.70
5	60	1.5	7	20	162.98
6	90	1.5	7	20	144.84
7	60	6	7	20	144.04
8	90	6	7	20	131.92
9	60	1.5	4	50	146.27
10	90	1.5	4	50	139.24
11	60	6	4	50	142.45
12	90	6	4	50	136.95
13	60	1.5	7	50	156.16
14	90	1.5	7	50	154.17
15	60	6	7	50	135.35
16	90	6	7	50	130.17
17	75	3.75	5.5	35	147.65
18	53.8	3.75	5.5	35	139.93
19	96.2	3.75	5.5	35	130.09
20	75	0.57	5.5	35	148.51
21	75	6.93	5.5	35	133.35
22	75	3.75	3.38	35	152.65
23	75	3.75	7.62	35	139.02
24	75	3.75	5.5	13.8	153.85
25	75	3.75	5.5	56.2	144.68

Table 2: Experimental design and results of methylene blue adsorption.

Compounds	CaO	SiO2	Al2O3	Fe2O3	SO3	K2O	MgO	Na2	LI
Raw clay	5.91	45.46	15.61	9.64	3.86	1.79	3.26	0.28	11.56
Activated clay	0.95	58.63	20.37	9.93	0.6	1.94	0.4	0.03	8.05

Table 3: Chemical analysis (wt %) and loss in ignition of raw and activated clay.

of the cation exchange capacity (CEC) is determined by the method using complex of copper ethylenediamine [13]. The mineralogical compositions were identified by Philips Panalytical X'Pert PRO X-ray powder diffractometer. The Infra Rouge spectra were obtained using a Perkin Elmer FTIR (the sample was diluted with potassium bromide: 50 mg of sample mixed with 350 mg of KBr). To obtain the DTA/ TGA curves, a differential calorimeter apparatus TG-DTA 92 SETARAM with an argon flow was used: the finely ground sample was introduced into a platinum crucible and placed in the furnace along with the reference sample (the temperature was increased from room temperature to 1000°C at a rate of 10°C/min, maintained at this temperature for 10 min then reduced again) [14].

Results and Discussion

Chemical analysis

The table 3 shows the chemical analysis and the loss on ignition of raw and activated clay.

The high loss on ignition is attributable to the presence of substantial volatiles, existing in clay minerals (mainly hydrous species), dolomite (carbonate), and organic matter [14].

The chemical analysis of raw moldier shows that the major elements are SiO_2 and Al_2O_3, which are both associated with the kaolinite structure [15].

The ratio of SiO_2 increase at activated sample: this increase is due to the formation of amorphous silica during activation. We note also a clear reduction in the calcium portion. Indeed, the composition of raw clay on CaO is 5.91 and for the activated sample is 0.95. So we can conclude that Ca^{2+} is a major exchangeable cation.

The increase of K_2O after activation is probably due to the existence of illite clay mineral: potassium is not involved in the cation exchange [16].

The ratio $\frac{SiO_3}{Al_2O_3 + Fe_2O_3 + MgO} = 1.59$ for the raw clay and 1.9 for the activated clay: this value is between 1 and 2, indicating that the clay is a mixture of 1/1 and 2/1 dioctahedral phyllosilicates [14]: our clay is a mixture of kaolinite and illite.

The overall structural formula of the clay is calculated: $[Al_{0.95}Fe_{0.56}Mg_{0.38}][Si_{3.52}Al_{0.48}]O_{10}(OH)_2Ca_{0.49}Na_{0.042}K_{0.18}$

X-Ray diffraction (XRD)

The diffractograms obtained shows that our clay is a well crystallized material.

Figure 1a and 1b show the X-ray diffraction pattern of respectively raw and activated clay. The raw clay sample contains illite as the major clay mineral associated with kaolinite; the peak at 9.97 Å is the wider peak which characterized the illite clay mineral, the two peaks at 7.14 and 3.33 Å are the characteristics peaks of kaolinite clay mineral. Accessory minerals such as calcite was detected in our natural clay, occur at 4.94 Å.

For the activated clay sample, we note: the disappearance of characteristic peak of kaolinite at d=7.14 Å: this disappearance is due to

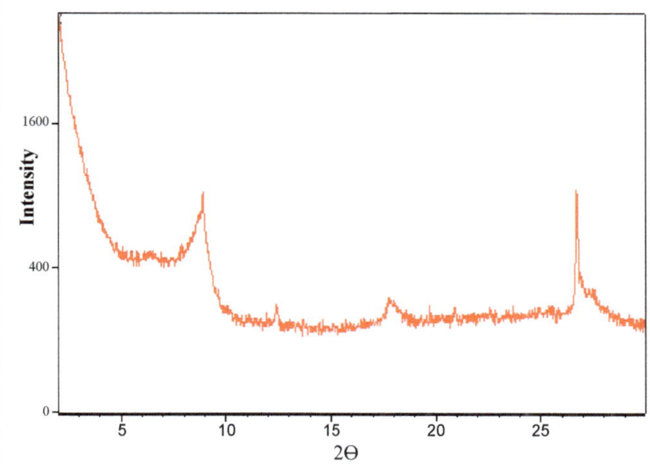

Figure 1a: Diffractogram (XRD) of raw clay.

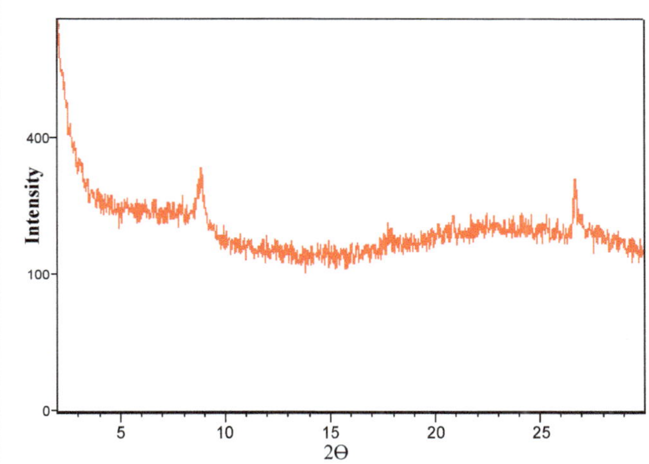

Figure 1b: Diffractogram (XRD) of activated clay.

the high temperature used during the acid activation [1]. We also noticed the disappearance of characteristic peak of calcite: this is explained by the cation exchange that occurs between the cations Ca^{2+} of clay and H^+ of the sulfuric acid which is proved by the chemical composition by the reduction of fraction of CaO of activated clay sample.

The two reflections at 9.97 and 3.33 Å correspond respectively to the phyllosilicate phase 1/2 of illite and 1/1 of kaolinite are still. However, there was a decrease in intensity which can be attributed to a change in the internal structure of the clay during the activation.

Infrared spectroscopy (IR)

The figure 2a and 2b below shows the IR spectrum of raw and activated clay.

The IR spectrum of raw and activated clay shows that the stretching band of the hydroxyl group of water of hydration H-O-H is around 3444 cm^{-1}, while the deformation is around 1637 cm^{-1}.

The band of stretching vibration of Al-Al-OH is at 3632 cm^{-1} and the strain is at 911 cm^{-1} indicates that this clay is dioctahedral [17].

The band at 1021 cm^{-1} is characteristic of vibration of Si-O of clay, while the deformation bands appear respectively at 525 cm^{-1} for Al-O-Si and at 470 cm^{-1} for the group Si-O-Si which increase on activated sample comparing with the raw sample's one: this result is confirmed by the chemical analysis.

The band that appears at 1385 cm^{-1} corresponds to the presence of carbonate (calcite ($CaCO_3$) or dolomite ($Ca,Mg((CO_3)_2)$) [18]. But, this band is reduced relative to that of natural clay: this is explained by the fact that cation exchange between Ca^{2+} cations of the clay and H^+ from the acid.

Thermal analysis (DTA-TGA)

Both DTA and TGA characterizations are carried out simultaneously with the same device.

The DTA and TGA curves of raw and activated clay are shown respectively on figure 3a and 3b.

The interpretation of the DTA-TGA curves of the raw clay sample (Figure 3a) leads to the following results: An endothermic peak at 160°C, due to the removal of moisture and interlayer water of the clay minerals (water of hydration) [14]. The endothermic peak at 580°C can be attributed to the elimination of constitution water resulting from the loss of hydroxyl groups from the clay structure (dehydroxylation of kaolinite). Weight loss is observed up to 799°C, in accordance with the

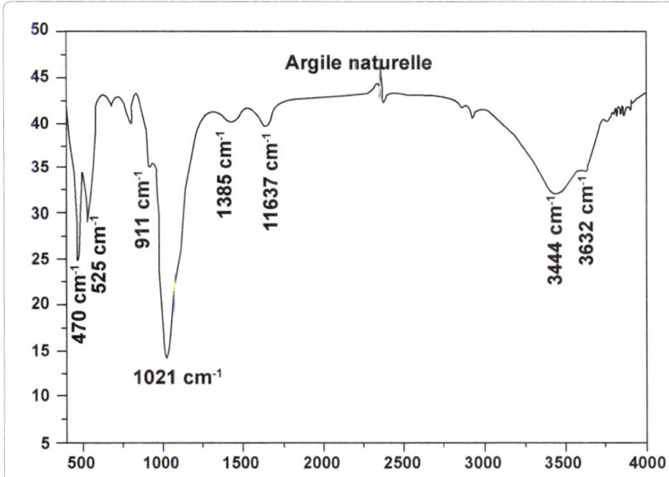

Figure 2a: Infrared spectra (IR) of raw clay.

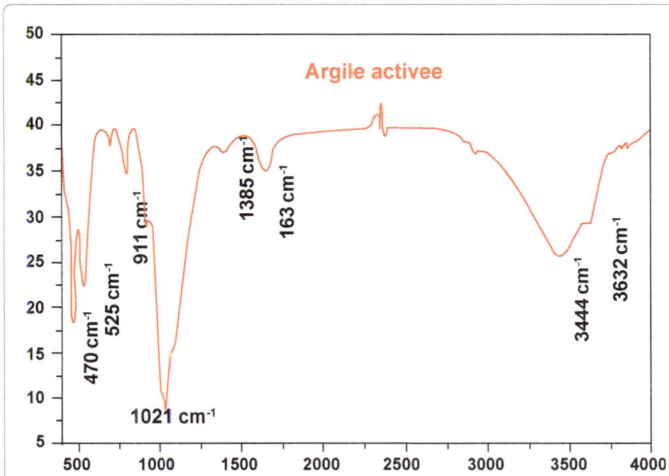

Figure 2b: Infrared spectra (IR) of activated clay.

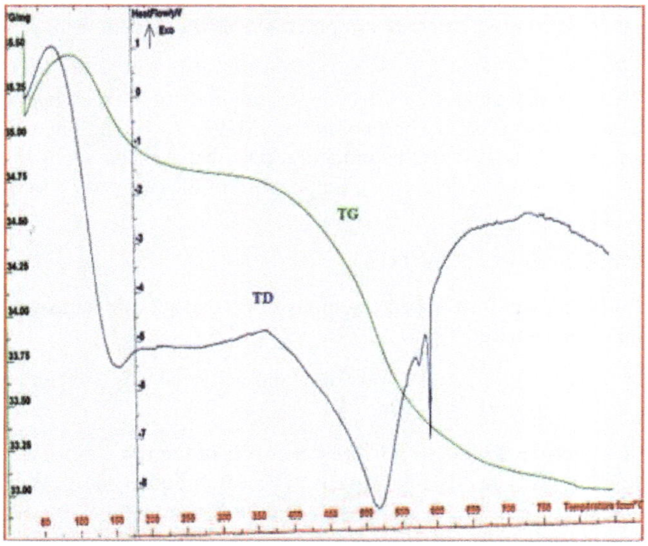

Figure 3a: Differential thermal and thermogravimetric analysis of raw clay.

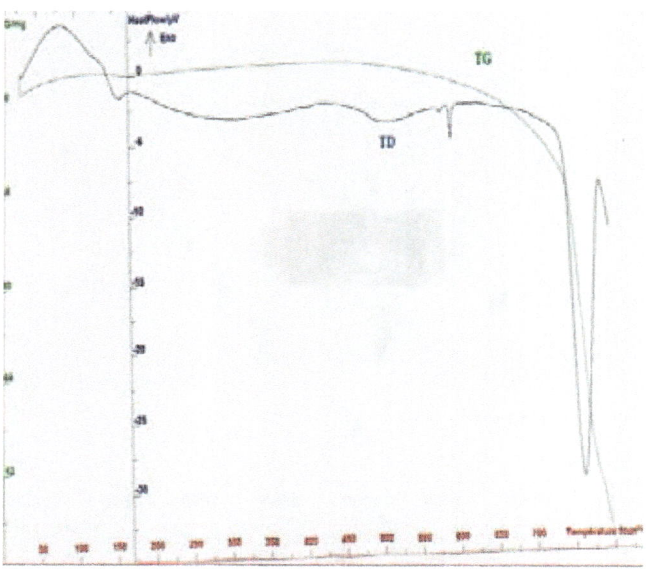

Figure 3b: Differential thermal and thermogravimetric analysis of raw clay.

decomposition of dolomite in the sample (decarbonation of clay).

For the activated clay (Figure 3a): the peak corresponding to the loss of the water of hydration is at 150°C and more significant than the raw clay's one. An endothermic peak appears at 520°C which correspond to the dehydroxylation of kaolonite with a high intensity compared with the raw clay's one. An endothermic peak appears at 600°C which attributed to the decarbonation of clay with a low intensity because of the decrease of the amount of CaO confirmed by the chemical composition and the IR spectrum.

Cation exchange capacity (CEC)

The value of the cation exchange capacity (CEC) of raw and activated clay is established by the method using the complex of copper ethylenediamine.

This low value shows that the clay studied is kaolinite: it is because the cation exchange capacity of kaolinite is generally between 3 and 15 meq/100 g of clay [19].

We note, from the table 4, the value of the cation exchange capacity of the activated clay is decreased compared to the raw clay. Indeed, this decrease is due to activation and shows that a portion of exchangeable cations were replaced during the activation by H^+ from sulfuric acid. Indeed, the surfaces of clay particles are negatively charged, and to balance these loads, the clay particles attract positive ions contained in the solutions that are exchangeable ions [20].

Determination of pH

The pH of clay is determined by the traditional method suggested by Oboth and coll [21] (Figure 4).

The knowledge of the pH change of a clay/water suspension is important for the environmental and industrial applications. Therefore, we have investigated the chemical behavior of clay/water suspension in natural and activated sample.

The table 5 shows that the pH of raw clay is 7.78 which give it a basic character. However; the pH of activated clay gives it an acidic character. Indeed, it has generally been observed [22], that high solution pH tends to disperse clay systems. This phenomenon is undoubtedly related to the dissociation of H^+ ions as the pH increases and the resultant increased negative charge on clay particles which increased repulsion between particles and the probability of the dispersion being stable. The reverse occurs at low pH where positive charges dominate the edges of particles and lead to enhanced flocculation by edge-to-face attraction. Low pH also affects the stability of the clay particles. Van Olphen [23] indicated that acid attacks clay particles at edges, thereby releasing Al^{3+} ions.

	CEC (meq/100g of clay)
Raw clay	6.22
Activated clay	5.11

Table 4: CEC of raw and activated clay.

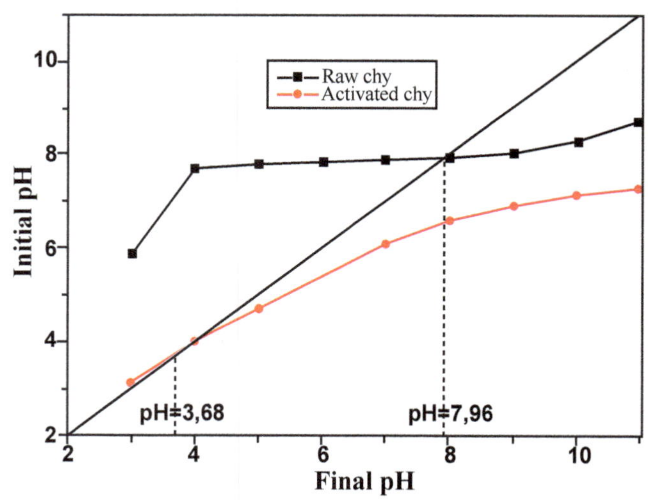

Figure 4: Isoelectric pH of raw and activated clay.

	pH
Raw clay	7.78
Activated clay	2.6

Table 5: pH of raw and activated clay.

Determination of isoelectric pH

The isoelectric pH is a method used to know the nature of the surface charge. To determine the isoelectric pH of our clay, we used the method developed by TJ Bondosz Keling and King in 1997 [21].

After activation of clay, the isoelectric pH decreases from 7.96 to 3.68. The acid treatment of the crude clay moves the PIE to acidic pH, which is directly related to the chemical modification of groups on the surface of these solids. The activated clay has indeed a negative charge on its surface for pH>IEP. Thus, clay minerals that are usually characterized as aluminum silicates exhibit negative charges in aqueous media, which results from the substitutions within the crystal lattice of Al^{3+} for Si^{4+} or Mg^{2+} for Al^{3+} [24]. As a result, negative charges are developed in the lattice to compensate for the so-called exchangeable cations, that is, Na^{+}, K^{+} and Ca^{2+} entering the crystal structure. If such clay minerals come into contact with water, some of these cations can easily dissociate, leading to a negatively charged surface. The electro neutrality of the system requires that the charged surface be surrounded by oppositely charged ions. The cloud of ions around the particle is described by the double-layer model, developed by Gouy, Chapman, Stern and Grahame [25].

Conclusion

The main objective of present study is chemical, mineralogical and thermal analysis of raw and activated clay taken from the region of Tataouine (southern Tunisia). Activated clays are obtained by acid activation using sulfuric acid and for different operating conditions.

Different techniques have been used to characterize raw clay and activated. We confirm that the red clay of Tataouine is a clay type illite kaolinite with a specific surface area BET equal to 44.87 m^2/g and 117 m^2/g by adsorption of methylene blue. After activation, this values increases to 163 m^2/g.

The influence of activation conditions for an experimental design has allowed highlighting changes in the specific surface area measured by adsorption of methylene blue according to the activation parameters.

Our study allowed us to define the optimal conditions for the largest surface area and which are: Activation time: 1.5 hours; activation temperature: 60°C; weight ratio: 7 and concentration of acid: 20%.

Acknowledgments

The author gratefully acknowledges the Research Unit of Applied Thermodynamics in the National Engineering School of Gabès ENIG for providing support of a project entitled: characterization of tataouine's raw and activated clay.

References

1. Amari A (2010) Adsorption des composés organo-volatils(COV) sur argiles tunisiennes naturelles et activées. Tunisie,Thèse de doctorat, ENIG.
2. Temuujin J, Jadambaa Ts, Burmaa G, Erdenechimeg Sh, Amarsanaab J, et al. (2004) Characterization of acid activated montmorillonite clay from Tuulant (Mongolia). Ceramics International 30: 251-255.
3. Novakovi´c T, Ro˘zi´c L, Petrovi´c S, Rosi´c A (2008) Synthesis and characterization of acid-activated Serbian smectite clays obtained by statistically designed experiments. Chemical Engineering Journal 137: 436-442.
4. Dali-Youcef Z, Bouabdasselem H, Bettahar N (2006) Élimination des composés organiques par des argiles locales. C R Chimie 9: 1295-1300.
5. Seghairi N, Koussa M, Achour S (2004) Effet de l'activation chimique de la bentonite Sur l'adsorption de substances humiques en eaux de minéralisation variable. Larhyss journal 3: 91-102.
6. Seghairi N, Achour S (2003) Influence du temps et du taux d'activation de la Bentonite sur l'adsorption de l'aniline. Larhyss Journal 2: 115-125.
7. Mak SY, Chen DH (2004) Fast adsorption of methylene blue on polyacrylic acid-bound iron oxide magnetic nanoparticles. Dyes Pigments 61: 93-98.
8. G¨urses A, Karaca S, Dogar C, Bayrak R, Ac¸ikyildiz M, et al. (2004) Determination of adsorptive properties of clay/water system: methylene blue sorption. J Colloid Interface Sci 269: 310-314.
9. Barton SS (1987) The adsorption of methylene blue by active carbon. Carbon 25: 343-350.
10. Kannan N, Sundaram MM (2001) Kinetics and mechanism of removal of methylene blue by adsorption on various carbons-a comparative study. Dyes and Pigments 51: 25-40.
11. Handreck GP, Smith TD (1988) Adsorption of methylene blue from aqueous solutions by ZSM 5-type zeolites and related silica polymorphs. J Chem Soc Faraday Trans 84: 4194-4201.
12. Giles CH, Silva APD (1969) Molecular sieve effects of powder towards dues. Measurement of porosity by dye adsorption. Trans Faraday Soc 65: 1943-1951.
13. Bergaya F, Vayer M (1997) CEC of clays: measurement by adsorption of a copper ethylendiamine complex. Appl Clay Sci 12: 275-280.
14. 14. Baccoura H, Medhioub M, Jamoussi F, Mhiri T, Daoud A (2008) Mineralogical evaluation and industrial applications of the triassic clay deposits, Southern Tunisie. Materials Characterization 59: 1613-1622.
15. Vieira CMF, Sa´nchez R, Monteiro SN (2008) Characteristics of clays and properties of building ceramics in the state of Rio de Janeiro, Brazil. Construction and Building Materials 22: 781-787.
16. Ayari F, Srasra E, Trabelsi-Ayadi M (2004) Application des modèles de Langmuir et Freundlich aux isothermes d'adsorption des métaux lourds par l'argile purifiée. J Phys IV France 122: 229-234.
17. Bekri-Abbes I, Srasra E (2008) Synthèse et caractérisation d'un hybride organique-inorganique à partir d'une argile tunisienne par une réaction à l'état solide. African journal of science and technology, Science and Engineering Series 1: 12-19.
18. Benzina M, Jarraya I, Eloussaief M, Kallel N, Yaacoubi A (2010) Valorization of Tunisian smectite clay: drying of the air application 1-5.
19. Azzouz FZ (2006) Contribution a 'étude de la stabilisation chimique de quelques argiles gonflantes de la région de Tlemcen. Mémoire de Magister, Faculté des Sciences de l'Ingénieur.
20. Wakim J (2005) Influence des solutions aqueuses sur le comportement mécanique des roches argileuses. France, Thèse de doctorat, Ecole nationale supérieure des mines de Paris.
21. Farhat I (2009) Préparation d'une argile pontée en vue de son utilisation comme catalyseur. Tunisie, Mastère de recherché en génie chimique-procédés, ENIG.
22. Mitchell JK (1976) Fundamentals of Soil Behavior. Wiley, New York 442.
23. Van Olphen H (1976) An Introduction to Clay Colloid Chemistry. (2nd edn), Wiley, USA.
24. Sedef D, Gulgun Y, Ertugrul Y, Emine K (2011) Zeta potential study of natural–and acid–activated sepiolites in electrolyte solutions. The Canadian journal of chemical engineering 90: 785-792.
25. Hunters RJ (1981) Zeta Potentia in Colloidal Science: principles and applications. Academic Press, London.

Experimental Study of Lactose Hydrolysis and Separation in Continuous Stirred Tank -Ultrafiltration Membrane Reactor

M. Pakizeh* and M Namvar-Mahboub

Department of Chemical Engineering, Faculty of Engineering, Ferdowsi University of Mashhad, P.O. Box 91775-1111, Mashhad, Iran

Abstract

In this study, lactose was hydrolyzed using β-galactosidase enzyme in a continuous stirred tank -Ultrafiltration (CSTR-UF) to produce galactose and glucose. The UF membranes of Molecular Weight Cut of 3 kDa of regenerated cellulose material were used to separate enzyme from products. Experiments were performed with 0.139 molar aqueous solution of lactose as feed. The effect of operating pressure ranging between 2 and 5 bar and time on rejection and permeate flux were studied. Results showed that the UF membrane rejects the enzyme completely. Also according to experimental data lactose concentration in permeate decrease with time due to concentration polarization and hydrolysis. It was found that rejection factor of lactose increases from 33 to77%, with time from 5 to 85 min.

Keywords: Hydrolysis of lactose; Reactor membrane; Ultrafiltration

Introduction

Whey disposal is one of the most important economical and environmental problems in diary industry. This liquid produced when milk is changed in to cheese and approximately it consists of 5 % lactose, 0.7 % protein and 93% water and salt. Lactose waste increases biochemical and chemical oxygen demand (BOD and COD) which is in contrast with legal standard for wastewater [1]. In addition, lactose as a disaccharide is scarcely digestible for some people due to lack of β-galactosidas enzyme in their body. Furthermore, high amount of lactose in diary product like ice cream, condensed milk, etc., lead to undesirable grainy texture. Thus it is necessary to remove lactose from diary product and waste streams through methods like hydrolysis or separation processes [1-3].

Lactose can be hydrolyzed by two principle method, using acid treatment at high temperature (above 150°C) or enzymatic catalysis. The second one is preferred because of its milder operating temperature and pH. β-galactosidas is suitable enzyme to perform hydrolysis, that change lactose to glucose and galactose monosaccharide [4-6]. There are several mechanisms to describe this reaction in presence of water such as those presented by Segal and Cha [7]. Some researches consider in detail, using enzymatic lactose hydrolysis in batch CSTR reactor as well as fixed bed reactor. In the comparison, superior performance of CSTR reactor was clearly established [8-10].

A lot of researchers focus on methods, recover valuable components from diary waste streams, like adsorption, NF, RO, UF membrane processes and chromatography [11-16]. In terms of lactose removal (342 Da) from diary product and waste streams, RO and NF processes are more efficient, but need higher operating pressure and energy consumption than UF. On the other hand, enzymatic hydrolysis has some problem such as the effects of galactose formation on hydrolysis completion, and the high enzyme cost. One possibility to ensuring protection of purified enzyme is offered by using membrane processes [17]. So, UF system in combination with hydrolysis reactor might be preferred due to enzyme separation via membrane [18-19].

In present study, CSTR-UF system was used to hydrolyze lactose and separate β-galactosidas enzyme (116 kDa) from glucose and galactose (180 Da). The UF membrane of MWCO of 3 kDa of regenerated cellulose (RC) material was used to separate enzyme from products. All experiments were carried out with 0.319 molar aqueous solution of synthesized lactose as feed. The effect of operating pressure ranging between 2 and 5 bar and time on rejection and permeate flux were studied.

Materials and Methods

Chemicals

β-galactosidase (activity of the enzyme at 37°C = 3000 lau/cc) and lactose were purchased from Sigma-Aldrich Company. In order to analyze the concentration of lactose, sulfuric acid (98%) and phenol (89%) were obtained from Merck Company.

Membrane module and experimental set-up

Experiments were performed in a dead-end filtration setup. A schematic of CSTR-UF system is shown in Figure 1. The stirred cell was fed from 3 L reservoir which was pressurized using nitrogen gas cylinder. Feed pressure is monitored by pressure gauges and Permeate side was connected to the atmosphere so its pressure was assumed approximately equal to 0 bar (gauge pressure).

The RC ultrafiltration membrane of 3 kDa (MWCO) was placed in the 150 mL stirred cell and supported with sintered stainless still disc.

Determination of lactose concentrations

Lactose concentration was determined by means of calibration plot

***Corresponding author:** M. Pakizeh, Department of Chemical Engineering, Faculty of Engineering, Ferdowsi, University of Mashhad, P.O. Box 91775-1111, Mashhad, Iran of absorbance value. In order to plot standard absorbance curve, five sample of lactose were diluted to prepared aqueous solution of 0.01, 0.02, 0.03, 0.04 and 0.05 g.L^{-1}. Then 2 mL of each sample was mixed with 0.1 mL phenol solution and 6 mL sulfuric acid for 10 min at room temperature. The absorbances of samples were measured via 490 nm UV-Vis, spectrometer (Perkin-Elmer, USA). By using absorbance data represented in Table 1, it is possible to determine lactose content in every permeate sample.

Governing equation

The most frequent parameters used to characterize membranes are founded on its performance such as flux and rejection.

Rejection

In order to evaluate membrane performance, rejection of lactose should be obtained. The observed rejection, R, was calculated using equation (1):

$$R = 1 - \frac{C_P}{C_f} \quad (1)$$

where C_p and C_f are the solute concentration in permeate and feed side, respectively.

Volumetric flux

Permeate volumetric flux is expressed as volume of fluid permeated in certain time step. To meet this value equation (2) was used:

$$J = \frac{1}{A}\frac{\Delta V}{\Delta t} \quad (2)$$

where J, volumetric flux (L.m^{-2}.h^{-1}), A, effective membrane area (m^2), ΔV, permeate volume (L) and Δt is sampling time (h).

Result and Discussion

Influence of pressure on permeate flux

In order to illustrate the effect of pressure on permeate flux, the operating pressure was controlled at different value in the range of 2 to 5 bar. Permeate flux verses pressure change is shown in Figure 2. It can be observed from the results that permeate flux increased with increasing operating pressure. Nevertheless, flux may not be increased proportional with pressure at high level because of pore compressing. It should also be noticed that Figure 2 didn't show any compaction due to pressure effect [15,20].

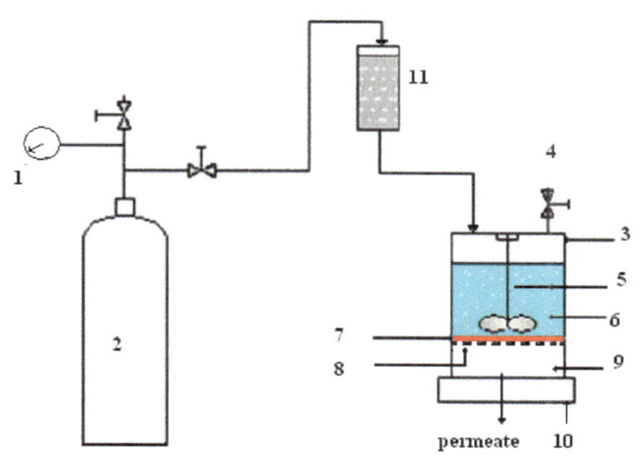

Figure 1: A schematic diagram of CSTR-UF setup.

Lactose concentration (g.L^{-1})	absorbance
0.01	0.130
0.02	0.155
0.03	0.175
0.04	0.190
0.05	0.207

Table 1: Calibration absorbance data.

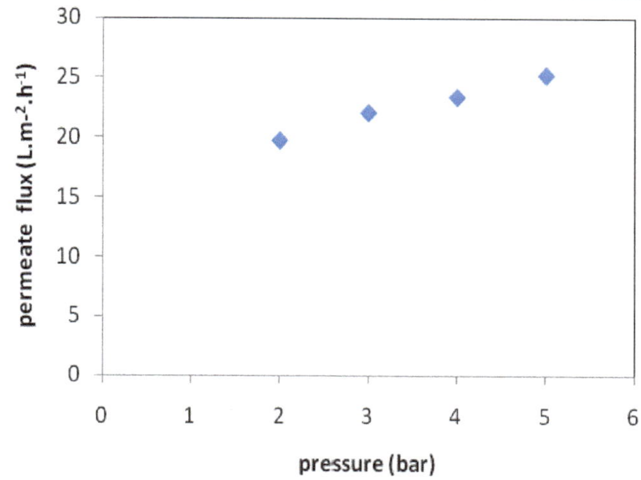

Figure 2: Influence of pressure on permeate flux.

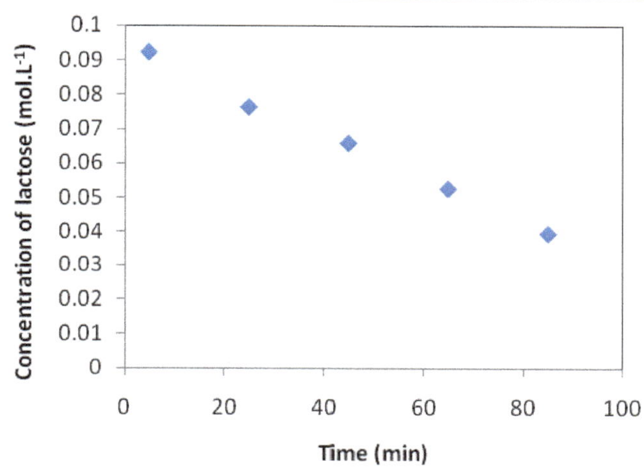

Figure 3: Influence of time on lactose concentration.

Influence of time on permeate concentration

To study the effect of time on lactose concentration at certain pressure, 2 bar, sample were collected from permeate at 5, 25, 45, 65 and 85 min. each sample were diluted thousand time and set for lactose amount determination. Test results are represented in Figure 3 and are shown that after 20 min lactose amount is decreased drastically from initial value of 0.139 mol.L^{-1} to 0.078 mol.L^{-1}, but further operation declined the value to 0.071 mol.L^{-1}. This phenomenon is caused because of lactose consumption in hydrolysis reaction by time spending. On the other hand, β-galactosidas as a high molecular weight protein are accumulated near membrane surface which cause resistance against lactose permeation. This behavior is known as the concentration polarization which can lead to fouling and plugging of membrane pore

[1,15]. Although, stirring minimized this effect by good mixing of bulk feed solution and turbulent flow near membrane surface [20].

Based on equation 1, rejection is enhanced with time at constant lactose concentration in feed. It should be mentioned that in all cases, sample appearance reveal that the UF membrane rejects the enzyme completely.

Influence of pressure on rejection of lactose

Effect of pressure rise between 2 and 5 bar on rejection of lactose is presented in Figure 4. Determinations of lactose amount on permeate side after 80 min showed that rejection is reduced from 0.77 to 0.57. This result suggest that increasing in permeate flux with pressure cause decrease of retention time of lactose in reaction zone. Thus lactose concentration on permeate side might be increased and rejection might be decreased. On the other hand, declining of rejection was slower at high pressure due to presence of enzyme molecules near membrane surface.

Effect of feed concentration on permeate flux

Figure 5 shows permeate flux verses pressure for both pure water and 0.319 molar of lactose as feed. This figure indicates that as lactose and enzyme added to feed, permeate flux decreased at all pressure, considerably. Base on this result, one can conclude that lactose and enzyme have resistance role on water permeation by deposition on top surface and pore wall of membrane [20]. Although, flux reduction is for the most part due to surface fouling of membrane by existence of enzyme.

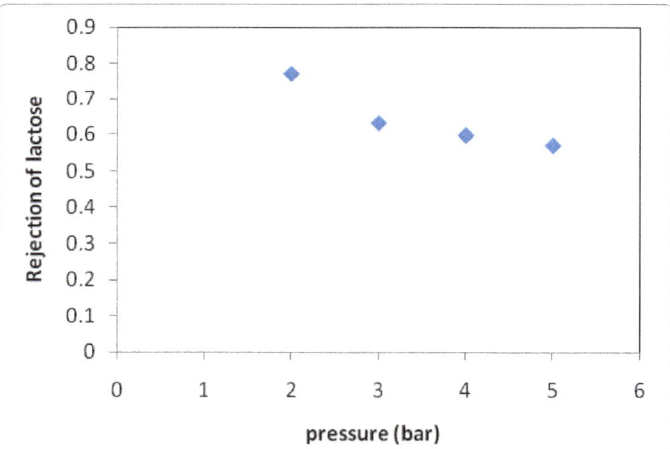

Figure 4: Influence of pressure on lactose rejection.

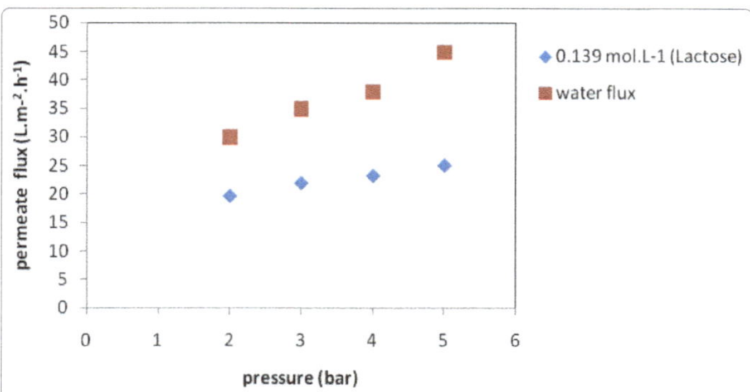

Figure 5: Permeate flux (L.m-2.h) vs. pressure (bar).

Conclusion

Lactose hydrolysis has been investigated using β-galactosidase enzyme in a continuous stirred tank -Ultrafiltration (CSTR-UF) to produce galactose and glucose. The major findings of the present study are summarized as follows:

- Increase in operation pressure as driving force, enhance permeate volumetric flux.
- Unchanged color of permeate samples reveal that no enzyme present in permeate and the UF membrane rejects the enzyme completely.
- Lactose concentration in permeate decrease with time due to concentration polarization and hydrolysis.
- It was found that rejection factor of lactose increases from 33 to77%, with time from 5 to 85 min due to drop lactose concentration.
- Fouling effect observed due to presence of lactose and enzyme in feed.

References

1. Cuartas-Uribe B, Alcaina-Miranda MI., Soriano-Costa E, Mendoza-Roca JA., Iborra-Clar MI, et al. (2009) A study of the separation of lactose from whey Ultrafiltration permeate using Nanofiltration. Desalination 241: 244-255.
2. Marwaha S, Kennedy JF (1988) Whey-pollution problem and potential utilization. Int J Food Sci Technol 23: 323–336.
3. Tragardh G (1991) Membrane applications in the food industry. Polymer Journal 23:521–529.
4. Illanes A, Ruiz A, Zuñiga M, Aguirre C, O'Reilly S, et al. (1990) Immobilization of lactase for the continuous hydrolysis of whey permeates. Bioprocess Eng 5: 257–262.
5. Bouhallab S, Touze C (1995) Continuous hydrolysis of caseinomacropeptide in a membrane reactor. dairy science and technology Lait 75: 251–258.
6. Grosová Z, Rosenberg M, Rebroš M (2007) Perspectives and applications of immobilised β-galactosidase in food industry: a review. Czech J Food Sci 26: 1–14.
7. Flashel E, Ratz F, Renken A (1982) The kinetics of lactose hydrolysis of the β-galactosidase from Aspergillus niger. Biotech bioeng 24: 2499-2518.
8. Mehaia MA, Alvarez J, Cheryan M (1992) Continuous hydrolysis of whey permeate lactose in a membrane reactor. International dairy journal 3: 179-192.
9. Prenosil JE, Hediger T (1988) Performance of membrane fixed biocatalyst reactors. I: Membrane reactor systems and modelling. Biotchnol Bioeng 31: 913-921.
10. Haider T, Husain Q (2009) Hydrolysis of milk/whey lactose by β galactosidase: A comparative study of stirred batch process and packed bed reactor prepared with calcium alginate entrapped enzyme. Chemical Engineering and Processing 48: 576–580.
11. Khider K, Akretche DE, Larbot A, (2004) Purification of water effluent from a milk factory by Ultrafiltration using Algerian clay support. Desalination 167: 147–151.
12. Hall CW (1995) Milk and Milk Products, Kirk-Othmer Encyclopedia of Chemical Technology. John Wiley & Sons.
13. Barba D, Beolchini F (2000) Minimizing water use in Diafiltration of whey protein concentrates. Sep Sci. Technol 35: 951–965.
14. Barba F, Beolchini D, Cifoni F, Vegli´o (2001) Whey protein concentrate production in a pilot scale two-stage diafiltration process. Sep Sci Technol 36: 587–603.

15. Baker RW (2005) Membrane Technology and application (2ndedn), John Wiley & Sons.

16. Pouliot Y, Wijers MC, Gauthier SF, Nadeau L (1999) Fractionation of whey protein hydrolysates using charged UF/NF membranes. J Membr Sci 158: 105–114.

17. Curcio S, Calabr`o V, Iorio G (2006) A theoretical and experimental analysis of a membrane bioreactor performance in recycle configuration. Journal of Membrane Science 273: 129–142.

18. Goulas AK, Grandison AS, Rastall RA (2003) Fractionation of oligosaccharides by Nanofiltration. J Sci. Food Agric 83: 675-680..

19. Chollangi A, Hossain Md M (2007) Separation of proteins and lactose from dairy wastewater. Chemical Engineering and Processing 46: 398–404.

20. Mhurchú BE, JN (2008) Dead-End and Cross flow Microfiltration of Yeast and Bentonite Suspensions: Experimental and Modeling Studies Incorporating The Use of Artificial Neural Networks, PhD Thesis, Dublin City University, School of Biotechnology.

Isolation, Purification and Characterization of β-Glucosidase from *Rauvolfia serpentina*

Om Prakash Verma[1]*, Abha Singh[1], Nirmala Singh[2] and Onkar Chaudhary[3]

[1]*Jacob School of Biotechnology & Bioengineering, SHIATS, Allahabad*
[2]*JDMC, KANPUR*
[3]*RTOU, ALLAHABAD*

Abstract

β-Glucosidase (EC 3.2.1.21) was extracted from different parts of *Rauvolfia serpentina* and was purified using ammonium sulphate fractional precipitation and Sephadex G-25 chromatography. The molecular weight of different bands in *Rauvolfia serpentina* was found in the range of 19.4KDa to 92.257KDa .The enzyme β-glucosidase has optimum pH 5.0 and the optimum temperature was found at 60°C while the thermal stability of enzyme β-glucosidase was found to be 30°C. The activation with $FeSO_4$ is due to Fe^{++} ions because $FeCl_3$ do not activate the enzyme.

Keywords: *Rauvolfia serpentine*; β-glucosidase; SDS PAGE; Enzyme activity

Introduction

β-Glucosidase (β-D-glucoside glucohydrolase, EC 3.2.1.21) comprises a heterogeneous group of enzymes that are able to cleave the β-glucosidic linkages of di- and/or oligosaccharides, or other glucose conjugates. β-Glucosidases are widely distributed and play pivotal roles in many biological processes, such as degradation of cellulosic biomass, hydrolysis of glycolipids, cyanogenesis, and modification of secondary metabolites [10]. In plants, β-glucosidase activity is involved in processes such as the compartmentalization and activity of phytohormones [17,18], defense mechanisms against microbes, insects, or parasitic plants [23], and they are also thought to have roles in lignification and cell wall decomposition. Several β-glucosidases of plant origin are highly substrate-specific, such as those having selective specificity toward saponin [16], hydrojuglone [9], cinnamyl alcohols [8], flavones [22] and isoflavones [15] β-Glucosidase occurs abundantly in young plants parts and localized in plastids [12].

Materials and Methods

The *Rauvolfia serpentina* was used as β -glucosidase source and was obtained from SHIATS, Allahabad. Plants were freshly harvested for use and processed immediately to avoid tissue breakage and loss of enzyme of activity. For the purification of β-glucosidase, ammonium sulfate, D-gluconolactone, imidazole, iodoacetate, EDTA, FCC reagent, TCA, paranitrophenol, β-D-glucopyranoside, standard BSA, acrylmide, bisacryl and glycine, TEMED, ammonium per sulfate were used.

Sephadex G-25 column chromatography

Pellets obtained after ammonium sulphate precipitation (30-70%) were dissolved in elution buffer and loaded on sephadex G. 25 column and fractions of volume 2 ml each were collected. The reading of protein was measured by spectrophotometer at 280 nm and activity was also estimated at 405 nm.

Electrophoretic methods

Crude enzyme preparations and chromatographic fractions were subjected to native PAGE. Samples were electrophoresed through 0.75 mm thick gel slabs (2cm 4% staking gel, pH 8.5 and 13.5 cm 10% resolving gel pH 8.8) using the alkaline gel system. The running buffer was 3.5 gm Tris and 14 gm Glycine, pH 8.3. For SDS PAGE, enzyme SDS buffer were mixed in ratio (4:1) v/v. Samples were heated in boiled water for 5 min. and were chilled in liquid nitrogen and. Thereafter, surface was loaded and electrophoresis was performed as described by [19].

Results and Discussion

The enzyme from each part of *Rauvolfia serpentina* was extracted and activity was estimated using spectrophotometer at wave length 405 nm. For the reaction, $_p$NPG was used as substrate. The reaction mixture was incubated for 15 min at room temperature. Change in OD was calculated and the activity was estimated (Table 1).

The activity of crude enzyme in *Rauvolfia serpentina* was observed in per gram. From data shown in Figure 1, it was observed that mature leaf 1 has the highest activity of crude enzyme in *Rauvolfia serpentina*. This indicates the higher requirement of β-glucosidase mediated reaction in corresponding plant parts.

Purification of β-glucosidase from crude protein extract

β-glucosidase was purified from crude protein extract by Ammonium sulfate precipitation and Sephadex G-25 column chromatographic procedure.

1. **Ammonium sulfate Precipitation:** $(NH_4)_2SO_4$ was mixed to crude enzyme under chilled conditions and kept for 2 hrs in refrigerator to develop the precipitate. There after centrifugation was done at 12,000 rpm, 4°C for 10 minutes and precipitate was discarded and supernatant processed further. For 0 to 30 % precipitation, weight of $(NH_4)_2SO_4$ = 7.79 gm, Volume of supernatant = 49 ml, for 30 to 70% precipitation,

***Corresponding author:** Om Prakash Verma, Jacob School of Biotechnology & Bioengineering, Shiats, Allahabad weight of $(NH_4)_2SO_4$ =12.201 gm, after centrifugation, pellet obtained was used for further process study.

2. Sephadex G -25 Column Chromatography: Pellet was dissolved in elution buffer and after loading protein on pre-equilibrated column was run and fractions were collected of volume 2 ml each, measured absorbance of protein at 280 nm and enzyme activity at 405 nm was estimated to be 60° C.

Characterization of enzyme

Molecular weight estimation of proteins in *Rauvolfia serpentina* from SDS gel: Total number of bands were recorded as per their distance migrated and Molecular weight determined from SDS gels of *Rauvolfia*. Graph was plotted for *Rauvolfia serpentina* (Figure 2).

In each plant, distance of marker was used as x-axis while log molecular weight was used as y-axis. Plant parts were selected in which there were a maximum number of bands. For each band distance was measured and logs molecular weight and molecular weight for each band was calculated. Total number of bands for each parts of plant was also calculated (Figure 3).

The SDS PAGE developed was analyzed to the protein bands and their molecular weight was determined in comparison to standard molecular weight marker. In *Rauvolfia serpentina* the maximum number of bands of protein was found to be five in fruit, and the minimum number of bands of protein was found one in old leaf and stem. The molecular weight of different bands in *Rauvolfia serpentina* was found in the range of 19.48 KDa to 92.257 KDa. (Table 2).

S. No.	Sample	Change in OD/10 μl	Units/ml/min	Activity (%)
1.	Fruit	0.03	200	118
2.	Flower	0.165	1100	64.7
3.	Flower stem	0.061	400	23.5
4.	Very young leaf	0.240	1600	94.1
5.	Young leaf	0.076	500	29.4
6.	Mature leaf 1	0.256	1700	100
7.	Mature leaf 2	0.095	633	37.2
8.	Old leaf	0.132	880	51.8
9.	Stem	0.035	230	13.5
10.	Root	0.025	160	9.4

Table 1: β glucosidase activity profile in *Rauvolfia Serpentina*

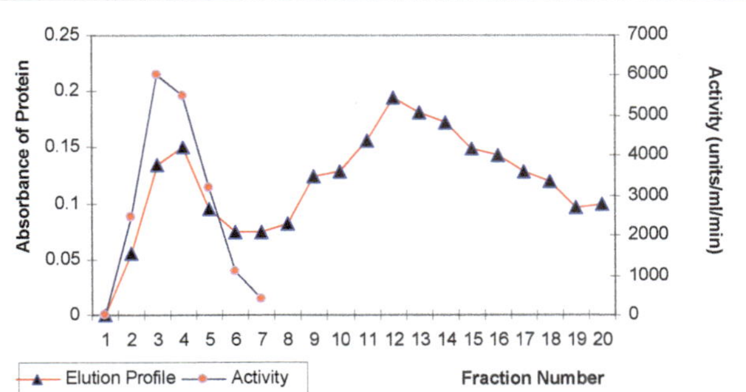

Figure 1: Elution profile and activity of enzyme in sephadex G-25 column chromatography.

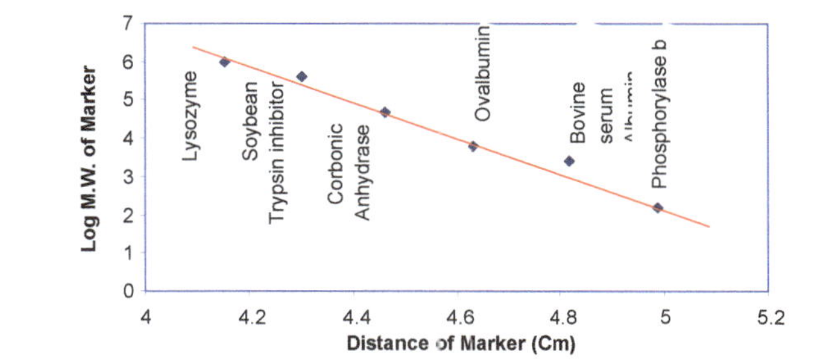

Figure 2: Molecular weight estimation of proteins in *Rauvolfia serpentina* from SDS gel.

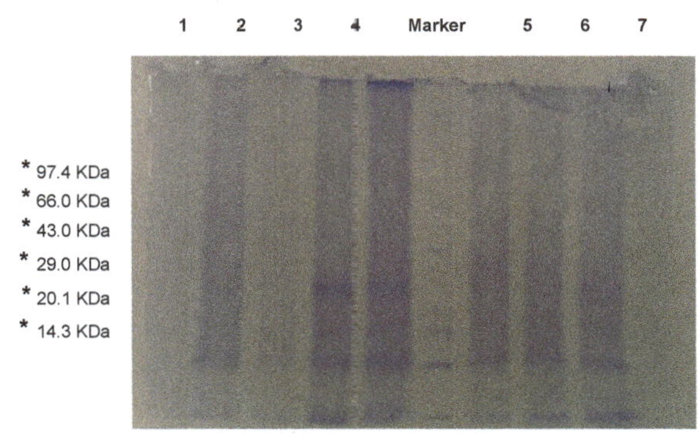

Figure 3: SDS PAGE gel of *Rauvolfia serpentina* in the presence of β mercaptoethanol.
Lane1 - stem, Lane2 - old leaf, Lane3 - mature leaf, Lane4 - young leaf, Lane5 - flower stem, Lane6 - flower, Lane7 - fruit.

S. No.	Log M.W. of marker	M.W. of marker (KDa)	Name of marker	Distance of marker (Cm)	Distance of protein band (Cm)	Log M.W. of protein	M.W. of protein (Da)
1.	4.988	97.4	Phosphorylase b	2.2	2.3	4.965	92,257.14
2.	4.819	66	Bovine serum Albumin	3.4	3.5	4.720	52,480.75
3.	4.633	43	Ovalbumin	3.8	4.1	4.595	39,355.01
4.	4.462	29	Corbonic Anhydrase	4.7	4.4	4.535	34,276.78
5.	4.303	20.1	Soybean Trypsin inhibitor	5.6	5.1	4.39	24,547.09
6.	4.155	14.3	Lysozyme	6.0	5.6	4.29	19,498.45

Table 2: Molecular Weight from SDS (Coomasie Staining) gel of *Rauvolfia serpentina*

Determination of optimum temperature

The pH 5.0 buffer was used for activity measurement using 20 mM substrate at temperature range 0 to 80°C. Thus optimum temperature was estimated. It is interesting to note that in case of *Rauvolfia serpentina* flower the β- glucosidase is active even beyond 60°C temperature (Figure 4). It is not so rare observation yet quite intriguing as it reflects its much stronger regulation at the cellular level, where increase in temperature at such a high level seems unlikely. On the other hand, it makes the reaction quite selective, if we extrapolate these results to in-vivo scenario. The optimum temperature in different plants were as Strawberry 60°C [24], *Leuconostoc mesenteroides* 50°C [6], Grape 45°C [20], Soybean 30°C [25] and maize 50°C [10].

Determination of optimum pH

After studying the kinetics, it was concluded that optimum pH was found in range of 4.5 to 5.0 as shown in Figure 5. The optimum pH for catalyzing β-glucosidase action of different plants ranged as strawberry - 4.0 [3], Lodgepole Pine - 5 to 6 in variety of substrate [8], *Leuconostoc mesenteroides* – 6.0 [2], *Dalbergia cochinchinensis* Thai Rosewood –5.0 [27], Grape –5.0 [1], soybean -6.0 [14], Maize –5.8 [10].

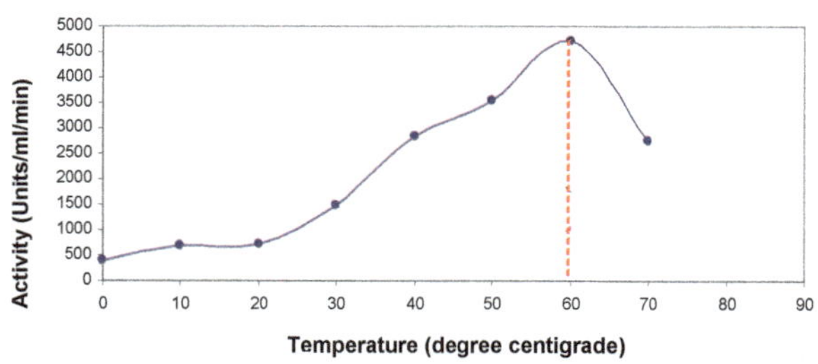

Figure 4: Determination of temperature optima for β-glucosidase.

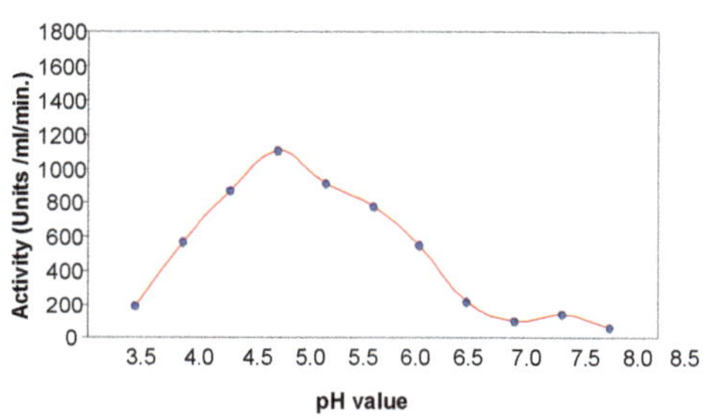

Figure 5: Determination of pH optima for β-glucosidase.

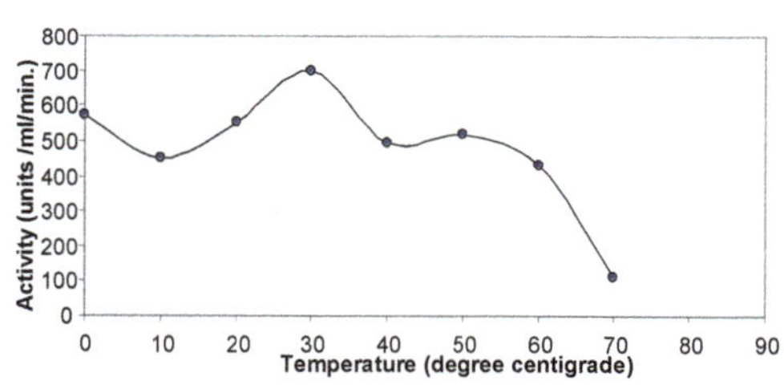

Figure 6: Thermal stability of enzyme.

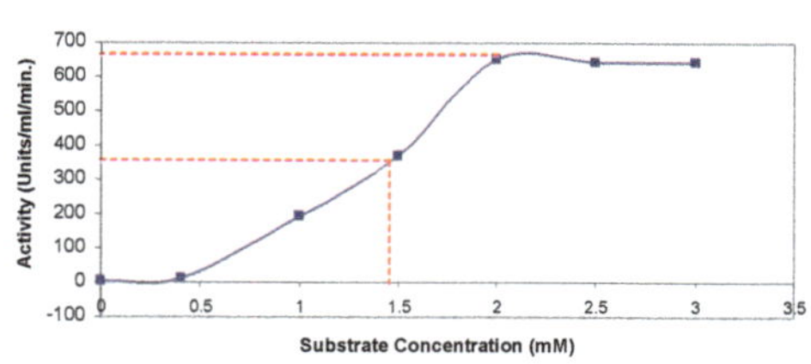

Figure 7: Effect of substrate concentration on β-glucosidase.

Determination of thermal stability

Thermal stability of β-glucosidase was determined. It showed highest activity display at temperature 60°C, while the activity was approximately 50% at temperatures below and above it. Interestingly, a substantial proportion (60%) remained even at 60°C, and around 20% even at 70°C as shown in Figure 6. It was clear that enzyme β-glucosidase has pH optima 5.0. The optimum temperature was found at 60°C (Figure 6). Thermal stability of enzyme β-glucosidase was found to be 30°C.

Effect of varying concentration of the substrate

The substrate para nitro phenyl β-D glycoside was used for measuring the effect of varying concentration of substrate. It was observed that β-glucosidase has V_{max} = 680 units / ml /min. and Km = 1.4 mM as shown in Figure 7. From the Michaelis - Menten curve, it was observed that β-glucosidase has Km= 1.4 mM and V_{max} = 680 units /ml/min (Figure 7). In *Catharanthus roseus* Km< 20mm and V_{max} = 243 nkatal per mg protein [21] In *Leuconostic mesenteroides* Km values were 3.7m mole^{-1} using linamarin as substrate while it was 0.07 m mole^{-1} using pN glycopyranoside substrate [13], while Km were 5.4 mM with pNPG substrate in *Dalbergia cochinchinensis* Thai Rosewood [27] In some other plants species, Km and V_{max} were known as following strawberry – Km = 18.5 mM, V_{max} = 474x10^{-6} IU/ml [7], papaya –Km = 0.11mM, V_{max} =9.51x10^{-6} IU/ml. [26], grape –Km= 1.81 mM, V_{max} = 43.9x10^{-6} IU/ml [5] and *Rauvolfia serpentina*, for Raucaffricine glucosidase Km = 1.3 mM, V_{max} =0.5n Kat /µg protein and for storictosidine glucosidase –Km= 1.8mM, V_{max}= 2.6p Kat/ µg protein [11]. In our study of kinetics, it was observed that D-glucono lactone highly inhibited the enzyme. The same result was concluded from some other plants *Dalbergia cochinchinensis* Thai Rose wood [27], grape [20]. In present study it was also found that D-glucono lactone is a strong inhibitor for glucosidase activity. It also implies that D-gluconolactone could be a completive inhibitor of the enzyme i.e., occupying the active site in place of substrates.

References

1. Bandaranayake H, Esen A (1996) Nucleotide sequence of a second β-glucosidase (glu2) cDNA from maize (*Zea mays L*). Plant Physiol 110: 1048.
2. Bennett RN, Hick AJ, Dawson GW, Wallsgrove RM (1995) Glucosinolate biosynthesis. Plant Physiol 109: 229-305.
3. Bothast RJ, Saha BC (1997) Ethanol production from agricultural biomass substrates. Adv App Microbiol 44: 261-286.
4. Cicek M, Esen A (1995) Cloning and sequencing of a cDNA coding for β-glucosidase (Dhurrinase) from *Sorghum bicolor* (L) Moench. Plant Physiol 109: 1497.
5. Cicek M, Esen A (1998) Structure and expression of a dhurrinase (β-glucosidase) from sorghum. Plant Physiol 116: 1469-1478.
6. Cicek M, Esen A (1999) Exp™ression of soluble and catalytically active plant (monocot) β-glucosidases in *E. Coli.* Biotechnol Bioeng 63: 392-400.
7. Day AJ, Canada FJ, Diaz JC, Kroon PA, Mclauchlan R, et al. (2000) Dietary flavonoids and isoflavone glycosides are hydrolysed by the lactase site of lactase phlorizin dydrolase . FEBS Lett 468: 166-170.
8. Dharmawardhana DP, Ellis BE, Carlson JE(1995) A β-glucosidase from lodgepole pine xylem specific for the lignin precursor coniferin. Plant Physiol 107:331-339.
9. Duroux L, Delmotte FM, Lancelin JM, Keravis G, Allemand CJ (1998) Insight into naphthoquinone metabolism; β-glucosidase catalysed hydrolysis of hydrojuglone β-D- glucopyranoside. Biochem J 333: 275-283.
10. Esen A (1992) Purification and characterization of maize (*Zea mays* L.) β-glucosidase. Plant Physiol 98: 174-182.

11. Esen A, Cokmus C (1990) The mechanism of substrate (aglycone) specificity in β- glucosidase is revealed by crystal structure of mutant maize β- glucosidase DIMBOA-DIMBOAGlc and dhurrin complexes. Proc Natl Acad Sci 97: 13555-13560.

12. Esen A, Stetler DA (1993) Subcellular localization of maize β-glucosidase. Maize Genet Coop News Lett 67: 19-20.

13. Gueguen Y, Chemardin P, Janbon G, Arnaud A,Galzy P (1996) A very efficient β-glucosidase catalyst for the hydrolysis of flavor precursors of wines and fruit juices. J Agric Food Chem 44: 2336-2340.

14. Warzecha H, Gerasimenko I, Kutchan TM, Stöckigt J (2000) Molecular cloning and functional bacterial expression of a plant glucosidase specifically involved in alkaloid biosynthesis. Phytochemistry 54:657-666.

15. Hósel W, Barz W (1975) β-Glucosidases from *Cicer arietinum L.* Purification and Properties of isoflavone-7-O-glucoside-specific β-glucosidases. Eur J Biochem 57:607-616.

16. Inoue K, Shibuya M, Yamamoto K, Ebizuka Y (1996) Molecular cloning and bacterial expression of a cDNA encoding furostanol glycoside 26-O-beta-glucosidase of Costus speciosus. FEBS Lett 389: 273-277.

17. Keresztessy Z, Brown K, Dunn MA, Hughes MA (2001) Identification of essential active-site residues in the cyanogenic β-glucosidase (linamarase) from cassava (*Manihot esculenta Crantz*) by site-directed mutagenesis. Biochem J 353: 199-205.

18. Keresztessy Z, Kiss L, Hughes MA (1994) Investigation of the active site of the cyanogenic beta-D-glucosidase (linamarase) from *Manihot esculenta* Crantz (cassava). I. Evidence for an essential carboxylate and a reactive histidine residue in a single catalytic center. Arch Biochem Biophys 314: 142-152.

19. Laemmli UK (1970) Cleavage of structural proteins during the assembly of the head of bacteriophage T4. Nature 277: 680-685.

20. Leah R, Kigel J, Svendsen I, Mundy J (1995) Biochemical and molecular characterization of a barley seed beta-glucosidase. J Biol Chem 270: 15789-15797.

21. Luijendijk TJC, Stevens LH, Verpoorte R (1998) Purification and characterisation of strictosidine β-d-glucosidase from *Catharanthus roseus* cell suspension cultures. Plant Physiol Biochem 36: 419-425.

22. Maier R, Carle R, Kreis W, Reinhard E (1993) Purification and Characterization of a Flavone 7-O-Glucoside-Specific Glucosidase from Ligulate Florets of *Chamomilla recutita*. Planta Med 59: 436-441.

23. Poulton JE (1990) Cyanogenesis in plants. Plant physiol 94: 401 – 405.

24. Poulton JE, Li CP (1994) Tissue Level Compartmentation of (R)-Amygdalin and Amygdalin Hydrolase Prevents Large-Scale Cyanogenesis in Undamaged Prunus Seeds. Plant Physiol 104: 29-35.

25. Rask L, Andréasson E, Ekbom B, Eriksson S, Pontoppidan B, et al. (2000) Myrosinase: gene family evolution and herbivore defense in Brassicaceae. Plant Mol Biol 42: 93-113.

26. Shen Z, Eisenreich W, Kutchan TM (1998) Bacterial biotransformation of 3 α(S) strictosidine to the monoterpenoid indole atkaloid valllesiachotamine. Phytochemistry 48: 293-296.

27. Svasti J, Srisomsap C, Surarit R, Champattanachai V, Sawangareetrakul P, et al. (1999) Isolation and characterization of an enzyme with β-glucosidase and β-fucosidase activities from dalbergia cochinchinensis Pierre. J BioChem 119: 585-590.

Synthesis of Zinc-Organic Frameworks Nano Adsorbent and their Application for Methane Adsorption

Mahnaz Aghajanloo[1], Ali Morad Rashidi[2]* and Mohamad Ali Moosavian[1]

[1]Oil and Gas Processing Centre of Excellence, School of Chemical Engineering, College of Engineering, Tehran University, Tehran, Iran
[2]Research Institute of Petroleum Industry, Nanotechnology Research Center, West Blvd., Azadi Sport Complex, Tehran, Iran

Abstract

In the present work, MOF-5 consisting of Zn_4O inorganic vertices was hydrothermally synthesized. Three-step activation treatment including preservation in initial solvent, solvent exchange and drying under evacuated condition was performed and it was found that activation is efficient by increasing the methane adsorption capacity of samples. The final product was characterized by Fourier transform infrared spectroscopy (FT-IR), X-ray diffraction (XRD), thermal gravimetric analysis (TGA), scanning electron microscopy (SEM) and pore size distribution (PSD). The corresponding pore size of MOF-5/DEF (sample A) and MOF-5/DMF (sample B) which was determined by nitrogen adsorption at 77 K, were equal to 31.27 Å and 42.65 Å and also BET surface area of samples were 2157 and 1532 m^2/g, respectively. Methane adsorption equilibrium for both samples was measured in a volumetric adsorption unit and it was found that the adsorbent (A) and (B) in this work have methane adsorption capacity of 18.15 and 11.36 mmol/g at temperature of 298 K and pressure of 36 bar, respectively. Four different models (Langmuir, Unilan, Toth and Sips) were used to correlate the experimental equilibrium data. The root mean square error values demonstrated that the Toth and Sips equation provide better models for correlating adsorption isotherms.

Keywords: Adsorption; Metal-Organic Framework; Methane storage

Introduction

The drawbacks of fossil fuels are well known as increasing the environmental pollution, changing the climate, and instability in oil market. Therefore, the use of alternative energy sources has recently attracted special attention [1,2]. Over the past two decades, considerable efforts have been allocated to develop new fuel technologies to meet transport fuel demand and reduce reliance on conventional fuels [3]. Natural gas (NG) has significant advantages over other hydrocarbon fuels because of its natural abundance and higher hydrogen-to-carbon ratio, which leads to higher thermal efficiency. In addition, it offers lower emission of carbon monoxide and nitrogen oxides. In practice, no particulates are produced compared to gasoline [4-6]. Among the possible ways to storage the natural gas (methane is the primary component of natural gas), its storage in the form of adsorbed gas (adsorption at moderate pressure of 35 bar) is superior to the current compression technology because it offers some advantages such as low cost, high energy efficiency and safety [7,8].

Porous materials provide a suitable base for satisfying storage demands [5]. In comparison with the same tank without adsorbent, efficient porous adsorbents have the capability of storing natural gas in tanks at lower pressure (40 bar) [9,10]. Traditional adsorbents such as zeolites, activated carbons, and MWCNTs have problems including low adsorption capacity in industrial applications. In recent years, hybrid porous solids such as metal organic framework (MOF) have emerged as promising materials for different applications including small molecule storage [11], separation [12,13], catalysis [13,14] and biological imaging [15,16]. In order to better understand the usage of MOFs for these applications, incremental attempt is being put forth to investigate the synthesis condition, physical properties, chemical properties and morphology of MOFs [17]. Metal-organic frameworks (MOFs) are an intriguing class of 3D organic-inorganic hybrid materials [18]. They are constructed by organic ligands and metal clusters linked with coordination bonds of moderate strength [19,20]. Recently, there has been significant progress towards building of MOF-n structures (n as an integer assigned in roughly chronological order) with desirable materials features, including ultra high porosity in the range of micro and nano pore sizes, unsaturated metal coordination sites and fascinating structures [21,22]. The MOF family also has high thermal stability with decomposition temperatures in the range of 300°C to 500°C. Due to superior properties of MOFs, some research works have also been done in order to increase the adsorption capacity of natural gas by the use of MOFs and expand the ANG (Adsorbed Natural Gas) technology [23,24].

Numerous studies on properties and chemical structures of MOFs have been done [25]. One of the most interesting structures of metal organic frameworks is MOF-5 or $Zn_4O(bdc)_3$ (bdc=1,4-benzenedicarboxylate), which is a highly porous material consisting of Zn_4O-cluster that has a decisive role in the final shape and the pore size of the crystals, linked to the terephthalate anion forming a cubic network (Figure 1) [26-28]. The initial findings of reversible gas adsorption of MOF-5 (also widely known as IRMOF-1) have made it one of the most studied metal–organic frameworks [29]. These MOFs have pore sizes ranging from microporous (<2 nm) to mesoporous (2-50 nm) [30]. Typically, the synthesis of MOFs is frequently performed by mass transfer methods (*e.g.* diffusion of amine in to a solution containing a metal salt and organic acid), hydrothermal and solvothermal methods [31].

In the present study, MOFs have been synthesized from two different solvents with same procedure. The samples of octahedral cluster of MOF-5/DEF and MOF-5/DMF are named as sample (A) and

***Corresponding author:** Ali Morad Rashidi, Research Institute of Petroleum Industry, Nanotechnology Research Center, West Blvd., Azadi Sport Complex, Tehran, Iran

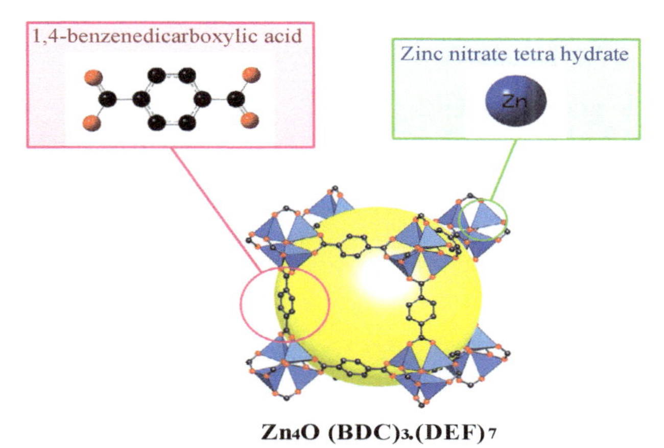

Figure 1: Crystal structures of MOF-5 (C, black; Zn, blue; O, red; gust molecule, yellow sphere and all H atoms have been omitted for clarity).

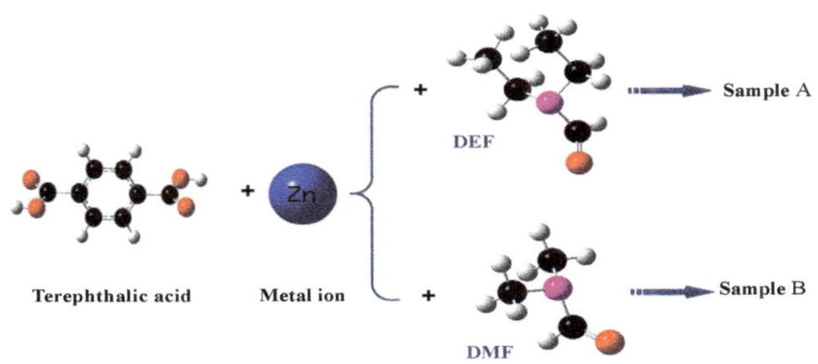

Figure 2: Reaction components for producing samples. (C,black; Zn,blue; O,red; N,pink and H;white).

sample (B), respectively (Figure 2). After vacating their pores by using a three-step activation process, methane adsorption measurements of samples were performed at temperature of 298 K and pressure up to 36 bar.

Experimental

Materials

The solvents N,N-diethylformamide (DEF, 99%) and N,N-demethylformamide (DMF, 99%) were purchased from Fluka. 1,4-Benzene dicarboxylic acid (H_2BDC) as a ligand and zinc nitrate tetrahydrate ($Zn(NO_3)_2{\cdot}4H_2O$) as a metal were purchased from Merck. All materials were used without further purification.

Synthesis of MOFs

During synthesizing the metal-organic frameworks, several factors should be considered. For instance, the type of solvent and its concentrations have essential role in process of the synthesis.

Synthesis of MOFs from DEF (sample A): Optimum conditions for the MOF synthesis were determined through experimental design. Consequently, for synthesis of MOF-5 at first, 0.34 g of terephthalic acid and 1.78 g of $Zn(NO_3)_2{\cdot}4H_2O$ were dissolved in 45 mL of pure N,N-diethylformamide (DEF) solvent. After sufficient stirring at ambient temperature and formation of a white precipitate, the mixed solution was transferred into a Teflon-lined autoclave and heated at 105°C for 24 h to yield MOF-5 as cubic crystals. It is important to note that heating slowly decomposes the mixed solution and helps to deprotonate the organic ligand. The yellow crystals formed on the wall of reaction container, were repeatedly washed with DEF and soaked in anhydrous chloroform for 72 h followed by filtration. Finally, the resulting solid product was dried under vacuum at 120°C for 8 h.

Synthesis of MOFs from DMF (sample B): The same procedure of sample (A) was used for synthesis of sample (B). The only difference was solvent type, which in this case, N, N-dimethylformamide (DMF) was used as solvent and the ligand concentration was doubled. After removal of sample from autoclave, a white product was filtered off and washed for three times with DMF, then treated with $CHCl_3$.

Activation method

Three step activation: (1) As-synthesized MOF-5 was washed with DEF/DMF three times (2) solvent exchange of DMF/DEF ↔ $CHCl_3$ was performed in which the sample was washed with a lower boiling point solvent for removal of the first solvent from the pores. For implementing this exchange, the mixture was immersed in $CHCl_3$ for 4 days and then the solvent removed under mild conditions. (3) After solvent exchange process, the sample was filtered and then the product was dried under vacuum at 120°C for 8 h.

Characterization

The structure of MOF-5, which was characterized by Fourier transform infrared spectrum of the samples, was recorded on a Bruker Vector 22 FT-IR spectrometer. The X-ray diffraction analysis was done on a Philips diffractometer with Co-K radiation source. The thermal stability of the sample was performed using thermogravimetric analysis (TGA/SDTA851e Mettler, Swiss) where 5.53 mg of sample was heated up to 700°C with a heating rate of 10°C/min in air atmosphere. To examine the morphology, the crystal structure and size of samples were analyzed by a CamScan S360 MV2300 scanning electron microscope (SEM) Cambridge with prior gold deposition. The nitrogen physical adsorption-desorption isotherm of MOF-5 adsorbent was measured at 77 K using a Micrometrics $ASAP_{2010}$ analyzer and pore size distribution was obtained by built-in software of the mentioned device. Prior to the measurement, the sample was out-gassed at 250°C for 3 h.

Methane adsorption measurement

To examine the methane adsorption capacity of MOF-5, we have used volumetric method under the equilibrium condition. A schematic diagram of experimental apparatus is shown in Figure 3. Before each experiment, 1 g of sample was loaded in the adsorption cell and de-gassing pretreatment process of the adsorbent was carried out by vacuum pump at 433 K for 1.5 h. After de-gassing, the system was cooled to ambient temperature and then the temperature was kept fixed by a water bath.

Methane (99.9% purity) adsorption measurements were performed at constant temperature of 298 K, which was made by water bath. For investigating the effects of solvent type and activation process on gas adsorption capacity of samples, the pressure was increased up to 36 bar. By using Helium (99.99% purity) as a test gas for measuring the dead volumes (including hollow space and the connected tubes), the exact pressure decrease because of gas adsorption was obtained. By applying the SRK equation of state and MATLAB software, the adsorption rate was calculated at each pressure.

Results and Discussion

FT-IR spectra

FT-IR spectra for both samples (Figure 4) confirmed the presence of all functional groups and MOF-5 bonds forming. Two sharp peaks

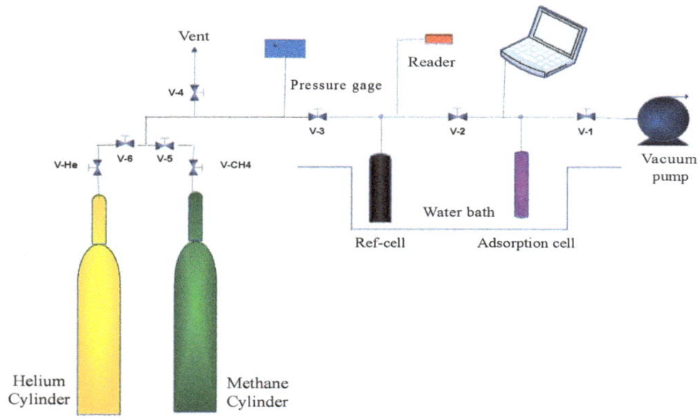

Figure 3: Setup for volumetric adsorption capacity measurement.

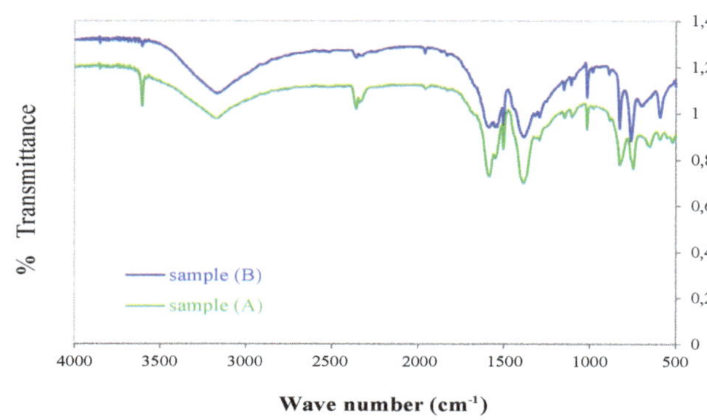

Figure 4: FTIR spectrum of MOF-5 for samples A and B.

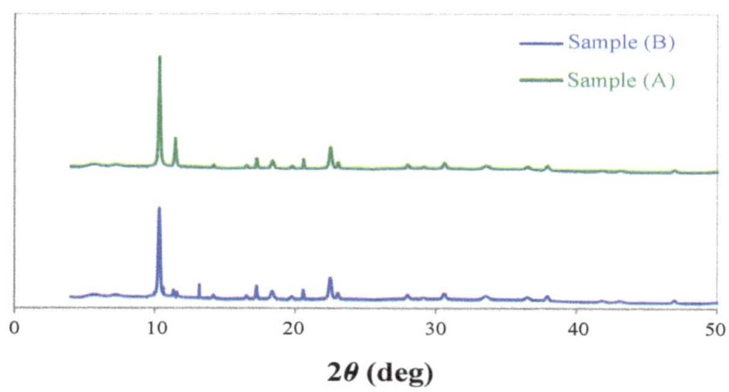

Figure 5: Powder x-ray diffraction pattern of MOF-5 for samples A and B.

at 1600 cm^{-1} and 1405 cm^{-1} in sample (A) and 1606 cm^{-1} and 1409 cm^{-1} in sample (B) were allocated to C-O linked to Zn. Small peaks, which take place in the range of 1330 cm^{-1} to 895 cm^{-1} in sample (A) and 1313 cm^{-1} to 827 cm^{-1} in sample (B), related to the C-H group present in the benzene ring of terephthalic acid. The bands between 2566 cm^{-1} and 3095 cm^{-1} in both samples which are due to methyl and $-CH_2-$ groups in DEF and DMF molecules, confirm the removal of guest molecules from the structure [32].

X-Ray diffraction

The X-ray diffraction of both sample A and B that were crystalline in nature, are represented in Figure 5. Sample (A) and (B) showed a sharp peak at 10.31° and 10.29°, respectively. Based on these results the crystals size of the sample A and sample B that have been obtained from scherrer equation, was 2491 Å and 2785 Å, respectively. The peak positions and relative intensities of synthesized samples were in good agreement with reported XRD patterns for MOF-5 [33].

Thermo-gravimetric analysis

Figure 6 shows the TGA result for activated MOF-5/DEF. The result indicates a thermal stability up to 400°C, where decomposition of the framework starts. The TGA curve experiences only one main weight loss around 450°C, meaning that guest molecules like DEF, were evacuated completely. Although due to the humidity, about 3.8 wt% loss was still observed on the activated sample curve.

Scanning electron microscopy

The morphology of MOF-5 (sample A and B) was studied by scanning electron microscopy (SEM) after gold deposition (Figure 7). Figure 7(A), shows three dimensional cube-like microcrystals of sample (A), synthesized under the same conditions with sample (B), but in DEF as a solvent and molar ratio of H_2BDC and zinc nitrate (H_2BDC:Zn) 1:6. Figure 7(B) shows sample (B) as hexahedron crystals from a simple reaction of zinc nitrate and H_2BDC in DMF and molar ratio of H_2BDC and zinc nitrate (H_2BDC:Zn) of 1:3. Moreover, comparing Figure 7(A) and (B), it is obvious that MOF/DEF crystalline is smaller than MOF/DMF.

Nitrogen adsorption-desorption isotherm

The nitrogen physical adsorption–desorption isotherm of MOF-5 was measured at 77 K by using a Micrometrics $ASAP_{2010}$ analyzer. Prior to gas adsorption analysis, the samples were heated at 250°C for 3 h. From N_2 adsorption isotherms, PSD was evaluated using the Barrett–Joyner–Halenda (BJH) method. At 77 K the isotherm (Figure 8) shows a typical type II isotherm based on the IUPAC classification which corresponds to mono- and multilayer physical adsorption.

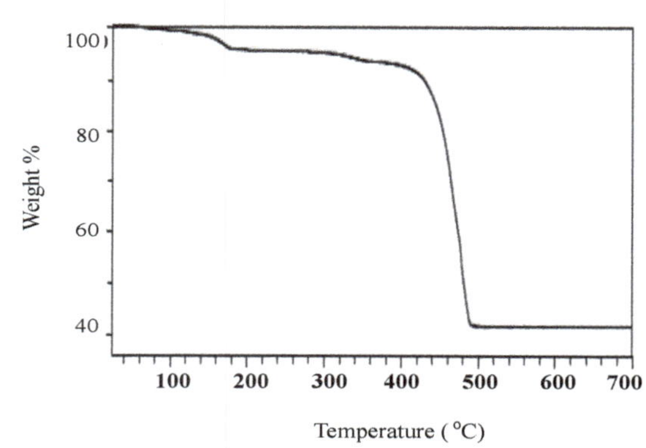

Figure 6: The TGA profile of activated MOF-5/DEF under air atmosphere.

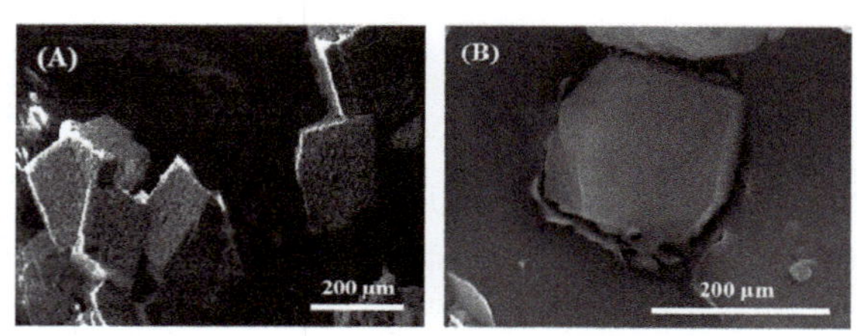

Figure 7: Morphology of MOF-5 crystals synthesized with, DEF (A) and DMF (B).

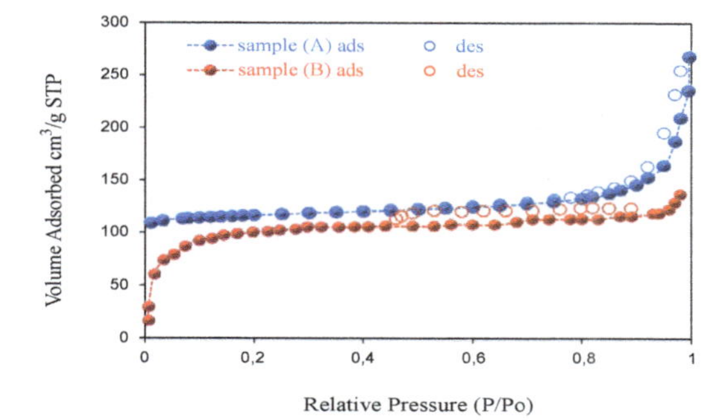

Figure 8: Adsorption-desorption isotherm of N_2 in synthesized materials.

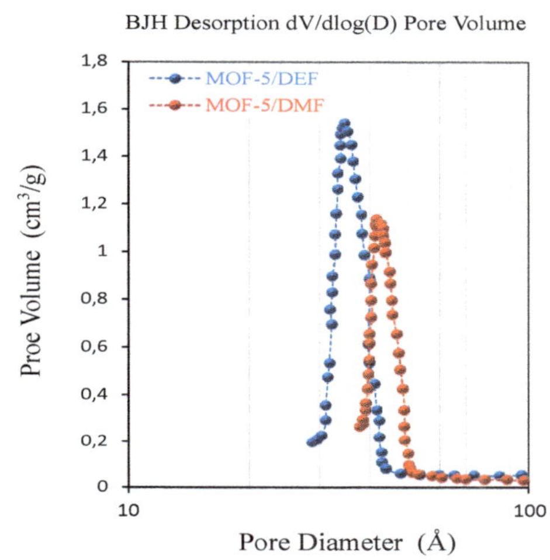

Figure 9: Pore size distribution.

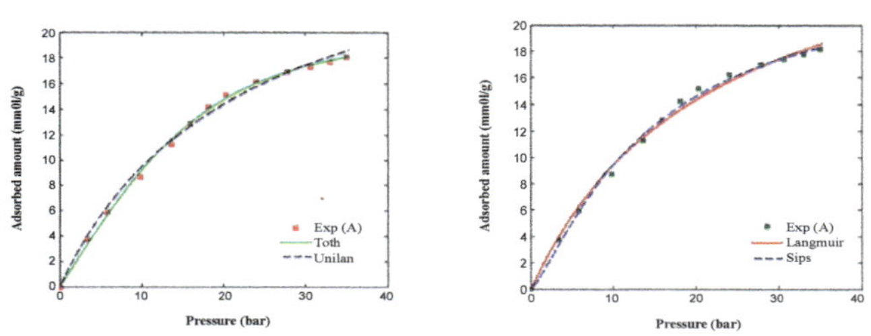

Figure 10: Comparison between experimental data of methane adsorption on MOF-5/DEF (sample A) and isotherms from Langmuir and Sips model (left), Toth and Unilan model (right) at 298 K.

Figure 9 shows the pore volume and pore size distribution of samples as an important parameter of adsorbents. Clearly, Figure 9 demonstrated that the pore size distribution of MOF-5/DEF displays sharper and higher peak. The BET surface area of sample A and B were 2157 and 1532 m^2/g, respectively.

Methane adsorption capacity of samples

Figures 10 and 11 compare the methane adsorption isotherms of two MOF-5 samples synthesized by two different solvents of dimethyl formamide (A) and diethyl formamide (B), at temperature of 298 K and different pressures in the range of 0-36 bar. The presence of solvent molecules in the crystal cavities prevents entering the methane gas in holes. Thus, the pore volume of the material and consequently the adsorption capacity are significantly reduced. To solve this problem, a three-step activation process was performed to remove the guest solvent molecules from the frameworks.

The chart shows that methane adsorption capacity in the sample (A) is higher in comparison with sample (B). The higher CH_4 adsorption capacity of MOF-5/DEF can be ascribed to its smaller crystal size, smaller mean pore diameter and larger pore volume.

Table 1 represents methane adsorption capacity of MOF-5 synthesized in this work with other adsorbents reported in the literature. The temperature in all works is at room temperature and the maximum pressure is around 30 bar. As it is indicated from this table our adsorbents have the highest methane storage capacity of 18.15 and 11.36 mmol/g for samples synthesized by DEF (A) and DMF (B) in comparison with the data reported in the literature.

Correlation of Isotherms: The data fitting was performed by using Langmuir, Toth, Unilan and Sips adsorption models, which are generally used for micro porous adsorbents. Table 2 shows the equations and correlated parameters. From this study, we conclude that Toth and Sips models lead to good correlation of the data for both samples. On the other hand, the Langmuir model and the Unilan model did not produce satisfactory results for correlating the isotherms of methane adsorption at 298 K.

Conclusions

In summary, MOF-5 was hydrothermally synthesized by employing two different solvents. These MOFs structure were obtained from the Zn(BDC) system by DEF (sample A) and DMF (sample B) as solvents. Resulting components were characterized by FT-IR, XRD, SEM, TGA and ASAP analysis. Three-step activation mechanism of MOF-5 including repeatedly washed sample with initial solvent, solvent exchange, and drying under vacuum, enhance the capacity for methane adsorption by removal of almost all guest molecules (solvent and

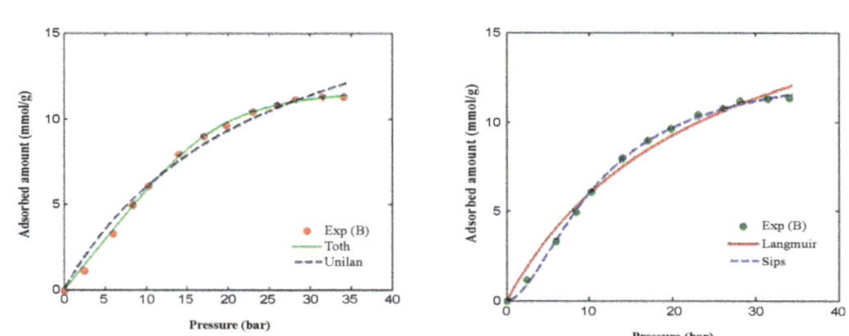

Figure 11: Comparison between experimental data of methane adsorption on MOF-5/DMF (sample B) and isotherms from Langmuir and Sips model (left), Toth and Unilan model (right) at 298 K.

Sample	Storage capacity Gravimetric (mmol/g)	Conditions pressure/ temperature	reference
Coconut shell-AC	7.4	35 bar / 300 K	[35]
CMK-3	5.2	35 bar / 298 K	[36]
MOF-5	14	30 bar / 298 K	[37]
SOC-MOF	4	27 bar / 298 K	[38]
MIL-53(Al) Al(OH)(bdc)	10.2	35 bar / 298 K	[39]
MOF-5/DMF	11.36	36 bar / 298 K	this study
MOF-5/DEF	18.15	36 bar / 298 K	this study

Table 1: Comparison of the methane adsorption capacity of MOF-5 with other adsorbent.

Isotherm Name	Isotherm equation	Isotherm constant	
		sample (A)	sample (B)
Langmuir	$Q_e = \frac{q_m bP}{1+bP}$	q_m=30.23	20.69
		b=0.045	0.041
		R-square=0.995	0.986
		RMSE=0.435	0.487
Unilan	$Q_e = \frac{q_m}{2s} ln\left(\frac{1+be^s P}{1+be^{-s} P}\right)$	q_m=30.24	20.44
		b=0.045	0.041
		s=0.0002	0.0008
		R-square=0.995	0.986
		RMSE=0.456	0.512
Toth	$Q_e = \frac{q_m bP}{(1+(bP)^n)^{\frac{1}{n}}}$	q_m=20.4	11.77
		b=0.049	0.05
		n=2.212	3.825
		R-square=0.998	0.998
		RMSE=0.258	0.158
Sips	$Q_e = \frac{q_m bP^{\frac{1}{n}}}{1+bP^{\frac{1}{n}}}$	q_m=24.22	13.21
		b=0.035	0.015
		n=0.792	0.575
		R-square=0.997	0.999
		RMSE=0.336	0.123

Table 2: Isotherm equations used to model the adsorption experimental data and their parameter.

water) and vacating the pores. The measurements of methane storage at temperature of 298 K and 36 bar, indicated that the sample (A) can adsorbed 18.15 mmol/g, so it showed better methane storage behavior than sample (B) with 11.36 mmol/g at the same pressure. Experimental data for the methane adsorption of both samples were modeled by isotherms of Langmuir, Toth, Sips, and Unilan equations. It was finally found that the Toth and Sips models provide better correlation of adsorption isotherms.

References

1. Farzad S, Taghikhani V, Ghotbi C, Aminshahidi B, Nemati Lay E (2007) Experimental and Theoretical Study of the Effect of Moisture on Methane Adsorption and Desorption by Activated Carbon at 273.5 K. Nat Gas Chem 16: 22-30.
2. Kumar A (2011) Adsorption of Methane on Activated Carbon by Volumetric Method. National Institute of Technology.
3. Lozano-Castello D, Alcaniz-Monge J, Casa-Lillo MA, Cazorla-Amoros D, Linares-Solano A (2002) Advances in the study of methane storage in porous carbonaceous materials. Fuel 81: 1777-1803.
4. Guan C, Su F, Zhao XS, Wang K (2008) Methane storage in a template-synthesized carbon. Sep Purif Technol 64: 124-126.
5. Zakaria Z, George T (2011) IJRRAS 9: 225-230.
6. Muto A, Bhaskar T, Tsuneishi S, Sakata Y (2005) Activated Carbon Monoliths from Phenol Resin and Carbonized Cotton Fiber for Methane Storage. Energy Fuels 19: 251-257.
7. He Y, Zhang Z, Xiang S, Fronczek FR, Krishna R, et al. (2012) A robust doubly interpenetrated metal-organic framework constructed from a novel aromatic tricarboxylate for highly selective separation of small hydrocarbons. See comment in PubMed Commons below Chem Commun (Camb) 48: 6493-6495.
8. Esteves IAAC, Lopes MSS, Nunes PMC, Mota JPB (2008) Sep Purif Technol 62: 281-296.
9. Nour UM, Tayeb AM, Farag HA, Awad SH (2009) IJET 9: 381-389.
10. Kuppler RJ, Timmons DJ, Fang QR, Li JR, Makal TA, et al. (2009) Potential applications of metal-organic frameworks. Coord Chem Rev 253: 3042-3066.
11. Murray LJ, DincÄƒ M, Long JR (2009) Hydrogen storage in metal-organic frameworks. See comment in PubMed Commons below Chem Soc Rev 38: 1294-1314.
12. Li JR, Kuppler RJ, Zhou HC (2009) Selective gas adsorption and separation in metal-organic frameworks. See comment in PubMed Commons below Chem Soc Rev 38: 1477-1504.
13. Seo JS, Whang D, Lee H, Jun SI, Oh J, et al. (2000) A homochiral metal-organic porous material for enantioselective separation and catalysis See comment in PubMed Commons below Nature 404: 982-986.
14. Ma L, Abney C, Lin W (2009) Enantioselective catalysis with homochiral metal-organic frameworks. See comment in PubMed Commons below Chem Soc Rev 38: 1248-1256.
15. Huxford RC, Rocca JD, Lin W (2010) Metal-Organic Frameworks as Potential Drug Carriers. Curr Opin Chem Biol 14: 262-268.
16. Farha OK, Hupp JT (2010) Tetratopic Phenyl Compounds, Related Metal-Organic Framework Materials and Post-Assembly Elaboration. US Patent 20100170395.
17. Kitagawa S, Kitaura R, Noro S (2004) Functional porous coordination polymers. See comment in PubMed Commons below Angew Chem Int Ed Engl 43: 2334-2375.
18. Papaefstathiou GS, MacGillivray LR (2003) Inverted metal–organic frameworks: solid-state hosts with modular functionality. Coord Chem Rev 246: 169-184.
19. Llewellyn PL, Bourrelly S, Serre C, Filinchuk Y, Férey G (2006) How hydration drastically improves adsorption selectivity for CO(2) over CH(4) in the flexible chromium terephthalate MIL-53. See comment in PubMed Commons below Angew Chem Int Ed Engl 45: 7751-7754.
20. Fromm KM, Sague JL, Mirolo L (2010) Coordination Polymer Networks: An Alternative to Classical Polymers? Macromol Symp 291-292: 75-83.
21. Spokoyny AM, Farha OK, Mulfort KL, Hupp JT, Mirkin CA (2010) Porosity Tuning in Carborane-Based Metal Organic Frameworks (MOFs) via Coordination Chemistry and Ligand Design. Inorg Chim Acta 364: 266-271.
22. Yaghi OM, O'Keeffe M, Ockwig NW, Chae HK, Eddaoudi M, et al. (2003) Reticular synthesis and the design of new materials. See comment in PubMed Commons below Nature 423: 705-714.
23. Makal TA, Li JR, Lu W, Zhou HC (2012) Methane storage in advanced porous materials. See comment in PubMed Commons below Chem Soc Rev 41: 7761-7779.
24. Jia Z, Li H, Yu Z, Wang P, Fan X (2011) Densification of MOF-5 synthesized at ambient temperature for methane adsorption. Mater Lett 65: 2445-2447.
25. Buso D, Nairn KM, Gimona M, Hill AJ, Falcaro P (2011) Fast Synthesis of MOF-5 Microcrystals Using Sol−Gel SiO2 Nanoparticles. Chem Mater 23: 929-934.
26. Li H, Eddaoudi M, OKeeffe M, Yaghi OM (1999) Design and synthesis of an exceptionally stable and highly porous metal-organic framework. Nature 402: 276-279.
27. Cheng SH, Liu S, Zhao Q, Li J (2009) Improved synthesis and hydrogen storage of a microporous metal–organic framework material. Energy Convers Manage 50: 1314-1317.
28. Lu CM, Liu J, Xiao K, Harris AT (2010) Microwave enhanced synthesis of MOF-5 and its CO2 capture ability at moderate temperatures across multiple capture and release cycles. Chem Eng J 156: 465-470.
29. Tranchemontagne DJ, Hunt JR, Yaghi OM (2008) Room temperature synthesis of metal-organic frameworks: MOF-5, MOF-74, MOF-177, MOF-199, and IRMOF-0. Tetrahedron 64: 8553-8557.
30. Zou R, Abdel-Fattah AI, Xu H, Zhao Y, Hickmott DD (2010) Storage and separation applications of nanoporous metal–organic frameworks. Cryst Eng Comm 12: 1337-1353.
31. Blanita G, Ardelean O, Lupu D, Borodi G, Mihet M, et al. (2011) Microwave Assisted Synthesis of Mof-5 At Atmospheric Pressure. Rev Roum Chim 56: 583-588.
32. Iswarya N, Kumar MG, Rajan KS, Balaguru RJB (2012) Synthesis, Characterization and Adsorption Capability of MOF-5. Asian J Sci Res 5: 247-254.
33. Kaye SS, Dailly A, Yaghi OM, Long JR (2007) Impact of preparation and handling on the hydrogen storage properties of Zn4O(1,4-benzenedicarboxylate)3 (MOF-5). See comment in PubMed Commons below J Am Chem Soc 129: 14176-14177.

34. Kaye SS, Dailly A, Yaghi OM, Long JR (2007) J. Amer Chem Soc 129: 14176-14177.

35. Prauchner MJ, Reinoso FR (2008) Microporous Mesoporous Mater 109: 581-584.

36. Zhou H, Zhu S, Honma I, Seki K (2004) Chem Phys Lett 396: 252-255.

37. Duren T, Sarkisov L, Yaghi OM, Snurr RQ (2004) Langmuir 20: 2683-2689.

38. Moellmer J, Celer EB, Luebke R, Cairns AJ, et al. (2009) Microporous Mesoporous Mater 129: 345-353.

39. Bourrelly S, Llewellyn PL, Serre C, Millange F, Loiseau T, et al (2005) Amer Chem Soc 127: 13519-13521.

Studies on Metal (Cu and Sn) Extraction from the Discarded Printed Circuit Board by Using Inorganic Acids as Solvents

Vijayaram R* and Dr. Chandramohan K

Department Of Chemical Engineering, Kongu Engineering College, Perundurai, Erode, India

Abstract

This Project deals with the experimental results of the leaching of metal like Cu and Sn from Printed Circuit Boards (PCB) from obsolete electronic devices by means of leaching using different combinations of acidic mixtures followed by precipitation. Printed circuit boards were dismantled, cut into small pieces, and fed into a Ball mill. The powder obtained was leached by using the aqueous solutions H_2SO_4, H_2SO_4+HCl, HCl, and HCl+HNO_3. The lowest values for the percentage of metal extraction were obtained with H_2SO_4 while the HCl+HNO_3 mixture exhibited an extraction of Sn and Cu in a maximum percentage. Precipitates were obtained at different pH values by neutralizing the leach liquors using NaOH. The HCl+HNO_3 leach system presented the highest recovery values from the powder feed as well as from the leach liquor. Comparing the results for Sn and Cu extraction after 120 minutes obtained with the various leach systems, 3.0 N HCl+1.0 N HNO_3 exhibited the highest percentage values for simultaneous Sn and Cu extraction (93.3%Sn and 92.7% Cu).

Keywords: Tin; Copper; Recycling; Electronic scrap; Extraction.

Introduction

More than millions of electronic devices have been installed over the past two decades. As new and more efficient electronic devices come onto the market, significant numbers of old electronic devices are being scrapped. Likewise, the number of obsolete electronic devices has also been growing. Due to the lack of specialized companies working with the recycling of obsolete electronic devices, such equipment is commonly scrapped in inappropriate disposal areas, together with domestic garbage, with no specialized recycling processes [1].

Printed circuit boards used in electronic devices are composed of different materials, such as polymers, ceramics, and metals, which render the process of scrappers even more difficult. The presence of metals, such as tin and copper, encourages recycling studies from an economic point of view. The Average Percentage availability of metal in PCB is shown in table 1. However, the presence of heavy metals turns this scrap into dangerous residues. This, in turn, demonstrates the need for solutions to this type of residue so as to dispose of it in a proper manner without harming the environment. The recycling of printed circuit boards from obsolete electronic devices is, at present, a fairly new activity, although opportunities are available for expansion in this area. For instance, gold, silver, tin, and copper, among other metals, can be recovered by means of the hydrometallurgical treatment of printed circuit boards [2]. The E-waste can be treated by thermal and non-thermal methods.

The Non-thermal method of treatments uses leaching as one of the main stages. Leaching is the process of extracting a soluble constituent from a solid by means of a solvent. In extractive metallurgy, it is the process of dissolving minerals from an ore or a concentrate, or dissolving constituents from metallurgical products.

Metals	Percent Availability
Copper	17-19
Aluminium	7-8
Zinc	0.1
Tin	3.25
Nickel	0.05
Iron	2.9
Lead	4.0
Silver	Traces

Table 1: Composition of PCB.

Lead, tin, and Copper were successfully recovered from alloy wire electronic scrap through acid/alkali leaching and a NaOH solution was added to the leach liquor for metals precipitation. Some researchers have also been working on the recycling of electronic scrap [3].

Recycling

The Recycling process can be done by two methods

Thermal processing

In Thermal processing the printed circuit boards can be recycled by applying enormous amount of heat. In this process technique three stages are involved. They are as follows

- Pyrolysis
- Hydration
- Metallurgy

Non-thermal processing

This is another way of recycling the printed circuit boards. This method is simple and easy to handle the feed materials. This process includes

- Dismantling or Disassembly
- Grinding or Crushing

***Corresponding author:** Vijayaram R, Department Of Chemical Engineering, Kongu Engineering College, Perundurai, Erode, India

- Separation
- Chemical Treatment

Materials and Methods

In this Project we have handled the non-thermal processing technique as shown in figure 1. The Printed Circuit Board was crushed in a ball mill and sieved to separate according to size of 0.2 mm. The undersized particle was separated which is feed to the Leaching mixture. The powdered PCB was dissolved in a leaching mixture of acids in different combination. Copper and Tin present in the PCB are allowed to dissolve in the acidic solutions. The amount of metals dissolved in the solution can be determined by using the AAS or by the volumetric analysis method. The concentrations of inorganic acids used in these tests are determined by volumetric analysis. The use of high concentrations of acids was allowed to control the experimental costs. About 10 gms of PCB powder was allowed to react with the 100ml of the leach solution of different concentration [4].

The experiments were carried out at 60 ± 2 °C. Samples of the leach liquor were collected at intervals of 10, 30, 60, and 120 minutes during the experiment. Leach liquor samples were collected, filtered and sent for chemical analysis to determine tin and copper concentration in aqueous solution [5].

Results and Discussions

Table 2 shows the percentage of copper and Tin Extracted from different leaching agents. The objective of the experiment is to determine the leaching ability of the inorganic acids at low concentration. Since the concentration of the acid is low, we can use the commercial grade as quite efficiently. Figure 2 and 3 shows the graphical representation of percentage metal (Cu and Sn) extracted versus reaction time. It was found that the leach liquor of concentrations 3.0 N HNO_3 and 3.0 N HCl+1.0N HNO_3 leach systems were the most efficient to recover Tin and Copper (93.3% and 92.7%) simultaneously.

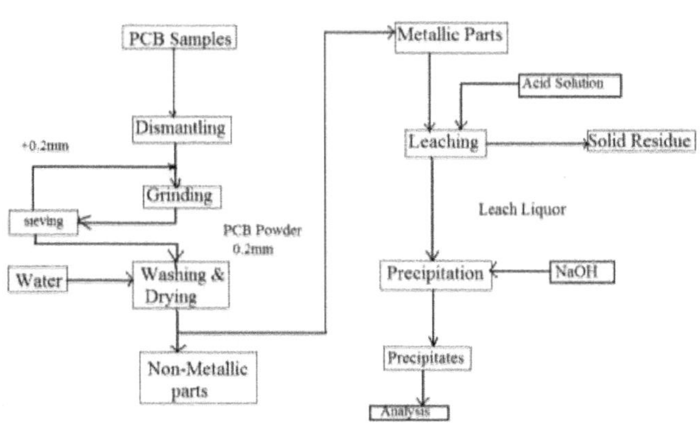

Figure 1: The block diagram of the process used in this work.

S.No	Leaching mixture	% Copper	% Tin
1	2.18 N H_2SO_4 + 3 N HCl	8.5	58.2
2	3 N HCl +1 N HNO3	92.7	93.3
3	3N HCl	23.8	85.1
4	2 N HCl + 2 N HNO_3	75.85	76.7
5	3 N HCl + 2N HNO_3	46.71	89.48
6	3 N HNO_3	86.95	93.3
7	2.5 N HCl + 2.5 N HNO_3	52.42	92.21

Table 2: Percentage of Cu & Sn obtained from leached acidic solution after 120 min.

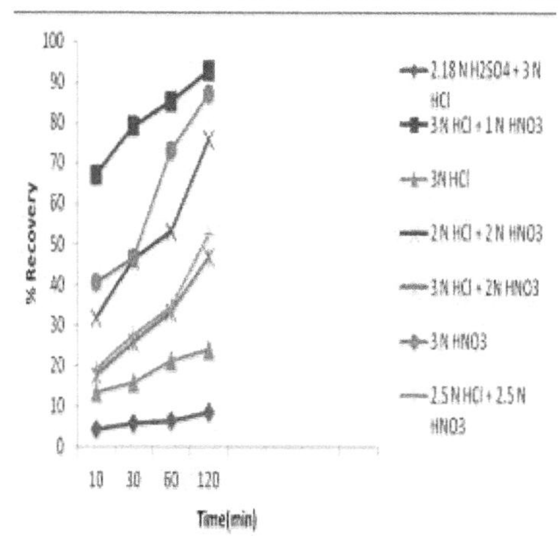

Figure 2: Cu recovery vs. reaction Time.

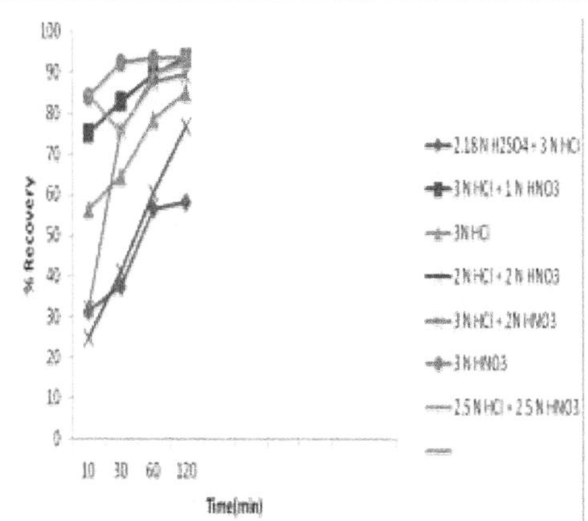

Figure 3: Sn recovery vs reaction Time.

Tin and copper content in the powder generated by the processing of PCBs from obsolete electronic devices is 3.25% Sn and 3.5% Cu. Thus the leach can be considered to be highly significant when compared to the content found in primary mineral sources.

The percentage of tin and copper extracted in the 2.18 N H_2SO_4+3 N HCl (58.2% and 8.5%) leach system showed the lowest results among the leach systems under study, while the 3.0 N HCl+1.0N HNO_3 (93.3% and 92.7%) system presented the highest results for simultaneous tin and copper extraction (using one stage leaching).

The precipitates obtained through the neutralization of the leach liquor from the 2.18 N H_2SO_4+3 N HCl systems exhibited the lowest results for tin and copper recovery from the powder feed and from the leach liquor. The 3.0 N HCl+2.0 N HNO_3 systems presented the highest values for simultaneous tin and copper recovery.

Concerning the morphology of the recovered copper, SEM images revealed dendritic growth of the deposits obtained from all three individual acid leaching solutions (Figures 4a, b, c). Specifically, in the case of the copper obtained from the sulphuric acid solution the recovered metal presented Fine dendritic structure with branches of about 80-100 μm (Figure 4a). On the other hand, copper deposited from nitric or hydrochloric acid

Figure 4: SEM images of recovered copper revealing the dendritic morphology from **(a) H_2SO_4 solution, (b) HNO_3 solution, (c) HCl solution.**

solutions demonstrated a more compact structure although a dendritic structure was conserved [6].

Conclusion

E-waste is an emerging issue, driven by the rapidly increasing quantities of complex end-of-life electronic equipment. The global level of production, consumption and recycling induces large flows of both toxic and valuable substances. Although awareness and readiness for implementing and Improvement is increasing rapidly, there are many obstacles to manage end-of-life products safely and effectively in industrializing countries [7]. Support securing economic efficiency and sustainability of e-waste management systems by optimizing the value added and improve the effectiveness of collection and recycling systems.

References

1. Gramatyka P, Nowosielski R, Sakiewicz P (2007) Recycling of waste electrical and electronic equipment JAMME 20: 535-538.
2. Li J, Shrivastava P, Gao Z, Zhang HC (2004) Printed circuit board Recycling: A State of Art survey. Ind Eng Dept 27: 33-42.
3. Havlik T, Orac D, Petranikova M, Miskufova A (2011) Hydrometallurgical treatment of used printed circuit board after thermal treatment. Waste Manag 31: 1542-1546.
4. Agarwal A, Perween M, Shubhajit P, Sahu KK (2011) Recovery of copper from PCB leach solution by solvent extraction. National Metallurgical Laboratory 20-22.
5. Veit HM, Bernardes AM, Ferreira JZ, Tenório JAS, Malfatti CDF (2006) Recovery of copper from printed circuit boards scraps by mechanical processing and electrometallurgy. Journal of Hazardous materials 137: 1704-1709.
6. Masavetas I, Moutsatsou A, Nikolaou E, Spanou S, Zoikis-Karathanasis A, et al. (2009) Production Of Copper Powder From Printed Circuit Boards By Electrodeposition. Global NEST Journal 11: 241-247.
7. Huang K, Guo J, Xu Z (2009) Recycling of waste printed circuit board: A review of current technologies and treatment status in china. Journal of hazardous materials 164: 399-408.

Modeling of Vapour Liquid Equilibrium Data for Thyme Essential Oil Based on UNIQUAC Thermodynamic Model

A. Varasteh[1*], H.A. Tahmasebi[2] and R. Gheshlaghi[3]

[1]*Chemical Engineering, Islamic Azad University, Shahrood Branch, Semnan, Shahrood, Iran*
[2]*Chemical Engineering Department, Islamic Azad University, Quchan Branch, Asian Road, Quchan, Iran*
[3]*Department of Chemical Engineering, Ferdowsi University of Mashhad, Mashhad, Khorasan, Iran*

Abstract

Vapour Liquid Equilibrium (VLE) data for natural ingredients and essential oils has not stucied at all yet. It is because of complexity of these mixtures. In this study, the equilibrium concentration of the mixture components was studied by using the UNIQUAC model at different pressure. This model is capable of determining temperature between 70°C to 170°C and the pressure limits 30 Kpa to 202 Kpa. The modeling's results showed thymol in the liquid phase were increased from 50% molar to 90% molar in constant pressure by increasing temperature from bubble point to dew point's temperature. Thymol concentration in the liquid phase was increased from 42% molar to 78% molar by increasing water concentration of the mixture at a constant pressure of 100 Kpa and constant temperature 177°C. Another obtaining result showed that thymol concentration in the triple mixture, which includes 30% molar water, 35% molar thymol and 35% molar Cymene in liquid phase, was increased from 60% molar to 82% molar at constant temperature of 177°C by decreasing pressure from 100 Kpa to 20 Kpa.

Keywords: Thyme; Essential oil; VLE; UNIQUAC; Modeling; Thermodynamic phases

Introduction

Thymus vulgaris plants of the family lamiaceae are fragrant [1]. Distribution of 215 species worldwide and 14 species of *Thymus vulgaris* has been reported in Iran [2]. This plant food is as traditional as the digestive, antispasmodic, antitussive, and due to the main compound thymol, in the food industry, pharmaceutical, health and beauty uses [3]. The aqueous extract, aqueous Ethanol and Propylene Glycol in the preparation of *Thymus vulgaris* shampoo, creams and ointments are used [4]. This plant is antifungal effects, antibacterial and ant parasite and its therapeutic efficacy for treatment of asthma, recurrent dry cough and bronchitis has been proved. This medical plant, syrups, pill, incense and extracts were prepared and has approved by the drug unit of supervision, ministry of health and medical education [5]. The treatment plant, flowering shoot and its leaves are dried. World wages species *Thymus vulgaris* (Thymus Daensis Celak) are distributed in different regions of Qazvin province. This plant is propagated by seed and can be divided. thymol is the main compound of *Thymus vulgaris* plant and the molar amount of thymol dittany have been reported about 40% to more than 70% of the world wages. Based on the results of extraction *Thymus vulgaris* world wages sample collected from the Hamedan which includes 26 compounds 95.1% of the thymol with 44.7%, Para Cymene with 18.6% and Alpha Trypen with 16.5% in most combinations [6]. Garden *Thymus vulgaris* every year in large areas of farmland in Europe, North Africa and the United states is grown. In Figure 1 and Table 1 the essential oils yield is 2 to 3 percent and thymol in the essential oil has been reported about 40 to 65 percent [7].

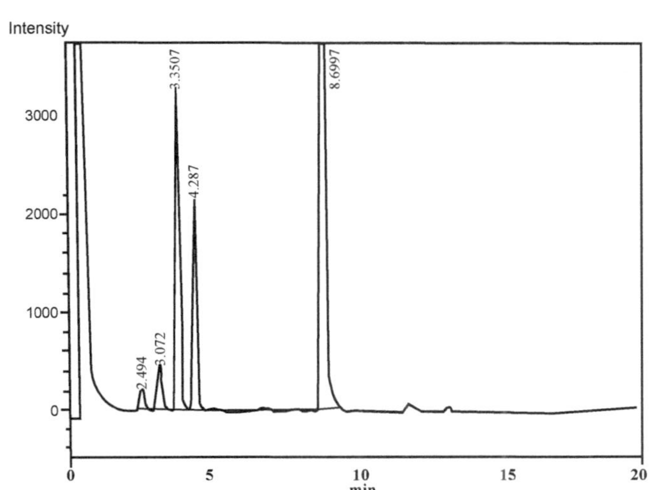

Figure 1: Chromatogram of the Thyme essential oil.

Gas chromatography

Chemical analysis was performed on a Hewlett–Packard 5890 series II gas chromatograph fitted with a HP-5 capillary column (25 m×0.20 mm, film thickness 0.33 μm). The carrier gas was hydrogen at a flow of 1.0 ml min-1 and a split ratio 1 100. The column temperature was programmed from 60–250°C at 2.0°C min-1; the injector temperature was 250°C and the detector (FID) temperature was 280°C [8].

Gas chromatography/mass spectroscopy

The analysis was performed on a Hewlett–Packard 5970 series II gas chromatograph coupled to a HP 5970 mass selective detector using a fused silica capillary column HP-5 (25 m×0.20 mm, film thickness 0.33 μm). The column temperature was programmed from 60-250°C

***Corresponding author:** A. Varasteh, Chemical Engineering, Islamic Azad University, Shahrood Branch, Semnan, Shahrood, Iran

NO	NAME	RI	Percent%
1	Tricyclene	925	0.53
2	β-Pinene	928	0.18
3	α-Tujene	931	0.33
4	Camphene	945	0.32
5	Myrcene	992	0.887
6	α-phlladrene	1005	0.13
7	α-Terpinene	1016	0.75
8	P-Cymene	1027	4.6
9	Limonene	1032	0.33
10	Carvacrol	1035	1.4
11	1,8-Cineole	1043	1.64
12	γ-Terpinene	1062	4.48
13	Cis-Sabinene Hydrate	1074	0.8
14	Terpinolene	1088	0.12
15	Trans-Sabinene Hydrate	1097	0.27
16	Borneol	1164	1.61
17	Terpinene-4-Ol	1188	0.57
18	Carvacrol/Methyl Ether	1246	4.27
19	Thymol	1297	73.61
20	β-Caryophylene	1416	0.99
21	Germacrene B	1487	0.64
22	α-Murolene	1499	0.94
23	β-Bisabolene	1509	0.17
24	δ-Cadinene	1525	0.16

Table 1: The type and composition of Thyme essential oil.

at 2.0°C min-1 using helium as the carrier gas at a flow rate of 1.0 ml min-1. The ion source temperature was 300°C and the electron energy 70 eV. Identification of components was based on comparison of their mass spectra with those found in the literature on retention indices, the mass spectrometry data bank (NIST) and a computer search of the Wiley library [8].

Theory

Modeling the vapour and liquid equilibrium

A number of important industrial processes such as distillation, absorption, and extraction, make two phases in contact with each other. Mass transfer between two phases occurs when the phases are not in equilibration. The amount and speed of transfer depend on the distance from the device and equilibrium. Quantitative analysis of mass transfer will require data and information about the equilibrium conditions (pressure, temperature and mole fraction of components). In the most industrial processes, coexistence phases are of kinds of liquid and vapour, although the liquid-liquid, vapour-solid and liquid-solid are also exist. This section is going to analyze and describe the qualitative and general behavior of the liquid and vapour phase in the essential oil distillation process, calculate the temperature, pressure and mole fraction of components for liquid and vapour equilibrium. Equilibrium is a static condition, which cause no change in the main properties of a device. This is because of the balance among all factors that may cause changes. In engineering practice, the equilibrium assumption is confirmed when the results would be satisfactory. For example, in the distillation column reboiler, it is generally assumed that the equilibrium is established between vapour and liquid phases. This is an approximation for certain evaporation, although this approximation in engineering calculations does not make an error. In a system with a constant amount of chemicals that contain vapour and liquid phases that are in close contact, there is no desire to change the system. Temperature, pressure and phase component are reached to final value and amount, and then remain constant. So the system is in equilibrium but at a molecular level, it is not a constant condition. Molecules forming one phase at a given moment are not the same as molecules as that phase in another moment. High speed molecules which are located near the boundary between the phases are overcome to the surface forces and go to another phase [9]. In vapour and liquid equilibrium condition the amount of molecules that leave the interface, and go to the vapour phase are equal with the number of another molecules component which returns to the liquid phase. Therefore, the concentration of each component in the vapour and liquid phase does not change during the time. Although theoretically, equilibrium required a long time, if the vapour liquid are placed vicinity and no reaction occurs between them, the equilibrium will come quickly (Figure 2).

According to thermodynamic relation, the equilibrium condition is the equality of chemical potential of component (i) in liquid and vapour phase in a constant temperature and pressure [10].

Macroscopic state of liquid vapour equilibrium is in constant temperature and pressure that the chemical potential of each component in liquid phase equal to chemical potential of each component in vapour phase [11].

$$(T,P) = cte\,\mu_i^V = \mu_i^L \tag{1}$$

On the other hand, with the help of the following thermodynamic relations, the chemical potential of each component in the liquid phase can associate to fugacity and activity coefficient of each component in the liquid phase. For complex mixture such as essential oils, using fugacity model for both vapour and liquid phase to predicting equilibrium data is necessary. Therefore, suitable activity model is used for liquid phase [12].

$$d\mu_i^V = RTd\ln f_i^V \tag{2}$$

$$d\mu_i^L = RTd\ln f_i^L \tag{3}$$

Thus, we have the equilibrium condition: $f_i^V = f_i^L$ (4)

If f_i^V, f_i^L values for each component are in the vapour and liquid phase, the fugacity which is a pressure unit is used for two non-ideal phases. Fugacity for vapour phase is as follows [13].

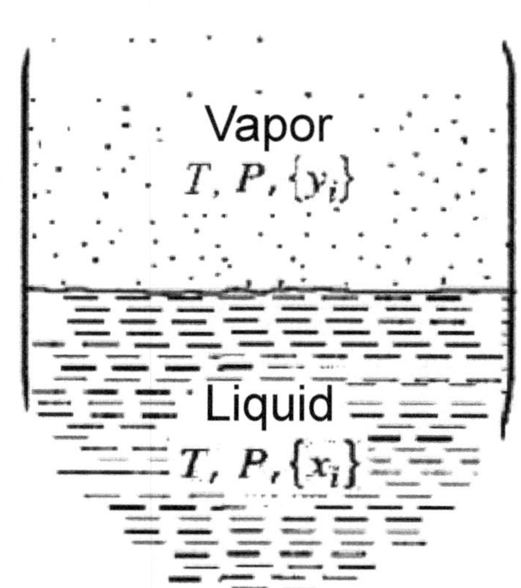

Figure 2: Schematic representation of VLE.

$$f_i^V = \varphi_i . y_i . P \tag{5}$$

The amount of fugacity in the liquid phase is calculated from the following equation:

$$f_i^L = \gamma_i . x_i . f_i \tag{6}$$

$$\varphi_i . y_i . P = \gamma_i . x_i . f_i \tag{7}$$

The required equilibrium information is usually expressed by the K value [14]: $K_i = y_i / x_i$ (8)

Where y_i is the mole fraction (i) in vapour phase and x_i is the mole fraction (i) in liquid phase. K_i is the equilibrium constant of component (i) at T temperature and P pressure [15]. By using thermodynamics, fugacity and activity coefficients can be related to the K value [16]: $K_i = \gamma_i f_i^0 / P\Phi_i$ (9)

γ_i is the activity coefficient of component (i) in the liquid phase and Φ_i is fugacity coefficient of component (i) in the vapour phase and P is the total pressure of the mixture. For compressible components, f_i^0 is pure fugacity of component (i) at T system temperature and P pressure [17]. Activity coefficients may be based on thermodynamic theory models, which based on rather strong scientific theory or based on an experimental model for example UNIQUAC, NRTL, UNIFAC, Wilson [18].

$$\frac{G^E}{RT} = g(X_1, X_2, \ldots, X_n) \tag{10}$$

$$\frac{G^E}{X_1 X_2 RT} = a + bX_1 + cX_1^2 + \ldots \tag{11}$$

$$G^E = RT \sum_i n_i \ln \gamma_i \tag{12}$$

$$RT \ln \gamma_i = \left(\frac{\partial n_i G^E}{\partial n_i}\right)_{T,P,n} \tag{13}$$

Equation UNIQUAC, with $\frac{G^E}{RT}$ and consists of two additive, the case course seeks to puts. The union sector g^c, for calculating the molecular size and shape differences. Other sector g^R, for calculating the molecule effects [19].

$$G = g^c + g^R \tag{14}$$

Function g^c only includes pure material parameters, while the function g^R includes the interaction of two parameters, which is formed for each pair of molecules. For a multi-components system [20]:

$$g^c = \sum_i x_i \ln\frac{\phi_i}{x_i} + 5\sum_i q_i x_i \ln\frac{\theta_i}{\phi_i} \tag{15}$$

$$g^R = -\sum_i q_i x_i \ln\left(\sum_i \theta_j \tau_{ji}\right) \tag{16}$$

$$\phi_i = \frac{x_i r_i}{\sum_j x_i r_i} \tag{17}$$

$$\theta_i = \frac{x_i q_i}{\sum_j x_j q_j} \tag{18}$$

(i) represents the material and (j) is an index. All calculations is for all materials and the value τ is equal to 1 for i=j [21].

In these equations, r_i (relative molecular size) and q_i (relative molecular level), are pure material parameter [22]. The temperature dependence of g^R at the above equation is determined by the following equations [23]: $\tau_{ij} = \exp-\left(\frac{u_{ji} - u_{ii}}{RT}\right)$ (19)

Interaction parameters in the UNIQUAC equation are $(u_{ji} - u_{ii})$. To obtain equation $\ln\gamma_i$ is as follows [24]:

$$\ln \gamma_i = \ln \gamma_i^c + \ln \gamma_i^R \tag{20}$$

$$\ln \gamma_i^c = 1 - j_i + \ln j_i - 5q_i\left(1 - \frac{j_i}{L} + \ln\frac{s_{ji}}{\eta_j}\right) \tag{21}$$

$$\ln \gamma_i^R = q_i(1 - \ln L_i) - \sum_j\left(\theta\frac{s_{ji}}{\tau_{ij}} - q_i \ln\frac{s_{ji}}{\eta_j}\right) \tag{22}$$

$$J_i = \frac{r_i}{\sum_j r_j x_j} \tag{23}$$

$$L_i = \frac{q_i}{\sum_j q_j x_j} \tag{24}$$

$$\theta = \sum_i q_i x_i \tag{25}$$

$$s_{ji} = \sum_m q_i \tau_{mj} \tag{26}$$

$$\eta_j = \sum_i s_{ji} x_i \tag{27}$$

$$\tau_{mj} = \exp\left(\frac{u_{ij} - u_{ij}}{RT}\right) \tag{28}$$

In these equations (i), is the material expression. (m) and (j) are counters. All calculations is for all materials and for m=j we have $\tau_{mj}=1$ [25].

All calculations have been performed using MATLAB software [26-28].

Values of parameters for the Margules, van Laar, and Wilson, NRTL, and UNIQUAC equations are given for many binary pairs by Gmehling [29].

Discussion

In this section, the data is shown in a graph for thymol component to achieve a better picture of the results. Diagrams show thymol concentration in the liquid phase according to temperature and pressure. This chart shows the change of thymol concentration in the liquid phase that is depending on temperature, and each curve represents the change of concentration in certain pressure.

As you see in Figure 3, the concentration of thymol in the liquid phase increases by increasing the temperature at constant pressure. At higher pressures, to achieve a certain amount of thymol concentration, the higher temperature is required. Due to sensitivity of plant

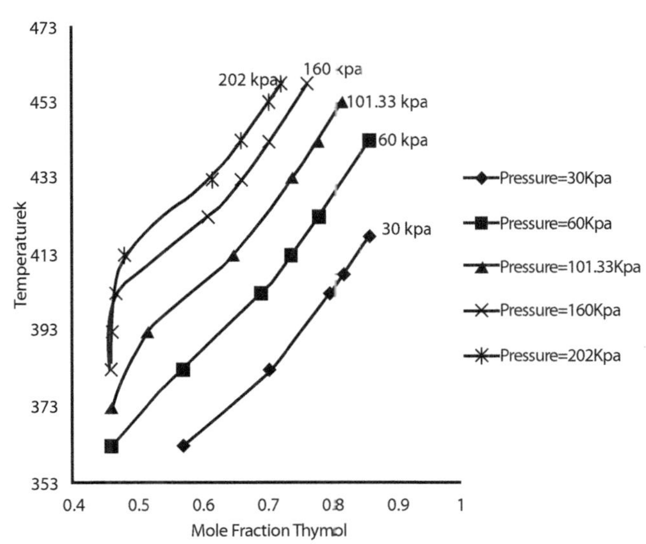

Figure 3: Thymol concentration changes in the liquid phase depending on temperature at constant pressure.

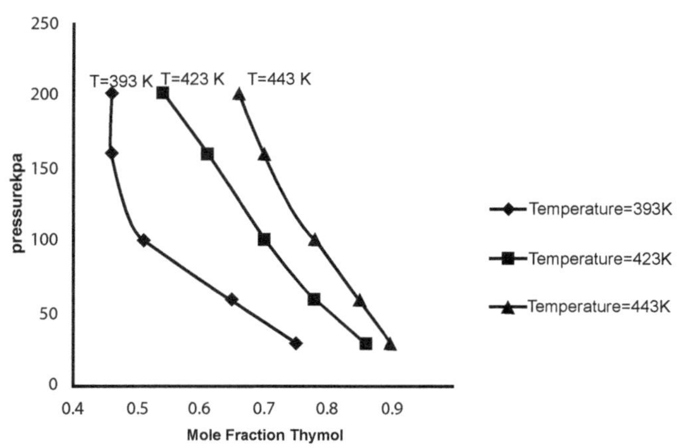

Figure 4: Thymol concentration changes in the liquid phase, depending on pressure at constant temperature.

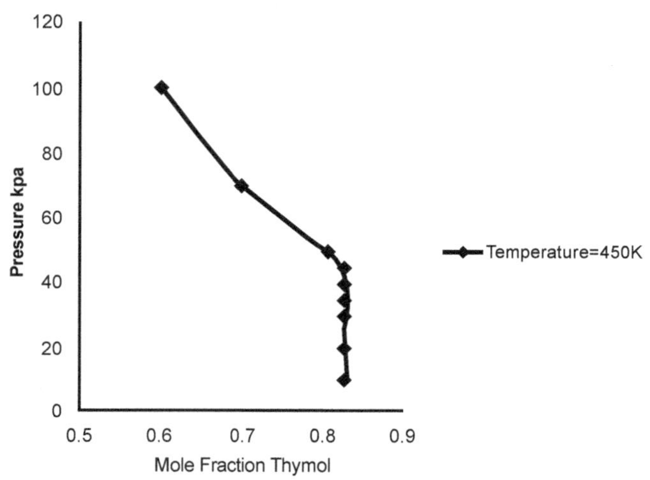

Figure 5: Thymol concentration changes in terms of pressure at constant temperature.

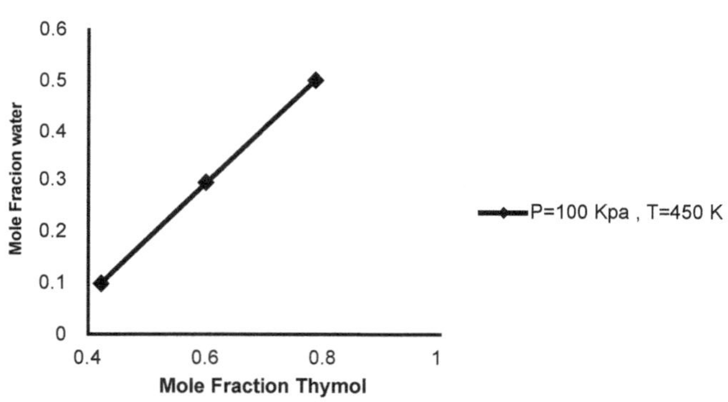

Figure 6: Thymol mole fraction changes depending on mole fraction of water changes.

compounds to temperature, low pressure is recommended. Change of thymol concentration in liquid phase depending on pressure at constant temperature.

Figure 4 shows the changes of thymol concentration in the liquid phase, depending on pressure in certain temperatures. This figure shows that by increasing pressure at constant temperature, the amount of thymol in the liquid phase decreases. In constant pressure at higher temperatures, the amount of thymol in the liquid phase increases. As it is indicated in the chart the highest concentration of thymol is in the lowest pressure.

In the triple mixture which includes 30% molar water, 35% molar thymol and 35% molar Cymene in liquid phase at constant temperature of 177°C by decreasing pressure from 100 Kpa to20 Kpa, thymol concentration is increased from 60% molar to 82% molar (Figure 5).

Thymol concentration in the liquid phase was increased from 42% molar to 78% molar by increasing water concentration of the mixture at a constant pressure of 100 Kpa and constant temperature 177°C (Figure 6).

Conclusions

According to the results of equilibrium data modeling and comparison on different pressure, temperature and water concentration, the following results were obtained.

1. By increasing temperature of the bubble point to dew point temperature at constant pressure, thymol concentration in the liquid phase is increasing.
2. By increasing temperature at constant pressure, the amount of thymol concentration in the liquid phase increases. At higher pressure, to achieve a certain amount of thymol concentration, the higher temperature is required. Due to sensitivity of plant compounds to temperature, low pressure is recommended.
3. By increasing pressure at constant temperature, the amount of thymol in the liquid phase decreases. At constant pressure and higher temperatures, the amount of thymol in the liquid phase increases.

References

1. Jam Zad, Thymus Research Institute of Forests and Rangelands 91: 1-17.
2. Akbarinia A, Sharifi-Ashabadi A, Sefidkon F (2009) Influence of harvest dynamics on herb and oil yield of Thymus kotschyanus and Thymus daenensis cultivated at two Sites. International workshop on medicinal and aromatic plant. ISHS Acta Horticulture 786.
3. Baigi RO (2005) Razavi, 1379: 88-99.
4. Askari F (2003) Essential oil composition of Thymus daenensis Celak. Journal of Essential oil Bearing Plants 6l: 123-125.
5. Naghdi Badi H, Yazdani D, Ali SM, Nazari F (2004) Effects of spacing and harvesting time on herbage yield and quality/quantity of oil in thyme, Thymus vulgaris L. Industrial crops and products 19: 231-236.
6. Avar N, Persuasive F, Review components of thyme essential oil world wages its flowering branches. Journal of Medicinal Herbs 13: 50-45.
7. Akbarinia A, Mirza M (2008) Identification of essential oil components of Thymus daenesis Celak. in field condition in Qazvin. Qazvin Agriculture and Natural Resources Research Center 12: 58-62.
8. Adams RP (1995) Identification of essential oil components by gas chromatography/mass Spectrometry. Allured Publ Corp.
9. Smith JM (2004) Introduction to chemical engineering thermodynamics 27: 584.
10. Pena DM, Colin CA, Compostizo A (1978) Isothermal liquid-vapour equilibria 1. The binary systems formed by toluene + methyl ethyl ketone, + methyl propyl ketone, and + methyl isobutyl ketone. J Chem Thermodyn 10: 337-341.
11. Mani JC, Cuony B (1993) Measurement of Vapor/Liquid Equilibria with an automatic lab reactor. Mettler Toledo publication.
12. Prausnitz JM (1969) Molecular Thermodynamics of Fluid-Phase Equilibria. Prentice-Hall, inc, Englewood Cliffs, New Jersey.

13. Renon H, Prausnitz JM (1968) Local compositions in thermodynamic excess functions for liquid mixtures. AIChE J 14: 135-144.

14. Debye P, Hückel E (1923) The theory of electrolytes. I. Lowering of freezing point and related phenomena. Physikalische Zeitschrift: 185-206.

15. Ohannes Jacobus van Laar (1860-1938) Dutch physical chemist.

16. Abrams DS, Prausnitz JM (1975) AIChE J 21: 116-128.

17. Wilson GM (1964) Vapor-Liquid Equilibrium. XI. A New Expression for the Excess Free Energy of Mixing. J Am Chem Soc 86: 127-130.

18. Redlich O, Kister AT, Turnquist CE (1952) Chem Eng Progr Symp Ser 48: 49-61.

19. McGlashan L, Rastogi RP, Tmns (1958) Faraday Soc 54: 496.

20. Van Ness HC (1995) Thermodynamics in the treatment of (vapor + liquid) equilibria. J Chem Thermodynamics 27: 113-134.

21. Van Ness HC (1995) Thermodynamics in the treatment of vapor/liquid equilibrium (VLE) data. Pure & Appl Chem 67: 859-872.

22. Eubank PT, Lamonte BG, Javier Alvarado JF (2000) Consistency Tests for Binary VLE Data. J Chem Eng Data 45: 1040-1048.

23. Van Ness HC, Abbott MM (1982) Classical Thermodynamics of Nonelectrolyte Solutions: With Applications to Phase Equilibria. McGraw-Hill, New York.

24. UNIQUAC Functional-group Activity Coefficients; proposed by Fredenslund Aa, Jones RL, Prausnitz JM.

25. AIChE J (1977) given detailed treatment in the monograph: Fredenslund Aa, Gmehling J, Rasmussen P, Vapor-Liquid Equilibrium using UNIFAC, Elsevier, Amsterdam 21: 1086-1099.

26. Reid RC, Prausnitz JM, Sherwood TK (1977) The Properties of Gases and liquids. (3rdedn), McGraw-Hill Book Co. New York, USA.

27. Hayden JG, O'Connell JP (1975) A Generalized Method for Predicting Second Virial Coefficients. Ind Eng Chem Process Des Dev 14: 209-216.

28. Sinor JE, Weber JH (1960) Vapor-Liquid Equilibria at Atmospheric Pressure. Systems Containing Ethy Alcohol, n-Hexane, Benzene, and Methylcyclopentane. J Chem Eng Data 5: 243-247.

29. Gmehling J, Onken U, Arlt W (1977) Vapor-Liquid Equilibrium Data Collection. DECHEMA, Chemistry Data Series. Frankfurt/ Main.

Simulation of Frontal Protein Affinity Chromatography Using MATLAB

Patricia Guerrero-German[1]*, Rosa Ma. Montesinos-Cisneros[2] and Armando Tejeda-Mansir[3]

[1]*Department of Chemical Engineering and Metallurgy, University of Sonora, L. Blvd Encinas and Rosales, Hermosillo, Sonora, Mexico*
[2]*Department of Mathematics, University of Sonora, L. Blvd Encinas and Rosales, Hermosillo, Sonora, Mexico*
[3]*Department of Scientific and Technological Research, University of Sonora, Luis Encinas and Rosales Blvd, Hermosillo, Sonora, Mexico*

Abstract

In this study a transport model that includes axial dispersion in the bulk liquid, pore diffusion, external film resistance and finite kinetic rate, was used to mathematically describe a frontal affinity chromatography system. The corresponding differential equations system was solved in a simple and accurate form by using the numerical method of lines implemented in MATLAB. The solution was compared with experimental data from literature and the analytic Thomas solution. The frontal affinity chromatography of lysozyme to Cibacron Blue Sepharose was used as a model system. A good fit to the experimental data was obtained with the simulated runs of the transport model using this methodology. This approach was used to perform a parametric analysis of the experimental frontal affinity system. The solution of the transport model results in simple way to predict frontal affinity performance as well a better understanding of the fundamental mechanisms responsible for the separation.

Keywords: Mathematical modeling; Simulation; Method of lines; Affinity chromatography; Frontal analysis

Nomenclature

- a specific area/volume of the adsorbent particle (m^{-1})
- c_0 initial protein concentration (mol/m^3)
- c protein concentration in the bulk liquid(mol/m^3)
- c_i protein concentration in the fluid of the pores(mol/m^3)
- C_D column diameter (m)
- d_p adsorbent particle diameter (m)
- D_{AB} protein diffusivity in free liquid (m^2/s)
- D_E = εiDieffective intraparticle protein diffusivity (m^2/s)
- D_i intraparticle protein diffusivity (free molecular diffusivity/pore tortuosity) (m^2/s)
- D_L column axial dispersion coefficient of the protein (m^2/s)
- F flow-rate (m^3/s)
- k_1 adsorption rate constant ($m^3/(mol\ s)$)
- k_2 desorption rate constant (s^{-1})
- k_f external film mass transfer coefficient (m/s)
- K_d equilibrium desorption constant, k_2/k_1
- L column length (m)
- nZ Node number in the axial direction
- nR Node number in the radial direction
- P protein molecule
- PS protein-active site complex
- q_i protein concentration in the adsorbed phase of the adsorbent particles (mol/m^3)
- q_m maximum equilibrium concentration,(mol/m3) of solid volume of adsorbent
- q_{ms} maximum equilibrium concentration, (mol/m3) of settled volume of adsorbent
- r radial distance in the adsorbent particle (m)
- r_m radius of adsorbent particle (m)
- S active site
- t time (s)
- v interstitial column velocity (flow-rate/bed porosity-column area) (m/s)
- z axial distance in the column (m)

Greek Letters

- ε bed porosity
- ε_i adsorbent particle porosity
- μ solution viscosity (g/m s)
- ρ solution density (g/m^3)

Introduction

High purity products are often required from the biotechnology industry; in order to achieve the demanded purity, several complex purification steps are needed. Chromatography is used both in analytical and preparative applications in pharmaceutical and biotechnology industry [1]. Among them, affinity chromatography is the preferred choice for the primary capture step. This purification method is based on the specific interaction between the biomolecule of interest and a ligand immobilized on a solid support [2].

A preparative chromatography process is often performed in packed bed columns. The packing material is a porous gel, i.e. small beads created by a matrix of a polymer structure. There are other kinds of column configurations, like expanded beds and moving beds, but this work is focused on the traditional packed bed column.

Frontal chromatography is a widely used process in the purification

***Corresponding author:** Patricia Guerrero-German, Department of Chemical Engineering and Metallurgy, University of Sonora. L. Blvd Encinas and Rosales, Hermosillo, Sonora, Mexico of biopharmaceuticals. Simulations of chromatographic processes are useful for studying and understanding complex column dynamics. When new chromatographic purification steps are planned or when an existing step is scaled up, it is common practice to perform numerous tedious chromatographic optimization experiments. As the target product molecules often are valuable and available only in small quantities, these experiments will be expensive to perform. Predicting the performance of the process by mathematical modeling and computer simulation minimizes the amount of experiments required. For this reason, simulation of chromatographic processes is becoming a valuable tool [3].

The scale-up and optimization of affinity chromatographic operations is of major industrial importance [4,5]. The development of mathematical models to describe affinity chromatographic processes, and the use of these models in computer programs to predict column performance is an engineering approach that can help to successfully attain these bioprocess engineering tasks [6]. An important requirement of this methodology is a thorough understanding of the fundamental mechanisms underlying such separations in order to develop realistic models based on basic physical and chemical principles or rate theories. The modeling in these areas often leads to a partial differential equation (PDE), or transport equation, which is parabolic in nature. Such equations may be linear, in which case analytic solutions are often achievable. However, many of the transport equations are non-linear and numerical simulations are used to understand the particular problem more fully [7].

Analytical solutions have been obtained through the rate limited breakthrough approach which considered that only one rate limiting step, i.e. either rate of interaction or rate of diffusion (pore or film) is controlling the overall adsorption mechanism. The non-dispersive flow model is used in the analysis. Chase [8] used the Thomas solution [9], which involves Langmuir reaction kinetics as the rate-limiting step, to predict the performance of affinity separations. More detailed solutions of affinity chromatography column behavior have been reported [10-14]. However, most of these methods are impracticable for the solution of transport models that include dispersed flow and finite reaction rates [5].

One of the most popular methods for solving evolutionary PDEs is the method of lines (MOL), for which efficient and effective integrating packages have been developed [15,16]. The MATLAB package has a set of strong ODE solvers, and an extensive functionality which can be used to implement the MOL [17].

The purpose of this work was to develop an accurate and simple solution of the transport model for an affinity column, using the method of lines and the MATLAB platform. The solution was compared with experimental data from the literature of the adsorption of lysozyme to Cibacron Blue Sepharose CL-6B and the Thomas analytic solution of the lumped parameter column model.

Frontal affinity chromatography model

During column operation in frontal mode the sample is fed continuously into the column. For a short time the solute in the feed is taken up almost completely, but after a while, solute breakthrough occurs and the effluent concentration increases with time. Much of the information needed to evaluate column performance is contained in these typical plots of effluent concentration versus time or breakthrough curves (BTC). These curves can be used to determine: (1) how much of the column capacity has been utilized, (2) how much solute is lost in the effluent, and (3) the processing time. This is precisely the performance information needed to optimize separation processing [18].

Many frontal affinity chromatography systems of industrial interest involve single-component adsorption. For this reason, in this study the frontal affinity model is based on the isothermal sorption of a single solute during flow through a porous fixed-bed of diffusive adsorbent particles with an average radius, r_m, and a porosity, εi, on which the ligand is immobilized. In this analysis, the feed protein concentration is, c_0, the protein solution in the system has a transient concentration, c(z, t), with a constant interstitial flow-velocity, v, through the column, with height, L, and a void-bed porosity, ε. The protein concentrations in the adsorbent fluid and solid phases are c_i and q_i, respectively (Figure 1). To achieve a mathematical description of a frontal affinity chromatographic process, two major phenomena must be included: matrix hydrodynamics must be assessed, as well as the nature of the binding process itself. In this study, a transport model that considers a dispersed flow in the column and three consecutive transport rate resistances to ideal equilibrium separation is used for the simulation of the frontal affinity chromatography system [5].

The Fickian convective dispersion in the column is characterized by the axial dispersion coefficient, D_L. The transport of protein is considered to involve the interfacial transport of protein to the outer surface of the adsorbent particles from the bulk liquid through the adsorbent surrounding stagnant film characterized by the coefficient, k_f, the diffusion in the pore fluid described by an effective diffusion coefficient, D_E, and the adsorption step of the protein with active sites on the surface of the adsorbent. The intrinsic adsorption rate can be described by different kinetic models. In this study, an adsorption-desorption model of the Langmuir type is used.

Due to the nonlinear equilibrium that characterizes affinity chromatography, adsorption behavior is best described by rate theories. This engineering approach to modeling involves the use of conservation equations, equilibrium lawsat interfaces, kinetic laws of transport and adsorption and initial and boundary conditions. To describe the concentration change of protein with time at the column exit, the following equation can be derived by a solute mass balance in the fluid phase:

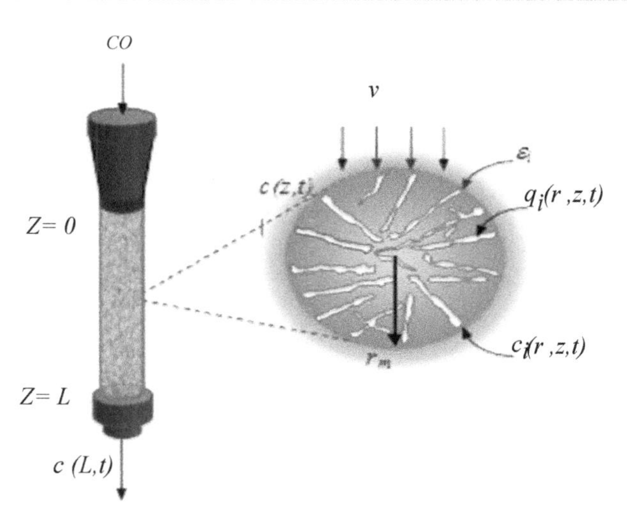

Figure 1: System conceptual: model of the affinity packed column with porous particle.

$$\frac{\partial c}{\partial t} - D_L \frac{\partial^2 c}{\partial z^2} + \upsilon \frac{\partial c}{\partial z} = -\frac{3}{r_m} \frac{(1-\varepsilon)}{\varepsilon} k_f \left(c - c_i\right)\Big|_{r=r_m} \quad (1)$$

The equation to describe the change of concentration of solute in the fluid of the adsorbent pores can be obtained by a solute mass balance in the particle,

$$\varepsilon_i \frac{\partial c_i}{\partial t} + \left(1-\varepsilon_i\right) \frac{\partial q_i}{\partial t} = D_E \left(\frac{\partial^2 c_i}{\partial r^2} - \frac{2}{r} \frac{\partial c_i}{\partial r} \right) \quad (2)$$

To describe the complex interactions between solute and affinity adsorbent, simplified models are often used [8,10,18]. In general, a second-order reversible adsorption reaction is considered, where the solute is assumed to interact with the adsorbent by a monovalent interaction and characteristic constant binding energy:

$$P + S \rightarrow PS$$

Where P is the protein in solution, S is the ligand adsorption site and PS is the protein-ligand complex.

The rate of adsorption for this type of interaction is usually represented as

$$\frac{\partial q_i}{\partial t} = k_1 c_i \left(q_m - q_i\right) - k_2 q_i \quad (3)$$

Where k_1 and k_2 are the adsorption and desorption rate constant, respectively. At equilibrium Eq. (3) gives the form of the Langmuir isotherm with equilibrium desorption constant $K_d = k_2/k_1$ and maximum adsorption capacity q_m.

Equation (1) requires one initial condition (IC) in t (since it first order in t), two boundary conditions (BCs) in z (since it second order in r). Equation (2) also requires two IC in t (one for c_i and one for q_i) and two BCs in r.

At the beginning of the operation there is no protein present in the system, therefore ICs:

$$\text{at } t = 0, \quad c = 0 \quad 0 \le z \le L \quad (4)$$

$$\text{at } t = 0, \quad c_i = 0 \quad 0 \le r \le r_m \quad (5)$$

$$\text{at } t = 0, \quad q_i = 0 \quad 0 \le r \le r_m \quad (6)$$

The Danckwerts boundary conditions [19] are used to account for dispersion at the entrance of the column and complete mixing with only convection flow at the end of the column and given by, respectively

$$\text{at } z = 0, \quad \varepsilon \upsilon c - \varepsilon D_L \frac{\partial c}{\partial z}\Big|_{z=0} = \varepsilon \upsilon c_0 \quad t > 0 \quad (7)$$

$$\text{at } z = L, \quad \frac{\partial c}{\partial z}\Big|_{z=L} = 0 \quad t > 0 \quad (8)$$

Due to particle symmetry,

$$\text{at } r = 0, \quad \frac{\partial c_i}{\partial r}\Big|_{r=0} = 0 \quad t > 0 \quad (9)$$

At the mouth of the particle pore,

$$\text{at } r = r_m, \quad k_f \left(c - c_i\right)\Big|_{r=r_m} = D_E \frac{\partial c_i}{\partial r}\Big|_{r=r_m} \quad t > 0 \quad (10)$$

The solution to Eqs. (1)-(10) gives c(z,t), c_i(r,z,t) and q_i(r,z,t) as a function of the independent variables r,z,t.

Numerical solution of the transport model

The detailed model (Eqs. 1-10), which is based on a set of partial differential equations (PDEs), was solved numerically using the method of lines (MOL). This means that the PDEs were discretized in space, resulting in a large set of ordinary differential equations (ODEs). The set of ODEs was solved using the MATLAB ODE solver, *ode15s*, which is a variable order (1-5) and variable step-length procedure that use implicit numerical differentiation formulas (NDF) and computes the solution over each time interval [17,20].

The Figure 2 shows the numerical solution procedure. Ads_Part was used as main program. System parameters, spatial domain in radial and axial direction, the MATLAB stiff integrator ode15s and the print instructions were incorporated in Ads_Part. For the integration of ODES, parameters AbsTol = 10^{-8} and RelTol = 10^{-5} were used. The program Ads_Part_ODE calls the routines dss004 and dss044 to obtain the first and second order spatial derivatives in the axial and radial direction, using five-point, fourth-order finite difference approximations [15,21].

Since the spatial differentiation dss routines compute numerical derivatives of one dimension (1D) array (vector), a conversion from two dimensions (2D) to 1D is first required at each value of z, to compute the change of concentration of solute in the fluid of the adsorbent pores. Then the first-order radial derivatives can be computed by call to dss004. Likewise, the second derivatives in r can be computed by called to dss044 [16]. nZ discretization points were used in the column axial domain and nR discretization points were used in the particle radius. A grid analysis was used to compare the breakthrough curve sharpness using a different number of discretization points. A significant decrease in curve dispersion was observed with an increase in the number of discretization points. The total number of grid points was nZxnR. Since Eq. (2) and Eq. (3) are function of the independent variables (r,z,t) and Eq. (1) is a function of the independent variables (z,t) there were (2×nZ×nR+nZ)ODEs programmed in the ODE routine.

The initial conditions, Eq. (4) is one of single dimension vector c(z)=0, the Eqs (5, 6) are two dimension vector c_i(r,z)=0 and q_i(r,z)=0 which are converted to initial conditions single dimension vector, (y_0) with (2×nZ×nR+nZ) rows. All the codes were incorporated in a MATLAB program that was run in a personal computer. The computational times to obtain a complete breakthrough curve vary between 0.8 min and 3 min.

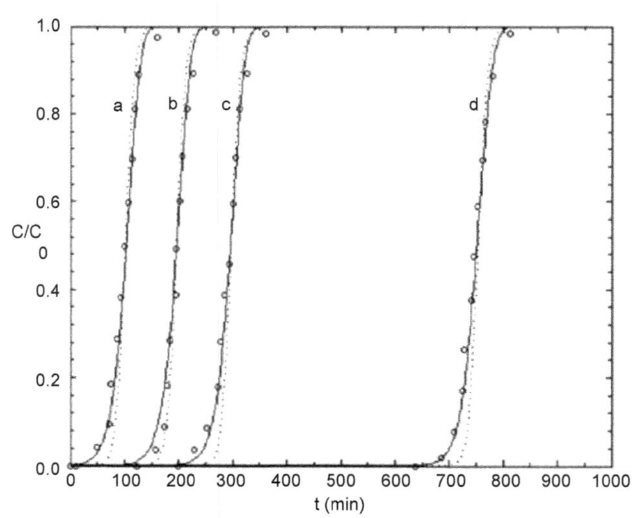

Figure 2: Numerical method of lines solution scheme of the frontal affinity transport model.

Lumped parameter model

The most general relation that has been developed to describe break through behavior involves Langmuir reaction kinetics as the rate-limiting step and non-dispersive convective flow through the column. It is known as the Thomas model [9]. Without mass-transfer effects on column performance, the overall rate of adsorption is only limited by the intrinsic adsorption kinetics. Another interpretation is that under mass-transfer limitations all effects of internal and external diffusion within and outside the beads as well as any dispersion in the column are lumped together with the kinetics. This approach is useful when mass-transfer resistance by pore-diffusion is relatively small. In this particular case, the analytical solution of the non-dispersive model is expressed as follows:

$$X = \frac{J(N/\Upsilon, N\Gamma)}{J(N/\Upsilon, N\Gamma) + [1 - J(N, N\Gamma/\Upsilon)]\exp[(1 - 1/\Upsilon)(N - N\Gamma)]} \quad (11)$$

Where

$$X = \frac{c}{c_0} \quad (12)$$

$$\Gamma = \frac{\varepsilon K_d \Upsilon (T-1)}{(1-\varepsilon) q_m} \quad (13)$$

$$\Upsilon = 1 + \frac{c_0}{K_d} \quad (14)$$

$$T = \frac{vt}{L} \quad (15)$$

$$N = \frac{(1-\varepsilon) q_m k_1 L}{\varepsilon v} \quad (16)$$

and J is a two-parameter function of α and β, given by:

$$J(\alpha, \beta) = 1 - e^{-\beta} \int_0^{\alpha} e^{-\xi} I_0\left(2\sqrt{\beta \xi}\right) d\xi \quad (17)$$

Where I_0 refers to the zero-order modified Bessel function of the first kind [22]. The analytical solution of Eqs. (11-17) (or Thomas model) was evaluated numerically for comparison with the numerical MATLAB solution and experimental data.

Input data for the study

The adsorption of lysozyme to Cibacron Blue Sepharose CL-6B was chosen as the model system. The values of the parameters utilized to conduct the simulation studies were obtained from the studies of [5,8,13] are presented in Table 1. To properly conduct this simulation, the experimental data were displaced one column residence time; because the time t in Chase's paper [8] is measured from the time at which non-adsorbing species exit the column. In this work, t is measured starting from the time at which the feed is introduced to the front of the bed. This last definition is commonly used in chromatography analysis because this time measurement is independent of the size of the non-adsorbing species, which is less ambiguous. The maximum adsorption capacity was calculated with respect to bed porosity, ε, and available volume to the protein as $q_m = 0.8 \times q_{ms} / [(1 - \varepsilon)(1 - \varepsilon_i)]$. In the analysis of the influence of bead diameter on the affinity process the mass-transfer coefficient was estimated using the Foo and Rice correlation [23].

$$Sh = 2 + 1.45 Re_p^{1/2} Sc^{1/3} \quad (18)$$

Where,

$$Sh = \frac{k_f d_p}{D_{AB}}, \quad Sc = \frac{\mu}{\rho D_{AB}}, \quad Re_p = \frac{d_p(\varepsilon v)\rho}{\mu} \quad (19)$$

Results and Discussion

The solution to the transport model for frontal affinity chromatography of lysozyme to Cibacron Blue Sepharose CL-6B was obtained using the MOL and MATLAB. This solution was compared with the experimental data and with the analytical solution of the lumped parameter model. Four column lengths were considered: 0.014, 0.027, 0.041 and 0.104 m.

The grid studied performed showed that in order to minimize the numeric dispersion, the number of nodes in the axial direction, nZ, was set to 40, 60, 80 and 100 for the 0.014, 0.027, 0.041 y 0.104 m column length simulations, respectively. The number of nodes in the radial direction was kept constant, nR=8, for all column lengths, since equal results were obtained with greater values of this parameter, an expenses of more computing time.

The results of the simulation of the breakthrough curve for the affinity column are shown in Figure 3. Taking into account the four column lengths, the average of the residual sum of squares between model calculations and experimental data were 0.0167 ± 0.0058 and 0.1146 ± 0.0776 for the MATLAB and the Thomas solution, respectively. A much better fit to the experimental data was obtained using MATLAB solution. In these computations, the kinetic parameter value was set to $k1=1.144\ m^3/(mol\ s)$, since in the transport model this is not a lumped parameter. The simulation runs with the Thomas model using the value for the lumped parameter $k_l=0.286\ m^3/(mol\ s)$ fitted fairly well to the experimental data.

Simulation studies

The MATLAB solution of the transport model was used to carry out a parametric analysis comparing frontal affinity curves from several computer simulations, in which one parameter was changed while the others were kept constant at the basic set of values in Table 1 and using a column length of L=0.014 m. Furthermore to describe in more detailed form the affinity chromatography process, e.g. detailing the protein dimensionless concentration profiles in the adsorbent pore

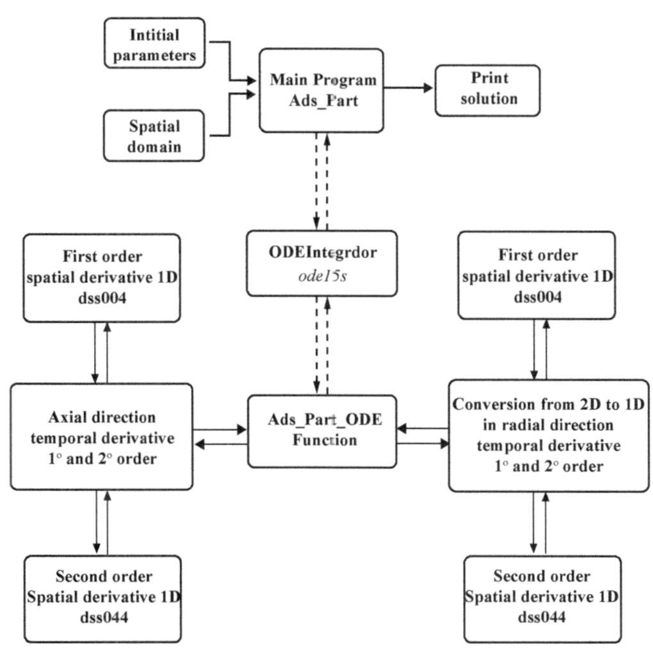

Figure 3: Transport and lumped kinetic parameter models compared with frontal affinity chromatography experimental data [8]. Operating conditions according to Table 1. (o) Experimental data. (—) Numerical method of lines solution of the transport model. (...) Lumped parameter model. The column lengths used were (a) 0.014m (b) 0.027 m, (c) 0.041m and (d) 0.104 m.

Variable	Value
Inlet protein concentration	$c_0 = 7.14 \times 10^{-3}$ mol/m³
Flow-rate	$F = 1.67 \times 10^{-8}$ m³/s
Column length	L = 0.014, 0.027, 0.041, 0.104m
Column diameter	$C_D = 0.01$m
Bed porosity	$\varepsilon = 0.39$
Bead porosity	$\varepsilon_i = 0.75$
Bead radius	$r_m = 5 \times 10^{-5}$ m
Axial dispersion	$D_L = 5.75 \times 10^{-8}$ m2/s
Film mass-transfer rate	$k_f = 6.9 \times 10^{-6}$ m/s
Solution viscosity	$\mu = 0.95$ g/m.s
Solution density	$\rho = 1.0 \times 10^{6}$ g/m³
Lysozyme diffusivity in free liquid	$D_{AB} = 1.06 \times 10^{-10}$ m²/s
Effective diffusion coefficient	$D_E = 5.3 \times 10^{-11}$ m²/s
Adsorption rate constant (Transport model)	$k_1 = 1.144$m³/(mol s)
Adsorption rate constant (Thomas model)	k_1=0.286m³/(mol s)
Equilibrium desorption constant	$K_d = 1.748 \times 10^{-3}$ mol/m³
Maximum adsorption capacity	$q_{ms} = 1.0$ mol/m³
Maximum adsorption capacity of solid gel	$q_m = 5.246$ mol/m³

Table 1: Base case data used in simulation studies of frontal affinity adsorption of lysozyme to Cibacron Blue Sepharose CL-6B [5,8,13].

liquid (average), c_{iAV}/c_0 and in the adsorbed phase (average), q_{iAV}/q_m as function of the dimensionless column length, Z, and the real time, t. The effect of inlet protein concentration and bead diameter is reported.

Upstream perturbations can initiate changes in process inlet concentrations that are important for study. The inlet protein concentration was changed using ± 40% variations of the $c_0 = 7.14\times10^{-3}$ mol/m³ base value. The corresponding curves are shown in Figure 4.

An increased inlet concentration gives an early and sharper break through curve (Figure 4). The concentration profiles are very symmetric suggesting the importance of both liquid film and pore diffusion mass transfer resistances in the adsorption process. The total column equilibration occurs in about 250, 150 and 120 min, respectively. The slope of the curves in the 0.5 region indicates the additional contribution of the dispersion to curve spreading.

As the inlet concentration increases, the driving-force for the transport process is also augmented. This results in a faster equilibration of the adsorbent beads. When the beads became equilibrated more rapidly, they will extract protein from the mobile phase for a shorter time, resulting in a sharper breakthrough curve. Hence, on this basis, it is more efficient to apply solute at high concentration. The utilization of the maximum capacity of the bed is greater at higher solute concentration as these conditions favor a greater extent of adsorption at equilibrium. As reported by Chase [8], when the dimensionless exit concentration of the column is plotted against the adsorbent applied to the column, an effect is only noticed on the shape and position of the breakthrough curve when the inlet concentration, c0, is comparable or smaller than the desorption equilibrium constant, K_d. The shape and position of the breakthrough curve becomes constant when $c_0 >> K_d$.

The inlet protein concentration effect on protein dimensionless concentration profiles in the adsorbent pore liquid (average), c_{iAV}/c_0, is shown in Figures 4. As the inlet concentration increases, the average protein concentration in the adsorbent pore-liquid, at the column entrance, reached faster the inlet concentration, (100, 75 and 50 min, respectively). The film mass transfer resistance and dispersion effect can also be observed.

The effect of the inlet protein concentration on the adsorbed phase (average), q_{AVi}/q_m, is shown in Figure 4. A greater degree of column saturation, q_{iAV}/q_m (0.71, 0.80 and 0.85 respectively) is observed in accordance with the Langmuir isotherm, and also a shorter equilibration time is needed (100, 60 and 40 min respectively).

The process parameter of most interest is the bead diameter. In the simulation studies the bead diameter was changed using ± 50% variations of the d_p=100 μm base value. The corresponding curves are shown in Figure 5.

A sharper breakthrough curve, consequently a greater operation capacity, is obtained as the bead diameter decreases (Figure 5). It can also be noted in the figure that this effect is less dramatic as the bead size decreases. As particle diameter decreases the initial adsorption rate increases markedly, since the diffusion time is decreased due to the shorter diffusion path. At the same time the area/volume ratio for a single particle ($3/r_m$) increases, giving an increased mass transfer area between the surrounding liquid phase and the bead. Both factors contribute to the increase in total adsorption rate.

The bead diameter effect on the profiles c_{iAV}/c_0 is shown in Figure 5. As particle diameter decreases, the film mass transfer resistance and dispersion effects are less marked. At the column entrance, the average protein concentration in the adsorbent pore liquid reached the column inlet concentration in shorter time as particle diameter decreases, (35, 75 and 135 min respectively).

The concentration profiles in the adsorbed phase (Figure 5) show that the degree of column saturation, q_{iAV}/q_m, was reached in shorter time as particle diameter decreases (30, 60 and 120 min respectively), although the particle diameter do not cause an effect on the degree of column saturation value.

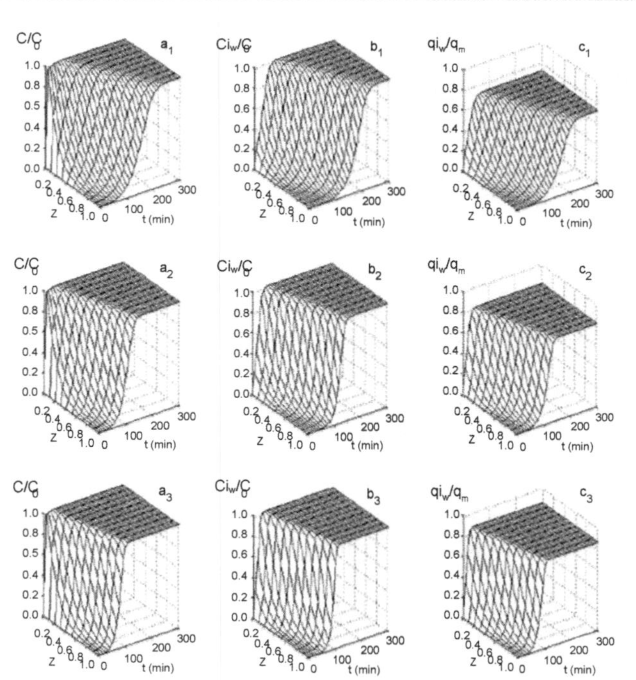

Figure 4: Influence of the inlet protein concentration on the breakthrough behavior shown as the variation of the dimensionless protein concentration in the bulk liquid, c/c_0, the variation of the dimensionless protein concentration in the adsorbent pore liquid (average), c_{iAV}/c_0, and of the dimensionless the adsorbed phase (average), q_{iAV}/q_m, with the dimensionless column length, Z_L, and the real time, t. (a_1, b_1 and c_1) $c_0 = 4.284 \times 10^{-3}$ mol/m³, (a_2, b_2 and c_2), $c_0 = 7.14 \times 10^{-3}$ mol/m³, and(a_3, b_3 and c_3) c0=9.996 × 10^{-3} mol/m³.

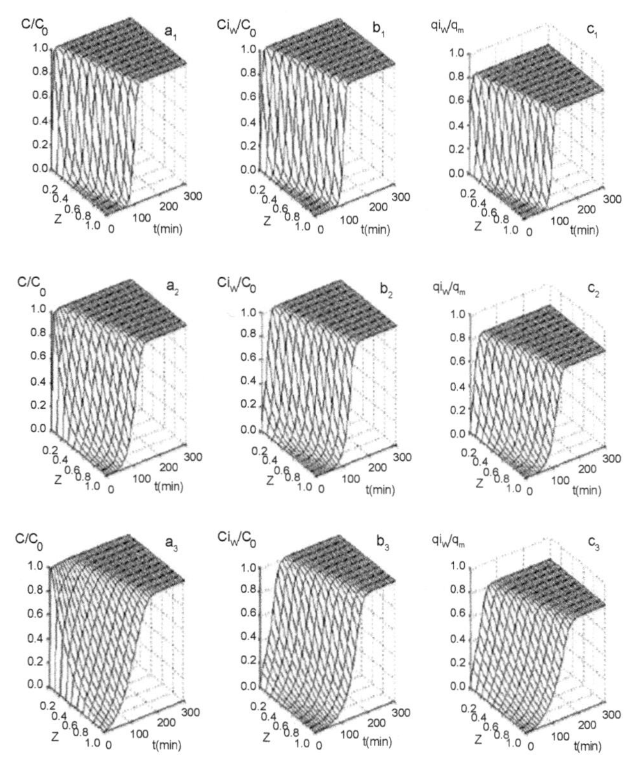

Figure 5: Influence of the bead diameter on affinity breakthrough behavior shown as the variation of the dimensionless protein concentration in the bulk liquid, c/c_0, the variation of the dimensionless protein concentration in the adsorbent pore liquid (average), c_{iAV}/c_0, and of the dimensionless the adsorbed phase (average), qiAV/qm, with the dimensionless column length, ZL, and the real time, t. (a_1, b_1 and c_1) dp = 50 μm,(a_2, b_2 and c_2) d_p = 100 μm and (a_3, b_3 and c_3) d_p = 150 μm.

Conclusions

The performance of frontal affinity chromatography of lysozyme to Cibacron Blue Sepharose CL-6B Sepharose was successfully described with a three-resistances and column dispersed flow model. Programming the model solution was quite simple using MOL in MATLAB platform. Additionally, parametric studies performed helped to gain qualitative information and to show the influence of both operation and system parameters on the affinity process. An early and sharper breakthrough curve and therefore a greater operating throughput of the affinity process was obtained as the column inlet concentrations were increased. In the bead-size parametric study, a sharper breakthrough curve and consequently a greater operating throughput was attained as the bead diameter decreases. This effect was less dramatic as the bead size decreases. The dynamic responses obtained are in concordance with theoretical predictions and show that the transport model can be used as a framework to provide a general description of almost all practical systems, when the appropriate basic experimental parameters and numerical solution are used, in order to obtain good simulation results. The MATLAB solution of the transport model permitted an accurate prediction of the frontal affinity performance and a better understanding of the fundamental mechanisms responsible for the separation. The influence of the adsorption properties of the protein from other components present in more complex mixtures should enable the work described here with a model system to be extended to more practical situations.

Acknowledgement

The authors gratefully acknowledge the support of this work by the National Council for Science and Technology of Mexico and by the Faculty Improvement Programme of the Secretariat of Public Education of Mexico, SEP and PROMEP the University of Sonora.

References

1. Persson P, Kempe H, Zacchi G, Nilsson B (2004) A methodology for estimation of mass transfer parameters in a detailed chromatography model based on frontal experiments. Chem Eng Res Des 82: 517-526.
2. Boi C, Dimartino S, Sarti GC (2007) Modelling and simulation of affinity membrane adsorption. J Chromatogr A 1162: 24-33.
3. Karlsson D, Jakobsson N, Axelsson A, Nilsson B (2004) Model-based optimization of a preparative ion-exchange step for antibody purification. J Chromatogr A 1055: 29-39.
4. Kempe H, Persson P, Axelsson A, Nilsson B, Zacchi G (2006) Determination of diffusion coefficients of proteins in stationary phases by frontal chromatography. Biotechnol Bioeng 93: 656-664.
5. Montesinos RM, Tejeda-Mansir A, Guzman R, Ortega J, Schiesser WE (2005) Analysis and simulation of frontal affinity chromatography of proteins. Sep Purif Technol 42: 75-84.
6. Liapis AI, Unger KK (1994) Highly Selective Separations in Biotechnology. (G. Street edn), Chapman and Hall, London.
7. Lee HS, Matthews CJ, Braddock RD, Sander GC, Gandola F (2004) A MATLAB method of lines template for transport equations. Environ Modell Softw 19: 603-614.
8. Chase HA (1984) Prediction of the performance of preparative affinity chromatography. J Chromatogr 297: 179-202.
9. Thomas HC (1944) Heterogeneous Ion Exchange in a Flowing System. J Am Chem Soc 66: 1664-1666.
10. Horstmann BJ, Chase HA (1989) Modelling the affinity adsorption of immunoglobulin G to protein A immobilised to agarose matrices. Chem Eng Res Des 67: 243-254.
11. Arve BH, Liapis AI (1987) Modeling and analysis of biospecific adsorption in a finite bath. AIChE Journal 33: 179-193.
12. Berninger JA, Whitley RD, Zhang X, Wang NHL (1991) A versatile model for simulation of reaction and nonequilibrium dynamics in multicomponent fixed-bed adsorption processes. Comput Chem Eng 15: 749-768.
13. Kempe H, Axelsson A, Nilsson B, Zacchi G (1999) Simulation of chromatographic processes applied to separation of Proteins. J Chromatogr A 846: 1-12.
14. Ruthven DM (1984) Principles of Adsorption and Adsorption Processes. New York: John Wiley and Sons.
15. Wouwer AV, Saucez P, Schiesser WE (2004) Simulation of Distributed Parameter Systems Using a Matlab-Based Method of Lines Toolbox: Chemical Engineering Applications. Ind Eng Chem Res 43: 3469-3477.
16. Schiesser WE, Griffiths GW (2009) A Compendium of Partial Differential Equation Models: Method of Lines Analysis with Matlab. New York : Cambridge University Press.
17. Shampine LF, Reichelt MW (1997) The MATLAB ODE suite. 1-33.
18. Arnold FH, Blanch HW, Wilke CR (1985) Analysis of affinity separations: I: Predicting the performance of affinity adsorbers. Chem Eng J 30: B9-B23.
19. Danckwerts PV (1953) Continuous flow systems: Distribution of residence times. Chem Eng Sci 2: 1-13.
20. Ashino R, Nagase M, Vaillancourt R (2000) Behind and beyond the MATLAB ODE Suite. Comput Math Appl 40: 491-512.
21. Schiesser WE, Silebi CA (1997) Computational Transport Phenomena: Numerical Methods for the Solution of Transport Problems. New York: Cambridge University Press.
22. Hiester NK, Vermeulen T (1952) Saturation performance of ion-exchange and adsorption columns. Chem Eng Prog 48: 505-516.
23. Foo SC, Rice RG (1975) On the prediction of ultimate separation in parametric pumps. AIChE Journal 21: 1149-1158.

Modeling of Cu Doped Cobalt Oxide Nanocrystal Gas Sensor for Methane Detection: ANFIS Approach

A. Ahmadpour[1], Z. Sheikhi Mehrabadi[1], J.R. Esfandyari[2]* and M. Koolivand-Salooki[3]

[1]Department of Chemical Engineering, Faculty of Engineering, Ferdowsi University of Mashhad, Khorasan, Iran
[2]The Technical teacher of saveh education office, saveh, Iran
[3]Department of Chemical Engineering, omideh beranch, Islamic Azad University, omideh, Iran

Abstract

In this paper, Nano-sized copper-cobalt compound oxide powders have been prepared by sol-gel technique with different mole ratios of Cu/Co (from 0.00 to 0.15); Detection of the methane gas, the most chemically stable hydrocarbon, is done. The structural properties and morphology of powders were studied by X-ray diffraction (XRD), Fourier transform infrared spectroscopy (FTIR) and Transmission electron microscopy (TEM). XRD analysis confirms that Co_3O_4 and $(CuO_{0.3}CO_{0.7})$ Co_2O_4 phases have been formed and mean grain size were decreased with increasing dopant (from 28 to 24nm). According to TEM images it was found that the particles have cubic morphologies with nearly uniform distribution. Then an adaptive neuro-fuzzy inference system (ANFIS) models have been utilized for prediction of sensitivity values of the corresponded sensor. The results of ANFIS model show that the independent predicted Sensitivity values (S) compared to the measured target values have a good agreement. And also, high coefficient were found ($R^2 > 0.98$) for the response.

Keywords: Nano particles; Cu/Co; ANFIS; XRD; TEM; FTIR

Introduction

Methane (CH_4) is a major component of the widely used natural fuel gases in the industry and the private household. Explosion will occur when its concentration in air reaches to 5-16% (v/v); therefore the monitoring of CH_4 in the industry is very important for safety reasons. In addition, it is also one of the greenhouse gases which are produced continuously by the anaerobic oxidation of organic materials, so it needs to be monitored. However, CH_4 is relatively difficult to detect at ambient conditions because its chemical reactivity is relatively low. The commonly used techniques for CH_4 detection include gas chromatography (GC) [1-9], semiconductor devices [10-21], optical fiber sensors [22], electrochemical methods [23-24], and biochemical methods [25, 26].

Many Nano-sized pure and doped semiconducting oxides have been used for gas sensors [27-30]. Use of Nano-materials is interested because of much greater surface area to the bulk ratio rather than course materials. In addition, when the grain size reduce to nanometer, especially when the dimensions become less than two times of the space-charge depth, a large fraction of atoms which will present in the surface and the surface properties become dominant. So, materials conduction type turns to surface conduction type [31-33]. The sensing mechanism depends on the changes in surface conductivity which is resulted from chemical reaction between target gases and the absorbed oxygen on the surface of the Nano-sized metal oxide. Consequently, it is important to enhance the rate of chemisorptions and surface reaction. Co_3O_4 is most active catalyst for oxidation of the methane [34]. So, it is a good candidate for the methane gas sensor. Addition of second component as doping is becoming one of the most interesting methods for optimization of the gas sensing properties in this type of sensors. These additives can act as catalysts in surface sites for adsorption of oxygen and detected gas or as promoters of the gas sensing characteristics [35]. The copper oxide has good catalytic activity for oxidation of the methane [34], so it was used as dopant for the methane gas sensor. For this reason, in this research, Nano-sized Cu- doped Cobalt oxides with different concentrations of the dopant have been prepared by the sol-gel method for the methane sensor application.

Material and Methods

The present paper covers an ANFIS modeling and experimental study of the gas sensing of the copper-cobalt compound oxide Nano particles. Experimental measurements have been made and Nano-sized powders have been prepared by the sol-gel technique with different mole ratios of the Cu/Co (from 0.05 to 0.15) for detection of the methane gas. The structural properties and morphology of the powders were studied by the X-ray diffraction (XRD), Fourier transform infrared spectroscopy (FTIR) and Transmission electron microscopy (TEM).

Adaptive network based fuzzy inference system

The ANFIS is a fuzzy Sugeno model placed in the framework of adaptive systems to facilitate learning and adaptation [36]. Sugeno type fuzzy inference systems (FIS) use a combination of least-squares and back propagation gradient descent methods, along with a hybrid learning algorithm to identify the membership function parameters and fuzzy IF–THEN rules based on a single output or singleton [37].

ANFIS represents a useful neural network approach for the solution of function approximation problems [38].

ANFIS structure For simplicity, it is assumed that the fuzzy inference system under consideration has two inputs and one output. The rule base contains two fuzzy if-then rules of Takagi and Sugeno's type [39] as follows:

***Corresponding author:** J.R. Esfandyari, The technical teacher of saveh education office, saveh, Iran

If x is A and y is B then z is f(x, y)

Where A and B are the fuzzy sets in the antecedents and z = f (x, y) is a crisp function in the consequent. f(x, y) is usually a polynomial for the input variables x and y.

Consider z= f (x, y) is a first-order Sugeno fuzzy inference system, which contains two rules.

Rule 1: If x is A_1 and y is B_1, Then: $f_1 = a_1 x + b_1 y + c_1$

Rule 2: If x is A_2 and y is B_2, Then: $f_2 = a_2 x + b_2 y + c_2$

ANFIS structure contains five layers excluding input layer.

Layer 0 is the input layer. It has n nodes where n is the number of inputs to the system.

Layer 1 is the fuzzification layer in which each node represents a generalized bell curve membership function value of a linguistic term as follows:

$$\mu_A(x) = \frac{1}{1 + \left| \frac{x - c_i}{a_i} \right|^{2b_i}} \tag{1}$$

Each node in Layer 2 provides the strength of the rule by means of multiplication operator. It performs AND operation.

Layer 3 is the normalization layer which normalizes the strength of all the rules according to the following equation:

$$O_{3,i} = \overline{w_i} = \frac{w_i}{w_1 + w_2} \quad i=1, 2 \tag{2}$$

Layer 4 is a layer of adaptive nodes. Every node in this layer computes a linear function where the function coefficients are adapted by using the error function of the multilayer feed-forward neural network.

Layer 5 is the output layer whose function is the summation of the net outputs of the nodes in Layer 4. The output f is computed as follows in Figure 1:

The first layer, every node i in this layer is an adaptive node with node function in Equation 1:

$$\mu_{Ai}(x) = e^{-((x-x^*)/\sigma^2)} \tag{3}$$

Where {x*} are σ premise parameters updated through hybrid learning algorithm and x is input variable. At least in the basic ANFIS method these parameters are not adjustable.

The second layer calculates the firing strength for each rule quantifying the extent which any input data belongs to that rule. The output of the layer is the algebraic product of the input signals as can be given as: [40-42]

$$O_{2,i} = \omega_i = \mu_{Ai}(x_1) \times .... \times \mu_{Ci}(x_n) \tag{4}$$

Figure 2 shows the ANFIS structure of T, Q and Ratio Cu/Co values as the three input parameters and two linguistic values (healthy or bruised) as the output parameters. After training the ANFIS model, predicted performance was tested.

Experimental

Nano-particles of copper-cobalt compound oxides were synthesized by sol-gel method. Systematically a proper amount of $CoCl_2.6H_2O$ and $CuCl_2.2H_2O$ dissolved in deionized water and ethanol. The initial solutions have been prepared with mole ratios of Cu/Co = 0.0, 0.05, 0.08, 0.10, 0.125 and 0.15. Then, citric acid (as complexing agent) was added in a mole ratio of $C_6H_8O_7/CoCl_2.6H_2O$ = 5, and primary sol was prepared by adding ethylene glycol (as polymerizing agent) in the mole ratio six time over cobalt cation and was stirred for 20 minutes. The clear mixed solution was refluxed at 120°C in oil bath for 6 hours. The final sol ensuing from reflux was kept at 80°C for 17 hours in oil bath to evaporate volatile compounds and a wet gel was formed. For removing the other residual volatile component the gel was heated again directly on hotplate at 180°C for 3 hours and then 220°C for 30 minutes. Finally the dry gel was annealed in 450°C for 1 hour to eliminate organic component and the copper-cobalt oxides Nano-powder was obtained. For gas sensing measurements, usually nanopowders screen-printed on the electrodes. So, using a component as binder, in many cases ethylene glycol, is necessary. Choi and Min [34] reported that sensitivity of a Co_3O_4 sensor decreases with increasing of the ethylene glycol. Accordingly, sensors were fabricated by pressing followed by hot treatment. The same amount of each powders were pressed into pellets at a pressure of 7 ton and then heated in furnace for 2 hours at 800°C with heating rate of 2°C/min. Silver paste was used as electrodes on the two sides of one surface of pellets for recording of the resistant change.

Phase formation and crystallinity were studied by X-ray diffraction using a D8 Advanced Bruker with Cu kα radiation (λ=0.15405 nm) in a 2θ range from 15° to 85° with a step size of 0.05°. The atomic link

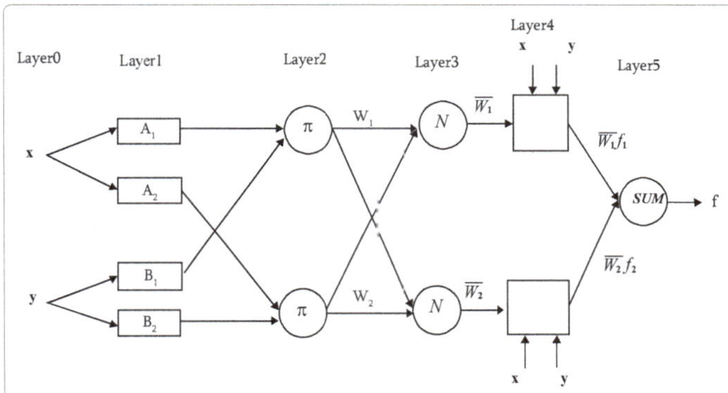

Figure 1: Basic ANFIS Structure.

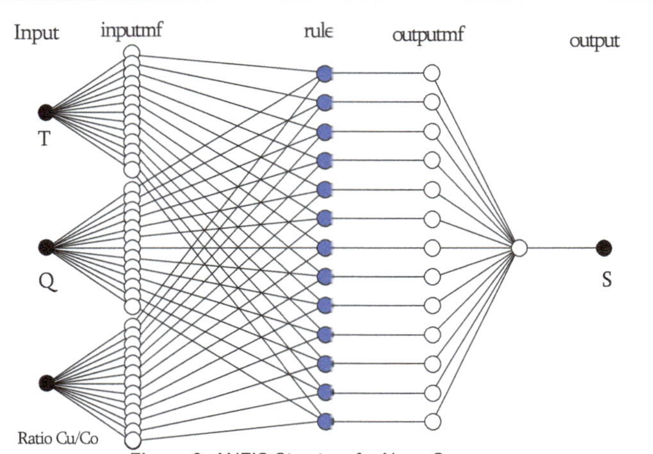

Figure 2: ANFIS Structure for Nano Sensor.

investigation was performed using Fourier transform infrared (FTIR) spectroscopy on a Bomem MB-154 spectrometer. Transmission electron microscopy (TEM) was carried out on a LEO 912 AB transmission electron microscope for determination of the particle shape and size. The prepared sensors were located in the testing chamber for gas sensing characterization. All sensitivity measurements were done in constant temperature in the range of 200-300°C and fixed quantity of sample gas of methane (3000 to 6000ppm). Electrical resistance of the sensor was monitored by a digital multimeter.

The sensitivity was defined as

$$(R_{gas}-R_{air})/R_{air}\times 100(\%)$$

Where R_{air} and R_{gas} are the electrical resistance in clean air and in presence of the gas at the same temperature, respectively.

Result and Discussion

Experimental results

The XRD patterns of copper-cobalt oxides nanopowders with different mole ratio of copper have been shown in Figure 3. It reveals polycrystalline structure consisting of cubic Co_3O_4 (a=8.08400Å) in all samples and $(CuO_{0.3}CoO_{0.7})\ Co_2O_4$ in which contain Cu.

Major XRD peaks are corresponded to (3 1 1) plane at 36.9° and it intensity does not change evidently by adding copper impurity in samples but it has become slight wider. Beside the strongest peak, also other peaks corresponded to (2 2 0) and (4 4 0) planes have got wider. It's clearly identified that copper impurity decreases the crystallinity and enhances the amorphous state that lead to the decreasing of the crystalline size. No other phase has been created in samples with different Cu/Co mole ratios and it demonstrates that formed phase are stable and its formation is independent of copper quantity.

Mean particle's size in major peaks is calculated using Scherrer's equation.

$$D=\frac{k\lambda}{\beta\cos\theta} \tag{5}$$

Where k is constant, depend on crystal morphology in the range of 0.89-1.39 and it is 0.9 in this calculation. λ is the x-ray wavelength, β is the line broadening at half the maximum intensity (FWHM) in radians and θ is the Brrag angle of XRD peak. The particle size values are given in Table 1. It was varied between 24 and 28 and decrease as Cu/Co mole ratio increase to 0.15.

Figure 4 shows FTIR spectra of copper-cobalt oxide Nano-particles with different mole ratios of Cu/Co = 0.0, 0.05, 0.10, 0.15 (S_1, S_2, S_4 and S_6 respectively).

The bands at 576 and 668 cm^{-1} are attributed to vibration of Co-O in the Co_3O_4. Any peak related to Cu-O could not be identified; it is probably Co-O peaks overlap Cu-O peak, and they have seen in very close wavelength. There is no peak that could be assigned to C-H which suggests that carbonate component removed completely from samples.

TEM images of cobalt oxide nanoparticles (S_1) and copper-cobalt oxide nanoparticles with Cu/Co = 0.15 (S_6) prepared by sol- gel method with different magnification are given in Figure 5.

It's found that particles have cubic morphologies with uniform distribution. The average particle size was 30 nm in S_1 and 15nm in S_6 which decreases by increasing of the dopant. It has a good agreement with result of Scherre's formula.

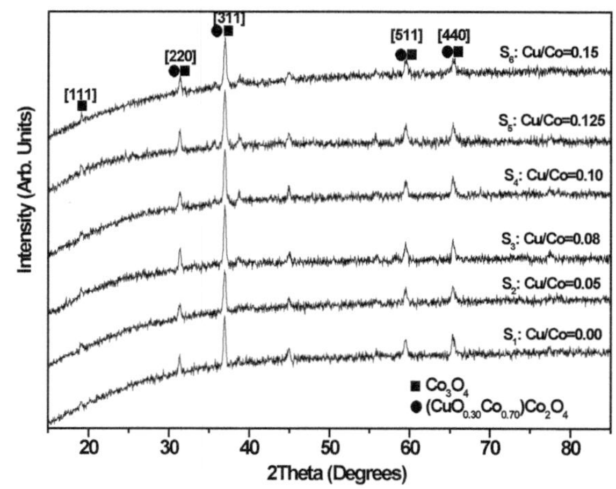

Figure 3: XRD patterns of Cu-Co oxides nano-particles with different Cu/Co mole ratio.

Sample	Mole ratio Cu/Co	2θ°	β	D(nm)
S_1	0.00	36.95	0.295	28
S_2	0.05	36.95	0.337	27
S_3	0.08	37.025	0.304	27
S_4	0.10	37.075	0.314	26
S_5	0.125	37.125	0.328	25
S_6	0.15	37.00	0.335	24

Table 1: The X-Ray Parameters and Particles Mean Size of Cu-Co Oxides Nano-Particles with Different Mole Ratios of Cu/Co Prepared By Sol-Gel Method.

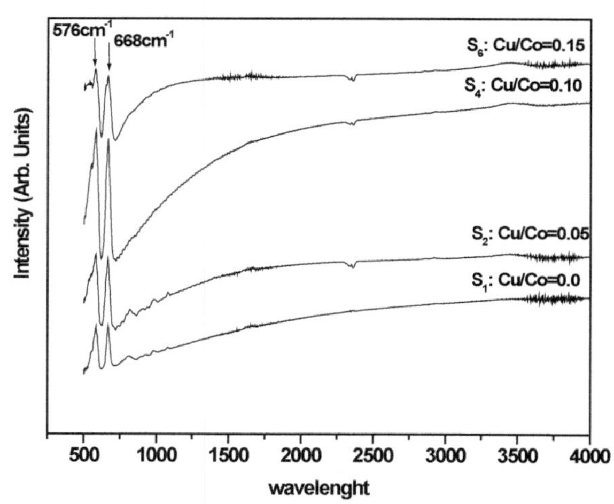

Figure 4: FT-IR Intensity Results for Cu-Co Oxides Nano-Crystals with Different Cu/Co Mole Ratios.

ANFIS comparative analyses and discussion

Two models of Sugeno, with an automatic extraction of data from FIS [GENFIS2] were used. The MATLAB software was adopted for comparison purposes. Moreover, the coverage threshold was fixes to 0.01. Table 2 shows experimental data and predicted data by ANFIS.

3D surface plots for the sensitivity values of nano sensor showing

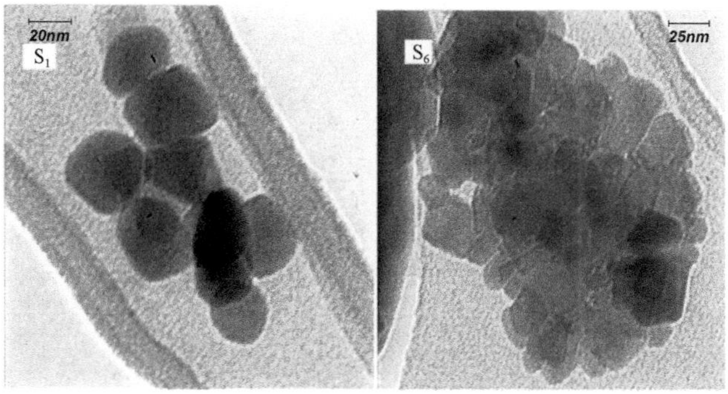

(a) (b)

Figure 5: TEM images for Cu-Co oxides nano-crystals with different Cu/Co mole ratios prepared by sol-gel method; (a) Cu/Co mole ratio=0.0, (b) Cu/Co mole ratio=0.15.

No.	T	Q (ppm)	Cu/Co	Experimental S	predicted S
Train data					
1	200	3000	0	0.1	0.103321
2	300	4500	0	7.2	7.154434
3	300	3000	0.05	2.5	2.418407
4	250	3000	0.05	1.3	1.340887
5	200	6000	0.08	4.1	4.164952
6	250	6000	0.08	5.3	5.377992
7	300	6000	0.1	23	22.28085
8	250	4500	0.1	5.9	5.87822
9	200	4500	0.125	2.9	2.870968
10	200	3000	0.15	2.2	2.177522
11	300	4500	0.15	23.9	21.86092
12	300	3000	0	0.4	0.392231
13	250	3000	0	0.2	0.20146
14	200	6000	0.05	3.8	3.538453
15	250	6000	0.05	4.4	4.177549
16	300	6000	0.08	19.7	20.136
17	250	4500	0.08	4.5	4.571422
18	200	4500	0.1	4.3	4.280648
19	200	3000	0.125	1.9	1.937442
20	300	4500	0.125	15.3	19.44232
21	300	3000	0.15	11.6	11.21666
22	250	3000	0.15	4.2	4.852579
23	200	6000	0	0.6	0.617481
24	250	6000	0	1.1	1.16363
25	300	6000	0.05	16.1	16.09997
26	250	4500	0.05	3.8	3.79999
27	200	4500	0.08	3.8	3.869347
28	200	3000	0.1	1.7	1.693322
29	300	4500	0.1	19.2	18.98301
30	300	3000	0.125	10.4	9.472365
31	250	3000	0.125	4.9	4.87564
32	200	6000	0.15	4.1	4.198856
33	250	6000	0.15	10.6	10.68818
34	200	4500	0	0.4	0.4
35	200	3000	0.05	1.1	1.08839
36	300	4500	0.05	12.5	12.65927
37	300	3000	0.08	5	4.86945
38	250	3000	0.08	2	2.02037
39	200	6000	0.1	5	4.953835
40	250	6000	0.1	7.4	7.39907
41	300	6000	0.125	25.3	25.58313
42	250	4500	0.125	7.6	7.646367
43	200	4500	0.15	4	3.961449
Test data					
44	300	6000	0	8.9	8.755444
45	250	4500	0	0.9	1.013
46	200	4500	0.05	3.2	3.11
47	200	3000	0.08	1.5	1.534764
48	300	4500	0.08	14.8	15.03857
49	300	3000	0.1	6.2	6.681882
50	250	3000	0.1	2.5	2.488226
51	200	6000	0.125	3.7	3.81
52	250	6000	0.125	8.9	8.854872
53	300	6000	0.15	31.1	29.97888
54	250	4500	0.15	8.9	9.01223

Table 2: Comparison of Experimental Data with Predicted Values for S.

the relationship between Temperature, Q, ratio Cu/Co and S are given in Figure 6:

A complete graphical comparison of experimental data used for testing or validating the predictions of ANFIS models are shown in Figures 7&8 for the train and the test data respectively. The coefficient of determination values (R^2), which quantifies the degree of agreement between experimental observations and numerically calculated values were found greater than 0.96 for all output variables are shown in Figure 9&10. R^2 was defined as below:

$$R^2 = \frac{\sum_{i=1}^{N}(x_i - \bar{x})(y_i - \bar{y})}{\sqrt{\sum_{i=1}^{N}(x_i - \bar{x})^2}\sqrt{\sum_{i=1}^{N}(y_i - \bar{y})^2}} \quad (6)$$

Table 3 reveals average relative error (ARE), absolute average relative error (AARE) and root mean square error (RMSE) for Nano Cu/Co respectively. ARE, AARE and RMSE are defined as below:

$$ARE = \frac{1}{N}\sum_{i=1}^{N}\left(\frac{X_{experimental(i)} - X_{caculated(i)}}{X_{experimental(i)}}\right) \quad (7)$$

$$AARE = \frac{1}{N}\sum_{i=1}^{N}\left(\left|\frac{X_{experimental(i)} - X_{caculated(i)}}{X_{experimental(i)}}\right|\right) \tag{8}$$

$$\text{RMSE} = \sqrt{\frac{\sum_{i=1}^{n}(X_{experimental(i)} - X_{caculated(i)})^2}{n}} \tag{9}$$

Conclusion

Copper-cobalt oxide Nano-sized powders with different mole ratio of Cu/Co from 0.0 to 0.15 were prepared by sol-gel technique for gas sensor application. Cubic Co_3O_4 and $(CuO_{0.3}CoO_{0.70})\,Co_2O_4$ phases were firmly formed. The mean grain size varied from 28 to 24 by changing Cu/Co mole ratios from 0.0 to 0.15. The gas sensing performances of sensors were investigated in presence of methane gas. The results have shown that optimal operating temperature is 300°C in all sensors. Doping of metal oxide to Co_3O_4 altered the sensing behavior by influence defect chemistry and decrease particle size. The performance of the ANFIS Prediction and experimental results was measured using the average relative error (ARE), absolute average relative error (AARE), root mean square error (RMSE) and the Coefficients (R^2) values. The developed ANFIS model showed a good regression analysis with the R^2 greater than 0.98. As the regression coefficients indicate the ANFIS approach could be considered as an alternative and practical technique to evaluate the sensitivity (S) based on the Q, temperature and ratio of Cu/Co with a high degree of accuracy.

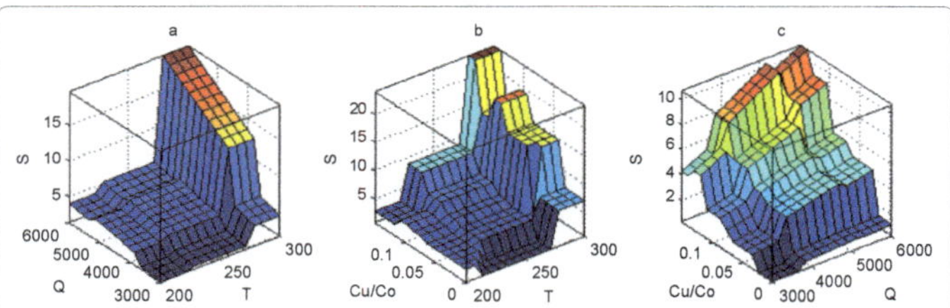

Figure 6: 3D surface plots for the sensitivity values of nano sensor.

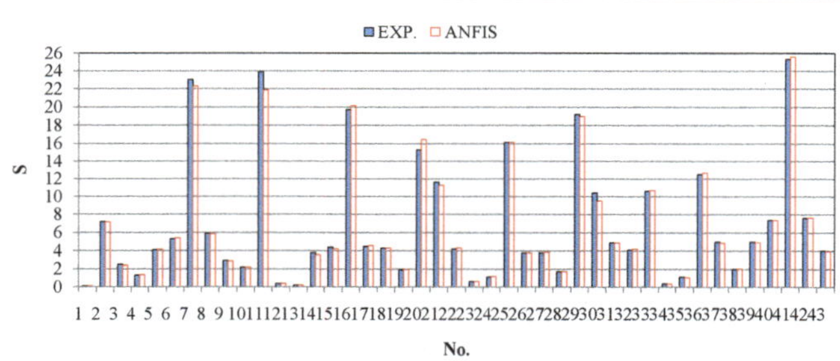

Figure 7: Comparison between the ANFIS modeling of this work and experimental data for train data.

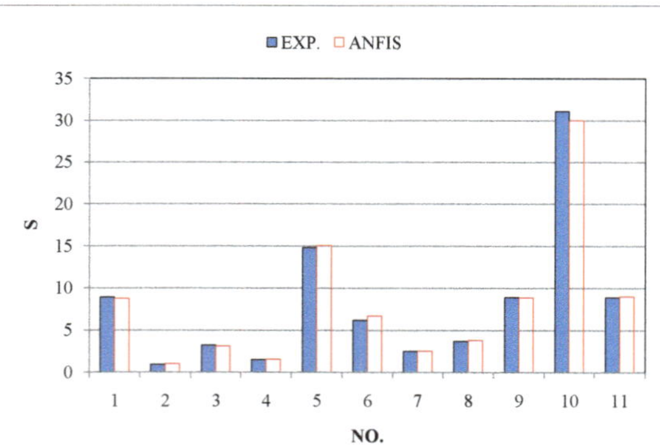

Figure 8: Comparison between the ANFIS modeling of this work and experimental data for test data.

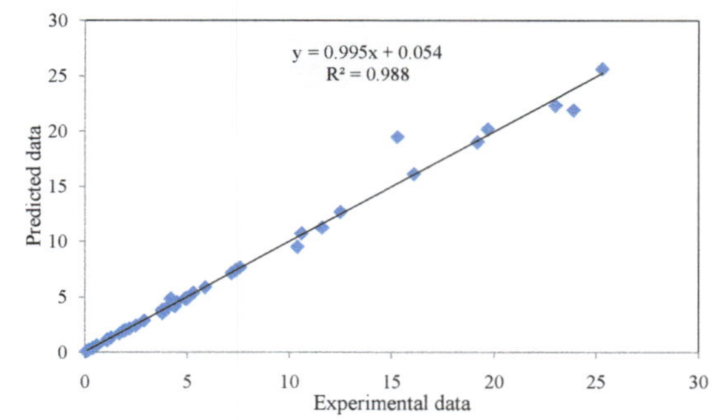

Figure 9: Comparison between of experimental data with predicted values for train data.

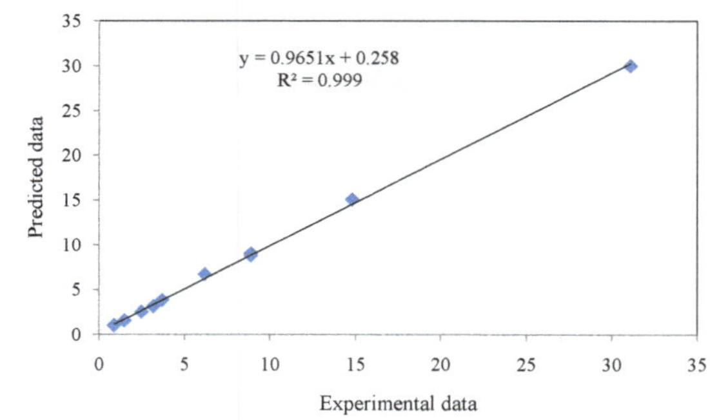

Figure 10: Comparison between of experimental data with predicted values for test data.

	Method	%ARE	%AARE	RMSE
Train	ANFIS	-0.56	2.98	0.74
Test	ANFIS	-1.77	3.41	0.38

Table 3: ARE, AARE and RMSE for nano sensor which modeled by ANFIS.

References

1. Hugon O, Sauvan M, Benech P, Pijolat C, Lefebvre F (2000) Gas separation with a zeolite filter, application to the selectivity enhancement of chemical sensors. Sens Actuators B 67: 235-243.
2. Ohta K, Terai H, Kimura I, Tanaka K (1999) Simultaneous Determination of Hydrogen, Methane, and Carbon Monoxide in Water by Gas Chromatography with a Semiconductor Detector. Anal Chem 71: 2697-2699.
3. Janicki W, Chrzanowski W, Zwan P, Namieśnik J (2003) Automated analyser for monitoring the contents of hydrocarbons in gas emitted from exploratory bore-holes in the gas and oil industry. J Autom Methods Manag Chem 25: 141-147.
4. Hansen TL, Schmidt JE, Angelidaki I, Marca E, Jansen JC et al. (2004) Method for determination of methane potentials of solid organic waste. Waste Manag 24: 393-400.
5. Forsyth DS (2004) Pulsed discharge detector: theory and applications. J Chromatogr A 24: 63-68.

6. Maris C, Chung MY, Lueb R, Krischke U, Meller R, et al. (2003) Development of instrumentation for simultaneous analysis of total non-methane organic carbon and volatile organic compounds in ambient air. Atmos Environ 37: 149-158.

7. Raksit A, Johri S (2003) Determination of dissolved volatile hydrocarbons in environmental aqueous samples by headspace-gas chromatography with flame ionization detection. Am Lab 35: 10-13.

8. Kamiński M, Kartanowicz R, Jastrzebski D, Kamiński MM (2003) Determination of carbon monoxide, methane and carbon dioxide in refinery hydrogen gases and air by gas chromatography. J Chromatogr A 989: 277-283.

9. Worthy DEJ, Levin I, Trivett NBA, Kuhlmann AJ, Hopper JF, et al. (1998) Seven years of continuous methane observations at a remote boreal site in Ontario, Canada. J Geophys Res 103: 15995-16007.

10. Seiyama T, Kato A, Fujiishi K, Nagatani M (1962) A new detector for gaseous components using semiconductive thin films. Anal Chem 34: 1502-1503.

11. Fleischer M, Kornely S, Weh T, Frank J, Meixner H (2000) Selective gas detection with high-temperature operated metal oxides using catalytic filters. Sens Actuators B 69: 205-210.

12. Faglia G, Comini E, Sberveglieri G, Rella R, Siciliano P, et al. (1998) Square and collinear four probe array and Hall measurements onmetal oxide thin film gas sensors. Sens Actuators B 53: 69-75.

13. Benounis M, Jaffrezic-Renault N, Dutasta JP, Cherif K, Abdelghani A (2005) Study of a new evanescent wave optical fibre sensor for methane detection based on cryptophane molecules. Sens Actuators B 107: 32-39.

14. Tournier G, Pijolat C (1999) Influence of oxygen concentration in the carrier gas on the response of tin dioxide sensor under hydrogen and methane. Sens Actuators B 61: 43-50.

15. Llobet E, Orts J, Vilanova X, Brezmes J, Correig X (1999) Selective methane detection under varying moisture conditions using static and dynamic sensor signals. Sens Actuators B 60: 106-117.

16. Pijolat C, Pupier C, Sauvan M, Tournier G, Lalauze R (1999) Gas detection for automotive pollution control. Sens Actuators B 59: 195-202.

17. Quaranta F, Rella R, Siciliano P, Capone S, Epifani M, et al. (1999) A novel gas sensor based on SnO_2 thin film for the detection of methane at low temperature. Sens Actuators B 58: 350-355.

18. Licznerski BW, Nitsch K, Teterycz H, Szecˇıowka PM, Wisniewski K (1999) Humidity insensitive thick film methane sensor based on SnO_2/Pt. Sens Actuators B 57: 192-196.

19. Schmid W, Barsan N, Weimar U (2003) Sensing of hydrocarbons with tin oxide sensors: possible reaction path as revealed by consumption measurements. Sens Actuators B 89: 232-236.

20. Gajdosik L (2002) The derivation of the electrical conductance/concentration dependency for SnO_2 gas sensor for ethanol. Sens Actuators B 81: 347-350.

21. Saha M, Banerjee A, Halder AK, Mondal J, Sen A, et al. (2001) Effect of alumina addition on methane sensitivity of tin dioxide thick films. Sens Actuators B 79: 192-195.

22. Llobet E, Rubio J, Vilanova X, Brezmes J, Correig X, et al. (2001) Electronic nose simulation tool centered on Spice. Sens Actuators B 76: 419-429.

23. Hammerichin O, Lund H, Hammerich O (2000) Organic Electrochemistry. (4thedn), Marcel Dekker, NY.

24. Lu Y, Li J, Han J, Ng HT, Binder C, et al. (2004) Room temperature methane detection using palladium loaded single-walled carbon nanotube sensors. Chem Phys Lett 391: 344-348.

25. Damgaard LR, Revsbech NP (1997) A microscale biosensor for methane containing methanotrophic bacteria and an internal oxygen reservoir. Anal Chem 69: 2262-2267.

26. Damgaard LR, Revsbech NP, Reichardt W (1998) Use of an oxygen-insensitive microscale biosensor for methane to measure methane concentration profiles in a rice paddy. Appl Environ Microb 64: 864-870.

27. Xiangfeng Ch, Donglia J, Yua G, Chenmoua Zh (2006) Ethanol gas sensor based on $CoFe_2O_4$ nano-crystallines prepared by hydrothermal method. Sens Actuators B 120: 177-181.

28. Lu F, Liu Y, Dong M, Wang X (2000) Nanosized tin oxide as the novel material with simultaneous detection towards CO, H2 and CH4. Sens Actuators B 66: 225-227.

29. Wang Sh H, Chou T Ch, Liu Ch Ch (2003) Nano Crystalline Tungsten Oxide NO_2 Sensor. Sens Actuators B 94: 343-351.

30. Chen Sh M, Wu MH, Thangamuthu R (2008) Preparation, Characterization, and Electrocatalytic Properties of Cobalt Oxide and Cobalt Hexacyanoferrate Hybrid Films. Electroanalysis 20: 178-184.

31. Zhang G, Liu M (2000) Effect of particle size and dopant on properties of SnO-based gas sensors. Sens Actuators B 69: 144-152.

32. Tan OK, Cao W, Hu Y, Zhu W (2004) Nano-structured oxide semiconductor materials for gas-sensing applications. Ceram Int 30: 1127-1133.

33. Tan OK, Cao W, Zhu W, Chai JW, Pan JS (2003) Ethanol sensors based on nano-sized a-Fe2O3 with SnO2, ZrO2 TiO2 solid solutions. Sens Actuators B 93: 396-401.

34. Choi SD, Min BK (2001) Co_3O_4-based isobutane sensor operating at low temperatures. Sens Actuators B 77: 330-334.

35. Korotcenkov GVM, Brinzari V, Tolstoy V, Schwank J, Cornet A, et al. (2004) Influence of Cu-, Fe-, Co-, and Mn-oxide nanoclusters on sensing behavior of SnO2 films. Thin Solid Films 467: 209- 214.

36. Jang JSR (1993) ANFIS: Adaptive-network-based fuzzy inference systems. IEEE Trans Syst Man Cybern 23: 665-685.

37. Ho YCh, Tsai ChT (2011) Comparing ANFIS and SEM in linear and nonlinear forecasting of new product development performance. Expert Syst Appl 38: 6498-6507.

38. Buragohain M, Mahanta Ch (2008) A novel approach for ANFIS modeling based on full factorial design. Appl Soft Comput 8: 609-625.

39. Takagi T, Sugeno M (1985) Derivation of fuzzy control rules from human operator's control actions, in: Proc IFAC Symp Fuzzy Inform. Knowledge Representation and Decision Analysis 55-60.

40. Meharrar A, Tioursi M, Hatti M, Boudghène Stambouli A (2011) A variable speed wind generator maximum power tracking based on adaptative neuro-fuzzy inference system. Expert Syst Appl 38: 7659-7664.

41. Sargolzaei J, Kianifar A (2010) Neuro-fuzzy modeling tools for estimation of torque in Savonius rotor wind turbine. Adv Eng Software 41: 619-626.

42. AliyariShoorehdeli M, Teshnehlab M, KhakiSedigh A (2009) Training ANFIS as an identifier with intelligent hybrid stable learning algorithm based on particle swarm optimization and extended Kalman filter. Fuzzy Sets Syst 160: 922-948.

Effect of O_2 Concentration on the Reaction Furnace Temperature and Sulfur Recovery Using a TSWEET® Process Simulator

Asadi S*, Hamed Mosavian MT and Ahmadpour A

Chemical Engineering Department, Faculty of Engineering, Ferdowsi University of Mashhad, Mashhad, P.O Box: 91775-1111, Iran

Abstract

In the present study, effect of a parameter for increasing the reaction furnace temperature and sulfur recovery of Claus sulfur recovery units (SRUs) is investigated by TSWEET process simulator based on Gibbs free minimization. The effect of oxygen enrichment on the reaction furnace temperature and sulfur recovery has been studied by proposed simulator. In the case of the effect of O_2 concentration, temperature of main burner increased monotonically. It is found that sulfur recovery (in the O_2 high concentrations) increases up to a maximum value and then decreases with increase of O_2 concentration, while this is not the case for O_2 low concentrations.

Keywords: Sulfur recovery; O_2 concentration; Reaction Furnace; Claus unit

Introduction

Sulfur recovery unit (SRU) is an important unit in natural gas processing units. It removes H_2S from acid gas feed before they can be released into the environment [1]. H_2S exists mainly as an undesirable by-product of gas processing units [2]. Different processes are applied to recover sulfur from H_2S. The most widely used process is Claus process [1,3]. It should be noted that several modifications were developed on the main Claus process in order to increase the overall sulfur recoveries.

The sulfur recovery requirements range from 97.5 to 99.8% for gas processing units and refining facilities processing 10 LT/d and greater of natural gas [4]. By increasing the number and type of beds in the case of rich acid gas feed, the sulfur recovery increases from 96.1% to 99.3%. In the case of lean acid gas feed, the sulfur recovery increases from 96.1% to 96.6%.

Recently, a number of studies have been studied the reaction furnace of Claus units. As the fraction of the AG splitter flow to main burner increased, initially the temperature of the main burner increased to a maximum temperature but it then decreased sharply as the flow was further increased. This was true for all three concentrations of oxygen [5]. However, if the concentration of H_2S, the H_2S/CO_2 ratio and the flow rate of air are increased, the temperature of the main burner increased [5].

In this paper, the simulation of Claus process was done, and Also effect of O_2 concentration on the reaction furnace temperature and sulfur recovery is investigated. For these purposes, intake air and acid gas feed are classified into three different categories in terms of their composition. The first type of intake air and acid gas feed contains 21 mol% of O_2 and 30 mol% of H_2S (usual feed and input air). The second type includes 50 mol% of O_2 and 50mol% of H_2S. The last type contains 85 mol% of O_2 and 90 mol% of H_2S (O_2 enriched air-H_2S reach feed).

Methodology

Research methodology consists of a review on O_2 concentration (in the tail gas ratio of 2.0) effect on the reaction furnace temperature and sulfur recovery of Claus unit using a process simulator called TSWEET.

Figure 1 shows the flow diagram of the Claus unit. In this process, the acid gas (acid gas + fuel gas) enters SRU and is divided into two streams in the AG splitter. A part of the feed stream is sent to the main burner, and another part is sent to the acid gas heater (for hydrolyzing sulfur components to H_2S). 1/3 of H_2S in the acid gas is oxidized to SO_2 at the main burner using outlet air of air blower.

$$H_2S+3/2O_2 \leftrightarrow SO_2 + H_2O \quad (1)$$

This combustion generates a large amount of heat. Further, the combustion products undergo Claus reaction between H_2S and SO_2.

$$2H_2S+ SO_2 \leftrightarrow 3/nS_n + 2H_2O \quad (2)$$

Where n is in the range of 6-8. Reaction (2) is exothermic and reversible, thus, processing under adiabatic conditions increases temperature.

The effluent gas from the main burner (reaction furnace) passes through the first pass waste heat boiler (1st pass WHB) to recover heat and produce high pressure steam [1,6]. The second pass of the waste heat boiler is where the redistribution of S_2 to S_8 is the primary reaction. It should be noted that, side reactions involving hydrocarbons and CO_2 in the acid gas feed can result in the formation of carbonyl sulphide (COS) and carbon disulfide (CS_2) in the main burner [7]. The effluent streams from the 2nd pass WHB and acid gas heater is cooled in condenser 1 to condense and recover the liquid sulfur. The effluent gas of the condenser 1 is preheated in the reheater 1 and is sent to the catalytic converters where Claus reaction occurs at lower temperatures. This leads to higher equilibrium conversion because Claus reaction is exothermic.

Typically, COS and CS_2 are also hydrolyzed in the first catalytic converter according to the following exothermic reactions [2]:

$$COS+H_2O \rightarrow CO_2+H_2S \quad (3)$$

$$CS_2+2H_2O \rightarrow CO_2+2H_2S \quad (4)$$

***Corresponding author:** Samer Asadi, Chemical Engineering Department, Faculty of Engineering, Ferdowsi University of Mashhad, Mashhad, P.O Box: 91775-1111, Iran

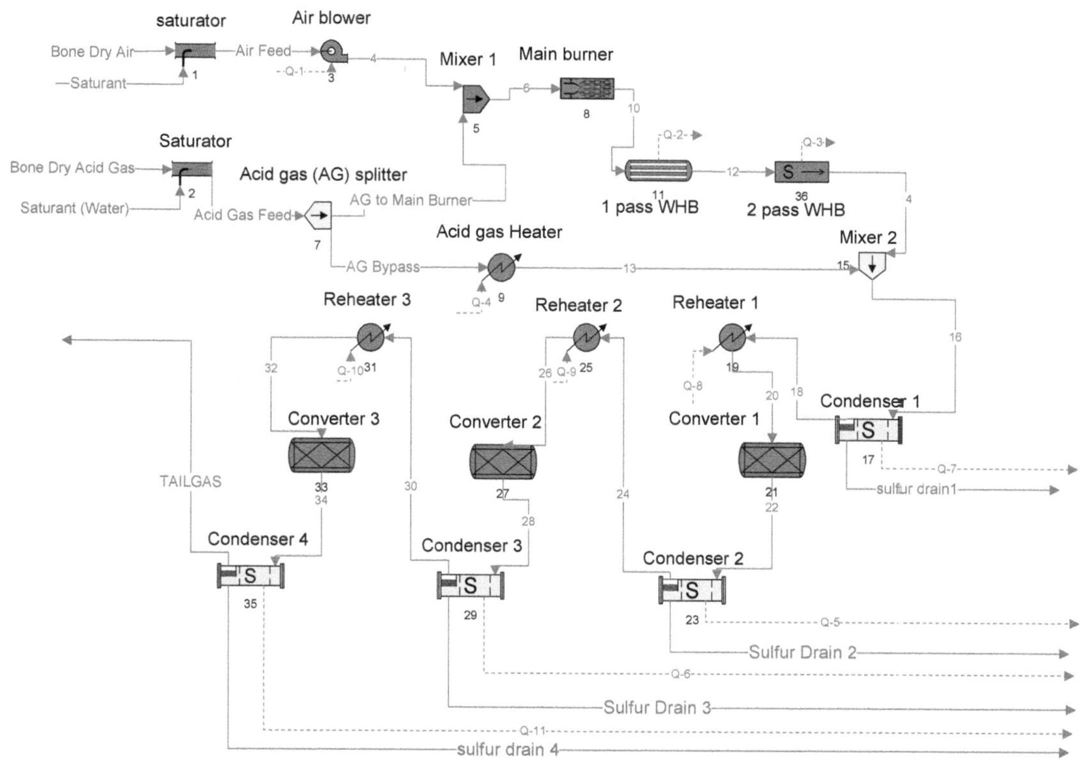

Figure 1: Flow diagram of simulated Claus process.

The first converter normally operates at temperature high enough to hydrolyze COS and CS_2.

Usually, the catalysts, based on γ- Al_2O_3 or TiO_2, are used in the process [8]. Sulphate is formed on the γ- Al_2O_3 catalyst when a mixture of H_2S/SO_2 or H_2S and SO_2, separately, passes on the catalyst at industrial conditions [9]. Sulfur is recovered after each catalytic stage by cooling converter effluent gases in sulfur condensers. Finally, effluent gas from the process is incinerated to SO_2 and vented to the atmosphere.

The simulation was performed using the TSWEET (copyright© 2007, Bryan Research & Engineering, Inc.) version 2.0 (PROMAX). PROMAX is a powerful and jack of all trades stream based process simulation pack. It is used by engineers around the world to design and optimize gas processing, refining, and chemical processes.

TSWEET assumes that the Claus beds operate at equilibrium conversions equal to 95% and it also assumes 4 lb of sulfur/100 lb moles of gas are entrained in the sulfur condensers [10]. This simulator uses three unit operations (main burner, 1st pass WHB, 2nd pass WHB) to model the burner/waste heat boiler [10,11]. The burner unit operation simulates the combustion of acid gas, with COS and CS_2 formation being calculated by NSER 1993 correlation [12]. This correlation predicts the concentration of COS, CS_2, CO, H_2, and/or S_2 out of the burner. The correlation appears to be most accurate when NH_3 presents.

The reaction furnace (modeled as the first pass of the waste heat boiler) is where large amounts of free sulfur, hydrogen and water recombine into H_2S and SO_2. The typical output temperature is about 600°C since all significant reactions other than sulfur redistribution have ceased in 600°C. The 2nd pass WHB is where the redistribution of S_2 to S_8 is the primary reaction, which the typically outlet temperature is about 315°C.

Results and Discussion

The first role of the reaction furnace in the Claus unit is to partially oxidize H_2S in the acid gas feed to SO_2 and it will also produce a significant amount of the total produced sulfur. A second role of the reaction furnace is to ensure of the destruction of other contaminants of the acid gas feed stream. This will ensure that these compounds do not break through the downstream process unit, where they can have a significant detrimental effect on the Claus catalyst activity. Therefore, it is critical to understand the effect of furnace operating parameter on the reaction furnace temperature and overall sulfur recovery.

There are different techniques to change the furnace temperature and sulfur recovery, including, O_2 concentration. The effect of this parameter on the reaction furnace temperature and sulfur recovery of the Claus unit is investigated by TSWEET simulator.

Oxygen enrichment

Many units are using oxygen enriched air in their SRUs for many reasons [9,13-15]. In most cases, oxygen enrichment is used to raise the throughput capacity of the plant. However, in some particular cases, oxygen enrichment has been used to improve the reaction furnace temperature. In some cases this has allowed for the processing of highly lean acid gas feed and in others it has utilized for better overall furnace and burner operation. The oxygen enrichment should be used with care. The accuracy control of the mixing of combustion gases is an essential factor in avoiding oxygen breakthrough into the catalytic converters. The poor concord in the combusting gases led to unreacted oxygen being accessible in the gas to further oxidize SO_2 to SO_3 either before the gas stream reached the catalyst bed or on the catalyst bed itself. This SO_3 then reacts with the alumina catalyst and depraves it by forming aluminium sulphate. The SO_3 may also react with H_2O to produce sulphuric acid, which is corrosive for steel equipments [16].

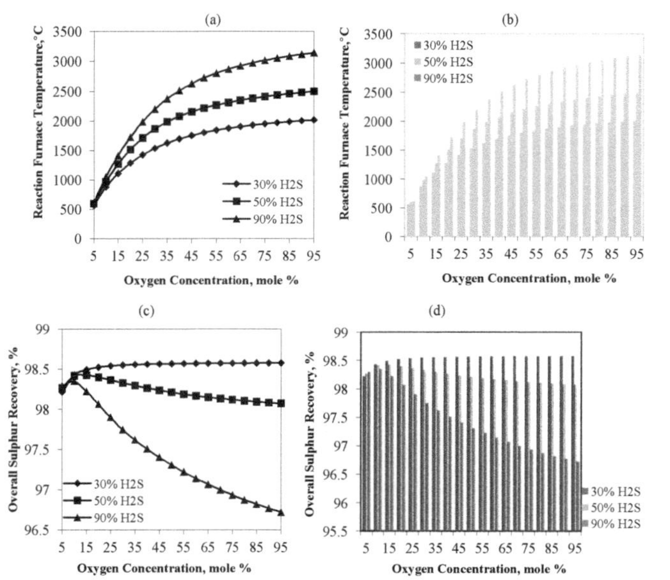

Figure 2: Effect of H_2S concentration on the reaction furnace temperature (a,b), and overall sulphur recovery (c,d).

As seen in the figures 2a and b, by increasing oxygen concentration, in tail gas ratio of 2.0 (optimal ratio) [5], reaction furnace temperature increases. Increased H_2S concentration in acid gas feed led to increase the reaction furnace temperature. In 30 mol% of H_2S, temperature of the main burner increased from 598°C to 2016°C. In concentrations of 50 and 90 mol% of H_2S, reaction furnace temperature was changed from 598°C to 2498°C and 624°C to 3143°C, respectively. Comparing these data with real ones indicates that when ratio of tail gas is 2.0, main burner temperature increases more. However, it is a general diagram because main burner temperature should be in the range of 700°C to 2000°C. The advantage of this study in comparison with other ones iTo investigate general changes of main burner temperature with respect to oxygen concentration.

To specify allowable range of oxygen concentration (in proportion to allowable range of reaction furnace temperature).

In this section the effect of oxygen concentration on sulfur recovery rate while changing the temperature of reaction furnace mentioned in the previous section has also been studied. Optimal temperature of reaction furnace is a temperature in which more sulfur recovery is obtained. So, in this study the changes in the rate of sulfur recovery were studied with respect to oxygen concentration in the intake air into the Claus unit. As seen in figures 2c and 2d, in 30 mol% of H_2S concentration, sulfur recovery was increased relatively however in 50% and 90% concentrations first it was increased to a maximum, and then decreased with a sharp slope. It is concluded that in 30 mol% of H_2S concentration, higher reaction furnace temperatures are required. However, in 50 and 90 mol%, optimum temperature was obtained in a specific concentration.

Also, as seen in figures 2c and 2d, as oxygen concentration changes in the range of 95-5%, sulfur recovery in 30 mol of H_2S concentration was increased from 98.2 to 98.57%. However, in 50 and 90 mol% of H_2S, sulfur recovery had maximums in concentrations of 10 and 15 mol%, respectively. Optimal temperatures of reaction furnace corresponding to the values of sulfur recovery were 1271°C and 1050°C, respectively.

It is also concluded that increasing the concentration of H_2S in acid gas feed leads to decrease in reaction furnace temperature.

Conclusion

In this study, using TSWEET simulator, effect of O_2 concentration on the sulfur recovery and furnace temperature of unit was investigated.

It is concluded that increasing concentration of O_2, temperature of main burner has increased monotonically (significant direct effect). As is found that sulfur recovery (in the O_2 high concentrations) increases up to a maximum value and then decreases with increase of O_2 concentration, while this is not the case for O_2 low concentrations. Whereas, it concluded that as oxygen concentration changes in the range 5-95%, sulfur recovery in 30% mol H_2S was increased from 98.2 to 98.57%.

References

1. Bishtawi RE, Haimour N (2004) Claus recycle with double combustion process. Fuel Process Tech 86: 245-260.
2. Zarenezhad B (2009) An Investigation on the most Important Influencing Parameters Regarding the Selection of the Proper Catalysts for Claus SRU Converters. J Indus Eng Chem 15: 143-147.
3. Polasek J, Bullin J (1993) Effect of Sulfur Recovery Requirements on Optimization of Integrated Sweetening, Sulfur Recovery, and Tailgas Cleanup Units. Proceedings of the Seventy-Second GPA Annual Convention, Tulsa, Gas Processors Association.
4. Blevins SK (1993) Proceeding of the Gas Conditioning Conference, Norman, Oklahoma.
5. Asadi S, Pakizeh M, Pourafshari chenar M (2011) An investigation of reaction furnace temperatures and sulfur recovery. Front Chem Sci Eng 5: 362-371.
6. Boussetta N, Lanoiselle JL, Bedel-Cloutour C, Vorobier E (2009) Extraction of Soluble Matter From Grape Pomace by High Voltage Electrical Discharges for Polyphenol Recovery: Effect of Sulphur Dioxide and Thermal Treatments. J Food Eng 95: 192-198.
7. Zarenezhad B, Hosseinpour N (2008) Evaluation of Different Alternatives for Increasing the Reaction Furnace Temperature of Claus SRU by Chemical Equilibrium Calculations. Appl Therm Eng 28: 738-744.
8. Zagoruiko AN, Matro YS (2002) Mathematical modelling of Claus reactors undergoing sulfur condensation and evaporation. Chem Eng J 87: 73-88.
9. Clarke D (2001) Laurance Reid Gas Conditioning Conference, Norman, Oklahoma.
10. Mattsson-Boze KW, Lyddon LG (1997) Using a Process Simulator to Improve Sulphur Recovery. Sulphur 37-41.
11. Lins VFC, Guimaraes EM (2007) Failure of a Heat Exchanger Generated by an Excess of SO_2 and H_2S in the Sulphur Recovery Unit of a Petroleum Refinery. J Loss Prevent Proc Ind 20: 91-97.
12. Monney WD, Svrcek WY, Behie LA (1993) Modeling the Modified Claus Process Reaction Furnace and the Implications on Plant Design and Recovery. Can J Chem Eng 71: 711-724.
13. Goar BG (1989) Proceedings of the Laurance Reid Gas Conditioning Conference 4-6.
14. Hyne JB, Goar BG (1996) Proceedings of the Laurance Reid Gas Conditioning Conference 3-6.
15. Kohl AL, Nielsen RB (1997) Gas Purification, fifth ed., Gulf Publishing Company, Houston.
16. Craig BD, Anderson DS (1995) Handbook of Corrosion Data. (2nd edn), ASM International, Ohio, USA.

A New Method for Demethylation of Lignin from Woody Biomass using Biophysical Methods

M Ferhan*, N Yan and M Sain

Center for Biocomposites and Biomaterials Processing (CBBP), Faculty of Forestry, University of Toronto, Toronto, ON, Canada

Abstract

Demethylation of softwood kraft lignin from woody biomass to improve the hydroxyl number in modified demethylated lignin and to produce lignin-based polyols was investigated using several biophysical techniques. Lignin is a tremendously under-developed natural polymer co-generated through papermaking and biomass fractionation. Molecular weights of lignins were analyzed by high-performance size-exclusion chromatography (HPSEC). Molecular weights of softwood kraft lignin (SKL) and demethylated lignin (DL) were determined as 1071 and 891 $gmol^{-1}$, respectively. For demethylation, iodocyclohexane in dimethyl formamide (DMF) was used under reflux conditions, with a resulting yield of 87%. Fourier transform infrared spectroscopy (FTIR) was used to determine characteristic absorption peaks of softwood kraft lignin and demethylated lignin. Significant spectral differences were noticed between the two types of lignin due to changes in chemical structure. Total hydroxyl numbers were determined by titration. Phosphorus nuclear magnetic resonance spectrometry (^{31}P-NMR) was employed to analyze the structure of lignin and different types of phenolic hydroxyl units. Nevertheless, further chemical and biological modifications within the lignin molecule are needed for various industrial applications to synthesize polyurethane foam by using chemically modified lignin-based polyols.

Keywords: Demethylation; Softwood kraft lignin (SKL); Demethylated lignin (DL); Hydroxyl numbers; FTIR spectroscopy; ^{31}P-NMR

Introduction

Lignin is described as an unsystematic, unstructured, 3-D polymeric complex system that does not possess a regular, precise structure with fixed recurring units. In lignin, the main functional groups are hydroxyl, methoxy, carbonyl, and carboxyl moieties in several amounts, whose components depend on the botanical source and the applied extraction processes [1]. In the plant kingdom, lignin is the most abundant polymeric organic material after cellulose [2]. The lignin biosynthetic pathway utilizes mainly three different types of phenylpropane units to make the lignin molecule. These include a guaiacyl precursor made from coniferyl alcohol and guaiacyl-syringyl precursors derived from coniferyl and sinapyl alcohol. Lignin from softwood (conifers) is mainly composed of guaiacyl units, while hardwood (angiosperms) lignin contains guaiacyl-syringyl units [3].

Both lignins contain mainly glycerol-aryl ether (β-O-4) linkages between short and linear chains of phenylpropane units [4]. The monomeric units of lignin are connected by a number of different C-C and ether linkages, which accounts for the complicated 3-D structure of lignin [2,5].

The different types of functional groups in a complex lignin macromolecule are presented in Figure 1, [3,6]. Lignins are the by-product of pulp and paper industries and are abundant in nature. The uses of lignin in different applications are as a dispersant agent for pesticides, emulsifiers, ion-exchange resins, water treatment agent, pesticide surfactants, heavy metal sequestrant, binders, animal foods, grinding aids, electrolytic refining, and tanning agents, or as a component of composites and copolymers [7-9].

To create value-added applications of lignin, medium- and long-term technologies mainly for the preparation of LMW compounds that can substitute for products produced by the petrochemical industry must be established [10]. Nevertheless, the structure of lignin is not completely recognized yet. Over the last two decades, many studies have been undertaken to understand the industrial applications of lignins [6]. The subjects of study by several research groups are lignin and their advanced applications, including, polyurethanes [10], acrylics [11], epoxies [12], and phenolic resins [13].

For polyurethane synthesis, the utilization of lignin as a macromonomer frequently follows two general methods: direct use of lignin without any major chemical modification, alone or in blends with other polyols [14], or (2) chemical modification, such as depolymerization, esterification and etherification reactions [10,15]. A broad range of lignin-based polyurethane materials (rigid foams, elastomers, sealants) has been produced and the resultant mechanical and thermal properties have been assessed [16].

Due to shared structural similarities, the industrially important organic compound vanillin can be produced from lignin, as has been reported by many scientists [17]. By catalytic degradation of lignin, many other industrially valuable small molecule chemicals can be produced [17]. However, to control for serious practical problems related to the enormous energy costs and basic purification processes, additional research is required to produce small molecule chemicals from this complex natural biopolymer. The second approach for the utilization of lignin is as a primary material in a diversity of green polymer products [18]. Nevertheless, the synthesis of commercial based

***Corresponding author:** M. Ferhan, Center for Biocomposites and Biomaterials Processing (CBBP), Faculty of Forestry, University of Toronto, Toronto, ON, M5S 3B3, Canada sustainable polymers requires high-purity initial materials having great potential and reproducible reactivity.

Lewis acids have been comprehensively studied for the demethylation of methoxyl groups in small molecule organic chemistry [19]. The methoxyl group in lignin can be converted to a more reactive hydroxyl group via demethylation [18,20]. In this study, we used iodocyclohexane as a hydriodic acid (HI) source in DMF. By increasing lignin hydroxyl-group contents and its reactivity, this treatment was found to be valuable, compared with other Lewis acids such as boron tribromide (BBr_3) [21], and trimethylsilyl iodide (TMSI) [22] which were not.

The aim of this study was to explore the demethylation of the lignin molecule and to characterize different hydroxyl groups present in commercially available technical and demethylated lignins using different biophysical methods. The hydroxyl contents are needed to characterize commercial polyols for the synthesis of bio-based polyurethane foam used for several applications as shown in Figure 3 and this concept was earlier presented by da Silva et al. [23].

Materials and Methods

Softwood kraft lignin (SKL) was provided by Krugar Wayagamack Inc. Quebec, Canada, contains 1.50% ± 0.02ash. The ash contents were determined by gravimetric method whilst sample ignition heating up to 525°C using Sybron-Thermolyne® bench top muffle furnace. Polystyrene standards were purchased from American Polymer Standards Corporation, Mentor, OH.

Analytical HPSEC

Polystyrene standards were used to determine the molecular weight of lignin. The calibration curve standards were used for several fractions of very low poly dispersity close to unity in HPSEC analysis. Polystyrene standards are commonly used for lignin molecular weight determination. Lignin fractions and polystyrene standards exhibited very comparable behavior throughout a large range of molecular weights in a regular HPSEC analysis of lignin [24]. The experimental conditions for HPSEC were: flow-rate 1.0ml/min, solvent DMF, sample concentration of 2mg/ml of DMF, volume injected variable from 100 µl to 200 µl, column temperature 80°C, RI detector temperature 40°C, UV-vis scanning range at 270-350 nm wavelength, Refractive Index

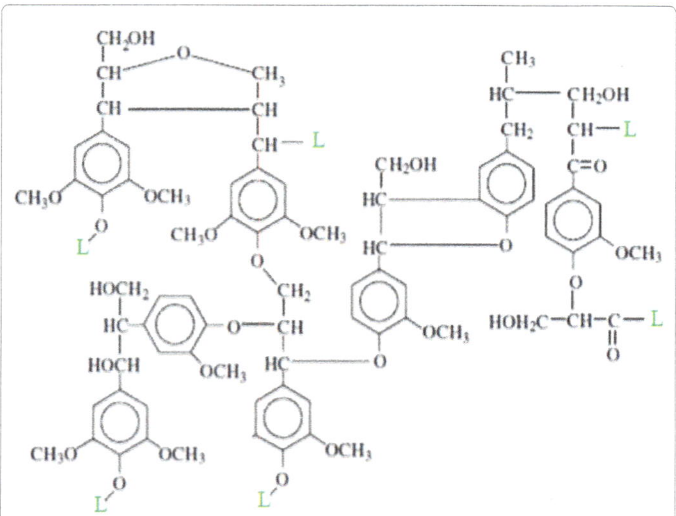

Figure 1: Partial lignin structure proposed by Adler [3,6].

scanning range at 400nm wavelength, Refractive Index detector-Waters 410 (Differential Refractometer), UV-vis detector-Waters M-490,GPC software- EMPOWER, HPLC system by Millipore-Waters and Styragel HR1 column 104A. Calibration curves were constructed with polystyrene standards.

Optimized Reaction Conditions for Demethylation

Reaction mixtures were prepared in a three neck round bottom flask and included 25mg of softwood kraft lignin (SKL) dissolved in 2.0 ml of dimethyl formamide DMF and 1.3 ml of iodocyclohexane [25]. The reaction mixture was purged with nitrogen to remove oxygen and setup with a reflux condenser. The reaction system was refluxed for 3h at 145°C, cooled, poured into water (20ml) and extracted with tetrahydro furan (THF) and toluene (20 ml x 3). The organic layer was washed with saturated aqueous sodium bisulphite ($NaHSO_3$) and brine solution and the samples were dried over oven-dry sodium sulphate (Na_2SO_4) and then filtered. The filtrate was concentrated at 82°C in a rotary evaporator to remove all organic solvents, including iodocyclohexane, and afterward dried in a vacuum desiccator to obtain demethylated lignin (13.7mg, 87% ± 0.51 yield). The process is illustrated by the general equation as shown in Figure 2.

FTIR Spectroscopy

FTIR spectra of softwood kraft lignin (SKL) and demethylated lignin (DL) were measured using a Bruker Tensor 27 FTIR spectrometer equipped with an IR-microscope and MCT detector (Karlsruhe, Germany). For spectral analysis, 2.0 mg of each lignin

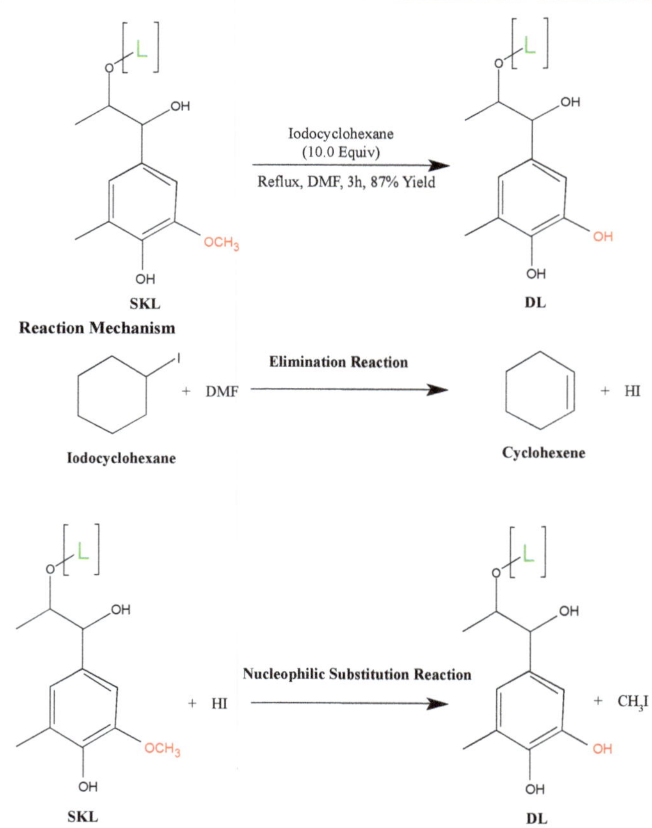

Figure 2: Reaction mechanism for demethylation process illustrating the replacement of methoxyl to hydroxyl group under optimized reaction conditions, where L indicates lignin molecule.

sample (SKL and DL) was mixed with200 mg of KBr. The transmission spectra (3500 to1000cm^{-1}) of SKL and DL were measured at room temperature. To prevent moisture contamination of the pellet, the following two steps were performed: SKL, DL and KBr were left at 40°C for 24 h under reduced pressure before pellet preparation, and (2) pellets were maintained under the same conditions for 12 h before FTIR analysis. By using Opus software, the baseline was corrected in the collected transmission spectra and stabilized according to the highest band.

Hydroxyl Number Determination by Esterification

Phthalation is one of the most common wet chemical methods for lignin hydroxyl group determination. The total hydroxyl content can be measured by a standard procedure with a phthalic anhydride, imidazole/dioxane mixture, followed by back-titration with a NaOH solution. This provides a simple method for determining total hydroxyl content. Using the esterification approach as reported by Kurimoto et al. [26,27] aided in the determination of the total OH value.

A mixture of 0.2-0.5 g of each lignin (SKL and DL) sample and 25 ml of phthalation reagent was separately heated at 110°C for 20 min. After cooling, 50ml of a dioxane/water solution (80:20 v/v) was added to the mixture and back titrated with 1N NaOH solution to get the equivalence point using a pH meter. The phthalation reagent is the combination of three components: 30g phthalic anhydride, 4.84 g imidazol, and 200 ml dioxane. The OH value was calculated based on esterification as follows:

Hydroxyl Value (mmol/g) = (B - A)N/W + acid value

where B is the volume of 1N NaOH solution used to titrate the blank solution (ml); A is the volume of 1N NaOH solution used to titrate the sample solution (ml); N is the normality of the NaOH solution; and W is the weight of the sample (g).

The acid values were also determined by titration with 1N NaOH. lg of each lignin (SKL and DL) was dissolved separately into 100ml of a dioxane/water solution (80:20 v/v) and titrated with 1N NaOH to determine the equivalence point. The acid value was calculated as follows:

Acid value=(C - B)N/W

where C is the volume of 1N NaOH solution used to titrate the sample solution(ml); and B is the volume of 1N NaOH solution used to titrate the blank solution (ml) (Tomoko et al. [28]).

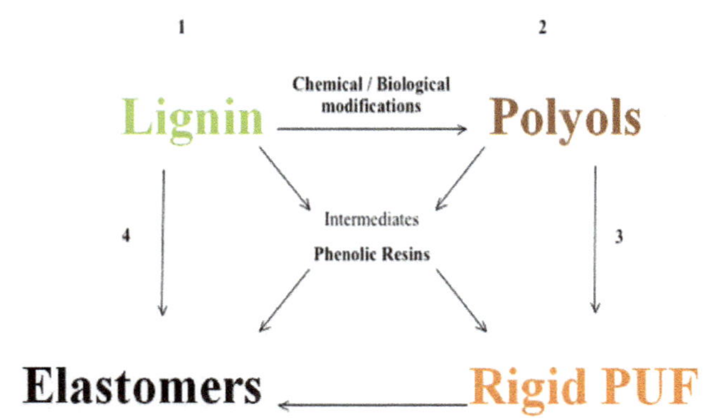

Figure 3: A general view for the synthetic routes of lignin mixing expended for different applications (1) direct use, (2) after chemical/biological modifications convert into phenolic resins (intermediates) and liquid polyol, (3) synthesis of rigid polyurethane foam (PUF) using polyols, (4) synthesis of rigid polyurethane elastomers using lignin based polyols. Previously, this concept was proposed by B.d Silva et al. [23].

Sample Preparation for ^{31}P-NMR Analysis

The lignin samples were phosphitylated as described by Hoareau et al. [29]. A combination of pyridine and $CDCl_3$ (1.6/1, v/v) was prepared and dried over molecular sieves. The above solution was further used for the preparation of relaxation reagent with 5 mg/ml of chromium (III) acetylacetonate [$Cr(C_5H_7O_2)_3$] solution and 5 mg/ml of cholesterol solution, which served as an internal standard.

Phosphitylation of lignin samples was carried out with 2-chloro-4,4,5,5-tetramethyl-1,3,2-dioxaphospholane. 20mg of each lignin were dissolved in 0.4 ml of DMF in a sealed vial with a Teflon-faced septum. Next, 0.25 ml of solvent solution was added, followed by the addition of 0.1 ml of internal standard and relaxation solution. The (0.1 ml) phosphitylation reagent was then added, and the sample was shaken to ensure homogeneity. After derivatization, the resulting solution was transferred to a 5-mm tube, and the ^{31}P-NMR spectrum was recorded by a Bruker Ultrashield Plus-400 MHz spectrometer.

Results and Discussion

The lignin macromolecule has been noted by many authors for its structural uniqueness and limited industrial usage [30-33]. Its complex structure and mass dispersion have also impeded the identification of its polymer properties.

Molecular Weight Determination by HPSEC

Polystyrene standards were used for the calibration of HPSEC columns and a calibration curve was obtained by applying a 3rd order polynomial equation to the polystyrene data. The calibration curve was then used to determine the molecular weights of lignins. The results demonstrate that the average molecular weight values (M_n) and the distributions of molecular weight (M_w) values obtained for both lignins were very similar, as shown in Table 1.

FTIR Spectra of SKL and DL

FTIR is one of the effective tools to determine the functional groups [4] and the structural properties of the lignin molecule. The peak assignments of IR bands and their relative transmittance in SKL and DL are listed in Table 2. The characteristic IR transmission peaks of SKL and DL are shown in Figure 4. In the fingerprint region, spectral differences were observed between SKL and DL in the range between (1600 to 1000 cm^{-1}). Due to aromatic skeletal vibration and guaiacyl ring breathing with carbonyl stretching, the dominating spectra were observed at band positions of 1603 and 1095 cm^{-1} in the SKL lignin.

Standards/ Lignins	Identification	Retention time (tr) minutes	M_w(Da)	M_n(Da)	(M_w/ Mn)
Polystyrene Standards	ps1	9.814	500	490	1.0
	ps2	8.682	610	600	1.0
	ps3	8.640	892	870	1.0
	ps4	8.086	1000	998	1.0
	ps5	6.682	3000	2980	1.0
	ps6	6.1	4020	4000	1.0
	ps7	6.065	5590	5570	1.0
	ps8	5.562	9145	9100	1.0
Lignins	SKL	7.916	1071	986	1.08
	DL	8.211	891	744	1.19

Table 1: HPSEC retention times of calibration curve standards.

As contrasted to SKL, the primary peak of DL was observed at bands 1664, 1389, and 1248 cm^{-1}. Large spectral differences were monitored in SKL and DL and we observed significant changes in lignin structure due to the displacement of methoxy to hydroxyl groups. These results are consistent with those reported by Kubo et al. [4] and Pandey et al. [34]. The different types of hydroxyl groups in SKL and DL were further characterized by ^{31}P-NMR.

Demethylation

The basicity of the solvent during demethylation plays a critical role in reaction efficiency. The production of hydriodic acid (HI), which is essential for demethylation, from iodoalkanes requires a strongly basic solvent. However, the solvent must not be so strongly basic as to neutralize HI. In situ, an elimination (E1) reaction produces HI, which afterward reacts with methoxyl groups of lignin by nucleophilic substitution (SN^2) reaction to produce methyl iodide (CH_3I) and demethylated lignin (DL). The optimized reaction conditions are shown in (Figure 2); the methoxyl groups were basically removed with 10.0 equivalents of iodocyclohexane to give the desired product, with a 87%±0.51 yield, without further purification. The method described here is more efficient than that which Canonica et al. [35] reported.

NMR Studies

The different types of hydroxyl groups in SKL and DL were investigated by ^{31}P-NMR. The spectra of SKL and DL are shown in Table 3 and Figure 5. For condensed phenolic units, the two sharp signals were observed at 146-145 ppm. Table 4 shows signal assignments for both lignins. The following spectral regions were integrated to acquire the different types of hydroxyl groups: aliphatic hydroxyl signals at 150 ppm, syringyl phenolic hydroxyls from 140-141ppm, guaiacyl phenolic hydroxyl signals at 137.5ppm, *p*-hydroxyphenyl phenolic hydroxyl at 132.5ppm, and carboxylic acid units from 131-130 ppm. The proportion of different hydroxyls can be computed by assimilating various signals of ^{31}P-NMR spectra, whereas the arrangement of different types of phenolic hydroxyl units in lignin can be seen in Figure 5. The phenolic peak of DL is higher than the peak of SKL due to the presence of different condensed phenolic hydroxyl units, as was previously reported by Carolina et al. [36]. By using the esterification method, the total hydroxyl values were determined containing phenolic

Serial No.	SKL Band position (cm^{-1})	DL Band position (cm^{-1})	Assignment
1	3424	3454	O-H stretching
2	2960	2937	C-H stretching
3	2920	2870	C-H stretching
4	2652	-	C-H stretching
5	-	1664	C=O stretching (unconjugated)
6	1603	-	Aromatic skeletal vibration+C=O stretching
7	1511	1501	Aromatic skeletal vibration
8	1455	1437	C-H deformation (CH_3 and CH_2)
9	-	1409	C-H in-plane deformation with aromatic ring stretching
10	-	1389	C-O of syringyl (S) ring
11	1264	1248	C-O of guaiacyl (G) ring
12	1206	-	C-C + C-O stretch
13	1095	1092	Aromatic C-H in-plane deformation in the guaiacyl ring

Table 2: Results of the IR Bands obtained in SKL and DL.

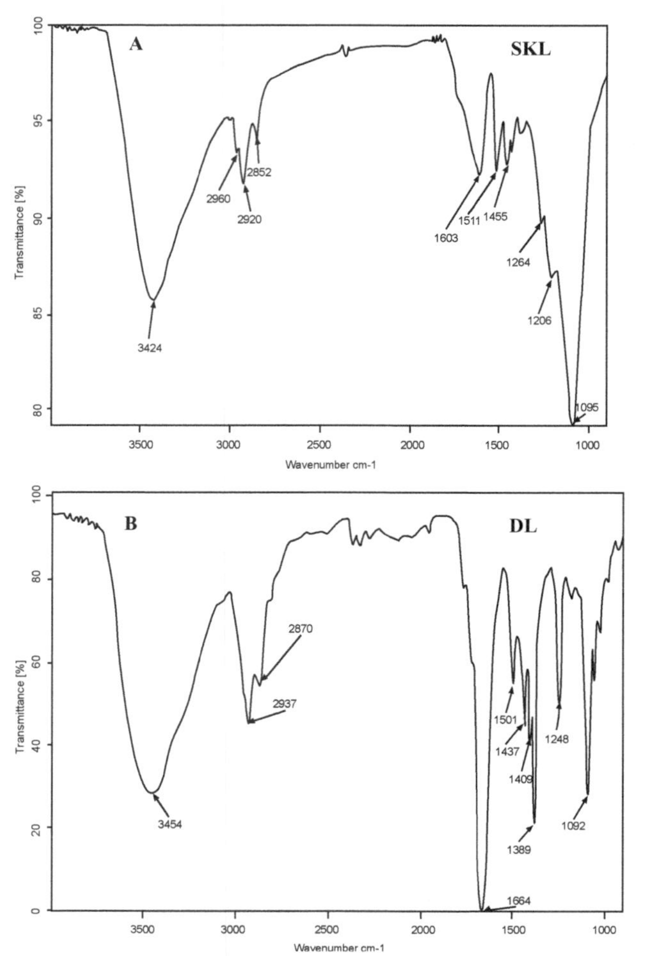

Figure 4: FTIR spectra of (A) softwood kraft lignin (SKL), and (B) demethylated lignin (DL).

Signal (ppm)	Assignment
150	Aliphatic-OH
146-145	Condensed phenolic units
140-141	Syringyl phenolic-OH
137.5	Guaiacyl phenolic-OH
132.5	p-Hydroxyphenyl-OH
131-130	Carboxylic acid units

Table 3: Signal Assignments of SKL and DL by ^{31}P-NMR spectrometry.

or sterically hindered hydroxyl groups; the values were 3.06 (±0.50) and 5.01 (±0.57) mmol/g for SKL and DL, respectively. In each lignin, the quantity of different hydroxyl groups was determined by ^{31}P-NMR method presented in (Table 4). The most abundant groups found in both lignins were carboxylic acid and guaiacyl-OH as compared to other hydroxyl units reported by Diao et al. [37].

Conclusions

The industrial production of goods mainly depends on sustainable growth, which involves the use of safe raw materials resources. Kraft lignin, as a large available biomass resource (by-product of pulp mills), is a very appealing and favourable raw material to produce biopolymers and biofuels for use in process unit operations. Economical affordability of lignin-based processes and advances in this field are likely to play a key role in the industrial development

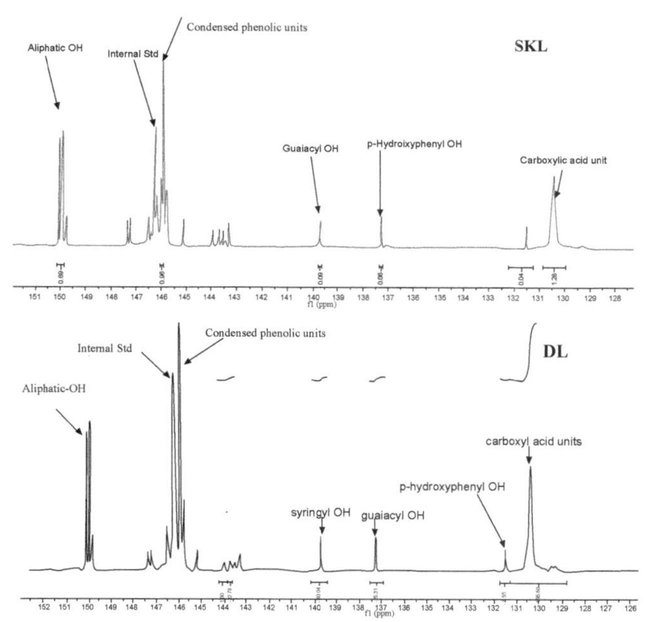

Figure 5: ^{31}P-NMR spectra of SKL and DL shows qualitative and quantitative characterization of different types of hydroxyl groups.

Hydroxyl Groups	SKL(mmol/g)	DL (mmol/g)
Carboxylic Acid	1.28	2.44
Guaiacyl-OH	0.96	1.32
Syringyl-OH	0.01	0.03
Aliphatic-OH	0.69	0.82
p-Hydroxyphenyl-OH	0.12	0.4

Table 4: Quantification of different hydroxyl groups of SKL and DL by ^{31}P-NMR spectrometry.

of lignin. HPSEC results showed that it is possible to use polystyrene standards for lignin molecular mass analyses. The fractions of different molecular masses have different chemical compositions and functional groups as indicated by FTIR analysis. Notable spectral differences were observed in SKL and DL due to changes in lignin structure and displacement of methoxyl to hydroxyl groups. The different types of phenolic hydroxyl units were investigated by ^{31}P-NMR. In summary, demethylation of the lignin molecule using iodocyclohexane as HI source in DMF has been developed. It was demonstrated that the total hydroxyl numbers (1.6-fold increase in DL as compared to SKL) indicate demethoxyl groups were removed successfully in high yields under optimized reaction conditions. The strategy offers a new alternative approach for protecting phenolic groups, which is helpful in regulating the optimum processing condition during the development of lignin based polyols and its application for the synthesis of bio-based polyurethane foams.

Acknowledgement

We are greatly thankful to the funding support from ORF-RE (Ontario Research Fund - Research Excellence program in Bark Biorefinery).

References

1. Sarkanen KV, Ludwig CH (1971) In Lignins: Occurrence, Formation, structure and Reactions. Wiley: New York.
2. Sjostrom E (1993) Wood Chemistry: Fundamentals and Application. Academic Press: Orlando.
3. Adler E (1977) Lignin chemistry-past, present and future. Wood Science Technol 11: 169-218.
4. Kubo S, Kadla J (2005) Hydrogen bonding in lignin: a Fourier transform infrared model compound study. Biomacromolecules 6: 2815-2821.
5. Bocchini P, Galletti GC, Seraglia R, Traldi P, Camarero S, et al. (1996) Matrix-assisted Laser Desorption/Ionization Mass Spectrometry of Natural and Synthetic Lignin. Rapid Commun Mass Spectrometry 10: 1144-1147.
6. Tejado A, Pena C, Labidi J, Echeverria JM, Mondragon I (2007) Physico-chemical characterization of lignins from different sources for use in phenol–formaldehyde resin synthesis. Bioresource Technol 98: 1655-1663.
7. Hu TQ (2002) Chemical Modification, Properties, and Usage of Lignin; Kluwer Academic/Plenum Publishers: New York.
8. Baker FS, Naskar NCGAK, Baker DA (2007) In FY 2007 Progress Report, ORNL; Oak Ridge National Laboratory (ORNL).
9. Liu Z, Luo X, Li Y, Li L, Huang Y (2009) Mater Sci Forum 571: 620.
10. Gandini A, Belgacem MN, Guo ZX, Montanari S (2002) Chemical Modification, Properties and Usage of Lignin. TQ Hu, TQ Edn, Kluwer Academic/Plenum, New York.
11. Chen R, Kokta BV, Valade JL (1980) Study on the graft copolymerization of lignosulfonate and acrylic monomers. Appl Polym Sci 25: 2211-2220.
12. Glasser WG, Leitheiser RH (1984) Engineering plastics from lignin. Polymer Bulletin 13: 73-95.
13. Sarkar S, Adhikari B (2004) Lignin-modified phenolic resin: Synthesis optimization, adhesive strength, and thermal stability. J Adhes Sci Technol 14: 1179-1193.
14. Yoshida H, Morck R, Kringstad KP, Hatakeyama H (1990) Kraft lignin in polyurethanes; II: Effects of the molecular weight of kraft lignin of the properties of polyurethanes from a kraft lignin-polyether-triol polymeric MDI system. J Appl Polym Sci 40: 1819-1832.
15. Nadji H, Bruzzese C, Belgacem MN, Benaboura A, Gandini A (2005) Oxypropylation of lignins and preparation of rigid polyurethane foams from the ensuing polyols. Macromol Mater Eng 290: 1009-1016.
16. Raquez JM, Deléglise M, Lacrampe MF, Krawczak P (2010) Thermosetting (bio)materials derived from renewable resources: A critical review. Prog Polym Sci 35: 487-509.
17. Doherty WOS, Mousavioun P, Fellows CM (2011) Value-adding to cellulosic ethanol: Lignin polymers. Ind Crop Prod 33: 259-276.
18. Meister JJ (2000) Polymer Modification: Principles, Techniques, and Applications. Meister JJ, Edn, Marcel Dekker: New York 2000: 67-144.
19. Bhatt MV, Kulkarni SU (1986) Cleavage of Ethers. Synthesis 249-282.
20. Meister JJ (2002) Modification of lignin. J Macromol Sci Poly Rev C 42: 235-289.
21. Westwood G, Horton TN, Wilker JJ (2007) Simplified polymer mimics of cross-linking adhesive proteins. Macromolecules 40: 3960-3964.
22. Yang Z, Pelton R (1998) The synthesis of poly(3,4-dihydroxystyrene) and poly [(sodium 4-styrenesulfonate)-co-(3,4-dihydroxystyrene)]. Macromol Rapid Commun 19: 241-246.
23. Borges da Silva EA, Zabkova M, Araújo JD, Cateto CA, Barreiro MF, et al. (2009) An integrated process to produce vanillin and lignin-based polyurethanes from *Kraft* lignin. Chemical Engineering Research and Design 87: 1276-1292.
24. Botaro VR, Curvelo AA (2009 Monodisperse lignin fractions as standards in size-exclusion analysis: comparison with polystyrene standards. J Chromatography A 1216: 3802-3806.
25. Zuo L, Yao S, Wang W, Duan W (2008) An efficient method for demethylation of aryl methyl ethers. Tetrahedron Letters 49: 4054-4056.
26. Kurimoto Y, Doi YS, Tamura (1999) Species effects on wood-liquefaction in polyhydric alcohols. Holzforschung 53: 617-622.
27. Kurimoto Y, Takeda M, Koizumi A, Yamauchi S, Doi S, et al. (2000) Mechanical properties of polyurethane films prepared from liquefied wood with polymeric MDI. Bioresource Technology 74: 151-157.
28. Tomoko U, Tatsuya A, Kokki S (2002) New method to determine the hydroxyl value in liquefied bark as Polyurethane material. J Wood Sci 48: 348-351.
29. Hoareau W, Trindade W, Siegmund B, Castellan A, Frollini E (2004) Sugar

cane bagasse and curaua lignins oxidized by chlorine dioxide and reacted with furfuryl alcohol : characterization and stability. Polym Degrad Stab 86: 567-576.

30. Bayerbach R, Nguyen VD, Schurr U, Meier D (2006) Characterization of the water-Insoluble fraction from fast pyrolysis liquids (pyrolytic lignin): Part III. Molar mass characteristics by SEC, MALDI-TOF-MS, LDI-TOF-MS, and Py-FIMS. J Anal Appl Pyrolysis 77: 95-101.

31. Pouteau C, Cathala B, Dole P, Kurek B, Monties B (2005) Structural modification of Kraft lignin after acid treatment: characterisation of the apolar extracts and influence on the antioxidant properties in polypropylene. Ind Crops Prod 21: 101-108.

32. Hirose S, Hatakeyama T, Hatakeyama H (2005) Glass transition and thermal decomposition of epoxy resins from the carboxylic acid system consisting of ester carboxylic derivatives of alcoholysis lignin and ethylene glycol with various carboxylic acids. Thermochim Acta 431: 76-80.

33. Sun XF, Sun RC, Fowler P, Baird MS (2004) Isolation and characterization of cellulose obtained by a two-stage treatment with organosolv and cyanamide activated hydrogen peroxide from wheat straw. Carbohydr Polym 55: 379-391.

34. Pandey KK (1999) A study of chemical structure of soft and hardwood and wood polymers by FTIR spectroscopy. J Appl Polym Sci 71: 1969-1975.

35. Canonica L, Rindone B, Santaniello E, Scolastico C (1972) A total synthesis of mycophenolic acid, some analogues and some biogenetic intermediates. Tetrahedron 28: 4395-4404.

36. Carolina AC, Maria FB, Alirio ER, Marie CBS, Thielemans W, et al. (2008) Lignins as macromonomers for polyurethane synthesis: A comparative study on hydroxyl group determination. J Appl Polym Sci. 109: 3008-3017.

37. Diao S, Feng Xu, Geng ZC, Sun RC, Jones GL, et al. (2010) Physicochemical characterization of extracted lignin from sweet sorghum stem. Ind Crops Prod 32: 21-28.

Experimental Studies on Hydrodynamics Characteristics of Co-current Three Phase Fluidization Using Different Sizes of Glass Beads and its Prediction by Fuzzy Logic

Anish T. Achankunju[1], Sivalingam A[2] and Kannadasan T[3]*

[1]*M.Tech Student, Department of Chemical Engineering, Coimbatore Institute of Technology, Coimbatore, Tamil Nadu, India*
[2]*Associate professor, Department of Chemical Engineering, Coimbatore Institute of Technology, Coimbatore, Tamil Nadu, India*
[3]*Head of the Department, Department of Chemical Engineering, Coimbatore Institute of Technology, Coimbatore, Tamil Nadu, India*

Abstract

The experiment was carried out with different sizes of Glass beads. That is Glass beads of diameter 0.2 cm, 0.4 cm, 0.6 cm. Water was used as a continuous phase and air was used as a dispersed phase. These three different sized Glass beads were used as solids for co-current studies. The three sections of the experimental setup viz. gas liquid disengagement section, test section and gas liquid distributed section. The experiment was conducted with Glass beads with varying liquid velocities and keeping constant gas velocity 0.16985138 cm/sec, 0.212314225 cm/sec, 0.304317056 cm/sec, 0.346779901 cm/sec. The graphs were plotted and their effects on individual phase hold-ups for various particle sizes were studied. Pressure drop of fluids in the fluidized bed was measured by using mercury manometer and by using ethyl acetate manometer. The phase holdups like solid, liquid, gas were determined, liquid holdup and solid holdup decreases with increase in liquid velocity whereas gas holdup increases with increases in liquid velocity. With the experimental data, simulation studies were carried out using Fuzzy Logic.

Keywords: Gas velocity; Glass beads; Phase holdup; Mercury manometer; Liquid velocity

Introduction

Fluidization is the operation by which solid particles are transformed into a fluid like state through suspension in a gas or liquid. A fluidized bed is formed when a quantity of a solid particulate substance (usually present in a holding vessel) is placed under appropriate conditions to cause the solid/fluid mixture to behave as a fluid [1]. This is usually achieved by the introduction of pressurized fluid through the particulate medium. This results in the medium then having many properties and characteristics of normal fluids; such as the ability to free-flow under gravity, or to be pumped using fluid type technologies. The resulting phenomenon is called fluidization. A fluidized bed demonstrates all the characteristics of a fluid. The static pressure at any height is approximately equal to the weight of bed solids per unit cross section above that level. The solids from the bed may be drained like a liquid through an orifice at the bottom or on the side. An object denser than the bulk of the bed will sink, while one lighter than the bed will float.

Materials and Methods

Experimental setup

The Perspex fluidized bed column used was of 5.4 cm I.D, 6 cm O.D and 160 cm height as shown in Figure 1. The liquid and gas flow rate were measured. The gas and liquid streams were merged and passed through a quick closing valve, a 0.3 cm thick perforated grid before entering the bed. The calming section and the grid ensured that the liquid and the gas were well mixed and evenly distributed into the particles. A tee joint at the top of the column allowed gas to escape and liquid to be recirculated to the reservoir. Pressure tapping were provided at the top and bottom of the test section.

Experimental procedure

The experiment is performed with water at 35°C and gas through the bed of glass beads with diameter 0.2 cm, 0.4 cm, 0.6 cm. The flow rates of both gas and liquid are regulated by the control valves. The liquid flow rate, manometer readings and the bed heights are observed. The gas velocity is kept constant and the liquid velocity is varied. After steady state is attained for each liquid velocity the fluidized bed height and manometer readings are noted. The same procedure is repeated for four different gas velocities. The effect on phase holdups, pressure drop and the bed porosity is studied for different sizes of glass beads at 35°C.

Fuzzy logic

A fuzzy logic (FL)-based expert system (ES) works based on the principle of fuzzy set theory, and it is a potential tool for dealing with imprecision and uncertainty. The FL-based ESs has been developed by various researchers, after realizing their potential in solving real world complex problems. The performance of an ES mainly depends on its KB, which consists of a data base and a rule base. Thus, designing a proper KB is very important, which is difficult too. In this paper fuzzy logic is used to predict gas holdup and liquid holdup for glass beads and gypsum particles for different sizes with varying liquid flow rates by keeping the gas flow to be constant.

Results and Discussion

The hydrodynamic parameters of interest in a three phase fluidized bed are: bed porosity (ε), solid holdup (ε_s), gas holdup (ε_g), Liquid holdup (ε_l) and fluidized bed pressure drop (ΔP) were calculated using

***Corresponding author:** Kannadasan T, Head of the Department, Department of Chemical Engineering, Coimbatore Institute of Technology, Coimbatore, 641014, Tamil Nadu, India

Ul (m/s)	Lf (m)	ε	ε_s	ε_g	ε_l	ΔP (N/m²)	$N_{Re,p,liq}$	$N_{Re,m,liq}$	$N_{Re,p,gas}$	$N_{Re,m,gas}$	$N_{Fr,liq}$	$N_{Fr,gas}$
1.21	0.127	0.2986	0.7014	0.0157	0.2828	10926.41	33.269	47.431	0.2058	0.2934	1.18	0.023
1.46	0.13	0.3148	0.6852	0.0385	0.2763	10986	39.923	58.262	0.2058	0.3004	1.66	0.023
1.7	0.132	0.3252	0.6748	0.053	0.2721	11271.62	46.577	69.018	0.2058	0.305	2.23	0.022
1.94	0.135	0.3402	0.6598	0.0741	0.2661	11319.63	53.23	80.671	0.2058	0.3119	2.84	0.022
2.18	0.137	0.3498	0.6502	0.0876	0.2622	11348.29	59.884	92.099	0.2058	0.3165	3.55	0.021
2.43	0.139	0.3591	0.6409	0.1007	0.2584	11501	66.538	103.826	0.2058	0.3212	4.32	0.021
2.67	0.141	0.3682	0.6318	0.1135	0.2548	11525.65	73.192	115.852	0.2058	0.3258	5.15	0.021
2.91	0.145	0.3857	0.6143	0.1379	0.2477	11569.58	79.846	129.969	0.2058	0.335	5.96	0.02
3.15	0.146	0.3899	0.6101	0.1438	0.246	11706.85	86.5	141.771	0.2058	0.3373	6.94	0.02
3.4	0.148	0.3981	0.6019	0.1554	0.2427	11726.07	93.153	154.768	0.2058	0.342	7.95	0.02
3.64	0.15	0.4061	0.5939	0.1667	0.2395	11744	99.807	168.064	0.2058	0.3466	9	0.02
3.88	0.155	0.4253	0.5747	0.1935	0.2317	11912.27	106.461	185.244	0.2058	0.3581	9.91	0.019
4.12	0.158	0.4362	0.5638	0.2089	0.2273	11933.78	113.115	200.631	0.2058	0.3651	10.97	0.019
4.37	0.165	0.4601	0.5399	0.2424	0.2177	11977.4	119.769	221.844	0.2058	0.3812	11.78	0.018
4.61	0.17	0.476	0.524	0.2647	0.2113	12133	126.422	241.265	0.2058	0.3928	12.74	0.017
4.85	0.177	0.4967	0.5033	0.2938	0.2029	12165.39	133.076	264.42	0.2058	0.409	13.56	0.017
5.09	0.183	0.5132	0.4868	0.3169	0.1963	12319	139.73	287.053	0.2058	0.4228	14.46	0.016
5.34	0.186	0.5211	0.4789	0.328	0.1931	12329.97	146.384	305.652	0.2058	0.4298	15.61	0.016
5.58	0.19	0.5312	0.4688	0.3421	0.1891	12473.52	153.038	326.417	0.2058	0.439	16.7	0.015
5.82	0.195	0.5432	0.4568	0.359	0.1842	12489.25	159.691	349.573	0.2058	0.4505	17.72	0.015
6.07	0.201	0.5568	0.4432	0.3781	0.1787	12506.28	166.345	375.342	0.2058	0.4644	18.66	0.015
6.31	0.207	0.5697	0.4303	0.3961	0.1735	12652	172.999	402.008	0.2058	0.4783	19.59	0.014
6.55	0.213	0.5818	0.4182	0.4131	0.1686	12666	179.653	429.571	0.2058	0.4921	20.53	0.014
6.79	0.218	0.5914	0.4086	0.4266	0.1648	12807.29	186.307	455.938	0.2058	0.5037	21.58	0.013
7.04	0.224	0.6023	0.3977	0.442	0.1604	12819.13	192.96	485.219	0.2058	0.5176	22.52	0.013
7.28	0.3	0.7031	0.2969	0.5833	0.1197	13042.59	199.614	672.255	0.2058	0.6931	18	0.01
7.52	0.307	0.7098	0.2902	0.5928	0.117	13047.87	206.268	710.872	0.2058	0.7093	18.78	0.01

Table 1.1: Hydrodynamic characteristcs for glass beads of dp = 0.2cm, Ug = 0.169851cm/s.

U_l (m/s)	L_f (m)	ε	ε_s	ε_g	ε_l	ΔP (N/m²)	$N_{Re,p,liq}$	$N_{Re,m,liq}$	$N_{Re,p,gas}$	$N_{Re,m,gas}$	$N_{Fr,liq}$	$N_{Fr,gas}$
Ul (m/s)	Lf (m)	E	εs	εg	εl	ΔP (N/m2)	NRe,p,liq	NRe,m,liq	NRe,p,gas	NRe,m,gas	NFr, liq	NFr, gas
1.21	0.14	0.3637	0.6363	0.0714	0.2922	9668.506	33.269	52.286	0.2573	0.4043	1.07	0.033
1.46	0.143	0.377	0.6229	0.0909	0.2861	9828.093	39.923	64.088	0.2573	0.413	1.51	0.032
1.7	0.147	0.394	0.606	0.1156	0.2783	9868	46.577	76.861	0.2573	0.4246	2	0.031
1.94	0.152	0.4139	0.5861	0.1447	0.2692	10037.75	53.23	90.829	0.2573	0.439	2.53	0.03
2.18	0.156	0.4289	0.571	0.1667	0.2623	10067.7	59.884	104.872	0.2573	0.4505	3.12	0.029
2.43	0.16	0.4432	0.5567	0.1875	0.2557	10094.29	66.538	119.512	0.2573	0.4621	3.75	0.029
2.67	0.164	0.4568	0.5432	0.2073	0.2495	10246.11	73.192	134.75	0.2573	0.4737	4.43	0.028
2.91	0.169	0.4729	0.5271	0.2308	0.2421	10272.8	79.846	151.481	0.2573	0.4881	5.11	0.027
3.15	0.18	0.5051	0.4949	0.2778	0.2273	10450.14	86.5	174.786	0.2573	0.5199	5.63	0.026
3.4	0.193	0.5384	0.4616	0.3264	0.212	10624.19	93.153	201.826	0.2573	0.5574	6.09	0.024
3.64	0.2	0.5546	0.4454	0.35	0.2046	10644.07	99.807	224.085	0.2573	0.5776	6.75	0.023
3.88	0.215	0.5856	0.4143	0.3953	0.1903	10679.2	106.461	256.951	0.2573	0.6209	7.14	0.021
4.12	0.22	0.595	0.4049	0.4091	0.186	10819.51	113.115	279.359	0.2573	0.6354	7.88	0.021
4.37	0.232	0.616	0.384	0.4397	0.1763	10840.76	119.769	311.926	0.2573	0.67	8.38	0.02
4.61	0.245	0.6364	0.3636	0.4694	0.167	10990.93	126.422	347.705	0.2573	0.7076	8.84	0.019
4.85	0.25	0.6436	0.3563	0.48	0.1636	11128.54	133.076	373.475	0.2573	0.722	9.6	0.018
5.09	0.256	0.652	0.348	0.4922	0.1598	11136.12	139.73	401.56	0.2573	0.7394	10.34	0.018
5.34	0.258	0.6547	0.3453	0.4961	0.1586	11138.53	146.384	423.969	0.2573	0.7451	11.25	0.018
5.58	0.265	0.6638	0.3361	0.5094	0.1544	11277.65	153.038	455.266	0.2573	0.7654	11.98	0.017
5.82	0.27	0.67	0.3299	0.5185	0.1515	11283	159.691	484.024	0.2573	0.7798	12.8	0.017
6.07	0.276	0.6772	0.3228	0.529	0.1482	11420.4	166.345	515.395	0.2573	0.7971	13.59	0.017
6.31	0.282	0.6841	0.3159	0.539	0.1451	11426.2	172.999	547.664	0.2573	0.8145	14.38	0.016
6.55	0.287	0.6896	0.3104	0.547	0.1425	11562.1	179.653	578.811	0.2573	0.8289	15.24	0.016
6.79	0.291	0.6938	0.3061	0.5533	0.1406	11697.1	186.307	608.615	0.2573	0.8404	16.16	0.016
7.04	0.298	0.701	0.2989	0.5638	0.1373	11703	192.96	645.514	0.2573	0.8607	16.93	0.015
7.28	0.304	0.7069	0.293	0.5724	0.1346	11707.7	199.614	681.218	0.2573	0.878	17.76	0.015
7.52	0.309	0.7117	0.2883	0.5793	0.1324	11711.5	206.268	715.503	0.2573	0.8924	18.66	0.015

Table 1.2: Hydrodynamic characteristics for glass beads of dp = 0.2cm, Ug = 0.212314cm/s.

U_l (m/s)	L_f (m)	ε	ε_s	ε_g	ε_l	ΔP (N/m²)	$N_{Re,p,liq}$	$N_{Re,m,liq}$	$N_{Re,p,gas}$	$N_{Re,m,gas}$	$N_{Fr,liq}$	$N_{Fr,gas}$
1.21	0.141	0.3682	0.6317	0.0496	0.3185	8247.7	33.269	52.66	0.3688	0.5837	1.06	0.067
1.46	0.145	0.3856	0.6143	0.0758	0.3097	8287.5	39.923	64.985	0.3688	0.6002	1.49	0.065
1.7	0.15	0.4061	0.5938	0.1066	0.2994	8329.9	46.577	78.43	0.3688	0.6209	1.96	0.063
1.94	0.157	0.4326	0.5673	0.1465	0.2861	8505.8	53.23	93.817	0.3688	0.6499	2.45	0.06
2.18	0.165	0.4601	0.5398	0.1878	0.2722	8551.4	59.884	110.922	0.3688	0.683	2.95	0.057
2.43	0.173	0.485	0.5149	0.2254	0.2596	8716.4	66.538	129.222	0.3688	0.7162	3.47	0.055
2.67	0.177	0.4967	0.5032	0.2429	0.2537	8732.6	73.192	145.431	0.3688	0.7327	4.1	0.053
2.91	0.2	0.5546	0.4454	0.33	0.2246	8932.3	79.846	179.268	0.3688	0.8279	4.32	0.047
3.15	0.215	0.5856	0.4143	0.3767	0.2089	8964.8	86.5	208.772	0.3688	0.89	4.72	0.044
3.4	0.23	0.6127	0.3873	0.4173	0.1953	8990.4	93.153	240.518	0.3688	0.9521	5.11	0.041
3.64	0.242	0.6319	0.3681	0.4462	0.1856	9268.3	99.807	271.143	0.3688	1.0018	5.58	0.039
3.88	0.248	0.6408	0.3591	0.4596	0.1811	9276	106.461	296.39	0.3688	1.0266	6.19	0.038
4.12	0.256	0.652	0.3479	0.4765	0.1754	9416.1	113.115	325.073	0.3688	1.0597	6.77	0.037
4.37	0.264	0.6625	0.3374	0.4924	0.1701	9424.8	119.769	354.951	0.3688	1.0929	7.36	0.036
4.61	0.276	0.6772	0.3227	0.5144	0.1627	9567.4	126.422	391.701	0.3688	1.1425	7.85	0.034
4.85	0.288	0.6907	0.3093	0.5347	0.1559	9577.7	133.076	430.243	0.3688	1.1922	8.33	0.033
5.09	0.297	0.7	0.2999	0.5488	0.1512	9716	139.73	465.873	0.3688	1.2295	8.91	0.032
5.34	0.302	0.705	0.2949	0.5562	0.1487	9719.6	146.384	496.273	0.3688	1.2502	9.62	0.031
5.58	0.309	0.7117	0.2882	0.5663	0.1453	9855.9	153.038	530.857	0.3688	1.2791	10.27	0.031
5.82	0.317	0.7189	0.281	0.5772	0.1417	9861.2	159.691	568.279	0.3688	1.3123	10.9	0.03
6.07	0.325	0.7259	0.274	0.587	0.1382	9997.6	166.345	606.897	0.3688	1.3454	11.54	0.029
6.31	0.331	0.7308	0.2691	0.5951	0.1357	10001.1	172.999	642.825	0.3688	1.3702	12.25	0.029
6.55	0.338	0.7364	0.2635	0.6035	0.1329	10136.6	179.653	681.666	0.3688	1.3992	12.94	0.028
6.79	0.344	0.741	0.2589	0.6104	0.1305	10139.8	186.307	719.462	0.3688	1.424	13.67	0.027
7.04	0.349	0.7447	0.2552	0.616	0.1287	10274.1	192.96	755.988	0.3688	1.4447	14.46	0.027
7.28	0.355	0.749	0.2509	0.6225	0.1265	10408.8	199.614	795.502	0.3688	1.4696	15.21	0.027
7.52	0.366	0.7566	0.2433	0.6338	0.1227	10414	206.268	847.489	0.3688	1.5151	15.75	0.026

Table 1.3: Hydrodynamic characteristics for glass beads of dp = 0.2cm, Ug = 0.304317cm/s.

U_l (m/s)	L_f (m)	ε	ε_s	ε_g	ε_l	ΔP (N/m²)	$N_{Re,p,liq}$	$N_{Re,m,liq}$	$N_{Re,p,gas}$	$N_{Re,m,gas}$	$N_{Fr,liq}$	$N_{Fr,gas}$
1.21	0.153	0.4177	0.5822	0.0522	0.3654	7866.8	33.269	57.142	0.4202	0.7217	0.98	0.08
1.46	0.165	0.4601	0.5398	0.1212	0.3389	7951.8	39.923	73.948	0.4202	0.7783	1.31	0.074
1.7	0.176	0.4938	0.5061	0.1761	0.3177	8136.2	46.577	92.024	0.4202	0.8302	1.67	0.07
1.94	0.185	0.5184	0.4815	0.2162	0.3022	8173.9	53.23	110.549	0.4202	0.8727	2.08	0.066
2.18	0.19	0.5311	0.4688	0.2368	0.2943	8319.9	59.884	127.728	0.4202	0.8963	2.56	0.065
2.43	0.205	0.5654	0.4345	0.2926	0.2727	8494.1	66.538	153.125	0.4202	0.967	2.93	0.06
2.67	0.218	0.5913	0.4086	0.3348	0.2565	8525.4	73.192	179.119	0.4202	1.0284	3.33	0.056
2.91	0.224	0.6023	0.3976	0.3526	0.2496	8796.6	79.846	200.78	0.4202	1.0567	3.86	0.055
3.15	0.232	0.616	0.3839	0.375	0.241	8812	86.5	225.28	0.4202	1.0944	4.37	0.053
3.4	0.245	0.6364	0.3635	0.4081	0.2282	8833.6	93.153	256.204	0.4202	1.1557	4.8	0.05
3.64	0.26	0.6573	0.3426	0.4423	0.215	8984.7	99.807	291.31	0.4202	1.2265	5.19	0.047
3.88	0.277	0.6784	0.3215	0.4765	0.2018	9004.6	106.461	331.048	0.4202	1.3067	5.54	0.044
4.12	0.286	0.6885	0.3114	0.493	0.1955	9144.4	113.115	363.167	0.4202	1.3491	6.06	0.043
4.37	0.293	0.6959	0.304	0.5051	0.1908	9281.8	119.769	393.941	0.4202	1.3822	6.63	0.042
4.61	0.308	0.7107	0.2892	0.5292	0.1815	9425.7	126.422	437.115	0.4202	1.4529	7.03	0.04
4.85	0.32	0.7216	0.2783	0.5468	0.1747	9435	133.076	478.048	0.4202	1.5095	7.5	0.038
5.09	0.33	0.73	0.2699	0.5606	0.1694	9573.2	139.73	517.636	0.4202	1.5567	8.02	0.037
5.34	0.342	0.7395	0.2604	0.576	0.1635	9581.1	146.384	562.005	0.4202	1.6133	8.49	0.036
5.58	0.357	0.7504	0.2495	0.5938	0.1566	9589.9	153.038	613.321	0.4202	1.6841	8.89	0.034
5.82	0.363	0.7546	0.2454	0.6005	0.154	9724.6	159.691	650.743	0.4202	1.7124	9.52	0.034
6.07	0.379	0.7649	0.235	0.6174	0.1475	9732.7	166.345	707.735	0.4202	1.7878	9.89	0.032
6.31	0.385	0.7686	0.2313	0.6233	0.1452	9867.1	172.999	747.697	0.4202	1.8161	10.53	0.032
6.55	0.392	0.7727	0.2272	0.6301	0.1426	10001.9	179.653	790.572	0.4202	1.8492	11.16	0.031
6.79	0.4	0.7773	0.2227	0.6375	0.1398	10137.1	186.307	836.584	0.4202	1.8869	11.76	0.031
7.04	0.408	0.7816	0.2183	0.6446	0.137	10140.5	192.96	883.791	0.4202	1.9246	12.37	0.03
7.28	0.416	0.7858	0.2141	0.6514	0.1344	10275.4	199.614	932.193	0.4202	1.9624	12.98	0.029
7.52	0.421	0.7884	0.2115	0.6555	0.1328	10409.1	206.268	974.844	0.4202	1.986	13.69	0.029

Table 1.4: Hydrodynamic characteristics for glass beads of dp = 0.2cm, Ug = 0.346779cm/s.

experimental data for both glass beads and gypsum particles [2,3]. The hydrodynamic characteristics like gas holdup and liquid holdup were predicted using fuzzy logic [4] (Tables 1-1.12).

Bed porosity (ε)

The results indicate that the expanded bed voidage depends on phase velocities, properties of solids and the other two phases [5-7]. As the gas phase Modified Reynolds number increases the bed voidage increases; the increase is high at low Modified Reynolds number, whereas it is less at higher Modified Reynolds numbers as shown in the figures. Higher flow rates of the phases lead to larger bed expansion

U_l (m/s)	L_f (m)	ε	ε_s	ε_g	ε_l	ΔP (N/m²)	$N_{Re,p,liq}$	$N_{Re,m,liq}$	$N_{Re,p,gas}$	$N_{Re,m,gas}$	$N_{Fr,liq}$	$N_{Fr,gas}$
1.21	0.131	0.3169	0.683	0.0076	0.3093	10034.61	66.5381	97.411	0.4116	0.6026	1.14	0.022
1.46	0.131	0.3169	0.683	0.0076	0.3093	10034.61	79.8457	116.893	0.4116	0.6026	1.65	0.022
1.7	0.131	0.3169	0.683	0.0076	0.3093	10158.5	93.1533	136.376	0.4116	0.6026	2.24	0.022
1.94	0.132	0.3221	0.6778	0.0151	0.3069	10176.86	106.4609	157.048	0.4116	0.6072	2.91	0.022
2.18	0.132	0.3221	0.6778	0.0151	0.3069	10176.86	119.7685	176.678	0.4116	0.6072	3.68	0.022
2.43	0.135	0.3372	0.6628	0.037	0.3001	10351.82	133.0762	200.771	0.4116	0.621	4.44	0.022
2.67	0.136	0.342	0.6579	0.0441	0.2979	10367.32	146.3838	222.484	0.4116	0.6256	5.34	0.022
2.91	0.137	0.3469	0.6531	0.051	0.2957	10507.25	159.6914	244.494	0.4116	0.6302	6.31	0.021
3.15	0.139	0.3562	0.6437	0.0647	0.2915	10535.47	172.999	268.736	0.4116	0.6394	7.29	0.021
3.4	0.141	0.3654	0.6346	0.078	0.2873	10687.23	186.3066	293.572	0.4116	0.6486	8.34	0.021
3.64	0.145	0.3829	0.6171	0.1034	0.2794	10734.22	199.6142	323.464	0.4116	0.667	9.31	0.02
3.88	0.15	0.4035	0.5965	0.1333	0.2701	10784.22	212.9219	356.926	0.4116	0.69	10.24	0.02
4.12	0.152	0.4113	0.5886	0.1447	0.2665	10929.06	226.2295	384.291	0.4116	0.6992	11.41	0.019
4.37	0.154	0.419	0.581	0.1558	0.2631	10945.92	239.5371	412.25	0.4116	0.7084	12.62	0.019
4.61	0.157	0.4301	0.5699	0.1719	0.258	11096.88	252.8447	443.629	0.4116	0.7222	13.8	0.019
4.85	0.16	0.4407	0.5592	0.1875	0.2532	11118.55	266.1523	475.902	0.4116	0.736	15	0.018
5.09	0.162	0.4476	0.5523	0.1975	0.2501	11132	279.4599	505.943	0.4116	0.7452	16.33	0.018
5.34	0.165	0.4577	0.5423	0.2121	0.2455	11279	292.7676	539.851	0.4116	0.759	17.6	0.018
5.58	0.167	0.4642	0.5358	0.2215	0.2426	11290.89	306.0752	571.231	0.4116	0.7682	19	0.018
5.82	0.169	0.4705	0.5294	0.2307	0.2397	11430.61	319.3828	603.205	0.4116	0.7774	20.45	0.017
6.07	0.174	0.4857	0.5142	0.2528	0.2328	11456.78	332.6904	646.929	0.4116	0.8004	21.55	0.017
6.31	0.176	0.4916	0.5084	0.2613	0.2302	11466.4	345.998	680.539	0.4116	0.8096	23.04	0.017
6.55	0.18	0.5029	0.4971	0.2777	0.2251	11613.41	359.3056	722.776	0.4116	0.828	24.3	0.016
6.79	0.185	0.5163	0.4836	0.2973	0.219	11633.98	372.6133	770.366	0.4116	0.851	25.42	0.016
7.04	0.19	0.529	0.4709	0.3157	0.2132	11652.51	385.9209	819.443	0.4116	0.874	26.56	0.015
7.28	0.196	0.5435	0.4565	0.3367	0.2067	11802.14	399.2285	874.469	0.4116	0.9016	27.55	0.015
7.52	0.203	0.5592	0.4407	0.3596	0.1996	11822.96	412.5361	935.89	0.4116	0.9338	28.4	0.014

Table 1.5: Hydrodynamic characteristics for glass beads of dp = 0.4cm, Ug = 0.169851cm/s.

U_l (m/s)	L_f (m)	ε	ε_s	ε_g	ε_l	ΔP (N/m²)	$N_{Re,p,liq}$	$N_{Re,m,liq}$	$N_{Re,p,gas}$	$N_{Re,m,gas}$	$N_{Fr,liq}$	$N_{Fr,gas}$
1.21	0.137	0.3469	0.6531	0.0146	0.3322	10172.9	66.5381	101.873	0.5145	0.7878	1.09	0.034
1.46	0.138	0.3516	0.6484	0.0217	0.3298	10437.2	79.8457	123.14	0.5145	0.7935	1.57	0.033
1.7	0.139	0.3562	0.6437	0.0287	0.3274	10701.6	93.1533	144.704	0.5145	0.7993	2.11	0.033
1.94	0.14	0.3608	0.6391	0.0357	0.3251	10716.86	106.4609	166.566	0.5145	0.805	2.74	0.033
2.18	0.145	0.3829	0.6171	0.0689	0.3139	11161.16	119.7685	194.079	0.5145	0.8338	3.35	0.032
2.43	0.147	0.3913	0.6087	0.0816	0.3096	11185.92	133.0762	218.617	0.5145	0.8453	4.08	0.031
2.67	0.15	0.4035	0.5965	0.1	0.3034	11346.04	146.3838	245.387	0.5145	0.8625	4.84	0.031
2.91	0.154	0.419	0.581	0.1233	0.2955	11386.99	159.6914	274.833	0.5145	0.8855	5.61	0.03
3.15	0.159	0.4372	0.5627	0.1509	0.2862	10923.47	172.999	307.403	0.5145	0.9143	6.38	0.029
3.4	0.165	0.4577	0.5423	0.1818	0.2758	10967.12	186.3066	343.541	0.5145	0.9488	7.13	0.028
3.64	0.168	0.4674	0.5326	0.1964	0.2709	11114.19	199.6142	374.772	0.5145	0.9661	8.04	0.027
3.88	0.173	0.4828	0.5172	0.2196	0.2631	11271.73	212.9219	411.655	0.5145	0.9948	8.88	0.027
4.12	0.177	0.4945	0.5055	0.2372	0.2571	11421.44	226.2295	447.496	0.5145	1.0178	9.8	0.026
4.37	0.182	0.5083	0.4916	0.2582	0.2501	11574.45	239.5371	487.204	0.5145	1.0466	10.68	0.025
4.61	0.186	0.5189	0.481	0.2741	0.2447	11592.36	252.8447	525.574	0.5145	1.0696	11.64	0.025
4.85	0.19	0.529	0.4709	0.2894	0.2395	11737.82	266.1523	565.133	0.5145	1.0926	12.63	0.024
5.09	0.193	0.5364	0.4636	0.3005	0.2358	11749.47	279.4599	602.759	0.5145	1.1098	13.71	0.024
5.34	0.197	0.5458	0.4542	0.3147	0.231	11893.27	292.7676	644.549	0.5145	1.1328	14.74	0.023
5.58	0.201	0.5548	0.4451	0.3283	0.2264	11906.9	306.0752	687.529	0.5145	1.1558	15.79	0.023
5.82	0.21	0.5739	0.4261	0.3571	0.2167	12063.97	319.3828	749.545	0.5145	1.2076	16.46	0.022
6.07	0.216	0.5857	0.4142	0.375	0.2107	12080.23	332.6904	803.084	0.5145	1.2421	17.36	0.021
6.31	0.224	0.6005	0.3994	0.3973	0.2032	12229.74	345.998	866.141	0.5145	1.2881	18.11	0.021
6.55	0.228	0.6075	0.3924	0.4078	0.1996	12238.7	359.3056	915.516	0.5145	1.3111	19.18	0.02
6.79	0.235	0.6192	0.3807	0.4255	0.1937	12253.2	372.6133	978.573	0.5145	1.3513	20.02	0.02
7.04	0.243	0.6318	0.3682	0.4444	0.1873	12398.66	385.9209	1048.024	0.5145	1.3973	20.76	0.019
7.28	0.256	0.6505	0.3495	0.4726	0.1778	12420.11	399.2285	1142.164	0.5145	1.4721	21.09	0.018
7.52	0.267	0.6649	0.3351	0.4943	0.1704	12435.81	412.5361	1230.949	0.5145	1.5353	21.59	0.017

Table 1.6: Hydrodynamic characteristics for Glass Beads of dp = 0.4cm, Ug = 0.212314 .

which in turn increases the bed voidage (Figure 2.1 and 2.2). Miura et al. reported increase of bed voidage with increase in superficial liquid velocities and superficial gas velocities. The influence of phase flow rates on ε seems to be considerable at low Modified Reynolds numbers, where it is marginal at high Modified Reynolds numbers.

Solid holdup (εs)

It has been observed that ε_s decreases steeply as the gas phase Modified Reynolds number increases as well as liquid phase Modified Reynolds number increases (Figure 2.3). Ik-Sang Shin et al. reported similar trends. These trends may be explained as follows; when either

U_l (m/s)	L_f (m)	ε	ε_s	ε_g	ε_l	ΔP (N/m²)	$N_{Re,p,liq}$	$N_{Re,m,liq}$	$N_{Re,p,gas}$	$N_{Re,m,gas}$	$N_{Fr,liq}$	$N_{Fr,gas}$
1.21	0.139	0.3562	0.6437	0.0143	0.3419	7442.59	66.5381	103.36	0.7375	1.1457	1.08	0.068
1.46	0.143	0.3743	0.6257	0.0419	0.3323	7609.31	79.8457	127.601	0.7375	1.1786	1.51	0.066
1.7	0.146	0.3871	0.6128	0.0616	0.3255	7637.36	93.1533	151.991	0.7375	1.2033	2.01	0.065
1.94	0.15	0.4035	0.5965	0.0866	0.3168	7796.18	106.4609	178.463	0.7375	1.2363	2.56	0.063
2.18	0.154	0.419	0.581	0.1103	0.3086	7825.62	119.7685	206.125	0.7375	1.2693	3.16	0.061
2.43	0.159	0.4372	0.5627	0.1383	0.2989	7984.4	133.0762	236.464	0.7375	1.3105	3.77	0.059
2.67	0.165	0.4577	0.5423	0.1697	0.2881	8017.77	146.3838	269.925	0.7375	1.3599	4.4	0.057
2.91	0.17	0.4736	0.5263	0.1941	0.2795	8169.43	159.6914	303.387	0.7375	1.4012	5.08	0.056
3.15	0.174	0.4857	0.5142	0.2126	0.2731	8186.87	172.999	336.403	0.7375	1.4341	5.83	0.054
3.4	0.18	0.5029	0.4971	0.2388	0.2641	8210.14	186.3066	374.772	0.7375	1.4836	6.53	0.052
3.64	0.185	0.5163	0.4836	0.2594	0.2569	8355.85	199.6142	412.696	0.7375	1.5248	7.3	0.051
3.88	0.188	0.524	0.4759	0.2712	0.2528	8365.46	212.9219	447.347	0.7375	1.5495	8.17	0.05
4.12	0.193	0.5364	0.4636	0.2901	0.2462	8509.16	226.2295	487.948	0.7375	1.5907	8.98	0.049
4.37	0.195	0.5411	0.4588	0.2974	0.2437	8514.76	239.5371	522.005	0.7375	1.6072	9.97	0.048
4.61	0.198	0.5481	0.4519	0.308	0.2401	8522.77	252.8447	559.482	0.7375	1.6319	10.94	0.048
4.85	0.202	0.557	0.4429	0.3217	0.2352	8662.08	266.1523	600.826	0.7375	1.6649	11.88	0.047
5.09	0.205	0.5635	0.4365	0.3317	0.2318	8669.25	279.4599	640.236	0.7375	1.6896	12.91	0.046
5.34	0.209	0.5719	0.4281	0.3445	0.2274	8807.77	292.7676	683.811	0.7375	1.7226	13.89	0.045
5.58	0.213	0.5799	0.4201	0.3568	0.2231	8816.31	306.0752	728.576	0.7375	1.7556	14.9	0.044
5.82	0.22	0.5933	0.4067	0.3772	0.2161	8830	319.3828	785.238	0.7375	1.8133	15.71	0.043
6.07	0.227	0.6058	0.3941	0.3964	0.2093	8972.33	332.6904	843.982	0.7375	1.871	16.52	0.042
6.31	0.235	0.6192	0.3807	0.417	0.2022	8985.12	345.998	908.675	0.7375	1.9369	17.26	0.04
6.55	0.24	0.6272	0.3728	0.4291	0.1981	8992.42	359.3056	963.701	0.7375	1.9781	18.22	0.039
6.79	0.25	0.6421	0.3579	0.452	0.1901	9136.17	372.6133	1041.035	0.7375	2.0605	18.81	0.038
7.04	0.258	0.6532	0.3468	0.4689	0.1842	9276.42	385.9209	1112.717	0.7375	2.1265	19.56	0.037
7.28	0.265	0.6623	0.3376	0.483	0.1793	9284.21	399.2285	1182.318	0.7375	2.1842	20.38	0.036
7.52	0.272	0.671	0.3289	0.4963	0.1747	9422.27	412.5361	1254.001	0.7375	2.2419	21.2	0.035

Table 1.7: Hydrodynamic characteristics for glass beads of dp = 0.4cm, Ug = 0.304317cm/s.

U_l (m/s)	L_f (m)	ε	ε_s	ε_g	ε_l	ΔP (N/m²)	$N_{Re,p,liq}$	$N_{Re,m,liq}$	$N_{Re,p,gas}$	$N_{Re,m,gas}$	$N_{Fr,liq}$	$N_{Fr,gas}$
1.21	0.143	0.3743	0.6257	0.0279	0.3462	7338.32	66.5381	106.334	0.8404	1.3431	1.05	0.086
1.46	0.145	0.3829	0.6171	0.0413	0.3415	7482.21	79.8457	129.386	0.8404	1.3619	1.49	0.085
1.7	0.148	0.3954	0.6046	0.0608	0.3345	7634.12	93.1533	154.073	0.8404	1.39	1.99	0.083
1.94	0.153	0.4152	0.5848	0.0915	0.3236	7799.56	106.4609	182.032	0.8404	1.437	2.51	0.08
2.18	0.156	0.4264	0.5735	0.1089	0.3174	7946.97	119.7685	208.802	0.8404	1.4652	3.12	0.079
2.43	0.16	0.4407	0.5592	0.1312	0.3094	7972.92	133.0762	237.951	0.8404	1.5027	3.75	0.077
2.67	0.164	0.4544	0.5456	0.1524	0.3019	7996.09	146.3838	268.29	0.8404	1.5403	4.43	0.075
2.91	0.168	0.4674	0.5326	0.1726	0.2947	8144.15	159.6914	299.818	0.8404	1.5779	5.14	0.073
3.15	0.172	0.4798	0.5202	0.1918	0.2879	8163.23	172.999	332.536	0.8404	1.6155	5.89	0.071
3.4	0.178	0.4973	0.5027	0.2191	0.2781	8316.58	186.3066	370.608	0.8404	1.6718	6.61	0.069
3.64	0.182	0.5083	0.4916	0.2362	0.272	8331.91	199.6142	406.004	0.8404	1.7094	7.42	0.067
3.88	0.189	0.5266	0.4734	0.2645	0.262	8484.34	212.9219	449.727	0.8404	1.7751	8.13	0.065
4.12	0.194	0.5388	0.4612	0.2835	0.2552	8499.66	226.2295	490.476	0.8404	1.8221	8.94	0.063
4.37	0.197	0.5458	0.4542	0.2944	0.2513	8637.09	239.5371	527.358	0.8404	1.8503	9.87	0.062
4.61	0.203	0.5592	0.4407	0.3152	0.2439	8653	252.8447	573.61	0.8404	1.9066	10.67	0.06
4.85	0.211	0.5759	0.424	0.3412	0.2346	8671.77	266.1523	627.595	0.8404	1.9817	11.37	0.058
5.09	0.216	0.5857	0.4142	0.3564	0.2292	8811.88	279.4599	674.59	0.8404	2.0287	12.25	0.057
5.34	0.218	0.5895	0.4104	0.3623	0.2271	8815.92	292.7676	713.257	0.8404	2.0475	13.32	0.056
5.58	0.223	0.5987	0.4012	0.3766	0.222	8825.49	306.0752	762.781	0.8404	2.0945	14.23	0.055
5.82	0.23	0.611	0.389	0.3956	0.2153	8967.71	319.3828	820.93	0.8404	2.1602	15.02	0.053
6.07	0.235	0.6192	0.3807	0.4085	0.2107	8975.87	332.6904	873.726	0.8404	2.2072	15.96	0.052
6.31	0.24	0.6272	0.3728	0.4208	0.2063	8983.47	345.998	928.008	0.8404	2.2541	16.9	0.051
6.55	0.247	0.6377	0.3622	0.4372	0.2004	9123.65	359.3056	991.809	0.8404	2.3199	17.71	0.05
6.79	0.253	0.6463	0.3536	0.4505	0.1957	9261.98	372.6133	1053.527	0.8404	2.3762	18.59	0.048
7.04	0.262	0.6585	0.34153	0.4694	0.189	9272.92	385.9209	1129.969	0.8404	2.4608	19.26	0.047
7.28	0.269	0.6674	0.3326	0.4832	0.184	9280.68	399.2285	1200.164	0.8404	2.5265	20.07	0.046
7.52	0.277	0.677	0.32303	0.4981	0.1787	9288.83	412.5361	1277.052	0.8404	2.6016	20.81	0.044

Table 1.8: Hydrodynamic characteristics for glass beads of dp = 0.4cm, Ug = 0.346779cm/s.

U_l (m/s)	L_f (m)	ε	ε_s	ε_g	ε_l	ΔP (N/m²)	$N_{Re,p,liq}$	$N_{Re,m,liq}$	$N_{Re,p,gas}$	$N_{Re,m,gas}$	$N_{Fr,liq}$	$N_{Fr,gas}$
1.21	0.126	0.2791	0.7208	0.0079	0.2712	12640.5	99.807	138.465	0.6175	0.8566	1.19	0.023
1.46	0.126	0.2791	0.7208	0.0079	0.2712	12640.5	119.769	166.158	0.6175	0.8566	1.71	0.023
1.7	0.126	0.2791	0.7208	0.0079	0.2712	12764.43	139.73	193.851	0.6175	0.8566	2.33	0.023
1.94	0.126	0.2791	0.7208	0.0079	0.2712	12764.43	159.691	221.544	0.6175	0.8566	3.05	0.023
2.18	0.126	0.2791	0.7208	0.0079	0.2712	12888.35	179.653	249.238	0.6175	0.8566	3.86	0.023
2.43	0.127	0.2848	0.7151	0.0157	0.2691	12915.6	199.614	279.128	0.6175	0.8634	4.72	0.023
2.67	0.127	0.2848	0.7151	0.0157	0.2691	12915.6	219.576	307.041	0.6175	0.8634	5.72	0.023
2.91	0.127	0.2848	0.7151	0.0157	0.2691	12915.6	239.537	334.954	0.6175	0.8634	6.8	0.023
3.15	0.128	0.2904	0.7095	0.0234	0.267	12941.38	259.499	365.724	0.6175	0.8702	7.92	0.023
3.4	0.128	0.2904	0.7095	0.0234	0.267	12941.38	279.46	393.857	0.6175	0.8702	9.19	0.023
3.64	0.128	0.2904	0.7095	0.0234	0.267	12941.38	299.421	421.989	0.6175	0.8702	10.55	0.023
3.88	0.129	0.2959	0.704	0.031	0.2649	13090.48	319.383	453.639	0.6175	0.877	11.91	0.023
4.12	0.129	0.2959	0.704	0.031	0.2649	13090.48	339.344	481.991	0.6175	0.877	13.44	0.023
4.37	0.13	0.3013	0.6986	0.0384	0.2629	13113.89	359.306	514.3	0.6175	0.8838	14.95	0.023
4.61	0.13	0.3013	0.6986	0.0384	0.2629	13113.89	379.267	542.872	0.6175	0.8838	16.66	0.023
4.85	0.131	0.3067	0.6933	0.0458	0.2609	13261.24	399.228	575.84	0.6175	0.8906	18.32	0.022
5.09	0.132	0.3119	0.688	0.053	0.2589	13407.91	419.19	609.247	0.6175	0.8974	20.04	0.022
5.34	0.134	0.3222	0.6777	0.0671	0.255	13448.04	439.151	647.93	0.6175	0.911	21.67	0.022
5.58	0.136	0.3321	0.6678	0.0808	0.2513	13484.62	459.113	687.491	0.6175	0.9246	23.34	0.022
5.82	0.139	0.3466	0.6534	0.1007	0.2458	13533.79	479.074	733.207	0.6175	0.945	24.86	0.021
6.07	0.141	0.3558	0.6441	0.1134	0.2424	13563.31	499.036	774.746	0.6175	0.9586	26.59	0.021
6.31	0.145	0.3736	0.6263	0.1379	0.2357	13615.8	518.997	828.594	0.6175	0.9858	27.97	0.02
6.55	0.147	0.3821	0.6178	0.1496	0.2325	13766.71	538.958	872.331	0.6175	0.9994	29.75	0.02
6.79	0.152	0.4024	0.5975	0.1776	0.2248	13947.06	558.92	935.41	0.6175	1.0334	30.94	0.019
7.04	0.155	0.414	0.5859	0.1935	0.2205	13974.82	578.881	987.939	0.6175	1.0538	32.55	0.019
7.28	0.159	0.4287	0.5712	0.2138	0.2149	14008.05	598.843	1048.38	0.6175	1.081	33.96	0.018
7.52	0.161	0.4358	0.5641	0.2236	0.2122	14023.24	618.804	1096.953	0.6175	1.0946	35.81	0.018

Table 1.9: Hydrodynamic characteristics for glass beads of dp = 0.6cm, Ug = 0.169851cm/s.

U_l (m/s)	L_f (m)	ε	ε_s	ε_g	ε_l	ΔP (N/m²)	$N_{Re,p,liq}$	$N_{Re,m,liq}$	$N_{Re,p,gas}$	$N_{Re,m,gas}$	$N_{Fr,liq}$	$N_{Fr,gas}$
1.21	0.128	0.2904	0.7095	0.0078	0.2826	10160.75	99.807	140.663	0.7718	1.0878	1.17	0.036
1.46	0.129	0.2959	0.704	0.0155	0.2804	10181.14	119.769	170.115	0.7718	1.0963	1.67	0.036
1.7	0.13	0.3013	0.6986	0.023	0.2782	10449.28	139.73	200.005	0.7718	1.1048	2.26	0.035
1.94	0.131	0.3067	0.6933	0.0305	0.2761	10592.73	159.691	230.336	0.7718	1.1133	2.93	0.035
2.18	0.132	0.3119	0.688	0.0378	0.274	10610.87	179.653	261.106	0.7718	1.1218	3.68	0.035
2.43	0.134	0.3222	0.6777	0.0522	0.2699	10769.84	199.614	294.513	0.7718	1.1388	4.48	0.034
2.67	0.136	0.3321	0.6678	0.0661	0.266	10800.96	219.576	328.8	0.7718	1.1557	5.34	0.034
2.91	0.138	0.3418	0.6581	0.0797	0.2621	10829.44	239.537	363.966	0.7718	1.1727	6.26	0.033
3.15	0.14	0.3512	0.6487	0.0928	0.2584	10855.61	259.499	400.011	0.7718	1.1897	7.24	0.033
3.4	0.141	0.3558	0.6441	0.0992	0.2565	10994.28	279.46	433.858	0.7718	1.1982	8.34	0.033
3.64	0.143	0.3648	0.6351	0.1118	0.2529	11017.74	299.421	471.441	0.7718	1.2152	9.44	0.032
3.88	0.145	0.3736	0.6263	0.1241	0.2495	11039.46	319.383	509.904	0.7718	1.2322	10.59	0.032
4.12	0.148	0.3863	0.6136	0.1418	0.2444	11069.2	339.344	552.982	0.7718	1.2577	11.72	0.031
4.37	0.152	0.4024	0.5975	0.1644	0.238	11104.31	359.306	601.335	0.7718	1.2917	12.79	0.03
4.61	0.156	0.4178	0.5821	0.1859	0.2319	11135.11	379.267	651.446	0.7718	1.3257	13.88	0.029
4.85	0.161	0.4358	0.5641	0.2111	0.2247	11168.67	399.228	707.712	0.7718	1.3682	14.91	0.029
5.09	0.165	0.4495	0.5504	0.2303	0.2192	11192.27	419.19	761.559	0.7718	1.4022	16.04	0.028
5.34	0.169	0.4625	0.5374	0.2485	0.214	11213.45	439.151	817.165	0.7718	1.4362	17.18	0.027
5.58	0.172	0.4719	0.528	0.2616	0.2103	11227.96	459.113	869.474	0.7718	1.4617	18.45	0.027
5.82	0.174	0.478	0.5219	0.2701	0.2079	11366.22	479.074	917.827	0.7718	1.4787	19.86	0.026
6.07	0.176	0.4839	0.516	0.2784	0.2055	11374.99	499.036	967.059	0.7718	1.4957	21.31	0.026
6.31	0.178	0.4897	0.5102	0.2865	0.2032	11383.35	518.997	1017.17	0.7718	1.5127	22.78	0.026
6.55	0.18	0.4954	0.5045	0.2944	0.2009	11520.79	538.958	1068.161	0.7718	1.5297	24.3	0.026
6.79	0.182	0.5009	0.499	0.3022	0.1987	11528.51	558.92	1120.03	0.7718	1.5467	25.84	0.025
7.04	0.183	0.5037	0.4963	0.306	0.1976	11661.83	578.881	1166.405	0.7718	1.5552	27.57	0.025
7.28	0.185	0.509	0.4909	0.3135	0.1955	11669.15	598.843	1219.813	0.7718	1.5722	29.19	0.025
7.52	0.188	0.5169	0.4831	0.3244	0.1924	11679.56	618.804	1280.914	0.7718	1.5977	30.67	0.024

Table 1.10: Hydrodynamic characteristics for glass beads of dp = 0.6cm, Ug = 0.212314cm/s.

U_l (m/s)	L_f (m)	ε	ε_s	ε_g	ε_l	ΔP (N/m²)	$N_{Re,p,liq}$	$N_{Re,m,liq}$	$N_{Re,p,gas}$	$N_{Re,m,gas}$	$N_{Fr,\,liq}$	$N_{Fr,\,gas}$
1.21	0.132	0.3119	0.688	0.0151	0.2968	8812.73	99.807	145.059	1.1063	1.6079	1.14	0.072
1.46	0.133	0.3171	0.6828	0.0225	0.2945	8828.36	119.769	175.389	1.1063	1.62	1.62	0.071
1.7	0.135	0.3272	0.6727	0.037	0.2902	8982.24	139.73	207.698	1.1063	1.6444	2.18	0.07
1.94	0.136	0.3321	0.6678	0.0441	0.288	8996.02	159.691	239.127	1.1063	1.6566	2.82	0.069
2.18	0.138	0.3418	0.6581	0.0579	0.2839	9021.85	179.653	272.974	1.1063	1.6809	3.52	0.068
2.43	0.14	0.3512	0.6487	0.0714	0.2798	9171.2	199.614	307.701	1.1063	1.7053	4.29	0.067
2.67	0.143	0.3648	0.6351	0.0909	0.2739	9203.82	219.576	345.724	1.1063	1.7418	5.08	0.066
2.91	0.145	0.3736	0.6263	0.1034	0.2701	9349.91	239.537	382.428	1.1063	1.7662	5.96	0.065
3.15	0.147	0.3821	0.6178	0.1156	0.2665	9368.5	259.499	420.011	1.1063	1.7906	6.9	0.064
3.4	0.15	0.3945	0.6054	0.1333	0.2611	9394.02	279.46	461.551	1.1063	1.8271	7.84	0.063
3.64	0.153	0.4063	0.5936	0.1503	0.256	9544.34	299.421	504.409	1.1063	1.8636	8.82	0.062
3.88	0.155	0.414	0.5859	0.1612	0.2527	9558.7	319.383	545.07	1.1063	1.888	9.91	0.061
4.12	0.158	0.4251	0.5748	0.1772	0.2479	9706.36	339.344	590.346	1.1063	1.9245	10.97	0.06
4.37	0.16	0.4323	0.5676	0.1875	0.2448	9846.74	359.306	632.984	1.1063	1.9489	12.15	0.059
4.61	0.163	0.4428	0.5571	0.2024	0.2403	9864.43	379.267	680.678	1.1063	1.9855	13.29	0.058
4.85	0.164	0.4462	0.5537	0.2073	0.2388	9870	399.228	720.899	1.1063	1.9976	14.63	0.058
5.09	0.167	0.4561	0.5438	0.2215	0.2346	10014.22	419.19	770.79	1.1063	2.0342	15.84	0.057
5.34	0.169	0.4625	0.5374	0.2307	0.2318	10152.75	439.151	817.165	1.1063	2.0585	17.18	0.056
5.58	0.171	0.4688	0.5311	0.2397	0.2291	10162.39	459.113	864.419	1.1063	2.0829	18.56	0.055
5.82	0.174	0.478	0.5219	0.2528	0.2251	10304.78	479.074	917.827	1.1063	2.1194	19.86	0.054
6.07	0.176	0.4839	0.516	0.2613	0.2226	10313.41	499.036	967.059	1.1063	2.1438	21.31	0.054
6.31	0.179	0.4926	0.5073	0.2737	0.2188	10325.62	518.997	1022.885	1.1063	2.1803	22.66	0.053
6.55	0.182	0.5009	0.499	0.2857	0.2152	10466.24	538.958	1080.029	1.1063	2.2169	24.03	0.052
6.79	0.184	0.5064	0.4936	0.2934	0.2129	10473.53	558.92	1132.338	1.1063	2.2412	25.56	0.051
7.04	0.187	0.5143	0.4856	0.3048	0.2095	10483.9	578.881	1191.9	1.1063	2.2778	26.98	0.05
7.28	0.19	0.5219	0.478	0.3157	0.2062	10623.18	598.843	1252.781	1.1063	2.3143	28.42	0.05
7.52	0.194	0.5318	0.4681	0.3299	0.2019	10635.43	618.804	1321.794	1.1063	2.3631	29.72	0.049

Table 1.11: Hydrodynamic characteristics for glass beads of dp = 0.6cm, Ug = 0.304317cm/s.

U_l (m/s)	L_f (m)	ε	ε_s	ε_g	ε_l	ΔP (N/m²)	$N_{Re,p,liq}$	$N_{Re,m,liq}$	$N_{Re,p,gas}$	$N_{Re,m,gas}$	$N_{Fr,\,liq}$	$N_{Fr,\,gas}$
1.21	0.135	0.3272	0.6727	0.0074	0.3198	8918.64	99.807	148.356	1.2606	1.8738	1.11	0.091
1.46	0.137	0.337	0.6629	0.0219	0.3151	8948.37	119.769	180.664	1.2606	1.9016	1.58	0.089
1.7	0.138	0.3418	0.6581	0.0289	0.3128	8962.29	139.73	212.313	1.2606	1.9155	2.13	0.089
1.94	0.141	0.3558	0.6441	0.0496	0.3062	9125.74	159.691	247.919	1.2606	1.9571	2.72	0.087
2.18	0.143	0.3648	0.6351	0.0629	0.3019	9274.56	179.653	282.865	1.2606	1.9849	3.4	0.086
2.43	0.145	0.3736	0.6263	0.0758	0.2977	9547.87	199.614	318.69	1.2606	2.0126	4.14	0.085
2.67	0.148	0.3863	0.6136	0.0945	0.2917	9578.87	219.576	357.812	1.2606	2.0543	4.9	0.083
2.91	0.15	0.3945	0.6054	0.1066	0.2878	9597.8	239.537	395.615	1.2606	2.082	5.76	0.082
3.15	0.153	0.4063	0.5936	0.1241	0.2822	9750.53	259.499	437.155	1.2606	2.1237	6.63	0.08
3.4	0.155	0.414	0.5859	0.1354	0.2785	9766.78	279.46	476.936	1.2606	2.1514	7.59	0.079
3.64	0.157	0.4215	0.5784	0.1465	0.275	9909.1	299.421	517.596	1.2606	2.1792	8.6	0.078
3.88	0.16	0.4323	0.5676	0.1625	0.2698	10057.94	319.383	562.653	1.2606	2.2209	9.6	0.077
4.12	0.164	0.4462	0.5537	0.1829	0.2632	10211.88	339.344	612.764	1.2606	2.2764	10.57	0.075
4.37	0.168	0.4593	0.5406	0.2023	0.257	10235.75	359.306	664.633	1.2606	2.3319	11.57	0.073
4.61	0.172	0.4719	0.528	0.2209	0.251	10257.27	379.267	718.261	1.2606	2.3874	12.59	0.071
4.85	0.175	0.481	0.5189	0.2342	0.2467	10528.86	399.228	769.252	1.2606	2.4291	13.71	0.07
5.09	0.178	0.4897	0.5102	0.2471	0.2425	10542.98	419.19	821.561	1.2606	2.4707	14.86	0.069
5.34	0.18	0.4954	0.5045	0.2555	0.2398	10551.87	439.151	870.353	1.2606	2.4985	16.13	0.068
5.58	0.183	0.5037	0.4963	0.2677	0.2359	10564.47	459.113	925.08	1.2606	2.5401	17.34	0.067
5.82	0.186	0.5117	0.4882	0.2795	0.2321	10834.24	479.074	981.126	1.2606	2.5817	18.58	0.066
6.07	0.192	0.5269	0.473	0.302	0.2248	10856.29	499.036	1054.974	1.2606	2.665	19.53	0.064
6.31	0.195	0.5342	0.4657	0.3128	0.2214	10866.36	518.997	1114.316	1.2606	2.7067	20.8	0.063
6.55	0.2	0.5458	0.4541	0.33	0.2158	11011.46	538.958	1186.845	1.2606	2.7761	21.87	0.061
6.79	0.208	0.5633	0.4366	0.3557	0.2075	11033.85	558.92	1280.035	1.2606	2.8871	22.61	0.059
7.04	0.216	0.5795	0.4204	0.3796	0.1999	11183.42	578.881	1376.741	1.2606	2.9981	23.36	0.057
7.28	0.224	0.5945	0.4054	0.4017	0.1927	11200.8	598.843	1476.963	1.2606	3.1092	24.11	0.055
7.52	0.229	0.6034	0.3966	0.4148	0.1885	11210.64	618.804	1560.262	1.2606	3.1786	25.18	0.054

Table 1.12: Hydrodynamic characteristics for glass beads of dp = 0.6cm, Ug = 0.346779cm/s.

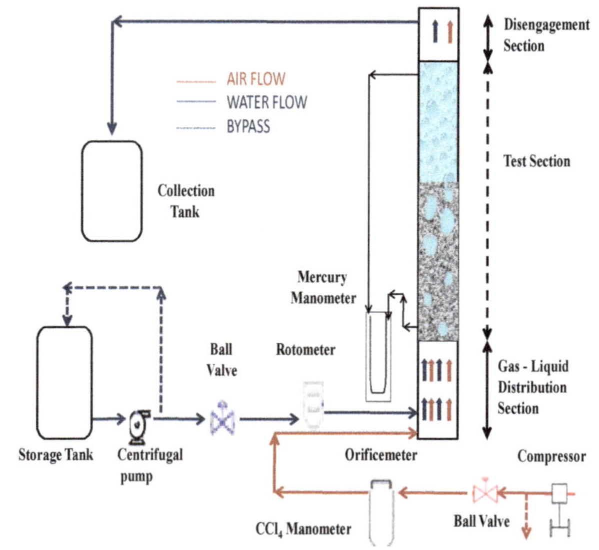

Figure 1: Schematic Diagram of the Experimental Setup.

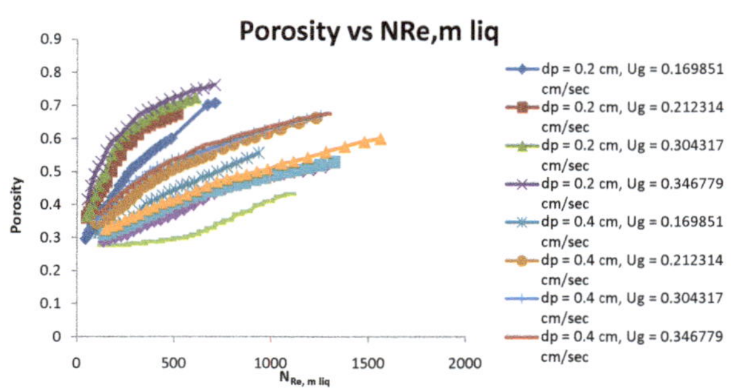

Figure 2.1: Porosity vs NRe,m liq.

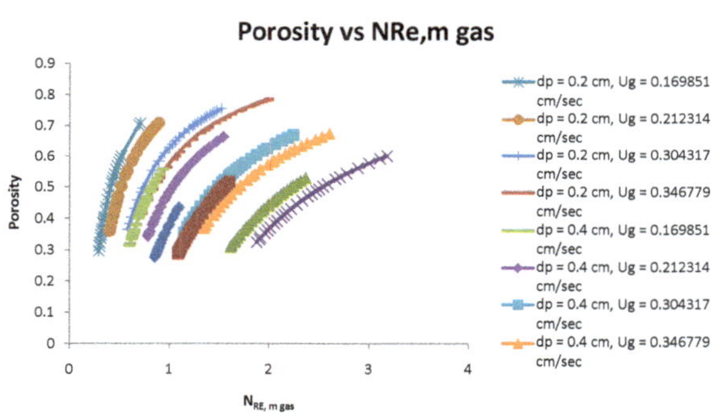

Figure 2.2: Porosity vs NRe,m gas.

liquid velocity or gas velocity is increased, there would be higher drag forces exerted on the solid particles, leading to more bed expansion; this results in reduction of solids holdup in the expanded bed. However, the decrease is considerable at low Modified Reynolds numbers and ε_s seem to be approaching a constant value at high gas phase Modified Reynolds numbers. Since $\varepsilon_s = (1- \varepsilon)$ and ε increase with Modified Reynolds number (gas or liquid) as explained in section 1, ε_s should decrease with an increase in Modified Reynolds number (Figure 2.4).

Gas holdup (ε_g)

The gas holdups are found to be increasing with the increase in gas

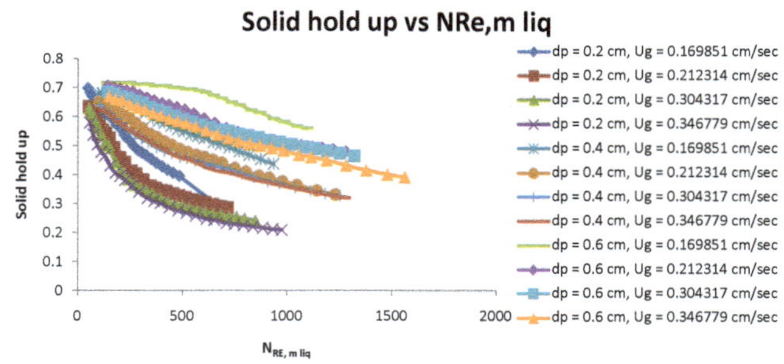

Figure 2.3: Solid hold up vs NRe,m liq.

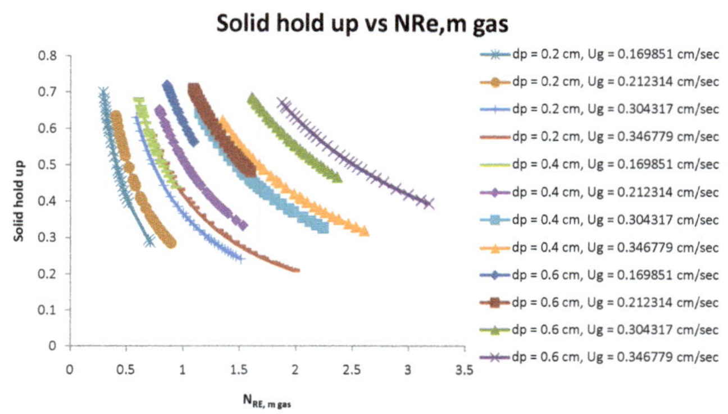

Figure 2.4: Solid hold up vs NRe,m gas.

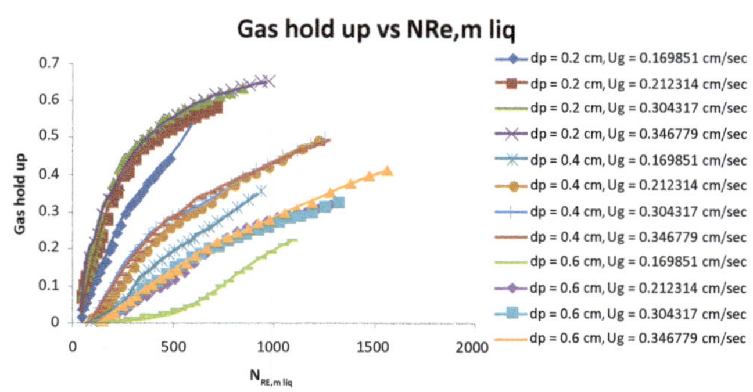

Figure 2.5: Gas hold up vs NRe,m liq.

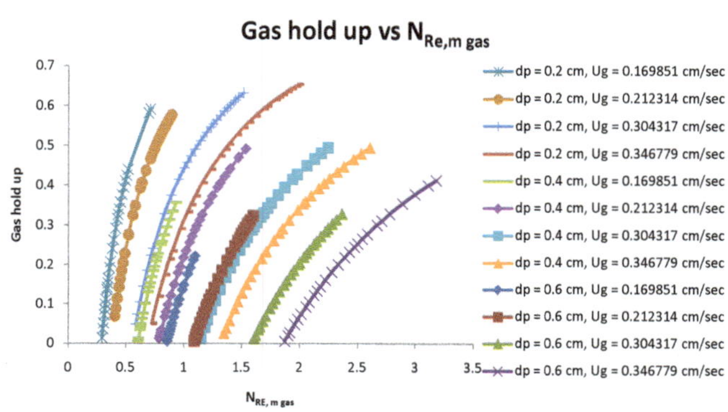

Figure 2.6: Gas hold up vs NRe,m gas.

phase Modified Reynolds number and also with the increasing liquid phase Modified Reynolds number as shown in the Figures 2.5 and

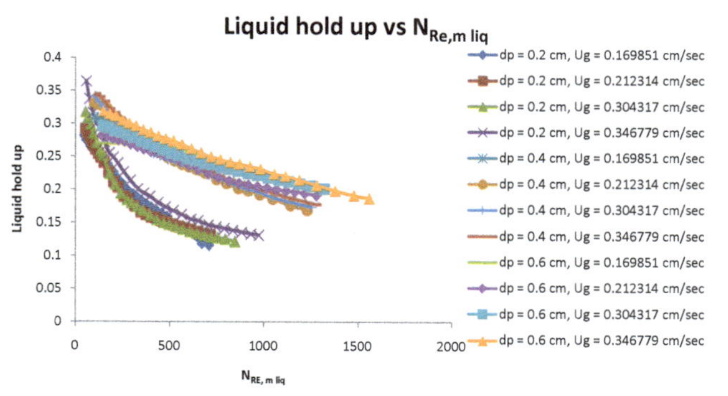

Figure 2.7: Liquid hold up vs NRe,m liq.

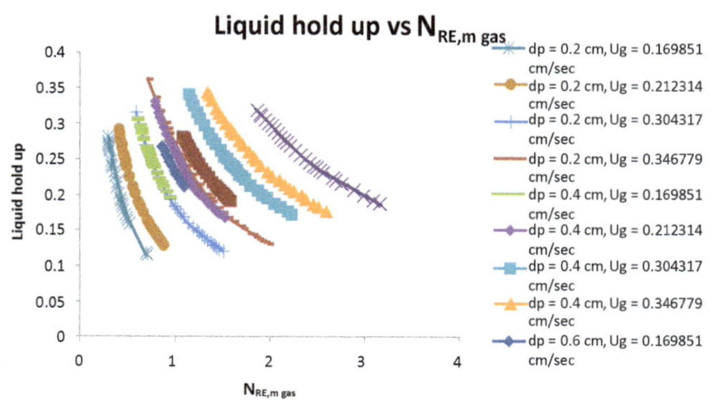

Figure 2.8: Liquid hold up vs NRe,m gas.

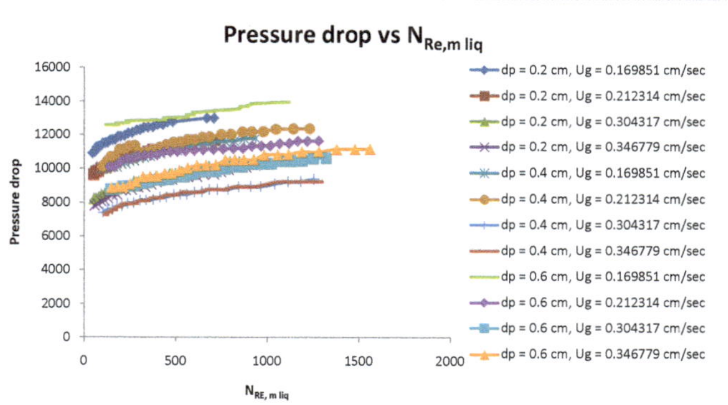

Figure 2.9: Pressure drop vs NRe,m liq.

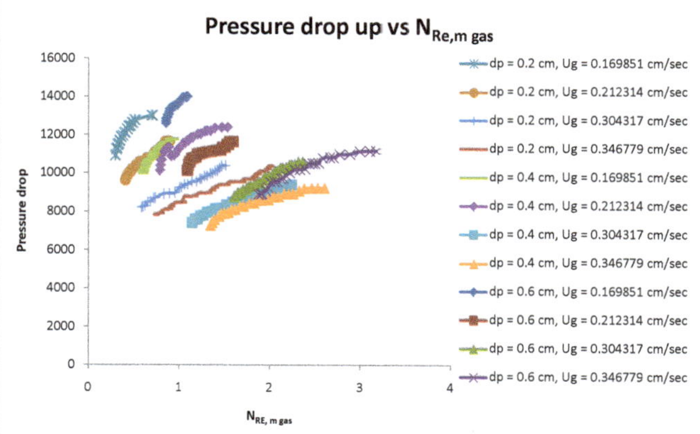

Figure 2.10: Pressure drop up vs NRe,m gas.

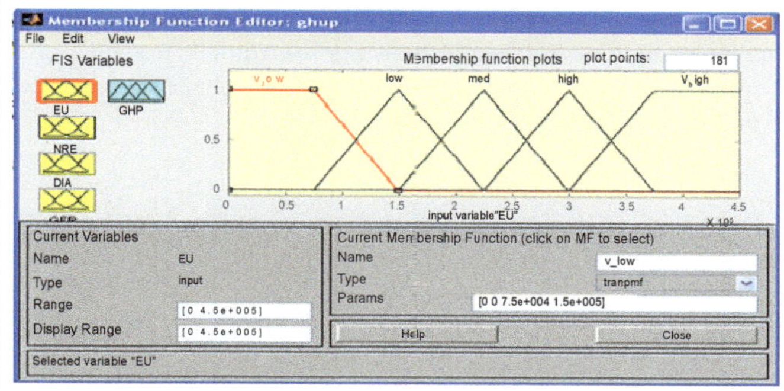

Figure 3.1: Membership Function for Input Variable EU.

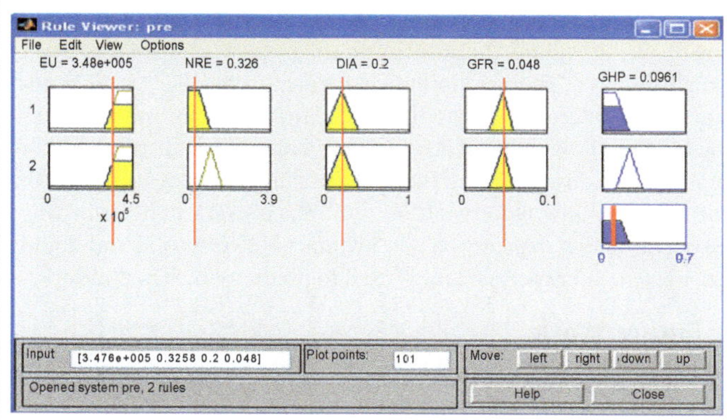

Figure 3.2: Rule viewers during training the expert system.

2.6; this increase is higher at low Modified Reynolds number (gas or liquid) and approaching a constant value at higher Modified Reynolds number. The gas and liquid phases are moving co-currently through the column; however the liquid phase velocities are higher than gas phase velocities obtained in this work. So, since the gas passage through the column is slower, more gas accumulation occurs in the column compared to liquid accumulation resulting in an increase in the gas holdup, gas holdup vs Modified Reynolds number follows similar trends for all the temperatures. The influence of temperature on phase properties causes a slight decrease of ε_g with an increase in temperature.

Liquid holdup (ε_l)

It is observed that ε_l is decreasing with increasing gas phase Modified Reynolds number as well as liquid phase Modified Reynolds number and approaching a constant value at higher Modified Reynolds numbers (Figure 2.7). The gas and liquid phases are moving co-currently through the column; however the liquid phase velocities are higher than gas phase velocities obtained in this work. So, since the liquid passage through the column is faster, less liquid accumulation occurs in the column and this accumulation decreases with increase in liquid velocity; this results in reduction in liquid holdup as the liquid phase Modified Reynolds number increases (Figure 2.8).

Pressure drop (ΔP)

The results show that the pressure drop increases with increase in Modified Reynolds number (both gas and liquid) (Figures 2.9 and 2.10). An increase is observed with lower liquid phase Modified

Reynolds number whereas the increase is gradual with high liquid phase Modified Reynolds number. Pressure drop is higher for large particle sizes.

Simulation

A training data set, which is required to design the optimal knowledge base of the fuzzy logic-based expert system [8-13], is collected by conducting experimental studies on hydrodynamics characteristics of co-current three phase fluidization [14] (as discussed in the earlier section) using fluidized bed for different combinations of parameters [15,16].

To develop a fuzzy logic-based expert system for the above problem, four parameters, namely Euler's number (Figures 3.1 and 3.2), Reynolds number, particles diameter, liquid and gas flow rates are considered as inputs where gas holdup and liquid holdup as two outputs [17]. Five linguistic terms are considered for each of the input and output variables, namely Very Low (V_LOW), Low (LOW), Medium (MED), High (HIGH) and Very High (V_HIGH). For simplicity, the shape of the membership function distributions is assumed to be triangular. As there are five linguistic terms for element size and shape ratio each, a total of 25 (i.e., 5.5) rules are to be considered.

Thus, a typical rule will look as follows:

IF EU is V_HIGH AND NRE is V_LOW AND DIA is LOW AND GFR is V_LOW, THEN GHP is MED

Conclusion

Improvement of mixing between liquid and solid particle in a fluidized bed can be achieved by a simple and effective method. Phase holdups, bed porosity and pressure drop in a gas-liquid-solid fluidized bed showed a marked variation with particle size and liquid flow rate at constant gas flow rate. Pressure drop is higher for large particle sizes. In the break-up regime, the gaseous phase forms a uniform dispersion of small bubbles. The gas holdup is greater in smaller particle and increase with increase in gas flow rate. Bed porosity increase with increase in gas flow rate and it is high in small particle. It is found that design calculations regarding three phase fluidized beds are based on phase holdups, flow pattern of the fluid phase and extent of mixing. The system mainly depends on good contact between solid and liquid. Fuzzy logic based expert system is used to predict the phase holdups.

For Future Work

Genetic algorithm may be used to tune fuzzy knowledge base to improve performance of the expert system.

References

1. Matsumoto T, Hidaka N, Takebayasi Y, Morooka S (1997) Axial Mixing and segregation in a gas-liquid-solid three-phase fluidized bed of solid particles of different sizes and densities. Chemical Engineering Science 52: 3961-3970.
2. Saez AE, Carbonell RG (1997) Hydrodynamic parameters for gas-liquid Cocurrent flow in packed beds. AIChE J 31: 52-62.
3. Tupin JL, Huntington RL (1967) Prediction of pressure drop for two-phase, two-component concurrent flow in packed beds. AIChE J 13: 1196-1202.
4. Jena HM, Roy GK, Meikap BC (2008) Prediction of Gas Holdup In A Three Phase Fluidized Bed From Bed Pressure Drop Measurement. Chemical Engineering Research and Design 86: 1301-1308.
5. Kim SD (1997) Heat and mass transfer in three-phase fluidized-bed reactor and overview. Chemical Engineering Science 52: 3639-3660.
6. Cho YJ, Kim SJ, Nam SH, Kang Y, Kim SD (2001) Heat transfer and bubble properties in three-phase circulating fluidized beds. Chemical Engineering Science 56: 6170-6115.
7. Charpentier JC (1976) Recent Progress in Two Phase Gas-Liquid Mass Transfer in Packed Beds. Chem Eng J 11: 161-181.
8. Comte MP, Bastoul D, Hebrard G, Roustan M, Lazarova V (1997) Hydrodynamics of a three-phase fluidized bed-the inverse turbulent bed. Chemical Engineering Science 52: 3971-3977.
9. Sato Y, Hirose T, Ida T, Kagaku K (1974) 38: 543.
10. Jena HM, Roy GK, Meikap BC (2009) Hydrodynamics of a gas–liquid–solid fluidized bed with hollow cylindrical particles. Chemical Engineering and Processing 48: 279-287.
11. Shah YT (1979) Gas-Liquid-Solid Reactor Design. McGraw-Hill, New York.
12. Han L, Al-Dahhan MH (2006) Gas–Liquid Mass Transfer In A High Pressure Bubble Column Reactor With Different Sparger Designs. Chemical Engineering Science 6: 131-139.
13. Luo B, Yan D, Ma YL, Barghi S, Zhu J (2007) Characteristics of Gas–Solid Mass Transfer in a Cocurrent Down flow Circulating Fluidized Bed Reactor. Chemical Engineering Journal 132: 9-15.
14. Miura H, Takahashi T, Kawase Y (2001) Effect of pseudoplastic behaviour of liquid in co-current three-phase fluidized beds on bed expansion. Chemical Engineering Science 56: 6047-6053.
15. Karr C (1991) Genetic algorithms for fuzzy controllers. AI Expert 26-33.
16. Cooper MG, Vidal JJ (1993) Genetic design of fuzzy logic controllers. In: Proceedings of the Second International Conference on Fuzzy Theory and Technology (FTT'93), Durham.
17. Ravish M, Kumar V (2011) A review on the applications of Fuzzy Logic technique related to the engineering applications. Indian Journal of Applied Research 1-35.

Removal of Uranium (VI) from Aqueous Solution using Low Cost and Eco-Friendly Adsorbents

Mohamed Ahmed Mahmoud[12*]

[1]*Reactors Material Treatment Department, Nuclear Material Authority, Kattamiya Road, Maddi, Cairo, Egypt*
[2]*Chemical Engineering Department, Faculty of Engineering, Jazan University, Jazan, Saudi Arabia*

Abstract

Adsorption process is an important technique in separation processes, which is used in heavy metals removal from aqueous solution. Orange peel has been investigated for the adsorption of uranium ions from aqueous solution. The parameters of pH, contact time, initial uranium ion concentration, temperature and adsorbent dose have been investigated in batch experiments. The adsorption kinetics were described using pseudo-second order model and the adsorption isotherms were described by Langmuir and Freundlich isotherms which indicate that the uranium adsorption onto orange peel fitted well with Langmuir isotherm.

Keywords: Uranium; Orange peel; Adsorption; Aqueous solution; Langmuir

Introduction

Several methods are used to remove uranium from liquid wastes. These include volume reduction followed by chemical, physical and biological treatment. However, these processes are both costly and ineffective, particularly when the concentration of uranium in liquid waste is a very low [1,2]. Adsorption process using agricultural material is the important technique in the wastewater treatment processes. Numerous researches have been carried out for using of low cost and eco-friendly adsorbents, such as peat [3,4], agricultural byproducts [5-7], and microbial biomass [8]. In the present work, the sorption of U (VI) onto orange peel carried out by studying the effect of various parameters such as pH, contact time, initial uranium concentration and adsorbent dosage in batch experiments. The kinetics and equilibrium of adsorption process are studied.

Experimental

Adsorbate

1000 mg /l of uranium solutions (Synthetic wastewater solutions) were prepared by dissolving a certain amounts of uranyl nitrate ($UO_2(NO_3)_2 \bullet 6H_2O$, Aldrich), in distilled water. The required concentrations were prepared by dilution solution. The concentrations of U (VI) in solution were determined spectrophotometrically using Shimadzu UV-VIS-1601 spectrophotometer using arsenazo (III) as complexing reagent at 510 NM against reagent blank [9,10].

Adsorbent

Orange peels, were collected from by-products of fruit industries. Orange peels were used without any treatment for the preparation of adsorbent, which were cut into small pieces, washed multiple times with distilled water and dried at 105°C. Then, milled and sieved by a size of 150-mesh size.

Methods

To study the influence of different parameters such as pH, contact time, initial metal ion concentration, temperature and dose of adsorbent for the removal of uranium, batch experiments were carried out by mixing 0.3 g of orange peels with 50 ml of adsorbate solution in 100 ml stoppered conical flask using a thermostatic water bath shaker of type (Julabo, Model SW -20°C, Germany), at 200 rpm. The influence of pH on the adsorption process was studied from pH 2.0 to 8.0. 0.1 M HCl or 0.1 M Na_2CO_3 solutions were used to adjust pH. Different concentrations of uranium ion solution from 25 to 200 ppm,, were used to study the influence of initial uranium ion concentration. The samples were taken and centrifuged at 5000 rpm for 5 min and were analyzed. Adsorption percent (%) was determined by the following equation: Adsorption percent (%)=$[(C_0 - C_e) / C_e] \times 100$ (1)

In addition, the adsorption capacity (qe) is determined according to the following equation:

$$q_e = (C_0 - C_e)\, V/M \qquad (2)$$

Where

C_0=initial concentration of solute (mg/L)

C_e=concentration of solute at equilibrium (mg/L)

V=volume of solution (L)

M=adsorbent dose (g)

Results and Discussion

Characterizations of adsorbent

Orange peel was studied using scanning electron microscopy (SEM) (Figure 1), showed that the orange peel has an irregular and porous surface, which have surface area of 0.98 m^2/g by BET method.

Effect of pH

The influence of pH on the removal of uranium ions onto orange peels was presented in Table 1. The percent of removal increases with

***Corresponding author:** Mohamed Ahmed Mahmoud, Assistant Professor, Reactors Material Treatment Department, Nuclear Material Authority, Kattamiya Road, Maddi, P.O. Box 530, Cairo, Egypt and Chemical Engineering Department, Faculty of Engineering, Jazan University, P.O. Box 114, Jazan, Saudi Arabia

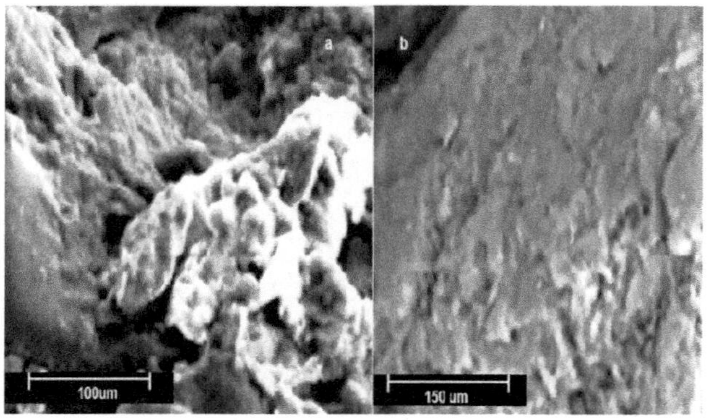

Figure 1a, 1b: Scanning electron micrographs (SEM) of orange peel.

Parameter		Ce (mg/L)	Adsorption percent (%)	q (mg/g)
pH:	2	61.10	38.90	6.483
	3	34.30	65.70	10.95
	4	4.50	95.50	15.91
(Condition: [U(VI)] = 100 mg L-1, V = 50 mL, mass = 0.3 g, 3 h, 28 °C)	5	9.55	90.45	15.07
	6	14.59	85.41	14.23
	7	16.77	83.23	13.87
	8	24.55	75.45	12.57
Time (min):	10	69.90	30.20	5.016
	20	49.79	50.21	8.368
	40	18.67	81.33	13.55
(Condition: [U(VI)] = 100 mg L-1, V = 50 mL, mass = 0.3 g, pH= 4 , 28 °C)	60	4.50	95.50	15.91
	80	4.50	95.50	15.91
	100	4.49	95.51	15.91
	120	4.50	95.50	15.91
Initial concentration (mg/l):	25	0.45	98.16	4.091
	50	1.05	97.72	8.158
	75	2.40	96.68	12.10
	100	4.50	95.50	15.916
Condition: pH= 4, V = 50 mL, mass = 0.3 g, 60 min, 28 °C)	125	29.54	76.36	15.91
	150	54.54	63.64	15.91
	175	79.60	54.51	15.90
	200	104.60	47.70	15.90
Temperature (°C):	28	4.50	95.50	15.91
Condition: [U(VI)] = 100 mg L-1, V = 50 mL, 60 min, pH= 4, mass = 0.3 g	40	4.51	95.49	15.91
	55	4.49	95.51	15.91
Adsorbent dose (g):	0.1	79.10	41.80	10.45
	0.2	41.32	96.04	14.67
Condition: [U(VI)] = 100 mg L-1, V = 50 mL, 60 min, pH= 4, 28 °C	0.3	4.50	95.50	15.91
	0.4	4.52	95.45	11.93
	0.5	4.49	95.44	9.551

Table 1: Adsorption data of adsorption of U (VI) onto orange peel.

increasing pH until reaches maximum at pH 4, then decreases with increasing pH. This is due to the competing of H^+ with uranium ions for the adsorption sites of orange peel, lead to decreasing the removal percent of uranium ions. However, after pH 4 the percent of removal decreases. This is due to the formation of stable complexes with carbonates such as UO_2CO_3, $[UO_2CO_3]^{2-}$ [11,12].

Effect of contact time

The influence of time on the adsorption of uranium ions was investigated from 10 to 120 min. Table 1 noted that the percent of removal increase with increasing time until reaching to a certain time 60 minutes. After 60 minutes, there is no significant change in the removal percent of uranium ions.

Effect of initial concentration of metal ions

Table 1, shows that the adsorption capacity (q_e) increases with increasing in equilibrium uranium ion concentration (C_e) until reaches the maximum q_e (15.91 mg/g), and then decreases with increasing C_e. This is due to saturation of adsorption sites of the adsorbent material.

Effect of adsorbent dosage

The adsorption percent at various doses of orange peel from 0.1to 0.5 g, is shown in Table 1. Increasing the adsorbent dose to 0.3g increase the adsorption percent of metal ions, which is due to the increasing in adsorption sites of adsorbent material resulting from increasing of surface area of adsorbent. However, further increase of adsorbent dosage does not afford exhaustive adsorption of U (VI). This may be due to overlapping of adsorption sites as a result of overcrowding of adsorbent particles. In addition, the amount of ions bound to the adsorbent and the amount of free ions remains constant even with further addition of the dose of adsorbent.

Effect of temperature

The effect of different temperature on uranium removal is carried out at different temperature (30, 40 and 55 °C), is shown in Table 1. The results show that the adsorption percent of uranium ions remain almost the same in the studied temperature range.

Equilibrium adsorption isotherm

Experimental results were studied using Langmuir and Freundlich isotherms [13-16], and these isotherms were analyzed by determination of correlation coefficients (r^2). The Langmuir isotherm supposes that the adsorbate molecules adsorbed as a monolayer on the surface of adsorbent material [15].

Langmuir model is represented as: $q_e = Q_0 b C_e / (1 + bC_e)$ (3)

Where q_e is the adsorption capacity of orange peel (mg/g), C_e is the equilibrium concentration of adsorbate (mg/L), b is the constant of Langmuir (L/mg) and Q_0 is the maximum monolayer coverage (mg/g).

By linearized Eq. (3): $C_e/q_e = 1/\, Q_0 b + C_e/\, Q_0$ (4)

Q_0 and b can be determined from the intercept and slope of the straight line resulting from Figure 2, respectively. The basic attributes of the Langmuir model can be expressed by dimensionless constant term (R_L) [17].

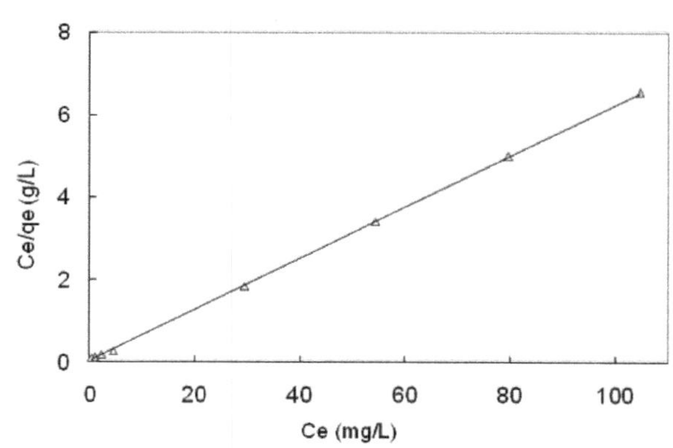

Figure 2: Langmuir isotherm of adsorption of uranium onto orange peels.

$R_L = 1/(1 + b C_0)$ (5)

When $0 < R_L < 1$ indicate that the adsorption process is favorable

Freundlich model is represented as: $q_e = K_F C_e^{(1/n)}$ (6)

Where K_F is the constant of Freundlich model [(mg/g)/(mg/l)] and n is a parameter refers to adsorption tendency.

By linearized Eq. (6): $\log q_e = \log K_f + 1/n \log C_e$ (7)

K_F and n can be determined from the intercept and slope of the straight line resulting from Figure 3, respectively. From Figures 2 and 3, and the result data in Table 2, show that the Langmuir isotherm model is the best fitting model because it gives a higher correlation coefficient (r^2=0.999) value than Freundlich isotherm model (r^2=0.682).

Adsorption kinetics

To study the adsorption kinetics of U (VI) ions onto the orange peel were tested using Pseudo-first-order and pseudo-second order model [18,19].

Pseudo- first -order model: The equation represents this model is:

$$\log(q_e - q_t) = \log q_e - \frac{k_1}{2.303} t \quad (8)$$

Where q_e and q_t (mg/g) are the adsorption capacity of U (VI) ions at equilibrium and at time t, respectively, k_1 is the pseudo first order constant (min^{-1}). The values of qe and k_1 were determined from the intercept and slope of Figure 4, respectively.

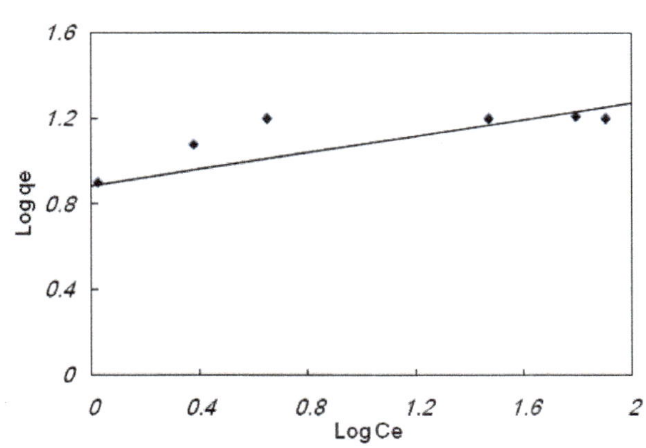

Figure 3: Freundlich isotherm of adsorption of uranium onto orange peels.

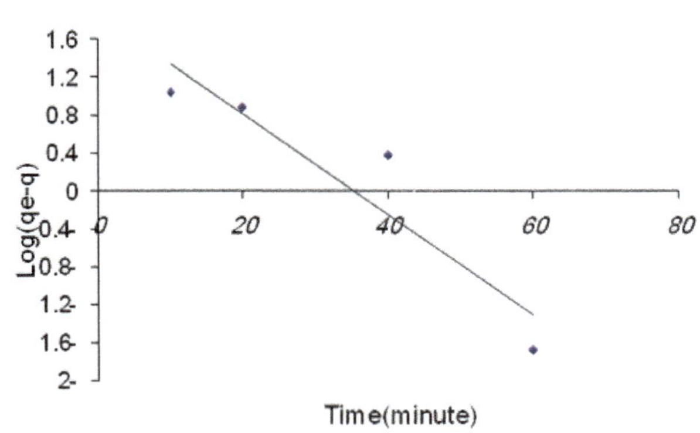

Figure 4: Pseudo-first order reaction for adsorption of U (VI) onto orange peel.

Pseudo-second-order model: The equation represents this model is:

$$\frac{t}{q_t} = \frac{1}{k_2 q_e^2} + \frac{1}{q_e} t \quad (9)$$

Where K_2 is the pseudo second order constant (g/mg min^{-1}). The values of q_e and k_2 were determined from the slope and intercept of Figure 5, respectively. From Figure 4 and 5, and the result data in Table 3, show that the pseudo second order model is the best fitting model because it gives a higher correlation coefficient (r^2= 0.980) than pseudo-first-order (r^2= 0.867).

Mechanism of adsorption

The orange peel can adsorb uranium ions through electrostatic interaction and/or ion exchange, complexation, or by a combination of all the processes Energy-dispersive X-ray spectroscopy (EDX spectra) of the orange peel before and after Adsorption of uranium ions are

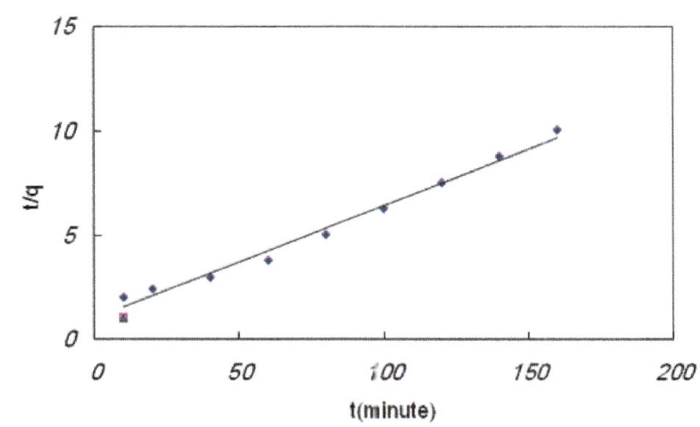

Figure 5: Pseudo-second order reaction for adsorption of U (VI) onto orange peel.

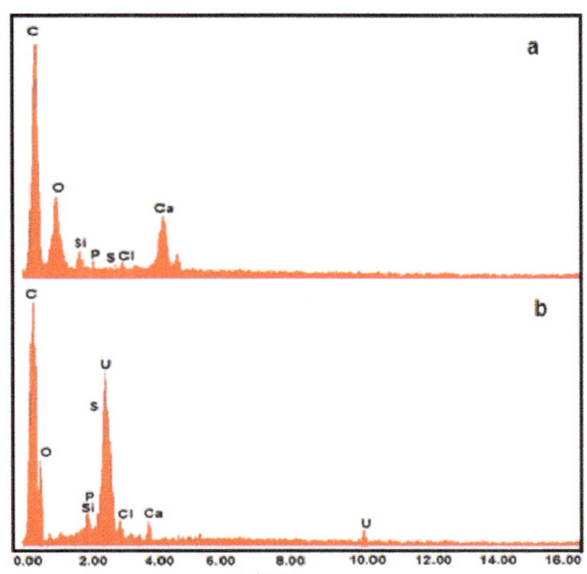

Figure 6: EDX analysis of orange peel (a) before adsorption (b) after adsorption.

Langmuir constant				Freundlich constant		
b	Q_0	R_L	r^2	K_f	n	r^2
1.477	16.12	0.0068	0.999	7.655	5.18	0.682

Table 2: Langmuir and Freundlich constant of adsorption system.

Experimental q_e (mg/g)	Pseudo-first-order kinetics			Pseudo-second-order kinetics		
	K_1 (1/min)	q_e (mg/g)	r^2	K_2(g/ mg.min)	q_e(mg/g)	r^2
15.91	0.119	73.62	0.867	0.0547	18.51	0.980

Table 3: Kinetics data of adsorption of U (VI) onto orange peel.

shown in Figure 6. It is noted that due to uranium ion adsorption by the orange peel, the peaks of Ca weaken with the concomitant appearance of uranium ion peaks, which indicates that there is an ion exchange between calcium and uranium ions, which indicate that ion exchange process is, appears higher than other processes mention above.

Conclusion

The results obtained indicate that orange peels can be used for removal uranium ions from liquid wastes. Performances of adsorption are strongly affected by pH, contact time, initial metal ions and adsorbent dose. The maximum removal of uranium ion was investigated at pH 4 and contact time 60 minutes. Langmuir model was more convenient than the Freundlich model for describing the adsorption conduct of orange peel. The adsorption of U (VI) ions onto orange peels can be described by the pseudo-second order equation.

References

1. Blázquez G, Hernáinz F, Calero M, Ruiz-Núñez LF (2005) Removal of cadmium ions with olive stones: the effect of some parameters. Process Biochem 40: 2649-2654.
2. Konstantinou M, Demetriou A, Pashalidis I (2007) Adsorption of hexavalent uranium on dunite. Global NEST J 9: 229-236.
3. Brown PA, Gill SA, Allen SJ (2000) Metal removal from wastewater using peat. Water Res 34: 3907-3916.
4. Ho YS, Mckay G (2000) The kinetics of sorption of divalent metal ions onto sphagnum moss peat. Water Res 34: 735-742.
5. Jaramillo J, Gomez-Serrano V, Álvareza PM (2009) Enhanced adsorption of metal ions onto functionalized granular activated carbons prepared from cherry stones. J Hazard Mater 161: 670-676.
6. Conrad K, Hansen HCB (2007) Sorption of zinc and lead on coir. Bioresour Technol 98: 89-97.
7. Febrianto J, Kosasih AN, Sunarso J, Ju YH, Indraswati N, et al. (2009) Equilibrium and kinetic studies in adsorption of heavy metals using biosorbent: A summary of recent studies. J Hazard Mater 162: 616-645.
8. Ahluwalia SS, Goyal D (2007) Microbial and plant derived biomass for removal of heavy metals from wastewater. Bioresour Technol 98: 2243-2257.
9. Onishi H (1989) Photometric determination of trace metals. Wiley, NY, USA.
10. Korkish J (1974) Modern method for the separation of rare metal ions. Pergamon Press: Oxford, UK.
11. Yusan SA, Mahmoud AA, Turkozu DA, Aycan HA, Aytas S, et al. (2010) Adsorption and thermodynamic behaviour of U(VI) on the Tendurek volcanic tuff. J Radioanal Nucl Chem 283: 231-238.
12. Man JK, Bo EH, Pil SH (2002) Precipitation and adsorption of uranium(VI) under various aqueous conditions. Environ Eng Res 7: 149-157.
13. Sciban M, Radetic B, Kevresan Z, Klasnja M (2007) Adsorption of heavy metals from electroplating wastewater by wood sawdust. Bioresour Tech 98: 402-409.
14. Yusan S, Doyurum AS (2008) Sorption of uranium (VI) from aqueous solutions by akaganeite. J Hazard Mater 160: 388-395.
15. Ahmad R (2004) Sawdust: Cost effective scavenger for the removal of chromium(III) ions from aqueous solutions. Water Air Soil Pollut 83: 163-183.
16. Saeed B, Amir L, Seyed JA, Mohammad FR, Akram H (2010) Uranium removal from aqueous solutions by wood powder and wheat straw. J Radioanal Nucl Chem 283: 289-296.
17. Bhatnagar A, Jain AK (2005) A comparative adsorption study with different industrial wastes as adsorbents for the removal of cationic dyes from water. J Colloid Interf Sci 28: 49-55.
18. Christian TCV, Fauduet H, Porte C, Delacroix A (2003) Removal of Cd(II) and Pb(II) ions, from aqueous solutions, by adsorption onto sawdust of Pinus sylvestris. J Hazard Mater 105: 121-142.
19. Monoj KM (2010) Removal of Pb(II) from aqueous solution by adsorption using activated tea waste. Korean J Chem Eng 27: 144-151.

Model Identification Using Identification Tool and Estimation of Optimum Control Parameters Using Relay Tuning Method for Bioreactor

Sunil A. Misal[1]*, V. S. Sathe[2], R.W.Gaikwad[1] and Dhirendra[1]

[1]*Department of Chemical Engineering, Pravara Rural Engineering College, Loni-413736, India*
[2]*Department of Chemical Engineering, Dr. BATU, Lonere, Dist-Raigad, India*

Abstract

A mathematical Model for Bioreactor was developed and the simulated experimental data was used in Identification tool of MATLAB. The transfer function was identified using identification tool. The first order plus time delay (FOPTD) transfer function was obtained for bioreactor. The FOPTD model was used in SIMULINK and a relay feedback test was conducted to obtain a symmetric relay response. Then the Ziegler et al. [1] optimum controller parameter method and Padmasree et al. [2] method were used for estimation of optimum control parameters for the bioreactor system.

Keywords: Model identification; Relay tuning; Control parameters; SIMULINK; Bioreactor

Introduction

Control plays a key role in the operation of chemical plants with respect to economical performance, safety and operability. To realize the role of control at a general level, the definition of the concept "chemical process" [3] is "Chemical process is control of (physico-chemical) phenomena for a purpose". This definition clearly shows the necessity of control: without control there would be no functional chemical process industry.

In a typical chemical plant there are hundreds of PID feedback loops. They are often poorly tuned because the choice of PID controller parameters requires professional knowledge by the user. The theory of controller tuning involves sophisticated mathematical manipulations, including complex-valued functions, differential equations, and integral transforms. It is thus not a surprise that the average process engineer repeatedly tunes controllers by trial-and-error methods. Because the PID controller has three tuning parameters, controller tuning by trial and error is a search in the three dimensional space. Evidently, optimal controller parameters are seldom instantly obtained by trial and error. Many modern controllers are equipped with various adaptive techniques such as self-tuning, on-line tuning and autotuning. These features provide easy-to-use controller tuning and have proven to be well accepted among process Engineers.

Identification of transfer function models from experimental data is essential for model based controller design. Often derivation of a rigorous mathematical model is difficult due to the complex nature of chemical processes. Hence, system identification is a valuable tool to identify low order models, based on input–output data. Astrom and Hagglund [4] have suggested the use of relay feedback test to generate sustained oscillations of the controlled variable and to get the ultimate gain (Ku) and ultimate frequency (ω_u) directly from the experiment. The relay feedback method has become very popular because it is time efficient as compared to the conventional method. Here, the Ziegler–Nichols continuous cycling method [1] is considered as a conventional method. The method is a closed loop method and the controller is a proportional one. The proportional gain is increased until sustained oscillations are obtained at the output for a unit step change in the set point. The gain (Ku) and the period of oscillation (Pu) in the output are noted and the controller is designed using Ziegler–Nichols tuning formulae as shown in Table 1. The conventional method requires several experiments. The relay feedback is a single shot experiment and the magnitude of oscillations can be varied. From the principal harmonics approximation, the ultimate gain and ultimate frequency are found. The ultimate gain and ultimate frequency are found from the principal harmonics approximation as given by equations (1) and (2).

$$\omega_u = \frac{2\pi}{p} \qquad (1)$$

$$k_u = \frac{4h}{\pi a} \qquad (2)$$

Luyben [5] has suggested the use of relay testing for identifying a transfer function model. Using the values of Ku and ω_u in the phase angle and amplitude criteria for first order plus time delay (FOPTD) model, we get the following two equations relating the three parameters.

$$\tau_d \omega_u - \tan^{-1}\left(\tau\omega_u\right) = -\pi \qquad (3)$$

$$\frac{K_u K_P}{\left(1+\tau^2\omega_u^{\,2}\right)^{0.5}} = 1 \qquad (4)$$

Since only Ku and ω_u are available, the additional information such as steady state gain or time delay should be known a priori in order to fit a typical transfer function model such as a first order plus time delay model. Many methods are proposed for obtaining the additional information. Li et al. [6] have suggested the use of two relay tests. The first one is a relay test on the system and the second one on the system with a known dynamic element (additional dead time) inserted in

***Corresponding author:** Sunil A. Misal, 1Department of Chemical Engineering, Pravara Rural Engineering College, Loni-413736, India the loop. However, the use of additional relay test is tedious and time consuming. An excellent review of relay feedback methods is given by Yu [7]. Shen et al. [8] have used a biased relay test for getting the model parameters using a single relay test. Friman and Waller [9] have proposed the use of two-channel relay test (one operating on process output and other on the integral of the process output). Scali et al. [10] have extended the technique suggested by Li et al. [6]; which assumed the delay of the process to be known, to estimate models with up to five parameters by means of a maximum of three relay tests with additional of different time delays. Luyben [5], Thyagarajan and Yu [11] have proposed a method that includes the effect of the shape factor in identifying the process model parameters.

Srinivasan and Chidambaram [12] have proposed a modified asymmetrical relay feedback method to get improved estimates of the parameters of the FOPTD model. Using a single asymmetrical relay test, a method is suggested for formulating additional equations to evaluate all the three parameters of the FOPTD model. The asymmetric relay method requires an extra parameter (γ, displacement in the relay height) and whose value is to be selected appropriately so that the calculation of the process gain and the estimate of ω_u are carried out accurately [8]. In the symmetric relay test such problems are not present. Vivek and Chidambaram [13] proposed a method to estimate the three parameters of an FOPTD model using a single symmetrical relay feedback method.

In present work, a mathematical Model for Bioreactor was developed and the simulated experimental data was used in Identification tool of MATLAB. The first order plus time delay (FOPTD) transfer function was obtained for bioreactor. The FOPTD model was used in SIMULINK and a relay feedback test was conducted to obtain a symmetric relay response.

Then the Ziegler-Nichols optimum controller parameter method and Padmasree et al. [2] method were used for estimation of optimum control parameters.

Materials and Method

A Bioreactor system was considered as shown in Figure 1. Biochemical reactors are cylindrical culture vessels used for the fermentation process in which anaerobic breakdown of complex organic materials by the action of anaerobic microorganism or free enzymes takes place. Materials such as carbon, nitrogen, oxygen, which are called substrate, and other nutrients are brought with the cell into the culture vessel (bioreactor) and converted within the cell via hundreds of reactions to the various constituents of the cell as well as to biochemical product. Bioreactors provide a controlled environment that is necessary to bring the better growth of microbes, and also maintain constant temperature according to the need of microbes. The method of identification of an unstable FOPTD system is applied to a nonlinear continuous bioreactor that exhibits output multiplicity. Initially, the system was assumed to be at the unstable steady state condition. The dilution rate (D) is considered as the manipulated variable in order to control the cell mass concentration (X) at the unstable steady state. A delay of 1h was considered in the measurement of X. For the given condition of the unstable operating point, the local linearized model is obtained as an unstable FOPTD with the parameters.

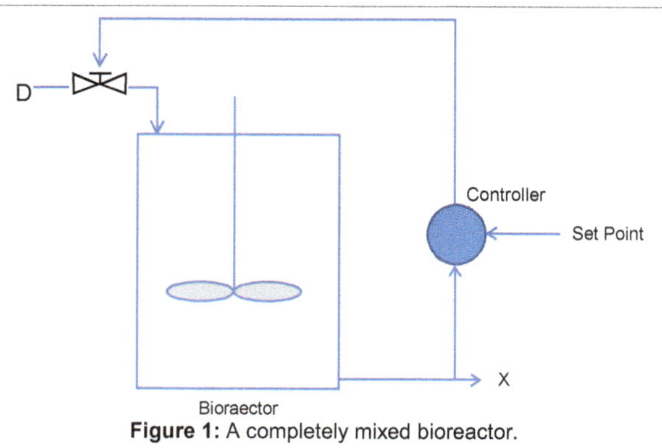

Figure 1: A completely mixed bioreactor.

Modeling of bioreactor

The dynamics of a completely mixed tank reactor were considered as shown in Figure 1. The influent flow rate is equal to the effluent (output) flow rate i.e. Fi=F [volume/time]. Hence, the volume V is constant. The influent has a substrate concentration Sin [mass/volume] and cell concentration X in [mass/volume]. The rate of accumulation of biomass is obtained from a mass balance. Assuming that the biomass has a specific growth rate μ. The total amount of produced biomass per time unit in a reactor with volume V is μVX. Since the reactor is completely mixed, the outflow concentration of biomass is equal to the concentration in the tank. The rate of change of biomass is then given as;

$$V\frac{dX}{dt} = \mu VX + FX_{in} - FX \tag{5}$$

Now defining the dilution rate;

$$D = \frac{F}{V} \tag{6}$$

The model [5] can be written as

$$\frac{dX}{dt} = \mu X + D(X_{in} - X) \tag{7}$$

For the substrate consumption we assume that the yield coefficient is Y. Paralleling, the procedure above for the substrate mass balance gives;

$$V\frac{dS}{dt} = FS_{in} - \frac{\mu}{y}VX - FS$$

Where S is substrate concentration.

Introducing the dilution rate from equation (6);

$$\frac{dS}{dt} = -\frac{\mu}{y}X + D(S_{in} - S) \tag{8}$$

Next it is considered the condition when μ is a Monod function;

$$\mu = \frac{\mu_0 S}{K_M + S + K_I S^2} \tag{9}$$

Where Km is substrate saturation constant and *KI* is substrate inhibition constant.

In the case the influent flow rate is different from the effluent flow rate; the volume variation in the reactor needs to be taken into account;

$$\frac{dV}{dt} = F_i - F \tag{10}$$

A mass balance for the biomass yields;

$$\frac{d}{dt}(VX) = \mu VX - FX \tag{11}$$

By applying chain rule and rearranging the above equation;

$$V\frac{dX}{dt} = \mu VX - F_i X \qquad (12)$$

Introducing dilution rate in equation (12);

$$\frac{dX}{dt} = (\mu - D)X \qquad (13)$$

Equations (8], (9) & (13) are the model equations. These equations were solved to get the transfer function of the system.

Result and Discussion

In the present control study the parameters used for a substrate inhibition model are $\mu o = 0.53\ h^{-1}$, Km = 0.12 g/l, $K_I = 0.4545$ l/g, y = 0.4, Xin = 4.0 g/l, & D = 0.3 h^{-1}

The transfer function between the controlled variable (Biomass concentration, X) and manipulated variables (dilution rate, D) obtained is given by equation;

$$G_p(s) = \frac{\exp(-s)}{0.45s+1} \qquad (14)$$

The model equations and transfer function obtained are used to obtain the simulated experimental data. The experimental data was used in Identification tool of MATLAB. The identification window and process model window are shown in Figure 2. The transfer function was identified by the identification tool and is given as

$$G_p(s) = \frac{1.03\exp(-1.5s)}{0.5s+1} \qquad (15)$$

The step response, frequency response, best fit model window, residual model, Poles & Zeros and noise spectrum are shown in Figure 3 to Figure 8 respectively.

The Simulink implementation of the relay auto tuning procedure is shown in Figure 9; whereas the Relay & process response obtained from the application of this automatic tuning procedure is depicted in Figure 10 and Figure 11.

The ultimate gain (Ku), period of oscillation (Pu) and frequency of oscillation (ω_u) were obtained from the relay response.

Hence, h =05; a =0.44; Pu =2.5; ω_u=2.51,

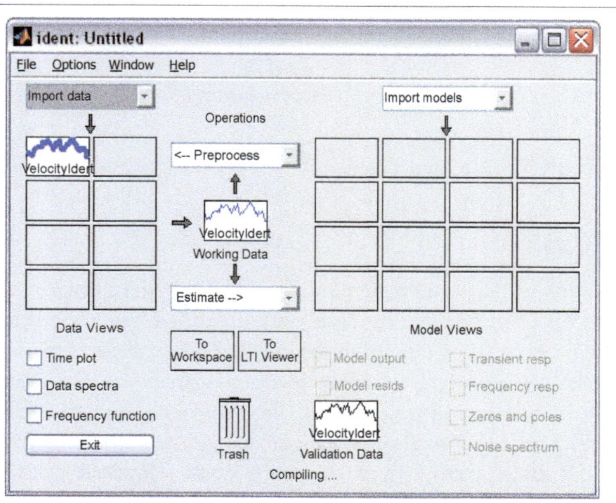

Figure 2: Identification window.

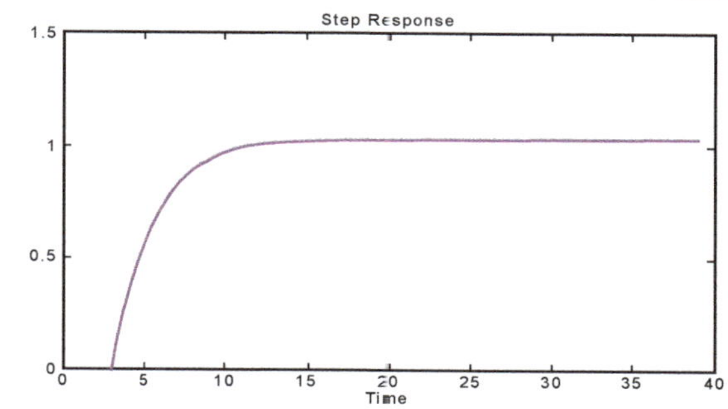

Figure 3: Step Response.

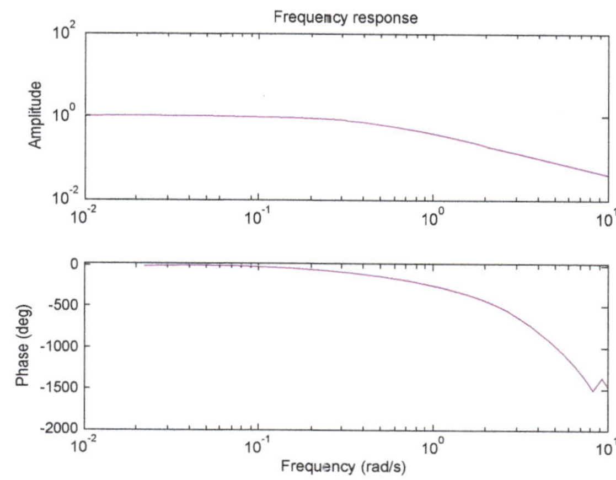

Figure 4: Frequency Response.

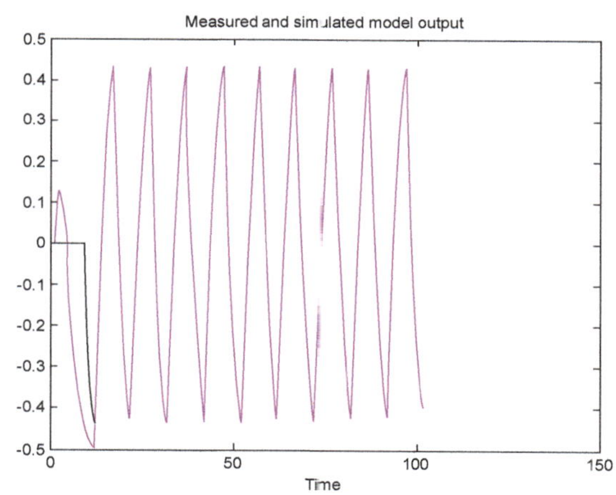

Figure 5: Best fit model.

Ku =1.44

The PID controller tuning parameters were calculated using Ziegler et al. [1] optimum controller parameter method as shown in Table 1.

Parameter

Then Padmasree et al. [2] method was used to calculate the PID parameters for FOPTD model as;

$$k_c k_p = \frac{\tau}{\tau_d} + 0.5\tau_d \qquad (16)$$

$$\tau_I = \tau + 0.5\tau_d \quad (17)$$

$$\tau_D = \frac{0.5\tau_d\left(\tau + 0.1667\tau_d\right)}{\tau + 0.5\tau_d} \quad (18)$$

The Simulink implementation of the closed-loop feedback control system is shown in Figure 12; whereas the closed-loop response obtained using a PID control system is depicted in Figure 13.

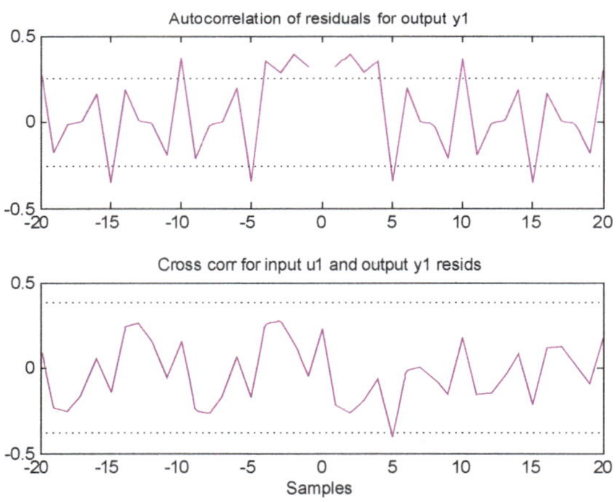

Figure 6: Residual model.

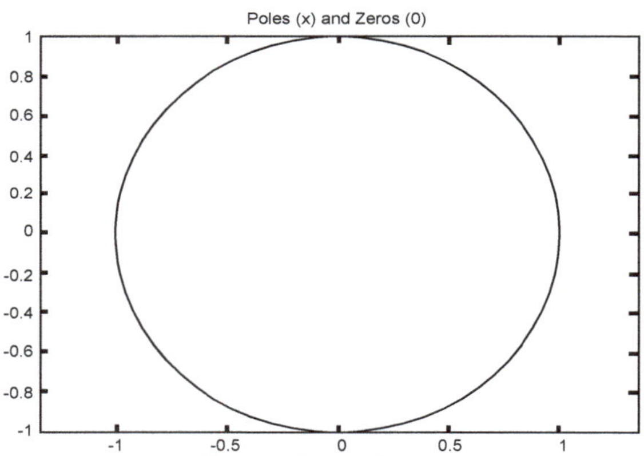

Figure 7: Poles & Zeros.

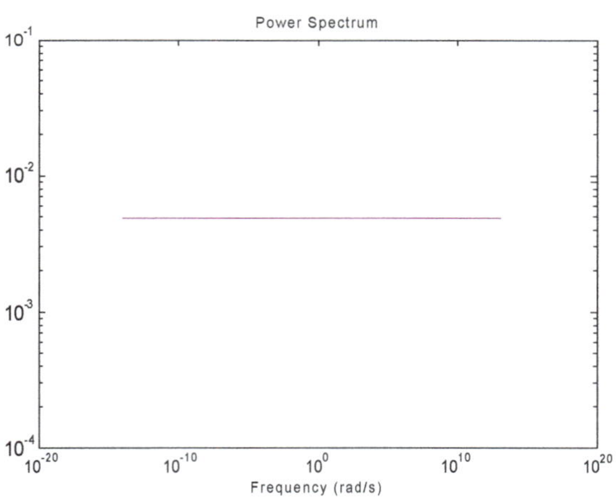

Figure 8: Noise Spectrum.

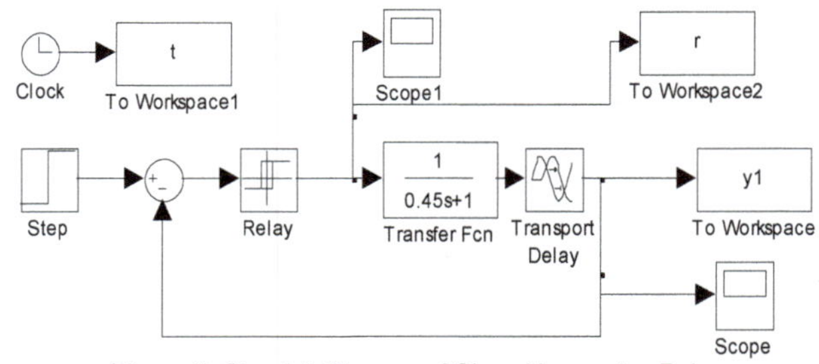

Figure 9: Simulink Diagram of Closed loop using Relay.

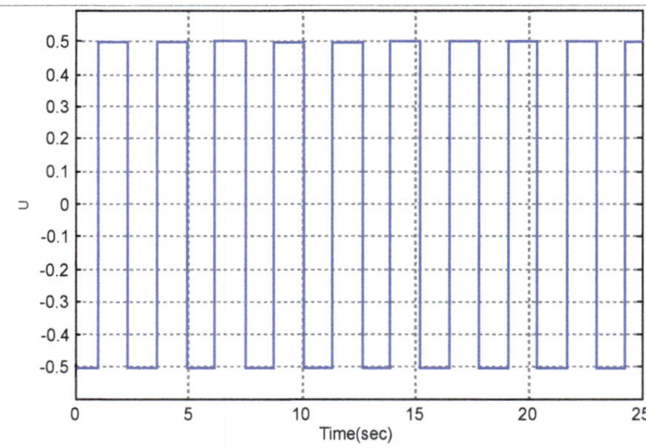

Figure 10: Relay Response.

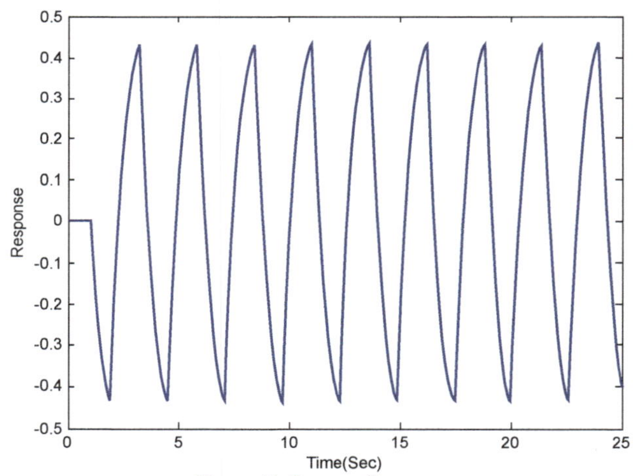

Figure 11: Process Response.

The PID parameters calculated by both the methods are shown in Table 2.

Conclusion

Simulated experimental data was used in Identification tool of MATLAB. The transfer function was identified using identification tool. The first order plus time delay (FOPTD) transfer function was obtained for bioreactor. A relay method is proposed, for estimating the ultimate gain and consequently the control parameters of FOPTD transfer function model of Bioreactor using a single symmetric relay test. The advantages of this method are: single relay feedback test and the PID parameters can be identified [13]. This method of identification is simple and will be of much industrial use. The present method can be

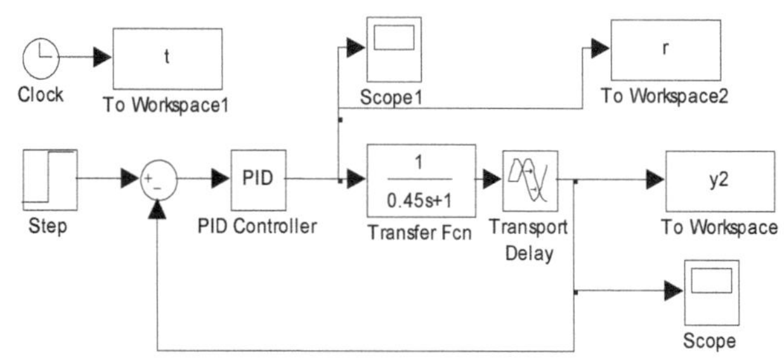

Figure 12: Simulink Diagram of Closed loop using PID.

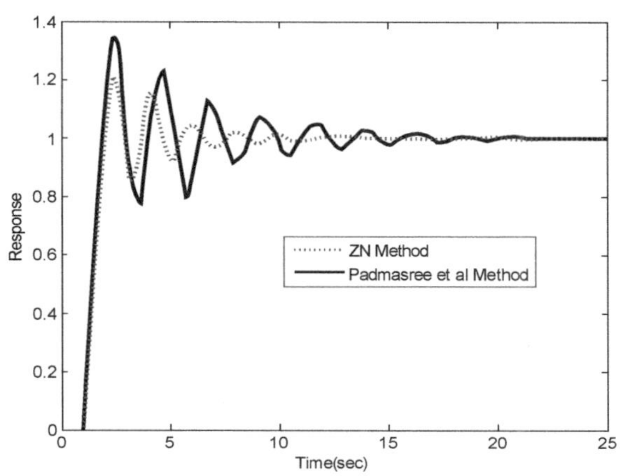

Figure 13: Process Response with tuned parameters.

Controller	Kc	τ_I	τ_D
P	0.5Ku	-	-
P+I	0.45Ku	$\frac{Pu}{1.2}$	-
P+I+D	0.6Ku	$\frac{Pu}{2}$	$\frac{Pu}{8}$

Table 1: Ziegler-Nichols optimum controller.

Controller	Kc	τ_I	τ_D
ZN method	0.36	0.8	0.31
Padmasree et al. method	0.35	0.95	0.38

Table 2: Control Parameters by both methods.

applied to higher order processes.

The control parameters obtained by Ziegler et al. [1] method and Padmasree et al. [2] method and compared and found close values for Bioreactor control.

References

1. Ziegler JG, Nichols NB, Rochester NY (1942) Optimum Settings for Automatic Controllers. Trans ASME 65: 433-444.
2. Padmasree R, Srinivas MN, Chidambaram M (2004) A simple method of tuning PI/PID controllers for unstable FOPTD systems. Comput Chem Eng 28: 2201-2218.
3. Pohjola VJ, Alha MK, Ainassaari J (1994) Methodology of Process Design. Comput Chem Eng 18: S307-S311.
4. Astrom KJ, Hagglund T (1984) Automatic tuning of simple regulators with specifications on phase and amplitude margins. Automatica 20: 645-651.
5. Luyben WL (1987) Derivation of transfer Function model for highly nonlinear distillation columns. Ind Eng Chem Res 26: 2490-2495.
6. Li W, Eskinat E, Luyben WL (1991) An improved autotune identification method. Ind Eng Chem Res 30: 1530-1541.
7. Yu CC (1999) Auto tuning of PID controllers: relay feedback approach. Berlin: Springer-Verlag.
8. Shen SH, Wu JS, Yu CC (1996) Use of biased relay feedback for system identification. AIChE Journal 42: 1174-1180.
9. Friman M, Waller KV [1994] Autotuning of Multiloop Control Systems. Ind Eng Chem Res 33: 1708-1717.
10. Scali C, Marchetti G, Semino D (1999) Relay and Additional Delay for Identification and Autotuning of Completely Unknown Process. Ind Eng Chem Res 38: 1987-1997.
11. Thyagarajan T, Yu CC (2003) Improved Autotuning Using the Shape Factor from Relay Feedback. Ind Eng Chem Res 42: 4425-4440.
12. Srinivasan K, Chidambaram M (2004) An improved Autotune identification Method. Chem Biochem Eng Q 18: 249-256.
13. Vivek S, Chidambaram M (2005) Identification using single symmetrical relay feedback test. Comput Chem Eng 29: 1625-1630.

A Simple Efficient Process for the Synthesis of 16-Dehydropregnenolone Acetate (16-DPA) – A Key Steroid Drug Intermediate from Diosgenin

Pritish Chowdhury*, Juri Moni Borah, Manobjyoti Bordoloi, Pradip K. Goswami, Aradhana Goswami, Nabin C.Barua and Paruchuri G. Rao

Natural Products Chemistry & Chemical Engineering Divisions, CSIR North East Institute of Science & Technology, Jorhat 785 006, India

Abstract

A three step efficient synthesis for the commercial production of 16-dehydropregnenolone acetate (16-DPA)- a potent steroid drug intermediate, from diosgenin, in an overall yield of 60%, has been developed, through generation of acetylonium ion (CH_3C^+=O) in optimum concentration from acetic anhydride using in the molar ratio of 1:3.5(diosgenin:Ac_2O) in a pressure reactor in a medium of a hydrocarbon solvent at a temperature in the range of 200°C with the corresponding in-built pressure of 5-6kg/cm^2 (Step 1) and an ultra sound irradiated (35 KHz) oxidation reaction (Step 2).

Keywords: Diosgenin; 16-dehydropregnenolone acetate (16-DPA); Steroid drug intermediate; Corticosteroids; Soft corticosteroids; Steroid-peptide conjugates; Hormones; Acetylonium ion

Introduction

Steroids play a vital role in human physiology and medicine. Although steroidal drugs have several side effects, these are still regarded as potent life saving drugs. Currently work on the synthesis of the modified steroid molecules to minimize their side effects in our body is gaining considerable importance. Development of such soft corticosteroids having less systemic side effects is therefore becoming important in steroid drug industry [1-3]. It is pertinent to note that, most of these steroid drugs including corticosteroids or soft corticosteroids, anabolic steroids, sex hormones and oral contraceptives etc. are synthesized from 16-dehydropregnenolone acetate (16-DPA) 1, barring a few which are obtained by total synthesis. The plant origin compounds solasodine 2 and its oxygen analogue diosgenin 3 (Figure 1) are the two main ingredients from which 16-DPA is generally synthesized [4-15]. Due to the favorable climatic condition of North East region of India, diosgenin yielding *dioscorea* species of plants, *viz., D. compositae* and *D. floribunda* grow abundantly in the region which led us to investigate commercial exploitation of these plants through large scale production of 16-DPA starting from diosgenin. Here we report a commercially viable three-step synthesis of 16-DPA 1 (Figure 2) starting from diosgenin 3 applying innovative synthetic strategies which involves the use of neither any high boiling solvents nor any toxic and costly catalyst in the crucial first step of acetolysis [16,17]. The hydrocarbon solvent used (commercial xylene: mixture of o-, p- & m-isomers: bp.138.5°C) in the step was also recovered fully for subsequent use.

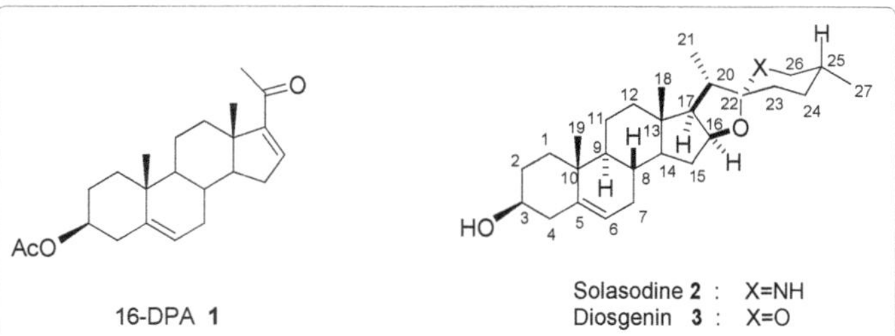

16-DPA **1**

Solasodine **2** : X=NH
Diosgenin **3** : X=O

Figure 1:

Synthesis of 16-dehydropregnenolone acetate (16-DPA) 1 from solasodine 2 and diosgenin 3 has been reported by several workers in various literatures [4-15]. For the conversion of this spirostanic compound, diosgenin (22-iso-5-spirostane-3β-ol) into 16-DPA (3β-acetoxy-pregn-5,16-diene-20-one) 1, it must be isomerized first to furostenol derivative pseudodiosgenin diacetate (3β,26-diacetoxy-furosta-5,20(22)- diene) 4, through acetolysis (Figure 2), which on subsequent oxidation of the 20(22) enolic double bond followed by hydrolytic degradation furnishes 16-DPA [4-15]. Chemical transformations of spirostanic compounds also find importance towards the synthesis of plant growth promoting substances including ecdysteroids and some furostenols having significant phytotoxicity [18]. Marker's degradation is the earliest known process for converting diosgenin into 16-DPA [5-7].

As mentioned earlier, acetolysis or isomerization (Step 1) is the crucial step in the process to get high yield of the final product. The condition of the isomerization reaction of diosgenin 3 as reported by Marker et al. involved the direct use of acetic anhydride at an elevated temperature and pressure in a sealed tube, while the use of high boiling anhydrides as solvent led to lower yield of the isomerized product pseudodiosgenin diacetate 4 [5-7].

Afterwards several modifications have been made including catalytic conversion to effect this isomerization process [9-13]. Gould et al. [9] observed that the conversion could be achieved at the boiling point of acetic anhydride in the presence of acetyl chloride or anhydrous $AlCl_3$. Similar conditions using acetyl chloride also have

***Corresponding author:** Pritish Chowdhury, Natural Products Chemistry & Chemical Engineering Divisions, CSIR North East Institute of Science & Technology, Jorhat 785 006, India

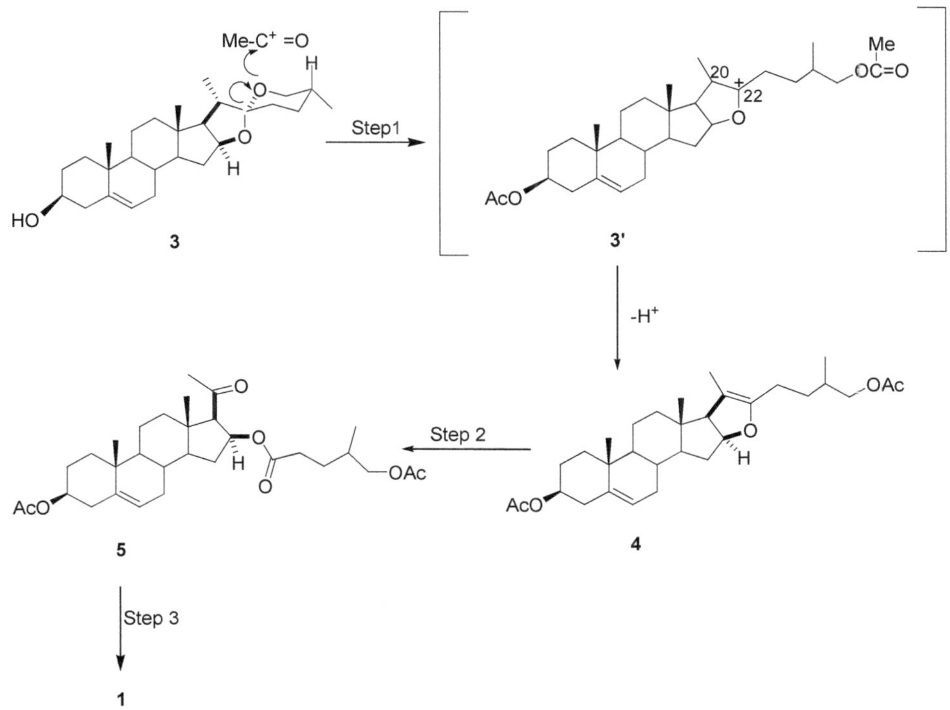

Figure 2: Reagents and Conditions: Step 1: Ac_2O, xylene, 200°C, 5-6Kg/cm^2, 91%; Step 2: CrO_3, CH_3COOH- H_2O, 0-15°C, sonication, 70%; Step 3: CH_3COOH, reflux, 90%.

been used for the transformation of $\Delta^{4,6}$-22-isospirostadiene-3-one to the corresponding 3β-acetoxy-3,5,7- triene [11].

The poor yield of the desired isomerized product pseudosapogenine diacetate (33-40%) is the main disadvantage of all these methods [5-7].To improve the yield of the isomerized product for its eventual conversion to this central steroid drug intermediate,16-DPA and its relatives several other workers have reported the isomerization of spirostanic compounds to furostenols in 70-84% yield by using various catalysts like pyridine and ammonium chloride, pyridinium hydrochloride, anhydrous titanium tetrachloride or aluminum chloride or p-toluenesulphonic acid etc. and also through the application of high boiling solvent like n-octanoic acid in the process [9-13]. Earlier we reported the application of magnesium iodide-diethyl ether and chlorotrimethylsilane as catalysts in this isomerization process giving fairly good yield of the isomerized product 4 [19]. For commercial production of 16-DPA from diosgenin, the use of both high boiling solvents and toxic & costly catalysts are not desirable and hence alternative methods to eliminate all these drawbacks are still sought after in steroid drug industry.

Results and Discussions

We argued that the isomerization of the isospirostane (diosgenin) 3 to furostenol (pseudodiosgenin diacetate) 4 is possibly due to the attack on the nucleophillic pyran oxygen of 3, by the highly reactive acetylonium ion (Me-C$^+$=O) generated *in situ* from acetic anhydride under different reaction conditions [19-23]. We further argued that in the case of Marker's degradation, the low yield of isomerized product 4 might be due to extensive decomposition of the starting material because of the indiscriminate formation of this highly reactive species under reaction condition, viz., heating diosgenin directly with acetic anhydride at 200°C in a sealed tube [5-7]. Considering this fact, we hit the idea of generating this acetylonium ion in an optimum concentration by heating diosgenin in a pressure reactor dissolving it in commercial grade of xylene containing low concentration of acetic anhydride (diosgenin: Ac_2O:: 1mol:3.5mol) at a temperature of 200°C to react selectively on the pyran oxygen of diosgenin 3 with concomitant acetylation of 3β- hydroxyl group [24] [Figure 1]. The opening of the spiroketal ring by acetylonium ion forms a carbonium ion at C-22 [intermediate 3] which undergoes favorable proton elimination from C-20 position to furnish the desired compound 4.The heating was continued for 10 h during which time the in-built pressure inside the reactor reached to a magnitude of around 5-6 kg/cm^2. Under this condition the yield of pseudodiosgenin diacetate 4 from diosgenin 3 was found to be more than 90% in a very clean reaction, a method so far not reported by any worker. In its ^{1}H NMR spectrum, the generation of the C-20(22) double bond was indicated by the presence of a singlet for the methyl on the double bond at 1.7 ppm while a broad singlet at 2.1ppm was appeared for C-3 & C-26 acetate protons. IR spectrum displayed the characteristic absorption band at 1735 cm^{-1} for the acetate functionalities. ^{13}C NMR and microanalyses were in conformity with its molecular structure and formula $C_{31}H_{46}O_5$. It is pertinent to note that the reaction proceeded without the aid of any catalyst. Moreover, the low cost aromatic hydrocarbon solvent i.e. xylene (commercial) used in the process could be recovered fully for reuse.

Oxidation of pseudodiosgenin diacetate 4 to Diosone 5 (Step 2) was performed by using chromium trioxide in a biphasic solvent medium containing acetic acid and water at a temperature in the range of (-) 10 to (+) 10°C and under ultrasound irradiation at a frequency in the range of 30-40 KHz [25]. Application of ultrasound irradiation reduced the amount of the oxidizing agent by about 40-60% than that required under normal oxidation condition. In its ^{1}H NMR spectrum, the compound 5 exhibited the newly generated C-20 methyl ketone signal at 2.2 ppm as a singlet. IR spectrum exhibited a band at 1715 cm^{-1} for this newly generated carbonyl functionality. ^{13}C NMR & microanalyses were in conformity with its molecular structure and formula $C_{31}H_{46}O_7$. Although some literature reveal the application of other oxidant like hydrogen peroxide in presence of catalyst/phase transfer catalysts etc [14]. CrO_3 is still regarded as an oxidant of choice, because of its high efficacy and low cost. The main advantage of carrying out oxidation under ultrasound irradiation is to minimize of the amount of chromium trioxide by about 40-60% as compared to normal oxidation process. Besides, the catalysts used with other oxidants could not be recovered for reuse making the process non-economical. Conversion of Cr (VI) to Cr (III) state by simply alcohol addition is another advantage of this oxidant for easy disposal through proper effluent treatment. Further work on complete replacement of chromium reagent in the oxidation process is in progress.

Hydrolytic degradation of diosone 5 to the final product 16-DPA 1 (Step 3) was easily achieved by refluxing it in acetic acid in almost quantitative yield. In its NMR spectrum the compound displayed C-16 olefinic proton at 6.4 ppm as a singlet and the other C-6 olefinic proton appeared at 5.3 ppm as a multiplet while C-3 acetate and C-20 methyl ketone protons appeared as two singlets at 2.0 and 2.2 ppm respectively. In its IR spectrum the band for α,β-unsaturated carbonyl system at C-20 appeared at 1675cm^{-1}. ^{13}C NMR and microanalyses were in conformity with its molecular structure and formula $C_{23}H_{32}O_3$. Although the same conversion may be effected by heating with a base such as sodium bicarbonate, sodium carbonate, potassium carbonate, potassium bicarbonate or other similar bases in an alcoholic solvent, use of acetic acid was preferred because of the simplicity of the operation and full recovery of acetic acid for subsequent use. In the process, the overall yield of 16-DPA 1 from diosgenin 3 was found to be more than 60% based on the 85-90% purity of commercial diosgenin.

The whole process was optimized by doing five experiments at 50gm per batch level giving consistent results for its eventual scaling-up to 3 kg per batch level at the pilot plant facilities [27,28] of the institute for commercialization of the process. Table 1 represents the mean, variance and standard deviation on the yield of 16-DPA (%) from diosgenin in 50 gm scale from five experiments while Table 2 represents same for the crop yield (in gm) of 16-DPA from diosgenin in 3 kg per batch level in three experiments.

The comparison of the present process with the existing processes is given in Table 3 which reflects the superiority of the present process with the existing ones.

No of Expts.	Yield of 16-DPA (%)	Mean (%)	Variance	Standard Deviation (%)
Exp. 1	61.2			
Exp. 2	58.3			
Exp. 3	64.4	61.2	4.964	2.228
Exp. 4	59.3			
Exp. 5	62.8			

Table 1: Percentage of Mean, Variance & Standard Deviation on the yield of 16-DPA (%) from Diosgenin in 50 gm scale.

No of Sets	Crop yield (in gm)	Mean (gm)	Variant	Standard Deviation (gm)
1`	1st crop: 911			
2	1st crop: 890	900.667	73.556	8.576
3	1st crop: 901			
1`	2nd crop: 342			
2	2nd crop: 403	372.00	620.667	24.913
3	2nd crop: 371			
1`	3rd crop: 131			
2	3rd crop: 138	132.667	14.889	3.859
3	3rd crop: 129			

Table 2: Variant & Deviation in crop yield (in gm) of 16-DPA from diosgenin in 3 kg per batch level.

Materials and Methods

General methods

Melting points were determined with an electro thermal melting point apparatus and are uncorrected. All the chemicals used were of reagent grade of E. Merck and were used without further purification. The progress of each of the reaction was monitored on Merck thin layer chromatography silica gel 60 F_{254} . IR spectra were recorded with a Perkin-Elmer model 2000 series FT-IR spectrometer for solutions in chloroform. Infrared absorbance is reported in reciprocal centimeters (cm^{-1}). ^{1}H and ^{13}C NMR spectra were recorded on a Bruker DPX (300MHz) spectrometer with tetramethylsilane (TMS) as internal standard on ppm scale(δ), Multiplicity of the resonance peaks are indicated as singlet(s), broad singlet(bs), doublet(d), triplet(t), quartet (q) and multiplet (m). Mass spectrometric analysis was performed by positive mode electro spray ionization with Bruker Esquire 3000 LC-MS instrument. Elemental analysis was carried out in Varian CHN analyzer.

Step 1: Acetolysis (isomerization) of diosgenin 3 to pseudodiosgenin diacetate 4

50 g (0.12 mol) of diosgenin **3** was charged with 40 ml of acetic anhydride (0.4 mol) and 150 ml of xylene in a pressure reactor vessel [24] and heating was started while stirring till the temperature of 250°C and corresponding pressure of 4-6 kg/cm^2 were reached. The reaction was carried for a period of 10 hours after attainment of desired temperature of 250°C. Heating was stopped and was allowed to cool under stirring for a period of one hour and the product was discharged through the discharge tube under positive pressure as the temperature came down to below 100°C. TLC analysis of the sample was carried out which showed the formation of only one major product. Solvent removal was done with a rotary vacuum evaporator under reduced pressure. The recovered solvent was kept for recycle. After the removal of the last traces of the solvent, solid material thus obtained was purified by crystallization from acetone to get pseudodiosgenin diacetate **4** in pure form. Yield: 55g as solid (91%).

Sl. No.	Items	Existing Process	Proposed Process	Comments
1	Raw materials	Diosgenin, Acetic anhydride, Catalysts like Pyridine hydrochloride, Titanium tetrachloride and anhydride Aluminium chloride etc.	Diosgenin, Acetic anhydride, commercial grade Xylene	Methods used by Glaxo and other companies are very old and based on the technology of 1952 but the present technology is developed through modern technology. In the present process acetic anhydride which is a reactant has been used not in large excess along with an aromatic hydrocarbon solvent to give very high yield of pseudodiosgenin acetate. The reaction being homogeneous, its easy to operate. Large excess of acetic anhydride lowers the yield of the above product.
	Catalyst i)Cost			Simple but efficient & cost effective process
	ii) Toxicity			In the present process the use of expensive and environmentally hazardous toxic catalysts has been completely eliminated.
2		Expensive: Pyridine hydrochloride : $200/Kg Titanium tetrachloride:$200/10gm Anhydrous Aluminum Chloride All these catalysts are environmentally hazardous, toxic and corrosive	No catalyst has been used	
	iii) Recovery & cost of catalysts	Catalysts used are not recoverable		Doesn't arise
3	Solvent used	High boiling solvent like n-octanoic acid (b.p. 237^0C) has been used by Glaxo	Medium boiling low cost aromatic hydrocarbon (b.p.< 150^0C) has been used	In the present process, no high boiling and toxic solvent like n-octanoic acid (corrosive nature) has been used. Besides the recovery of such solvent would be high energy intensive.
4	Yield of pseudodiosgenin diacetate(2)	Max. 87% (using n-octanoic acid)	92%	Yield under present process is better without using such high boiling solvent
5	Oxidation process	Oxidant in acetic acid	Oxidant in solvent	The oxidation has been done in a non toxic solvent which is recoverable
6.	Yield of diosone(3)	55%	70-75%	The better yield
7.	Yield of 16-DPA(over all)	42-56%	55-62%	Better process
8	Effluent generation	Step 1: Toxic effluent Step 2: Chromium salts	Step 1: nil Step 2: Chromium salts	In the present process the use of ultra-sound in the oxidation step 2 reduces the amount of chromium reagent by about 30-40% which would minimize amount of chromium salts in the effluent.

Table 3: Comparison of the present process with the existing processes.

Mp: 98^0 C (lit.[19] mp. 97-98^0);

IR ($CHCl_3$): ν =1735(broad), 1375, 1250, 1025 cm^{-1}

^{1}H NMR(300MHz): δ=0.81(s, 3H, H-18), 0.97(d,3H,J=6.3Hz, H-27), 1.2(s,3H,H-19), 1.7(s,3H,H-21), 2.1(bs, 6H, acetate protons), 3.9(d,2H, J=4.1Hz, H-26), 4.6 (m,2H, H-3 & H-16), 5.4(m,1H, H-6) ppm;

^{13}C NMR : δ= 34.2; 35.1; 78.6(C-3); 39.5; 155.0(C-5); 142.3(C-6); 32.0; 29.0; 46.2; 35.8; 20.9; 39.5; 40.5; 55.4; 34.7; 80.6(C-16); 62.0(C-17); 16.5; 15.2; 41.7; 14.5; 109.3(C-22); 29.4; 28.2; 30.9; 66.9(C-26); 17.1; 21.4(C-CH_3COO); 178.2(C-OCOCH_3) ppm; Mass spectrum (m/z) : 498 (M^+),

Anal. Calcd. for $C_{31}H_{46}O_5$: C, 74.66; H, 9.3; Found: C, 74.34; H, 9.27;

Step 2: Oxidation of pseudodiosgenin diacetate 4 to Diosone 5:

Oxidation under normal condition

Preparation of the oxidant solution:

25 g (0.25 mol) of chromium trioxide (CrO_3) was dissolved in 25 ml of water and 10 ml of glacial acetic acid to get a clean solution which was precooled to 0-5°C.

Addition of oxidant solution:

50 g of Compound 4 (0.1 mol) was dissolved in 100 ml of dichloromethane and 100 ml of glacial acetic acid and 25 ml of water. The mixture was cooled to 0-5°C and the oxidant solution prepared as above was added to it drop wise keeping the temperature of the reaction mixture below 5-7°C till the addition was over. After the addition of the oxidant solution was complete, cooling was discontinued and the temperature of the reaction medium was allowed to rise up to 15°C and stirred at that temperature for a period of another 25 minutes. When the TLC indicated the completion of the reaction, a solution 5 g of sodium chloride in water (200 ml) and methanol (10 ml) were added and the stirring was continued for further 20 minutes.

Keto ester diosone 5 thus formed was extracted from the reaction mixture with 1,2-dichloroethane. The organic layer after separation was subjected to distillation under reduced pressure to recover the solvent. A gummy residue was obtained which was purified by column chromatography using petroleum ether and ethyl acetate as eluent to get pure diosone **5**. Yield: 37.5 g as gum (70%).

IR ($CHCl_3$): ν = 1735(broad), 1715, 1375, 1250, 1100 cm^{-1}

^1H NMR (300 MHz) : δ = 0.81(s,3H, H-18), 0.97(d, 3H,J=6.3Hz, H-27), 1.2(s, 3H,H-19), 2.0(bs, 6H, acetate protons), 2.2 (s, 3H, H-21), 3.9(d,2H, J=4.1Hz, H-26), 4.6 (m,2H, H-3 & H-16), 5.4(m,1H, H-6) ppm;

^{13}C NMR : δ = 34.2; 35.1; 78.6(C-3); 39.5; 155.0(C-5); 142.3(C-6); 32.0; 29.0;46.2; 35.4; 20.9; 39.5; 40.5; 55.4; 34.7; 82.6(C-16); 62.0(C-17); 16.5; 15.2; 217(C-20); 19.2; 173.5(C-22); 29.6; 28.7; 30.3; 71.3(C-26); 177.0(C-$OCOCH_3$); 21.4(C-CH_3COO) ppm.

Mass spectrum (m/z): 530 (M^+).

Anal. Calcd. for $C_{31}H_{46}O_7$: C, 72.06; H, 9.36; Found: C, 72.10; H,9.26;

Oxidation under ultra sound irradiation

Preparation of the oxidant solution

12.5 g (0.125 mol) of chromium trioxide (CrO_3) was dissolved in 25 ml of water and 10 ml of glacial acetic acid to get a clean solution which was precooled to 0-5°C.

Addition of oxidant solution:

50 g (0.1 mol) of compound 4 was dissolved in 100 ml of dichloroethane and 100 ml of glacial acetic acid and 25 ml of water. The mixture was cooled to 0-5°C and the oxidant solution prepared as above was added to it drop wise keeping the temperature of the reaction mixture below 5-7°C till the addition was over. The reaction mixture was sonicated at 35 KHz from the time of addition of chromium trioxide solution. After the addition of the oxidant solution was complete, cooling was discontinued and the temperature of the reaction medium was allowed to rise up to 15°C and stirred at that temperature for a period of another 25 minutes still under sonication. The rest of the operation was same as described in the case of oxidation under normal condition as above.

IR, NMR and Mass spectra of the product were superimposable with those of the compound **5** isolated under normal oxidation condition.

Yield: 38 g as gum (70%).

1.2.1.7. Step 3: Hydrolytic Degradation of diosone 5 to 16-DPA 1:

50 g (0.94 mol) of diosone 5 was allowed to reflux directly in 200 ml of glacial acetic acid for a period of 2 hr. The reaction was monitored on TLC. After completion of the reaction acetic acid was recovered by distilling under reduced pressure with a rotary evaporator. The residue was thoroughly washed with cold water and was subjected to exhaustive extraction with petroleum ether (bp. 60-80°C) when a yellow solution of 16-DPA 1 was obtained leaving a black residue. The yellow solution after distillation gave a crude yellow product which on recrystalisation from ethanol (up to 3 crops can be collected) furnished creamy white crystals of 16-DPA 1. The structure of 16-DPA 1 was confirmed by mixed melting point determination and by comparison of ^1H & ^{13}C NMR spectral data (recorded for both authentic & present 16-DPA) with those of the authentic material supplied by M/S Glaxo-India Ltd., Mumbai.

Yield: 30g as solid (90%).

Mp: 172°C (lit.[3-6, 9] mp 169-175°);

$[\alpha]_D$: (-) 39.4° (c=1.0, $CHCl_3$) [lit [6,9] $[\alpha]_D$: -37.7to -41.2°(c=1% in $CHCl_3$)];

IR ($CHCl_3$): ν = 1735, 1675, 1375, 1250, 1140, 950 cm^{-1}

^1H NMR (300MHz): δ= 1.0 (s, 3H, H-18), 1.2 (s,3H, H-19), 2.0 (s,3H, acetate protons), 2.2 (s, 3H, H-21), 4.6 (m, 1H, H-3), 5.4 (m, 1H, H-6), 6.4 (m, 1H, H-16) ppm;

^{13}C NMR : δ = 34.2; 35.1; 78.6(C-3); 39.5; 155.0(C-5); 142.3(C-6); 32.0; 29.0; 46.2; 35.4; 20.9; 39.5; 40.5; 55.4; 34.7; 122.9(C-16); 144.5(C-17); 16.7; 15.2; 178.9(C-$OCOCH_3$); 170.5(C-20); 21.3 ppm.

Mass spectrum (m/z): 356(M^+). Anal.Calcd.for $C_{23}H_{32}O_3$: C, 77.49; H, 9.05;

Found: C, 77.41; H, 9.03;

Scaling-up experiments with 3 kg batch level:

Step 1: Acetolysis (isomerization) of diosgenin 3 to pseudodiosgenin diacetate 4:

3 kg of diosgenin 3 (88% pure by GLC) was charged with 2.4 L of Acetic anhydride and 9.0 L of xylene in the Reactor `A' [27]. and heating was started through circulation of hot oil while stirring till the temperature of 250°C and corresponding pressure of 4-6 kg/cm^2 were reached. The reaction was carried for a period of 10 hours after attainment of desired temperature of 250°C. Heating was stopped and was allowed to cool under stirring for a period of one hour and the product was discharged through the discharge tube under positive pressure as the temperature came down to below 100°C. TLC analysis of the sample was carried out which showed the formation of only one major product. Solvent removal was done with a 10 L capacity Buchi Rotavapor under reduced pressure. The recovered solvent was kept for recycle. After the removal of the last traces of

the solvent, solid material thus obtained was directly subjected to the next step of oxidation (Step 2). A small amount of material was purified by preparative TLC (Pet.ether: EtOAc:: 1:12) to get pseudodiosgenin diacetate 4 in pure form. The product was directly compared with that obtained by laboratory scale experiment (TLC, IR, NMR & Mass spectra are superimposable).

Step 2: Oxidation of pseudodiosgenin diacetate 4 to Diosone 5:

Oxidation under normal condition

Preparation of the oxidant solution:

1.5 Kg of chromium trioxide (CrO_3) was dissolved in 1.5 L of water and 600ml of glacial acetic acid to get a clean solution which was precooled to 0-5°C.

Addition of oxidant solution:

Entire amount of pseudodiosgenin diacetate 4 obtained in Step 1 was transferred to the Reactor 'B' along with 6.5L of dichloroethane, 6.5L of glacial acetic acid and 1.5 L of water [28]. The mixture was cooled to 0-5°C by circulating ice cold water through outer jacket of the reactor. The oxidant solution prepared as above was added to it drop wise keeping the temperature of the reaction mixture below 5-7°C till the addition was over. After the addition of the oxidant solution was complete, cooling was discontinued and the temperature of the reaction medium was allowed to rise up to 15°C and stirred at that temperature for a period of another 30 minutes. A solution 300 g of sodium chloride in water (10 L) and methanol (500 ml) were added and the stirring was continued for further 30 minutes.

Keto ester diosone 5 thus formed was extracted with 1,2-dichloroethane (10L) by transferring the material to a separating unit (a 7 ft long glass column having a diameter measuring 10 inches). The organic layer after separation was subjected to distillation under reduced pressure using a rotavapor as described earlier to recover the solvent. A gummy residue thus obtained was directly subjected to the next step of hydrolytic degradation (Step 3). A small amount of the product was purified by preparative TLC (Pet.ether:EtOAc::1:4) to get the keto ester 5 in pure form. The product was directly compared with that obtained in laboratory scale experiment (TLC, IR, NMR & Mass spectra were superimposable).

1.3.3. Step 3: Hydrolytic Degradation of diosone 5 to 16-DPA 1

Crude diosone 5 obtained in Step 2 was dissolved in 10 L of commercial grade acetic acid and was transferred to the Reactor 'A' [27]. The reaction mixture was heated to reflux by circulating hot oil through the outer jacket of the reactor for a period of 3 hrs. The reaction mixture was cooled by stopping heating and acetic acid was recovered through distillation under reduced pressure using a rotavapor for reuse. The residue was thoroughly washed with cold water and was subjected to exhaustive extraction with 7L of petroleum ether (bp.60-80°C) when a yellow solution containing 16-DPA 1 was obtained leaving a black residue. The yellow solution after distillation under reduced pressure furnished a crude yellow product which on recrystallization from ethanol furnished creamy white crystals of 16-DPA 1. From mother liquor another 2 crops of pure 16-DPA could be collected.

First crop: 911 gms (98% purity by GLC).

Second crop: 341 gms (95% purity by GLC).

Third crop: 130 gms (92% purity by GLC).

Overall yield > 60% (based on 85-90% purity of commercial diosgenin).

The product obtained was directly compared with 16-DPA obtained by laboratory scale experiment (TLC, IR, NMR & Mass spectra were superimposable). Mixed melting point remained undepressed.

Conclusion

The present work provides a simple efficient way to synthesize 16-DPA from diosgenin. 16-DPA is still regarded as a potent intermediate for the synthesis of various life saving steroidal drugs and hormones including steroid- peptide conjugates having unique biological properties [26]. Other diosgenin related sapogenins, *viz.*, sarsasapogenin etc. could also be converted to the relatives of 16-DPA using same route as reported here. Further Step 2 and Step 3 (Figure 2) could be performed in the same vessel after completely decomposing the excess oxidant with alcohol. Economic viability is also another advantage of the process to have a high potential for its commercial production.

References

1. Bodor N, Buchwald P (2003) Retrometabolism-based drug design and targeting. In: Abraham, D.J. ed. Drug Discovery and Drug Development: Burger's Medicinal Chemistry and drug Discovery. John Wiley & Sons, NY 2:533-608.
2. Bodor N, Buchwald P (2000) Soft drug design: general principles and recent applications. Med Res Rev 20:58-101.
3. Pavesio CE, DeCory HH (2008) Treatment of ocular inflammatory conditions with loteprednol etabonate. Br. J. Ophthalmol 92:455-459.
4. Asolkar LV, Chadha YR (1979) In: Diosgenin and other steroid drug precursors. Publications & Information Directorate, CSIR, New Delhi.
5. Marker RE, Wagner RB, Ulshafer PR, Wittbecker EL, Goldsmith DPJ, et al. (1947) Steroidal sapogenins. J Am Chem Soc 69: 2167-2230.
6. Marker RE, Rohrman E (1939) Sterols LXXXI Conversion of sarsasapogenin to pregnanediol-3(α),20(α). J Am Chem Soc 61:3592- 3593.
7. Marker RE, Rohrman E (1940) Sterols LXXXVIII Pregnanediols from sarsasapogenin.J Am Chem Soc 62:518-520.
8. Bandhoria P, Gupta VK, Gupta DK, Jain SM, Varghese B (2006) Crystal structure of 3β-acetoxy-pregna-5,16-diene-20-one (16-DPA). J Chem Cryst 36: 161-166.
9. Gould DH, Staeudle H, Hershberg EB Catalytic Isomerisation of spirostanes to furostenols. J Am Chem Soc 74: 3685-3688.
10. Dauben WG, Fonken GJ (1954) Isomerization of Isospirostanes to Furostenols with pyridinium hydrochloride as the catalyst. J Am Chem Soc 76: 4618-4619.
11. Cameroon AFB, Evans RM, Hamlet JC, Hunt JS, Jones PG, et al. (1955) Studies in the synthesis of cortisone Part XII Improvements in the conversion of sapogenins into pregnan-20-ones.J Chem Soc 77:2807- 2816.
12. Dauben WG, Eastham JF, Micheli RA, Takemura KH, Mandell L, et al. (1953) The preparation of $\Delta^{5,7}$ -steroidal dienes. J Am Chem Soc 75: 3255- 3258.
13. Micovic IV, Ivanovic MD, Piatak DM (1990) Simplified preparation of 16-dehydropregnenolone acetate Synthesis.591-592.
14. Goswami A, Kotoky R, Rastogi RC, Ghosh AC (2003) A one-pot efficient process for 16-dehydropregnenolone acetate. Org Proc Res Develop 7:306-308.
15. Manosroi A, Manosroi J, Buddhasukh D, Sripalakit P,Maier R, et al. (2006) Synthesis of cyproterone acetate.US Pat No 11/438193:816-819.
16. Chowdhury PK, Bordoloi M, Barua NC,Sarmah HP, Goswami PK, et al. (1998) Process for the production of 16-dehydropregenolone acetate from diosgenin US Pat No. 5,808,117.

17. Chowdhury PK, Bordoloi M, Barua NC, Sarmah HP, Goswami PK, et al. Process for the production of 16-dehydropregnenolone acetate (16-DPA) from Diosgenin Ind Pat No. 183826.

18. Anielka RA, Margarita RA, and Iglesias-Arteaga MA (2008) An unexpected and useful E-ring oxidative cleavage in furostanes. ARKIVOC 274-281.

19. Borah P, Chowdhury PK (1998) Catalytic Isomerization of isospirostanes to furostenols. Ind J Chem *37B* 408-410.

20. Chowdhury P, Sharma RP, Baruah JN(1983) A new method for enol acetylation. Tetrahedron Lett *24*: 3383-3384.

21. Chowdhury PK (1990) Magnesium-Iodine-Diethyl Ether: An efficient system for deoxygenation of oxiranes to olefins. J Chem Res 192-193.

22. Chowdhury PK (1992) Magnesium Iodide-Diethyl Ether- Acetic Anhydride: A convenient system for deprotection of methyl thiomethyl ethers. J Chem Res 68.

23. Chowdhury PK (1993) Magnesium Iodide-Diethyl Ether- Acetic Anhydride: A new and efficient acylating system for primary, secondary, tertiary alcohols & phenols. J Chem Res 338.

24. Pressure reactor: Make: Berghof (with attached pressure gauge & temperature meter), Eningen, Germany: Art No. 570014, Serial No.86, 265, 12, Input: 220V, 50Hz, 1800 W

25. Ultrasonic Bath: Model: Powersonic 405 Bath Size (mm) : 300x155x150D Power output: 350W

26. Rivera DG, Pando O, Coll F (2006) Synthesis of peptidomimetic-spirostane hybrids Ugi reaction: a versatile approach for the formation of peptide–steroid conjugates Tetrahedron. 62: 8327-8334.

27. Reactor `A' for Acetolysis (Step 1): 25 L capacity SS Reactor Vessel with an outer Jacket for hot oil circulation and Stirrer. Mak.

28. Reactor `B' for Oxidation (Step 2):50 L capacity Glass Lined Reactor Vessel with an outer Jacket for cold water circulation and Stirrer: Make : Gujrat Machinery Manufacturing Ltd. India.

Experimental Study of Volumetric Mass Transfer Coefficients in Slurry Bubble Column Reactor

Saba Adnan Ghani[1]*, Rafei J. Yaqub[2] and Suhaeb S. Saleh[1]

[1]*Department of Chemical Engineering, University of Tikrit, Iraq*
[2]*Department of Petroleum Engineering, University of Tikrit, Iraq*

Abstract

Scaling up of Fisher-Tropsh requires suitably accurate estimations of hydrodynamic and mass transfer parameters as a function of column size and slurry concentration. The present study focuses on gas holdup, volumetric mass transfer and local mass transfer coefficients in a slurry bubble column employing a C_9-C_{11} Iraqi paraffin oil as a liquid phase, alumina and silica particles as a solid phase and oxygen was used as a gas phase because the properties of the paraffin oil are similar to those of Fisher-Tropsch wax under actual operating conditions in the industrial slurry bubble column and alumina and silica serve as carrier for the active cobalt catalyst utilized in the Fischer-Tropsch process. The experimental work was carried out using two rectangular (0.1 m length, 0.02 m width and 0.95 m height) and (0.1 m length, 0.02 m width and 2.6 m height) slurry bubble columns. The superficial gas velocity was varied in the range of (1-6 cm/s) for short column and (7-12 cm/s) for long column. For all experiments the height of liquid phase was maintained at (75 cm) for short column and (160 cm) for long column from the gas distributor and solid loading varied in 0, 5, 10, 15% vol. The experimental results showed that the gas holdup increases linearly with superficial gas velocity at both homogeneous and heterogeneous regimes but the rate of increasing is slower at the heterogeneous flow and also the volumetric mass transfer coefficient increases with superficial gas velocity for both regimes. And the results showed that the gas holdup and volumetric mass transfer coefficient decreased with increasing solid loading.

Keywords: Slurry bubble column; Hydrodynamics; Mass transfer coefficient

Introduction

The use of Bubble columns (BCs) is widely spread through a number of industries and will become especially important in the future in large scale conversion of natural gas and synthesis gas to fuels and chemicals. It is the future use of BCs that will severely test the current know-how applied to their design and operation. Hence, a more fundamental understanding of the fluid dynamics and transport processes in BC is highly desirable in order to reduce the uncertainties in scale-up and optimize the design and performance of these reactors. As one of the key reactor design parameters, the volumetric gas-liquid mass transfer coefficient, K_La, is directly affected by the hydrodynamics, phase mixing, and physical properties. In the ongoing effort to reduce uncertainty, there has been a large body of mass transfer research covering various specific topics of Luo et al. [1], Fan et al. [2], Behkish [3] and Vandu and Krishna [4]. Al-Zaharani and Daous [5] measured the volumetric mass and heat transfer coefficient inside a bubble column employing a single gas nozzle was measured and the effect of adding the solid phase was investigated. They showed that the use of a single nozzle helps to obtain fine and almost mono sized bubbles, and they investigate the effect of gas superficial velocity, solid concentration and static liquid height on mass and heat transfer coefficient in a bubble column using air-water and a cation resins as the solid phase. It was found that both volumetric mass and heat transfer coefficient increased with superficial gas velocity. An addition of 5% wt of suspended solids into the bubble column enhanced the volumetric mass transfer coefficient was observed when the superficial gas velocity greater than 0.056 m/s did not contribute to any further change in the values of this coefficient.

Vandu et al. [6] reported the results of an experimental study of the gas holdup, ε_g, and volumetric mass transfer coefficient, K_La, in a bubble column slurry reactor of 0.1m diameter operated at ambient temperature and pressure conditions. The superficial gas velocity U was varied in the range (0-0.4 m/s), Air was used as the gas phase. The liquid phase used was C_9-C_{11} paraffin oil of porous catalyst (alumina catalyst support). With increasing slurry concentrations, ε_G is significantly reduced due to enhanced bubble coalescence.

The objective of this study is to investigate the hydrodynamics and mass transfer parameters in a rectangular slurry bubble column reactor simulating the Fischer-Tropsch synthesis under actual conditions using Iraqi paraffin oil via hydrodynamics effect and their influence on superficial gas velocity and volumetric mass transfer coefficient.

Experimental work

The experiments were carried out using two rectangular slurry bubble columns, which were made of Perspex to enable visual observation and photographic recording of the bubbles. One of them has dimensions of 0.1 m length, 0.02 m width and 0.95 m height and, the other of (0.1 m length, 0.02 m width and 2.6 m height) (Figure 1). The base plate of the column has one off-center orifice for liquid or slurry outflow and a central orifice for gas inflow through a distributor, which was supplied by oxygen cylindrical through check valve and a humidifier. The two columns had two pressure tapes located along the column, for short column distance between two tapes is 45cm, and for long column is 140 cm. The two pressure taps were connected to pressure difference transmitter by a plastic tube of 1cm inside diameter

***Corresponding author:** Saba Adnan Ghani, Department of Chemical Engineering, University of Tikrit, Iraq placed at the bottom of the column. Normal Iraqi paraffin oil (C_{10}-C_{13}) mixture cut supplied by Arab Company for Detergent Chemicals is used as the liquid phase. Alumina and silica powders manufactured by "GMBH", Germany were used as solid phases. The properties of the catalyst carrier are similar to those envisaged for used in the commercial Fischer Tropsch reactor. Oxygen gas is used as a gas phase; with a purity of 97%. Nitrogen gas with a purity of 99% is employed to flash out the dissolved oxygen. Oxygen dissolved meter (type; LTT-Lutron, DO-5512 SD, Taiwan), and a high speed video camera (Sony, model No: HDR-XR550E, wide-angle is 26.3mm, Japan) were connected online. Pressure difference method was used to determine the pressure drop between two taps on the column by using a pressure transducer type (ST- 3000 Transmitter- "YAMATAKE", Japan). The slurry concentration, ε_s was varied in the range 0, 5, and 10% vol. All experiments are conducted at ambient pressure and temperature conditions, and in a batch mood operation with respect to the liquid phase. The sensor constant value corresponds to an inherent delay in readings obtained as a result of the fact that the oxygen sensor has a finite response time and it is obtained as previously described Han and Al-Dahhan [7]. Sufficient time was given in each experimental run for the oxygen saturation concentration in the liquid, C^*_L to be reached. When saturated concentration is achieved, the volumetric mass transfer coefficient k_La can be calculated using eq. below:

$$K_La = K_La_L(1-\varepsilon_g)(1-f_s\varepsilon_s) \quad (1)$$

where: K_La_L = the volumetric mass transfer coefficient per unit volume of liquid, ε_g = gas holdup, f_s = the volume fraction of the solid particles. ε_s = solid holdup. Windows Media Player program was used to examine the flow pattern inside the column.

Results and Discussion

Effect of superficial gas velocity and solid loading on K_La

The effect of superficial gas velocity on the volumetric mass transfer coefficient, K_La at different solid concentration is shown in Figure 2 and 3 and Table 1 for short and long columns and two types of solid loading. These figures indicate that the volumetric mass transfer coefficient, K_La increases rapidly from 0.01 to 0.12 sec^{-1} at

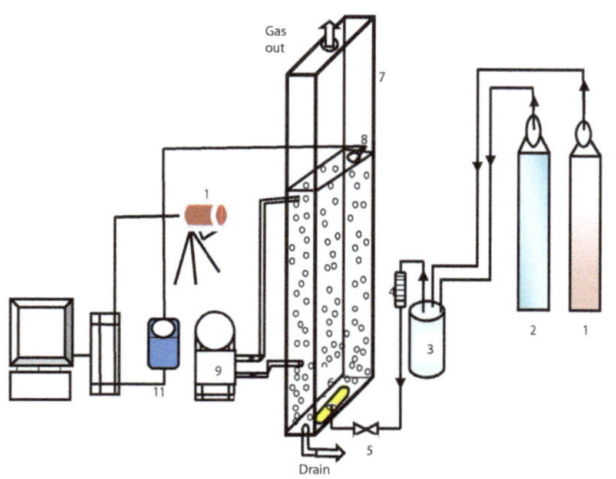

Item No.	Description	Item No.	Description
1	O_2 cylinder	7	Slurry bubble column reactor
2	N_2 cylinder	8	Oxygen dissolve sensor
3	Humidifier	9	pressure difference transmitter
4	Gas flow meter	10	High speed video camera
5	Check valve	11	Oxygen Dissolved Metter(ODM)
6	Distributor	12	Computer

Figure 1: Schematic diagram of the bubble column apparatus used in the experimental work.

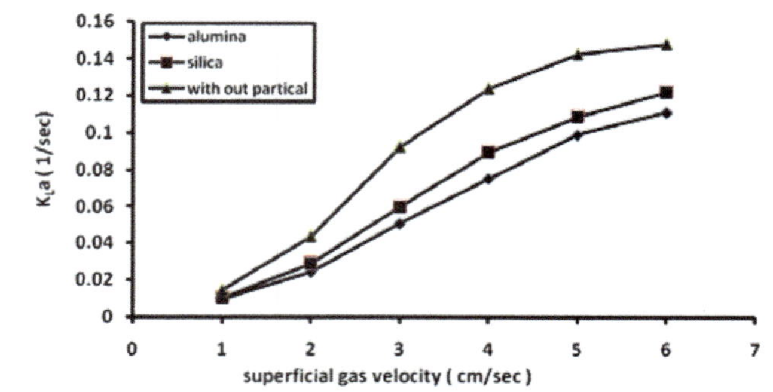

Figure 2: Effect of superficial gas velocity on mass transfer coefficient for long column at (5% vol.) slurry concentration.

zero loading with increase of superficial gas velocity (1-4) cm/sec, a less increase was realized in the case of using silica particles in the same range of velocity increasing (0.028-0.088 sec^{-1}), and alumina exhibited the lowest value increase with velocity graduation (0.01-0.074 sec^{-1}) but the increasing trend was reduced at high superficial gas velocities where it reached just 0.15, 0.12 and 0.107 (all in sec^{-1}) for no loading, silica and alumina respectively at 6 cm/sec. The reason for this increasing is mainly attributed to the increase in the interfacial area. As the superficial gas velocity increases, the mean bubble diameter decreases. However, the bubble size decreases up to a certain point and then remains constant even if the superficial gas velocity is increased further. Thus, the increase in the interfacial area can be explained mainly as a result of the higher superficial gas velocity. An increase in the superficial gas velocity also enhances the turbulence induced by the gas flow, which increases K_L. The increase in the interfacial area and K_L at high superficial gas velocities predominates over the decrease in the mass-transfer coefficient because of shorter gas-liquid contacting time and results in an increase in K_La. These results are in agreement with results of Al-Zahrani and Daous [5], Behkish [3], Abdul-Rahman and Alazzi [8] and Asgharpour et al. [9]. The volumetric mass transfer coefficient, K_La decreases also in the range of (0.108- 0.0983) sec^{-1} with increasing the alumina concentration from 5-10%, this behavior can be attributed to the bubble size distribution in the column under these conditions. Although the volume fraction of gas bubbles is reduced at lower solid concentrations, this reduced volume mainly consists of a large population of small gas bubbles, contributing enormously to the gas-liquid interfacial area. When increasing solid concentration above certain value, however, the relative change of the volume fraction of small gas bubbles slightly decreases, whereas the population of these small bubbles sharply decreases due to coalescence, resulting in a sharp decrease of the gas-liquid interfacial area and subsequently K_La. These results are in agreement with results of Behkish [3], Vandu and Krishna [4] and Abdul-Rahman and Alazzi [8] and in disagreement with results of Vandu et al. [6] and Hashemi et al. [10]

Effect of superficial gas velocity and solid loading on gas holdup (εg)

The effect of superficial gas velocity on the gas holdup, ε_g at different solid concentration is shown in Figure 4 and 5 and Table 1 for short column and long column. These figures indicate that the gas holdup, ε_g increases with increasing of superficial gas velocity for both columns, the gas holdup increases sharply as U_g increases. A decrease in the slope of variation curve signifies the onset of bubble clustering and hence the

Type of column	U_g (cm/sec)	Pressure drop (mbar)	Gas holdup (-)	K_La_L (1/sec)	K_La (1/sec)
Silica, short column	1	67	0.034586	0.003484	0.008396
	2	65	0.0625	0.025331	0.021373
	3	64	0.090909	0.049638	0.040613
	4	63	0.112426	0.090606	0.072378
	5	62	0.132948	0.118559	0.092517
	6	61	0.152542	0.133584	0.101886
Alumina, short column	1	80	0.034586	0.00415	0.003606
	2	79	0.0625	0.02005	0.016917
	3	77	0.090909	0.04229	0.034601
	4	76	0.112426	0.0796	0.063586
	5	75	0.132948	0.10846	0.084636
	6	74	0.152542	0.12899	0.098382
Silica, long column	7	142	0.116022	0.05774	0.045937
	8	140	0.130435	0.080335	0.062871
	9	136	0.139785	0.13289	0.102883
	10	134	0.148936	0.16665	0.127647
	11	132	0.162304	0.20004	0.150815
	12	130	0.175258	0.24998	0.185552
Alumina, long column	7	172	0.111111	0.05333	0.042664
	8	169	0.123288	0.0741	0.058468
	9	166	0.135135	0.1158	0.090136
	10	163	0.144385	0.1546	0.11905
	11	161	0.153439	0.18839	0.143535
	12	159	0.157895	0.2011	0.152413

Table 1: Experimental results for 10% loading in long and short column using alumina and silica.

onset of the churn-turbulent bubble regime. As U_g further increases, bubble collisions intensify, leading to more bubble clustering. Larger bubble clusters rise faster and tend to aggregate in the center of the bubble column. As they rise, the momentum of these clusters is transferred to smaller bubbles and hence increases their rising velocity. The clustering effect contributes to a decrease in the rate of increase of the gas holdup with increasing U_g. This maximum point corresponds to the onset of bubble coalescence. These larger bubbles have a much higher rise velocity, causing a decrease in the gas holdup. As the gas velocity further increases, the gas holdup increases solely due to the higher through flow of the gas. At ambient pressure, the sharp increase in gas holdup at low U_g is not observed. The regime transition is believed to occur at U_g lower than the minimum measurement condition. Therefore, these figures show that the gas holdup is mainly dependent on the superficial gas velocity and solid concentration. These results are in agreement with results of Vandu et al. [6], Behkish [3], Hashemi et al. [10], Abid and Jameel [11], Sulaymon et al. [12] and Asgharpour et al. [9]. These figures show also that the gas holdup, ε_g decreases with the increase in the solid concentration for both columns. As seen from these figures that this decreasing is more rapid at high solid concentration and high superficial gas velocity. The reason is mainly due to the increase in the apparent viscosity. Actually, the influence of increasing liquid viscosity and solid concentration on the hydrodynamics is the same qualitatively; this increase is accompanied by significant increase of average bubble size which results in an increase in bubble rise velocity and decrease in the gas holdup. These results are in agreement with results of Li and Prakash [13], Mena et al. [14], Vando and Krishna [4] and Vandu et al. [6] and Alwasiti et al. [13]

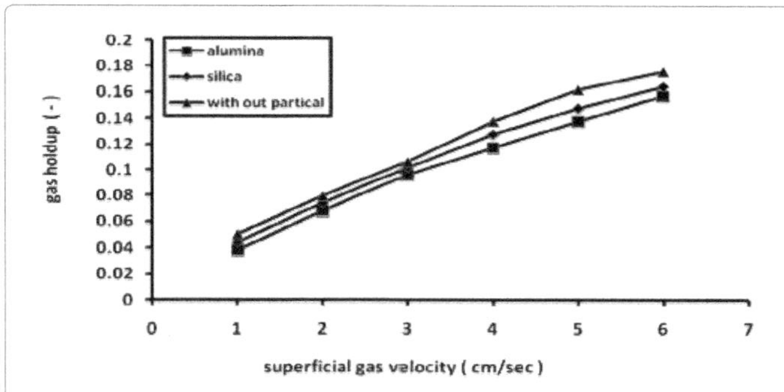

Figure 4: Effect of superficial gas velocity on gas holdup for short column at (5% vol.) slurry concentration.

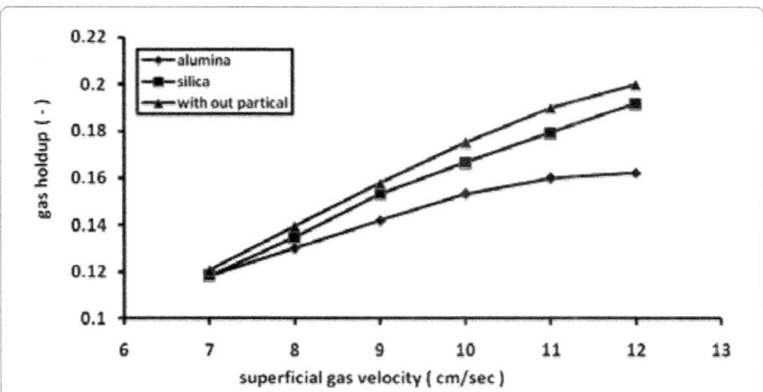

Figure 3: Effect of superficial gas velocity on gas holdup for long column at (5% vol.) slurry concentration.

Conclusions

The gas holdup increases from (0.01315 to 0.1758) with increasing superficial gas velocity from (1 to 6 cm/s) and increases from (0.101124 to 0.2) with increasing superficial gas velocity from (7 to 12 cm/s), and

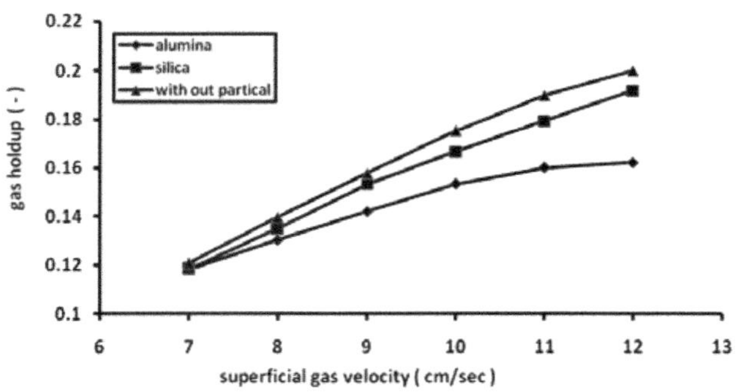

Figure 5: Effect of superficial gas velocity on gas holdup for long column at (5% vol.) slurry concentration.

decreases with increasing solid concentration. The volumetric mass transfer coefficient increases from (0.002826 to 0.148043 s^{-1}) with increasing superficial gas velocity from (1 to 6 cm/s) and increases from (0.024593 to 0.239051 s^{-1}) with increasing superficial gas velocity from (7 to 12 cm/s), and decreases with increasing solid concentration and it is higher for silica loading than alumina loading.

References

1. Luo X, Lee DJ, Lau R, Yang G, Fan LS (1999) Maximum Stable Bubble Size and Gas Holdup in High-Pressure Slurry Bubble Columns. AIChE Journal 45: 665-680.
2. Fan LS, Yang GQ, Lee DJ, Tsuchiya K, Luo X (1999) Some Aspects of High-Pressure Phenomena of Bubbles in Liquid and Liquid-Solid Suspensions. Chemical Engineering Science 54: 4681-4709.
3. Behkish A (2005) Hydrodynamic and Mass Transfer Parameters in Large-Scale Slurry Bubble Column Reactors. Doctor of Philosophy thesis, University of Pittsburgh, USA.
4. Vandu CO, Krishna R (2004) Volumetric mass transfer coefficients in slurry bubble columns operating in the churn-turbulent flow regime. Chemical Engineering and Processing 43: 987-995.
5. Al-Zaharani A, Daous M (2002) Measurement of Heat and Mass Transfer Coefficients in a Bubble Column. The 6th Saudi Engineering Conference. KFUPM, Dhahran, 2: 537.
6. Vandu CO, Koop K, Krishna R (2004) Volumetric mass transfer coefficient in a slurry bubble column operating in the heterogeneous flow regime. Chemical Engineering Science 59: 5417-5423.
7. Han L, Al-Dahhan MH (2007) Gas–Liquid Mass Transfer in a High Pressure Bubble Column Reactor with Different Sparger Designs. Chemical Engineering Science 62: 131-139.
8. Abdul-Rahman A, Alazzi NJ (2010) Gas Hold-Up and Volumetric Liquid-Phase Mass Transfer Coefficient in Solid-Suspended Bubble Columns with Draught Tube. Eng & Tech Journal 28: 1464-1483.
9. Asgharpour M, Mehrnia MR, Mostoufi N (2010) Effect of Surface Contaminants on Oxygen Transfer in Bubble Column Reactors. Biochem Eng J 49: 351-360.
10. Hashemi S, Macchi A, Servio P (2009) Gas–Liquid Mass Transfer in a Slurry Bubble Column Operated at Gas Hydrate Forming Conditions. Chemical Engineering Science 64: 3709-3716.
11. Abid MF, Jameel FS (2009) Scale Effects on the Hydrodynamics of Bubble Column. Eng Tech Journal 27.
12. Sulaymon AH, Mohammed TJ, Jawad AH (2010) Hydrodynamic Characteristics of Three-Phase Non Newtonian Liquid-Gas - Solid Fluidized Beds. Emirates Journal for Engineering Research 15: 41-49.
13. Li H, Prakash A (2000) Influence of Slurry Concentrations on Bubble Population and Their Rise Velocities in a Three-Phase Slurry Bubble Column. Powder Technology 113: 158-167.
14. Mena PC, Ruzicka MC, Rocha FA, Teixeira JA, Drahos J (2005) Effect of solids on homogeneous–heterogeneous flow regime transition in bubble columns. Chemical Engineering Science 60: 6013- 6026.

Removal of Copper from Phosphoric Acid by Adsorption on Tunisian Bentonite

Abdennebi N*, Bagane M and Chtara C

Chemical Engineering Department, National engineering school of Gabes (ENiG), University of Gabes, Tunisia

Abstract

This experimental study was articulated around the efficiency of the technique of adsorption as a method for the elimination of copper from phosphoric acid by using bentonite as a support. Several parameters such as the effect of the initial concentration of the pollutant, the isotherms and the retention capacity of bentonite, was investigated. The adsorption was modeled by simples' models of Langmuir, Freundlich. The results obtained have proved that the adsorption of copper follow the Freundlich Model.

Keywords: Copper; Phosphoric acid; Bentonite; Adsorption

Introduction

Several studies carried out have shown that the clay has a very important role as a powerful adsorbent for removing heavy metals contained in phosphoric acid, due to their high availability and relatively lower costs compared to industrial adsorbents. A study was conducted by Mellah and Jaffar [1], on the adsorption of heavy metals from phosphoric acid by an Algerian activated bentonite. Some tests of zinc (II), cadmium (II) and chromium (III) adsorption contained in 5.5 M phosphoric acid pretreated (30% P_2O_5) were performed. The results showed a removal of 88% for zinc (II) and 89% for Cd (II) and chromium (III) of the initial concentrations (Table 1). In the same context, another study by Omri et al. [2] in the purification of Tunisian phosphoric acid, using various types of clays in their natural state or modified for the adsorption of cadmium, fluorine, carbon and organic chlorides. The experimental results show good fixation of these impurities on these adsorbents. For cadmium is of about 50%, fluorine and chloride about 98%, and finally for organic carbon can close to 85%. According to data in a study developed by El-Bayaa et al. [3] for the elimination of iron (III) ions and of uranium (VI) ions from 5 M phosphoric acid (52% P_2O_5) through silica sand. The maximum removal of iron (III) ion was 84% with a dose of 60 g of adsorbent/l and for uranium (VI) ions was 75% with a dose of 200 g of adsorbent/l. El-Sofany et al. [4] studied the cadmium, copper, zinc and lead removals from phosphoric acid by the use of impregnated charcoal. The removal rate was 43.7% for the (Cd), 36.74% for the (Zn), 27.45% for the (Cu) and 9.12% for (Pb).

The objective of this study is to achieve the possibility of the maximum elimination of copper from phosphoric acid by adsorption on bentonique Tunisian clay [5]. The effect of the initial concentration of copper, the different isotherms and the retention capacity of bentonite to remove copper.

Materials and Methods

Preparation of solutions and used reagents features

Main solutions: We have prepared a solution of 1M phosphoric acid from a standard stock solution (Fischer, Analar Normapur, France) of 85% purity with a density of 1.692 and a molecular weight of 98 g/mol, while injecting the adequate amount of copper sulfate ($CuSO_4$, $5H_2O$) with a molar mass 249.69 g/mol. The various studied solutions are prepared by adding copper to the main solution of 1M phosphoric acid, as a range of the metal concentrations in the North African phosphoric acid.

Elements studied	Cu (ppm)
Algerian rock [5]	6-26
Tunisian rock [6]	1-3
Marocan rock [7]	35

Table 1: Phosphoric acid composition (28% P_2O_5).

Experimental protocol: Our experimental study is based on the purification of contaminated phosphoric acid solutions assigns a range of initial copper concentration of 3 to 32 ppm. The first step was started by mixing the appropriate amount of clay (0.5 to 3 g) with a volume of 100 ml for each of the acid solution to be treated and then is subsequently stirred at 25°C and with a speed of 450 rpm for 120 minutes. After centrifugation step in a centrifuge-type Grosseron, MICRO 22, for 10 minutes at 6000 rpm, the solutions were filtered in a Buchner vacuum filter while using filters based on a cellulose membrane 0, 45 microns.

The analysis of the filtrate is made by an atomic absorption apparatus air-acetylene flame of mark (VARIAN AA140 Atomic Absorption Spectrophotmeter) lamp cathode (copper) is used to determine the adsorption capacity and the removal efficiency [6].

Bentonites tested: The clay that we used is a rich montmorillonite bentonite and from south-eastern Tunisian deposits (Hicha El Hamma). It is a calcium, grey bentonite and in powder form. Its specific surface area is around 70 m^2/g. Table 2 presents some characteristics.

Description of the adsorption tests on bentonite

The tests were realized in a batch reactor, where the acid solutions at different initial copper concentrations (3 to 32 ppm) are brought into contact with increasing doses of clay ranging from 0.5 g/100 ml acid to 3 g/100 ml acid. Tests were stirred at a speed of 450 rpm and maintained at a temperature of 25°C for two hours. When equilibrium is reached, samples will be centrifuged, filtered and analyzed.

***Corresponding author:** Abdennebi N, Chemical Engineering Department, National engineering school of Gabes (ENIG), University of Gabes, Tunisia

Chemical compound	specific area (m2/g)	CaO	SiO2	Al2O3	Fe2O3	SO3	K2O	MgO	Na2O	MnO	P2O5	TiO2
Composition (%)	70	2,52	48,92	19,31	13,13	0,28	1,06	2,53	1,21	0,05	0,19	1,63

Table 2: Physico-chemical characteristics of bentonite.

Model	Langmuir			Freundlich		
Dose (g/l)	R2	Qm (mg/g)	KL (l/mg)	R2	n	KF (l/mg)
5	0,9537	0,5903	0,0163	0,9962	2,156	0,0623
15	0,9083	0,3405	0,0472	0,9953	1,329	0,0212
30	0,7886	0,1257	0,0108	0,9904	0,7305	0,0077

Table 3: Parameters of Langmuir and Freundlich isotherms.

During our tests, some parameters were studied to see their effects on the performance of the adsorption: the influence of dose of clay added the effect of treatment by the clay on the concentration of phosphoric acid and the effect of the initial copper content [7].

Results and Interpretation

Effect of the amount of bentonite

We followed the evolution of the removal efficiency of copper with initial concentrations of copper (5-32 mg/l) while varying each time the dose of bentonite (5-30 g/l). Measuring the residual copper content of the treated samples was made after 120 minutes contact with the bentonite.

From the results shown in Figure 1, we can see that the removal efficiency of copper increased with increasing dose of bentonite introduced whatever the initial concentration of the copper present in the phosphoric acid. Note that for small amounts of clay (5 and 15 g/l), the best yields (14-20%) are for initial copper concentrations of 3 ppm. But for large dose of clay (30 g/l), the best yield obtained (43%) is for high copper concentration (32 ppm). On the adsorption phenomenon produced, it can be justified by Basta and Tabatabai's [8] proposed the following mechanism for the adsorption of metals:

$$M^{2+} + H_2O \square MOH^+ + H^+$$

$$MOH^+ + X^- \square XMOH$$

X: surface; M: metal.

Effect of initial copper content

We conducted tests for initial copper contents varying from 3 to 32 ppm doses of clay (5 to 30 g/l) to see its effect on the removal efficiency.

The results shown in Figure 2, we can see that the performance is gradually improving with the initial copper content for a high clay dose (3 g/l). But for both low doses (5 and 15 g/l) we see that the efficiency decreases with the initial copper content.

These results demonstrate that treatment with bentonite is not very effective for a low initial concentration of copper. The removal of copper was 23%, obtained for an initial concentration of 3 ppm copper. While for high initial concentrations (32 ppm) was used a high dose of bentonite to get a good removal efficiency of about 43%.

Treatment effect of bentonite on the concentration of phosphoric acid

We made the assays acid before and after treatment to see its influence on the rate of P_2O_5. We studied the effect of the dose of clay on the loss of P_2O_5 and found that there's a slight decrease (0.03%) on the initial concentration of phosphoric acid to a low dose of bentonite (5 g/l) and decreased more for larger doses (15 and 30 g/l), which varies between 0.05 and 0.07% (Figure 3).

Adsorption isotherms of copper

To study of the isotherms of adsorption of copper by bentonite, volumes of 100 ml of different initial concentrations of pollutants from (3 to 32 mg/l) are contacted with a mass of 0.5 g of the adsorbent. The fixed amount of copper per gram of adsorbent (mg/g) is given by the following relationship:

$$Qe = (C_0 - C_e) * \frac{v}{m}$$

Where Qe the fixed amount of copper in mg per gram of adsorbent, C_0 and C_e: are respectively the initial and instantaneous concentrations of copper (mg/l), V is the volume of the solution, m the mass of adsorbent used (g). Figure 4 shows the isotherm of copper for clay dose of 5 mg/l.

From Figure 4, we observed that the amount of adsorbed copper increases more or less rapidly at low concentrations in solution, then decreases to reach a plateau corresponding to a saturation adsorption sites, resulting in a monolayer adsorption. The isotherm obtained is L-type according to the classification of Giles [9].

The results show that the adsorption capacity for bentonite decreases with the solid/liquid ratio of 5 to 15 g/l and increases with the initial contents of Copper. For a high dose of bentonite (30 g/l) of the maximum adsorption capacity (0.47 mg / g) is obtained for an initial copper concentration of 32 mg/l and 0.28 mg/g for a dose of 5 g/l (Figure 5).

Linearization of adsorption isotherms: We have adopted the classic models of Langmuir and Freundlich isotherms adsorption, which allow linear regression to obtain the values of their constants (Q_m, K_L, 1/n, K_F), to better interpret the results of adsorption.

The Langmuir model [10] is based on the following assumptions:

Forming a single layer of adsorbate on the surface of the adsorbent, the existence of the defined sites of adsorption, the surface is uniform with no interaction between the adsorbed molecules.

Freundlich model [11] is based on an empirical equation reflects a change in energy with the adsorbed amount. This distribution of the interaction energies can be explained by heterogeneity of adsorption sites. Unlike the Langmuir model, the Freundlich equation does not provide higher adsorption which restricts its application to dilute media limit. However, this model admits the existence of interactions between the adsorbed molecules [12].

In fact, we remember different linearization equations for each model:

Langmuir relation:

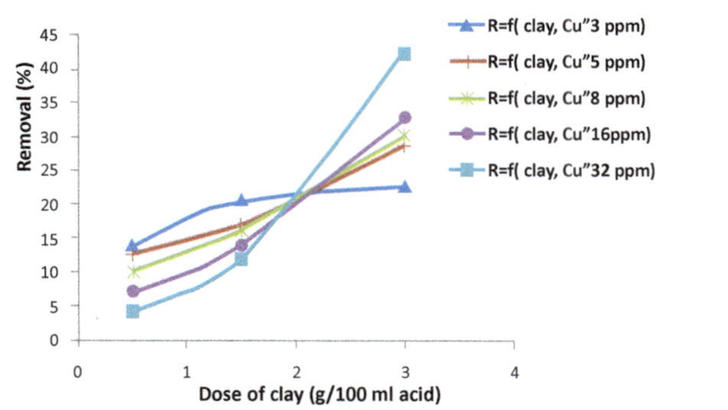

Figure 1: Effect of the dose of bentonite on the removal efficiency of copper.

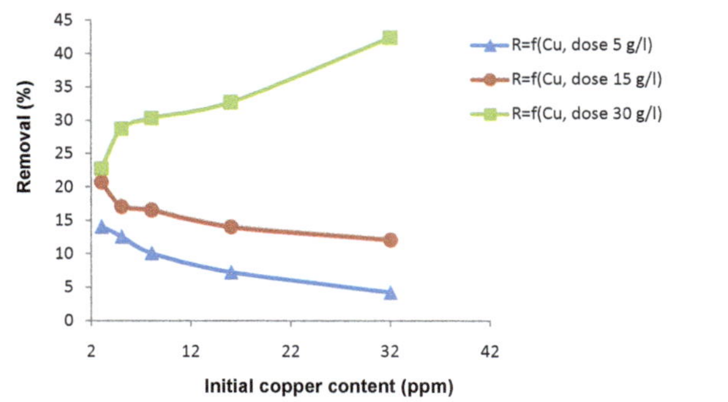

Figure 2: Variation of removal efficiency as a function of the initial copper content.

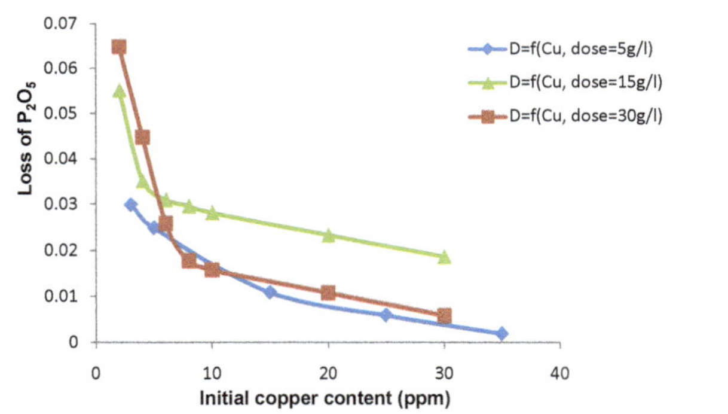

Figure 3: Effect of clay dose on loss of P_2O_5.

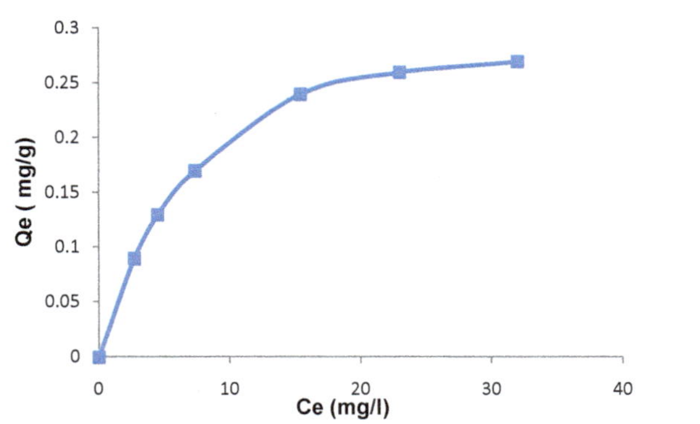

Figure 4: Isothermal adsorption of Cu2+ on the bentonite clay to a dose of 0.5 g to 25°C.

$$\frac{Ce}{Qe} = \left(\frac{1}{Q_m}\right) * Ce + \frac{1}{(K_L * Q_m)}$$

Freundlich relation: $\ln Qe = \left(\frac{1}{n}\right) \ln Ce + \ln K_F$

With:

C_e: Concentration in liquid phase at equilibrium (mg/l)

Q_e: Quantity adsorbed at equilibrium (mg/g)

Q_m: maximum adsorption capacity of the monolayer (mg/g)

K_L: Parameter Langmuir (l/mg)

K_F: Distribution coefficient Freundlich on the total sorption capacity of the solid (l/mg) and

n: constant characterizing the Freundlich adsorption affinity.

Figures 6 and 7 show the performance of the isotherms in their linearized forms.

From Figure 6, we note that the linearization of the adsorption isotherms of copper bentonite is satisfactory with good correlation coefficients. We can say that the Freundlich model is adequate for a good description of the adsorption isotherms. The various constants derived equations of the lines obtained are summarized in Table 3.

Note that according to the R^2 values of different models studied the adsorption of copper in phosphoric acid follows the Freundlich model.

The application of the linearized forms of Langmuir and Freundlich laws has verified that the latter model was applicable and that the removal efficiency of copper varies in the same direction with the Freundlich model.

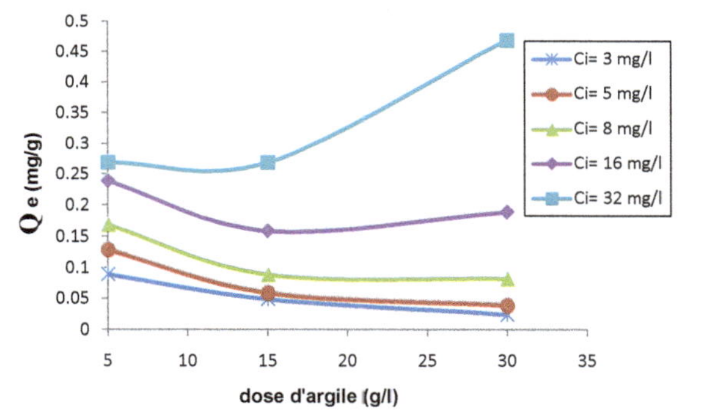

Figure 5: Effect of solid/ liquid ratio on the adsorbed amount of Cu^{2+} for different initial concentrations of copper at 25°C.

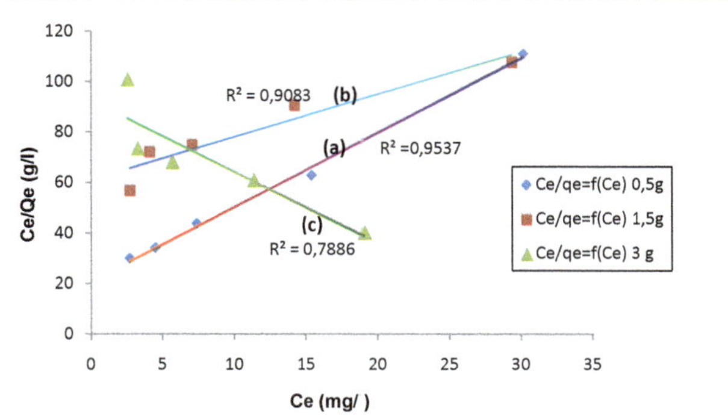

Figure 6: Modeling of adsorption isotherms of copper by the bentonite according to the Langmuir model at 25°C.

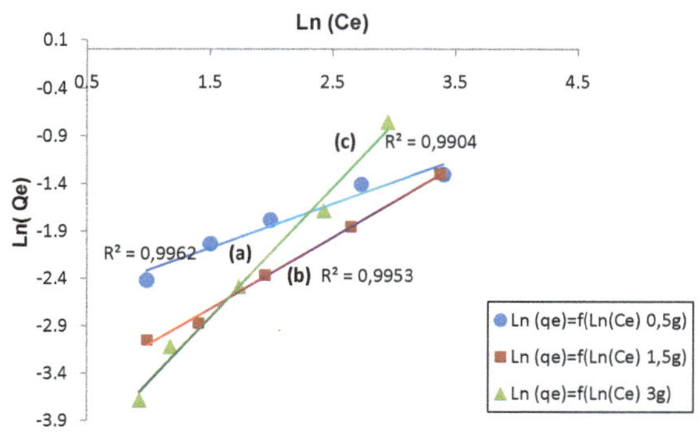

Figure 7: Modeling of adsorption isotherms of copper by the bentonite according to the model of Freundlich at 25°C.

Conclusion

Bentonite that we used in this study was moderately effective for the removal of copper from phosphoric acid. She gave us a yield of 23% removal for low initial concentrations of copper in varying doses of 5 to 15 g/l and a yield of 43% for an initial content of 32 ppm copper and a dose of 30 g of clay. The maximum retention capacity of copper to an initial copper concentration of 32 mg/l is 0.47 mg/g for a clay dosage of 30 g/l and 0.28 mg/g to a dose of 5 g/l.

The adsorption isotherms of copper on the bentonite are satisfactorily described by the Freundlich model, while the Langmuir model cannot describe our experimental results over the entire concentration range studied.

References

1. Mellah Abdelhamid, Benachour Jafar (2007) Adsorption of heavy metals from industrial phosphoric acid by an Algerian activated bentonite. Modeling. Annals of Chemistry 32: 487-504.
2. Hayet Omri, Rekik Salma Moalla Narjès Batis Valuation of clays in the purification of industrial phosphoric acid (INSAT). Centre Urbain Nord, Tunis.
3. El-Bayaa AA, Badawy NA, Gamal AM, Zidan IH, Mowafy AR (2011) Purification of wet process phosphoric acid by decreasing iron and uranium using white silica sand. J Hazard Mater 190: 324-329.
4. El-Sofany EA, Zaher WF, Aly HF (2009) Sorption potential of impregnated charcoal for removal of heavy metals from phosphoric acid. J Hazard Mater 165: 623-629.
5. Souad A, Nabiev M, Mimoun H (2000) Utilisation du diŽthyldithiocarbonate pour la purification de l'acide phosphorique des mŽtaux lourds par extraction liquide-liquide.
6. Ayadi Abdel Monem (1993) Elimination du cadmium dans l'acide phosphorique. Informate Chemicals No. 345. February.
7. Berthélémé S (2008) Acide phosphorique. Guide d'intervention chimique. Octobre, Saint Thonan.
8. Basta NT, Tabatabai MA (1992) Effect of cropping systems on adsorption of metals by soils. Soil Sci 153: 108-114.
9. Giles CH, McEwan TH, Nakhwa SN, Smith D (1960) A System of Classification of Solution Adsorption Isotherms, and its Use in Diagnosis of Adsorption Mechanisms and in Measurement of Specific Surface Areas of Solids. Journal of Chemical Society 93: 3973-3994.
10. Langmuir L (1918) The adsorption of gases on plane surfaces of glass, mica and platinum. J Am Chem Soc 40: 1361-1403.
11. Freundlich H, Hatfield SH (1926) Colloid and Capillary Chemistry. Metheum, London.
12. McKay G (1996) Use of Adsorbents for the Removal of Pollutants from Wastewaters. CRC Press.

Particulate Solids Removal from Synthetic and Real Turbid Water and Wastewater by Electro coagulation Using Vertical Expanded Al Anode

E. M. Elsayed, T. M. Zewail* and A.A. Zaatout

Chemical Engineering Department, Faculty of Engineering, Alexandria University, Alexandria, Egypt

Abstract

The performance of an innovative cell design in removing particulate solids from synthetic turbid water by electrocoagulation has been investigated. The cell consists of two concentric vertical cylindrical Al-electrodes; the outer electrode is a cylindrical solid Al cathode, whereas the inner electrode is an expanded cylindrical Al anode. The distance between the two electrodes is 1 cm apart. The performance of the present cell has been measured in terms of % removal of particulate solids and power consumption. The effect of key parameters such as electrolysis time, current density, pH, initial particulate solid concentration, NaCl concentration and solution velocity on the % removal of particulate solids has been investigated. The results revealed that as current density increases the % removal increases. Whereas as NaCl concentration and solution velocity increase, the % removal decreases. The optimum pH corresponding to the maximum % particulate solids removal and minimum power consumption is 10. At the optimum condition an energy consumption of 0.48 kW.h is required for the removal of one kg of particulate solids. Kinetic study of the present time data revealed that the rate of the removal follows second-order kinetic model. The performance of the present cell has been proven to be efficient for the treatment of real raw and wastewater.

Keywords: Particulate solids removal; Turbidity removal; Water treatment; Wastewater; Electrocoagulation; Expanded electrodes; Aluminum electrode; Kaolinite clay removal

Abbrevations: A: Active anode area cm^2; C0: The initial clay concentration ppm; C: The clay concentration at time t ppm; I: Applied Current A; k: Second order rate constant (mg.s/l)-1; t: Time of electrolysis min;

Introduction

One of the major challenges facing mankind today is providing clean water to a vast majority of the population around the world. The quality and quantity of available water are two particularly critical problems facing not only the third-world countries but also the whole world. Rivers; canals and other water bodies are being constantly polluted due to indiscriminate discharge of industrial effluents as well as other anthropogenic activities and natural processes [1]. Particulate solids pollutants in particular are present in water in two main forms: suspended particles and colloids. Suspended particles; such as sand; vegetable matter and silts; range in size from very large particles down to particles with a typical dimension of 10 μm. Colloids are very fine particles; typically ranging from 10 nm to 10 μm. There are two types of colloids: hydrophilic colloids and hydrophobic colloids. Hydrophobic colloids, including clay and non-hydrated metal oxides, whereas hydrophilic colloids like soap [2].

The presence of such particulate solids such as algae; clays; silts; colloids and organic particles in water often causes it to get turbid. Turbidity is one of the principal physical characteristics of water; excessive turbidity; or cloudiness; in drinking water is aesthetically unappealing; and may also represent a health concerns as it can provide food and shelter for pathogens. If not removed; turbidity can promote re-growth of pathogens in the distribution system; leading to waterborne disease outbreaks; which have caused significant cases of gastroenteritis. Although turbidity is not a direct indicator of health risk; numerous studies showed a strong relationship between removal of turbidity and removal of protozoa [3].

The conventional methods for the removal of turbidity causing particulate solids involve the bulk addition of inorganic coagulants (e.g. aluminum or ferric salts) which are responsible for combining small particles into larger aggregates. As the conventional chemical coagulation has advantages such as it is simply operated; it has also major problems such as high amount of chemicals needed and high amount of sludge produced [4-6].

Electrocoagulation process is a promising technology that can be used for the removal of both suspended solids and colloidal particles. Electrocoagulation is the electrochemical production of destabilization agents that brings about charge neutralization of negative charged surfaces of colloidal or suspended particles. Once charge neutralized; the particles bond together like small magnets to form a largeflocs [7]. These flocs can be easily separated either by gravity settling or by electroflotation where the evolved gas bubbles at the cathode would enhance flocs growth and their separation by flotation.

Electrocoagulation process offers some advantages over traditional chemical treatment: less coagulant ion is required; less sludge is formed; and electrocoagulation equipment is very compact; thus; suitable for installation where the available space is rather limited. Moreover; The EC process has the advantage of removing the smallest colloidal particles; because the applied electric field sets them in faster motion; thereby facilitating the coagulation [8].

***Corresponding author:** Taghreed Zewail, Chemical Engineering Department, Faculty of Engineering, Alexandria University, P.O Box 21544; Alexandria, Egypt

Recently EC has been applied successfully to treat lignin and phenol [9] kaolin-polluted wastewater [10] heavy metals [11] chemical mechanical polishing wastewater [12] oil emulsions [13] and fermentation wastewater [14]. All previous studies have investigated the performance of EC for removal of pollutants using traditional parallel plate electrode cell. There is scant attention has been given to the use of innovative cell designs.

The aim of the present study is to investigate the performance of a new EC cell configuration for the removal of kalointe as source of particulate solids from synthetic turbid solution. The present EC cell consists of two vertical cylindrical Al sheets separated by 1 cm gap. The outer cylinder is solid Al sheet cathode; whereas the inner cylinder is a screen mesh Al anode. The small electrode gap could minimize the IR drop; while the mesh screen anode would provide high specific surface area.

Optimum conditions of electrocoagulation process for the kaolinite clay removal were studied in relation to the following key parameters: i)-Electrolysis time; ii)-Current density; iii)- pH; iv)- Clay concentration; v)- NaCl concentration and vi)- solution velocity. In addition the performance of the present study was investigated to treat real turbid raw and waste water effluent.

Material and Methods

Materials

Figure 1 shows a schematic diagram of the batch experimental setup. It consisted mainly of a plexi glass container; a cell and an electric circuit. The cell consisted of a solid cylindrical Al cathode of 12 cm diameter and 20 cm height which was insulated from the container side and was supported on the inner wall of the container. The anode was a cylindrical Al anode screen of 10 mesh number; 10 cm diameter and 15 cm height. Four baffles were supported on the cathode to keep a fixed distance of 1cm between the anode and the cathode to avoid short circuit of the cell. The electric circuit consisted of a D.C power supply; a multi-range ammeter connected in series to measure the cell current and a voltammeter connected in parallel to measure the cell voltage.

The study was conducted on a synthetic turbid solution using kaolinite clay. The kaolinite clay was supplied from LECICO for ceramics company; Alexandria; Egypt. Table 1 shows the chemical composition of main oxides present in the used clay. All the regents and chemicals used in the present work are analytical reagent grade.

Figure 2, shows a schematic diagram of the flow system experimental setup. It included beside to the above mentioned components a flow system. The flow system consisted of 5 liters storage tank; a 0.12 hp plastic pump to deliver the clay solution to the EC cell and a filter screen was supported in the overflow weir to prevent the recirculation of the sludge to the storage tank.

Methods

Before each run the electrodes were pickled in a 10% HCl for 10 mins to remove any oxide. A stock of 500 ppm of the kaolinite clay solution was prepared using distilled water and few drops of 1N NaOH were added to the solution as peptizer to ensure deflocculation of clay. Batch experiments were carried out by placing 1.5l of freshly prepared clay solution in the cell. The pH of solution was measured before and after each experiment using a microprocessor pH meter (HANNA pH 211; Italy). NaCl was used as an electrolyte. Current was adjusted to predetermined values using a D.C power supply (BK Precision model 1686) whereas cell voltage was recorded during the run using an avometer (EXCEl DT9205A; China). EC progress was followed by taking a sample from the cell every 2 minutes from the same location in the cell and each sample was left for 10 mins before analysis to allow settling. The absorbance of each sample was measured using a spectrophotometer (ZUZI spectrophotometer 4201/20; Spain) at a wave length 340 nm which corresponding to maximum absorbance. Initial pH of the solution was adjusted either using 30% caustic soda (NaOH) or concentrated sulfuric acid (H_2SO_4).

In case of studying the effect of solution velocity; the flow rate of the solution was estimated by measuring the time of delivering a known volume. The variation of flow rate was done by adjusting the main and/or the bypass valve.

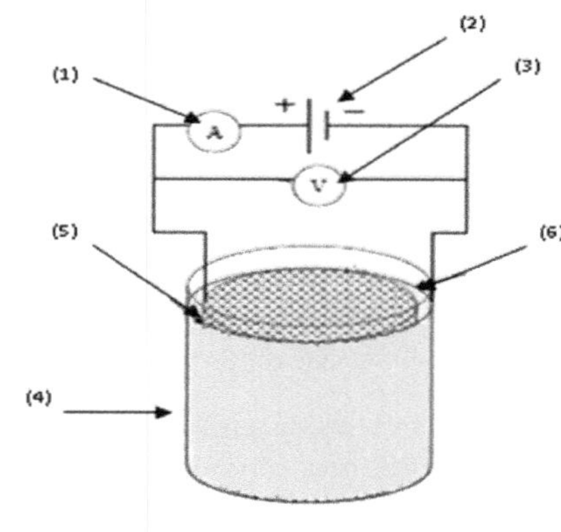

Figure 1: Electrocoagulation cell.

(1) Ammeter.
(2) D.C power supply.
(3) Voltmeter.
(4) 2 litreplexi glass container.
(5) Al screen anode.
(6) Al sheet cathode.

Components	Percentage (%)
SiO_2	46.6
Al_2O_3	36.5
H_2O	13.9
Impurities	3

Table 1: Chemical composition of the kaolinite clay.

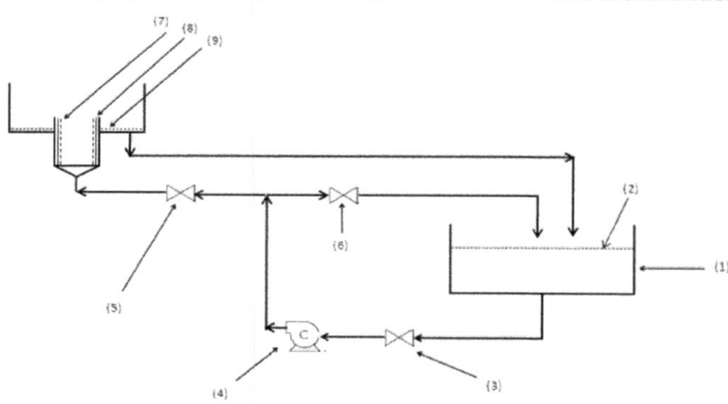

Figure 2: Continuous electrocoagulation system.

(1) Five liter Storage tank.
(2) Contaminated solution.
(3) Drain valve.
(4) Plastic Pump (0.12 hp).
(5) Main valve.
(6) Bypass valve.
(7) Mesh screen Al anode.
(8) Cylindrical Al cathode.
(9) Sludge filtration screen.

Results and Discussion

The effectiveness of the present cell was measured in terms the % of clay removal using the following equation:

$$\% \text{ clay removal} = (C_0 - C)/C_0 \quad (1)$$

Where; C_0: The initial clay concentration; C: The clay concentration at time t.

Effect of electrolysis time

Figure 3, depicts the effect of electrolysis time on % clay removal at different current densities. The % clay removal increases rapidly to reach about 77.4 to 94.5 % in a 2 mins (reactive stage) then it approaches almost a constant removal rate (stable stage) at approximately 12 mins. Rahmani [4]; and Kilic and Hosten [15] have observed the same trend. Whereas Holt et al. [16,17]; have indicated that in the removal of clay or colloidal matters by EC there are three operating stages: i) lag (where no significant change in clay removal takes place); ii) reactive stage (in which most of clay removal occurs) and finally iii) stable stage (where clay removal is approximately constant). The absence of a lag stage in the present study can be attributed to the high anode surface area. This high area can release sufficient Al^{+3} into the solution for effective destabilization of the colloidal particles. Mollah et al. [6] and Canizares et al. [10] indicated that the main mechanisms responsible for the clay or colloidal matters removal during EC are:

i. Charge neutralization of negatively charged particles of the clay through the electrophoresis migration of those particles towards the anode under the influence of electrical field.

ii. Charge neutralization of the negatively charged clay particles by adsorption of monomeric and polymeric hydrolysis species of Al^{+3}.

iii. The charge neutralized clay particles coalesce upon collision as a result of their Brownian movement or their electrophoretic migration to form a larger floc which can be easily removed.

iv. Sweep flocculation which takes place as the $Al(OH)_3$ reaches its minimum solubility and precipitates and during its precipitation clay particles may be enmeshed between $Al(OH)_3$ and removed from the bulk.

v. The coalesced clay particles may adhere to H_2 bubbles generated at the cathode and float to the surface and can be easily removed.

It should be noted that all the above mechanisms contribute to a certain degree in the removal of clay during reactive stages. However; during stable stage; the concentration of clay particle decreases and charge neutralization through electrophoretic migration diminishes; on the other hand the pH reaches neutrality and $Al(OH)_3$ precipitates. Thus sweep flocculation or enmeshment is the main dominant removal mechanism in that stage [17].

A time of 12 mins was chosen as the optimum electrolysis time for further experiments.

Effect of current density

Figure 4, shows the effect of current density on the % clay removal and power consumption. It is noted that as current density increases from 10.13 to 29.1 A/m^2; the % clay removal increases slightly from 96.9 to 99.1 % and power consumption increases from 1.05 to 6.03 kW.hr/kg. The increase in the % removal with increasing current density may be attributed to the increase of the electrophoretic motion under the applied electrical field; beside to the increase of the concentration of Al^{+3} ions generated at the anode according to Faraday's law. The increase of both electrophoretic motion and concentration of Al^{+3} ions enhance charge neutralization of clay particles which is believed to be the main mechanism responsible for their removal. Moreover; as current density increases hydrogen bubble generation rate increases and bubble size decreases. Which could improve floc formation and growth and their subsequent flotation [18]. The optimum current density of 10.13 A/m^2 was selected for further experiments based on the minimum power consumption.

Effect of NaCl concentration

Figure 5, shows the effect of NaCl concentration on % clay removal and power consumption. It is well seen that as NaCl concentration increases from 0.25 to 1 g/l the % clay removal decreases from 99.39 to 92.6 % while the power consumption decreases to a minimum value and then increases once again. This effect of NaCl concentration on % clay removal may be attributed to that the increase of NaCl concentration

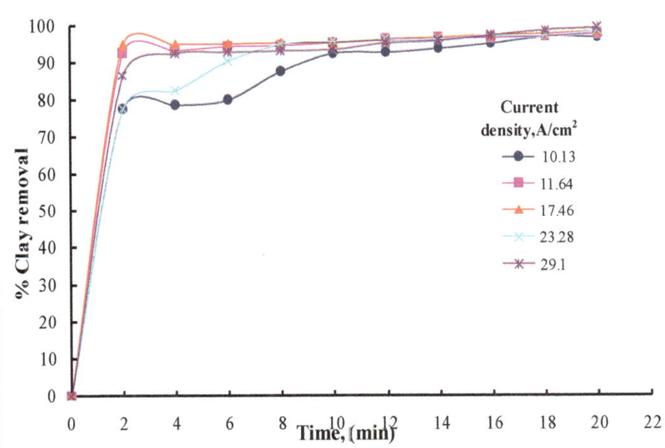

Figure 3: Effect of electrolysis time on the % clay removal [Clay concentration= 500ppm, pH=10.52, NaCl concentration=1g/l].

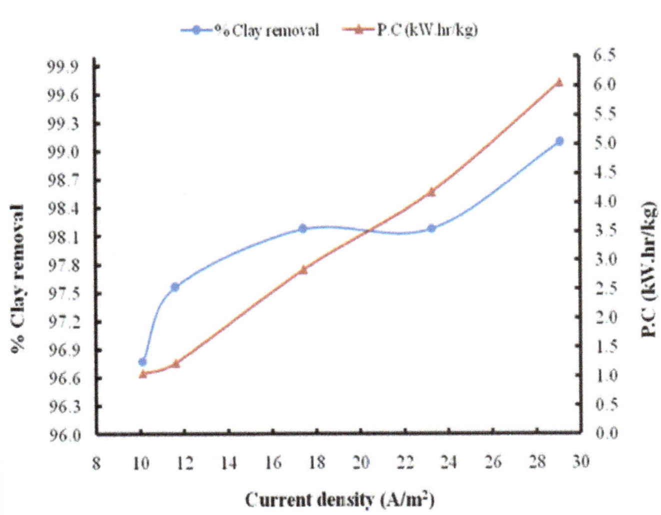

Figure 4: Effect of current density on the % clay removal [Clay concentration= 500ppm, pH=10.52, NaCl concentration=1g/l, Electrolysis time=12 min].

causes an increase in the solution ionic strength which could affect the electrophoretic mobility of the clay particles. In addition; the Cl^- could retard the electrophoretic mobility of the clay particles as it has a higher mobility than the clay particles towards the anode. Fouad et al. [19] found that the demulsification of a crude oil emulsion by electrophoresis decreases with increasing NaCl concentration. They attributed their results to that the competing chloride migrates to the anode surface inpreference to the negatively charged clay particles. Lung Chou et al. [20] have indicated that an excess amount of NaCl in the solution is detrimental to removal of turbidity. The present trend is in agreement with the work of Can et al. [21]. However; Emamjomeh and Sivakumar [22] and Merzouk et al. [23] showed a trend in contrast to the present work.

The optimum NaCl concentration of 0.75 g/l was selected for further experiments based on the minimum power consumption.

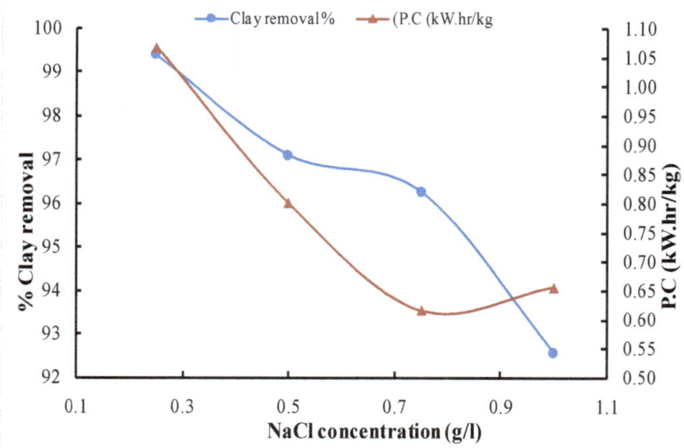

Figure 5: Effect of NaCl concentration on the % clay removal [Clay conc. = 500ppm, pH=10.50, Current density=10.13 A/m2, Electrolysis time=12 min].

Effect of initial pH

Figure 6, shows the effect of initial pH on % clay removal and power consumption. It can be noted that the initial pH has a little effect on the % clay removal as the % removal ranges between 93.44 to 97.8 % within the investigated pH range (4-11). Maximum % clay removal (97.8) corresponding to pH 10. Solak et al. [24] found that maximum removal of suspended solids by EC on Al anode is at pH 9. Kilic and Hosten [15] found that the highest % turbidity removal occurred near pH =9 where the zeta potential approached the isoelectric point of kaolinite particles.

The insignificant effect of pH on the clay removal may be explained by the presence of different mechanisms which contribute to the clay removal as stated previously.

The optimum pH was found to be 10 based on maximum % clay removal and minimum power consumption.

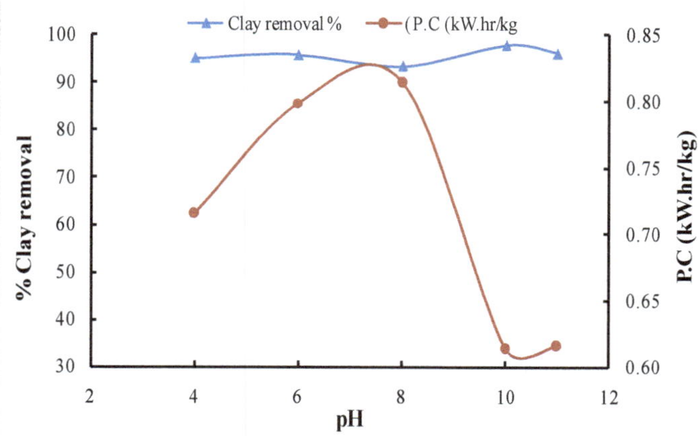

Figure 6: Effect of initial pH on the % Clay removal [Clay concentration = 500 ppm, Current density=10.13 A/m2, NaCl concentration=0.75 g/l, Electrolysis time = 12 min].

Effect of initial clay concentration

Figure 7, shows the effect of initial clay concentration on both % clay removal and the power consumption. It is well noted that as the initial clay concentration increases from 250 to 700 ppm; the % removal decreases slightly from 98 to 95. 6% and power consumption decreases from 1.17 to 0.53 kW.hr/kg. The insignificant effect of initial clay concentration on the removal may be due to that as the initial clay concentration increases; the chance for coalescence by electrophoretic migration increases at the same time the adsorption capacity of $Al(OH)_3$ consumed rapidly for fixed current density and the removal by sweep coagulation decreases and due to the presence of such two opposing effects; the net result is the slight dependence of % removal on the initial clay concentration .

Merzouk et al. [18,23] showed a decline in % turbidity removal or clay removal as the initial concentration increases.

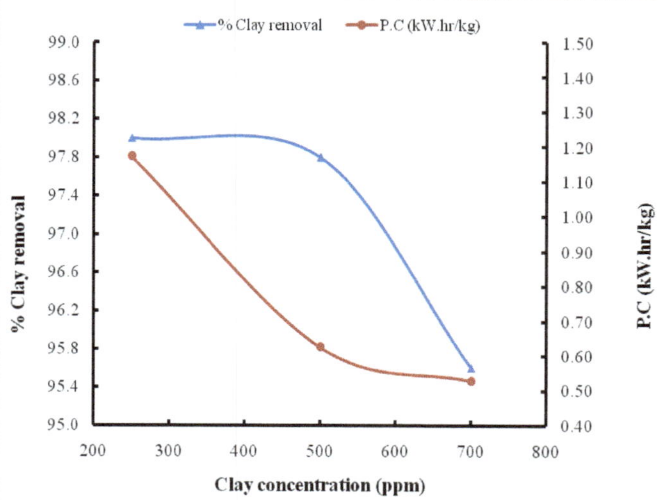

Figure 7: Effect of initial concentration on the % clay removal [Current density=10.13 A/m2, pH=10, NaCl concentration =0.75g/l, Electrolysis time=12 min].

Effect of solution velocity

Figure 8, shows the effect of solution velocity on % clay removal and power consumption. It is noticed that as the solution velocity increases from $2.83*10^{-3}$ to$8.51*10^{-3}$ m/s; the % clay removal decreases from 90.53 to 84.44. Matteson et al. [25] have indicated that vigorous stirring rate will hinder the coagulation because particles are removed from the region near the anode before they have chance to charge neutralized by the generated ions. At the same time if flocs are formed; they will not have the sufficient time to allow flocs aggregation to form larger flocs. Moreover; large flocs may be broken to smaller flocs due to vigorous stirring. Emamjomeh and Sivakumar [22] showed the same trend of solution velocity on % removal. A velocity of $4.25*10^{-3}$ m/s was chosen as the optimum velocity. It should be mentioned that under the optimum investigated operating conditions the corresponding power consumption was found to be 0.48 0.48 kW.hr/kg.

Kinetics of electrocoagulation

There are many models to describe some of the physical processes taking place in the electrocoagulation. In the present study it was assumed that the process undergoes a second order reaction rate. Hence; the rate of electrocoagulation can be expresses by the following equation:

$$-[dc/dt]=kC^2 \quad (2)$$

By integration yields:

$$[(1/C)-(1/C_0)]=kt \quad (3)$$

Where:

C: Clay concentration at time t; C_0: Initial clay concentration; k: second order rate constant; t: electrolysis time.

Figures 9-11, show the plot of 1/C versus time at different conditions. The data fit the second order kinetic model to a fair extent. The present results are in agreement with kinetic model proposed by Matteson [25]. According to their model it is proposed that the ions generated by the electrodissolution of the anode neutralize the superficial charge of the colloidal particles; and that the efficiency of the process is a function of the electrophoretic velocity of colloidal particles toward the anode and not of the Al generation rate. The electrocoagulation rate is represented by a second order rate equation that incorporates the electrophoretic migration of particles toward the anode. Figure 12 shows the effect of electro-generated Al dose on % clay removal. It is well noted that the Al dose has insignificant effect on the % clay removal which could also support the proposed kinetic model.

Case study

The efficiency of turbidity removal from real wastewater and raw water obtained from El-Maamoura water treatment plant (Alexandria; Egypt) were examined using the present cell. Table 2 and 3 shows the main characteristics of wastewater and raw water.

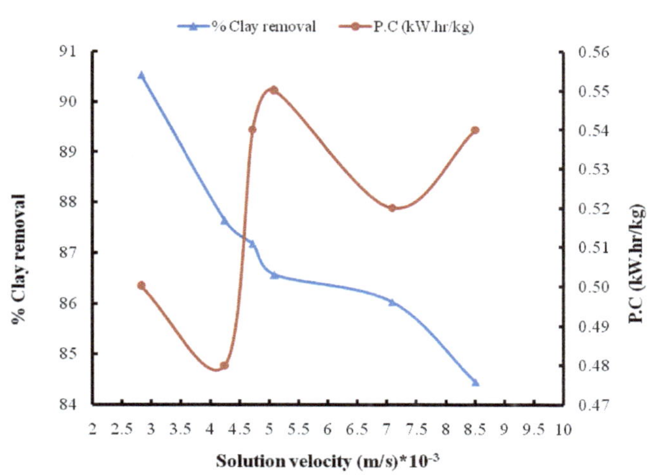

Figure 8: Effect of solution velocity on the % Clay removal [current density=33.85 A/m2, pH=10,NaCl concentration =0.75g/l, Electrolysis time=12 min].

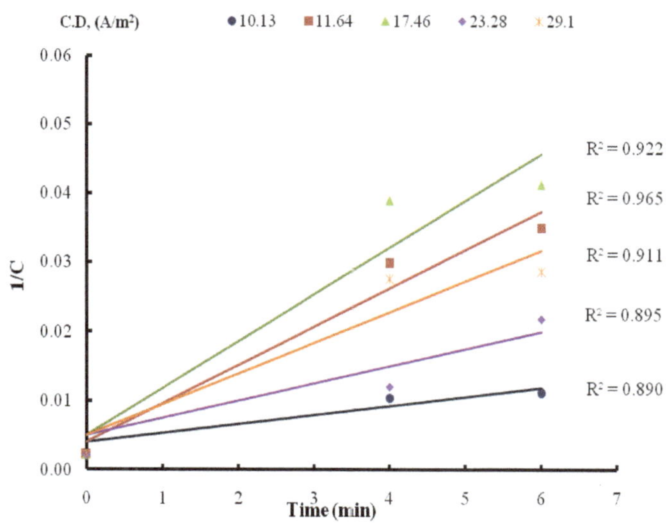

Figure 9: (1/C) vs. time at different current densities [Clay concentration= 500ppm, pH=10.52, NaCl=1g/l].

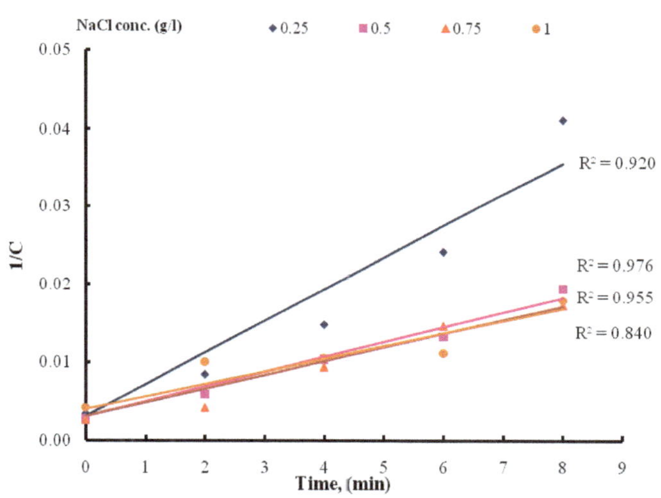

Figure 10: (1/C) vs. time at different NaCl concentrations [Clay concentration= 500ppm, pH = 10.52, C.D=10.13 A/m2].

Figure 13 and 14 show % turbidity removal versus time for the wastewater and raw water respectively. Figure 13 shows that there is a region at which % turbidity is approximately constant with the increase of electrolysis time. The presence of such constant % turbidity removal with time may be attributed to the presence of interfering dissolved solids which can affect the ionic strength of the solution and may also affect the zeta potential of the colloidal particles. A 91 % turbidity removal and 43% COD reduction were achieved after 12 minutes of electrolysis at current density of 33.85 A/m² for the wastewater. Whereas a 96.3% turbidity removal ; a 100% algae removal and 49% COD reduction were achieved at 12 minutes of electrolysis and current density of 33.85 A/m² for the raw water.

Conclusions

The present work has investigated the removal of clay from synthetic solution in both batch and flow electrocoagulation systems.

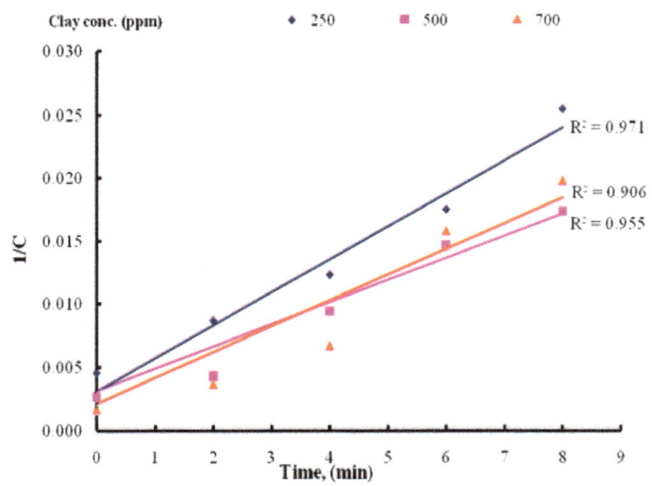

Figure 11: (1/C) vs. time at different solution concentrations [pH=10, NaCl=0.75 g/l, C.D=10.13 A/m2].

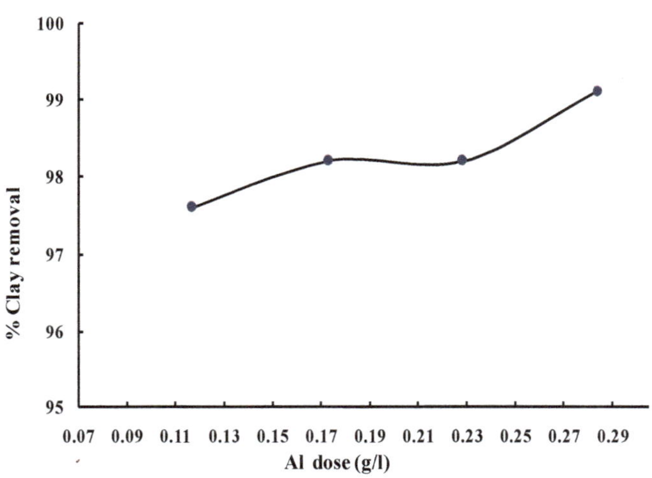

Figure 12: % clay removal versus Electro-generated Al dose [clay concentration=500 ppm, time= 20 mins, pH=10, NaCl concentration=0.75 g/l].

Parameter	Before electrocoagulation	After electrocoagulation
Turbidity, NTU	245.4	22.12
pH	7.27	7.74
Temp, °C	14.7	16
COD, mg/l	26.1	14.9
TDS, mg/l	1410	1387
Dissolved Al, mg/l	0.084	3.37

Table 2: Characteristics of wastewater influent before and after electrocoagulation.

The performance was expressed in terms of percentage clay removal and power consumptions calculations. Different parameters have been studied

The following conclusions have been withdrawn:

- % Clay removal increases as both current density and electrolysis time increase.
- As the NaCl concentration increases the % clay removal decreases.
- The optimum pH corresponding to the maximum % clay removal and minimum power consumption is 10
- The initial clay concentration has a slight effect on the % clay removal.
- The % clay removal was found to decrease with increasing the solution velocity.

Parameter	Before electrocoagulation	After electrocoagulation
Turbidity, NTU	112.7	4.15
pH	8.03	8.31
Temp, °C	16.4	17
COD, mg/l	14.84	7.58
TDS, mg/l	1258	1118.4
Algae number, Cell/ml	396746	0
Dissolved Al, mg/l	0.325	8.5

Table 3: Characteristics of raw water influent before and after electrocoagulation.

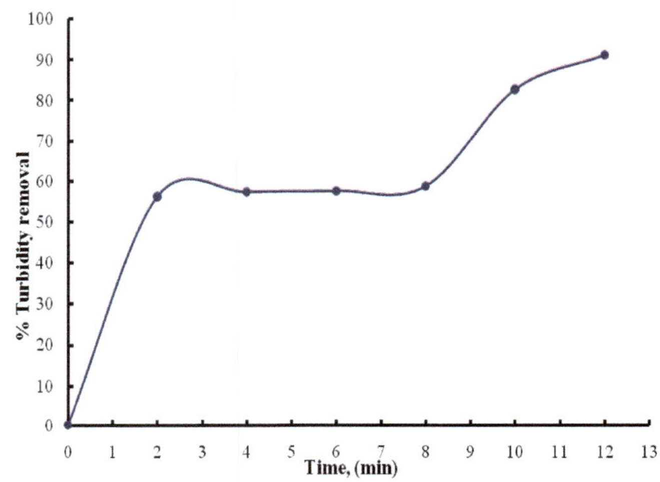

Figure 13: % Turbidity removal versus time for wastewater influent [NaCl conc.=0.75 g/l, C.D = 33.85 A/m2, Solution velocity=4.25*10-3, pH=10].

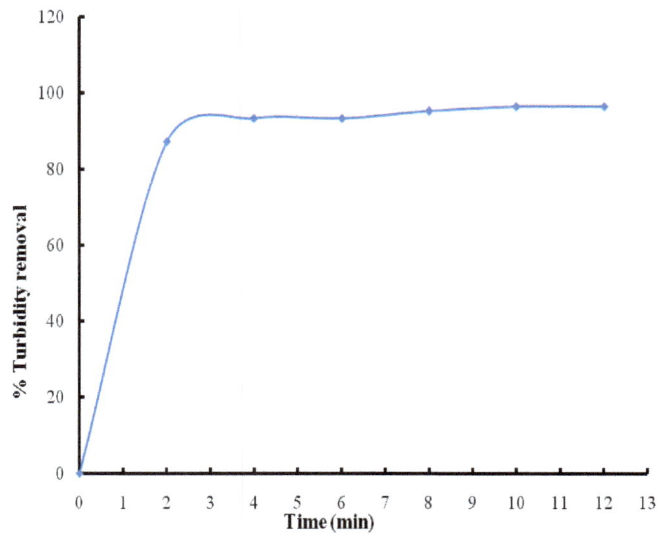

Figure 14: % Turbidity removal versus time for raw water influent [NaCl conc. =0.75 g/l, C.D = 33.85 A/m2, Solution velocity = 4.25*10-3, pH=10].

- Power consumption (PC) is a very important economical parameter in Electrocoagulation. The results showed that as the current density increases; PC increases whereas as NaCl concentration increases; PC decreases.
- At optimum operating conditions of the present flow cell a 0.48 kW.h is required per one kg of clay removed.
- A kinetics study revealed that the rate of clay removal follows second-order reaction kinetics.
- Treatment of raw water influent by the present flow system showed promising results in terms of turbidity removal. It has been found that under optimum conditions 96.3% turbidity removal; 49% COD reduction and 100 % algae removal have been achieved.
- Treatment of wastewater influent by the present flow system showed that under the optimum conditions 91% turbidity removal and 43% COD reduction has been achieved.
- The present new cell configuration has proven to be efficient in terms of technical and economic aspects for the removal of suspended and colloids from raw and wastewater.
- In view of the ability of the present cell to remove algae completely; it may be used as a pretreatment of feed water of reverse osmosis plants to reduce biological fouling.

References

1. Mollah M, Schennach R, Parga J, Cocke D (2001) Electrocoagulation (EC)-science and applications. J Hazard Mater B 84: 29-41.
2. Koohestanian A, Hosseini M, Abbasian Z (2008) The Separation Method for Removing of Colloidal Particles from Raw Water. J Agric & Environ Sci 4: 266-273.
3. DeZuane J (1997) Handbook of drinking water quality, 2nd edn. New York: John Wiley & sons.
4. Rahmani A (2008) Removal of water turbidity by the electrocoagulation method. J Res Health Sci 8: 18-24.
5. Niam M, Othman F, Sohaili J, Fauzia Z (2007) Removal of COD and turbidity to improve wastewater quality using electrocoagulant technique. The Malaysian J of Analyt Sci 11: 198-205.
6. Mollah MYA, Morkovsky P, Gomes J, Kesmez M, Parga J, et al. (2004) Fundamentals, Present and future perspectives of electrocoagulation. J Hazard Mater B 114: 199-210.
7. Emamjomeh M, Sivakumar M (2009) Review of pollutants removed by electrocoagulation and electrocoagulation/flotation processes. J Environ Manag 90: 1663-1679.
8. Linares-Hernandez I, Barrera-Diaz C, Roa-Morales G, Bilyeu B, Urena-Nunez F (2009) Influence of the anodic material on electrocoagulation performance. Chem Eng J 148: 97-105.
9. Uğurlu M, Gürses A, Doğar C, Yalçin M (2008) The removal of lignin and phenol from paper mill effluents by electrocoagulation. J Environ Manage 87: 420-428.
10. Canizares P, Martınez F, Rodrigo M, Jimenez C, Saez C, et al. (2008) Modelling of wastewater electrocoagulation processes Part I. General description and application to kaolin-polluted wastewaters. Sep Purif Technol 60: 155-161.
11. Heidmann I, Calmano W (2008) Removal of Zn(II), Cu(II), Ni(II), Ag(I) and Cr(VI) present in aqueous solutions by Aluminum electrocoagulation. J Hazard Mater 152: 934-941.
12. Drouiche N, Ghaffour N, Lounici H, Mameri M (2007) Electrocoagulation of chemical mechanical polishing wastewater. Desalin 214: 31-37.
13. Cañizares P, Martínez F, Rodrigo MA, Jiménez C, Sáez C, et al. (2008) Modelling of wastewater electrocoagulation processes part II. Application to dye-polluted wastewaters and oil-in-water emulsions wastewaters. Sep Purif Technol 60: 147-154.
14. Gadd A, Ryan D, Kavanagh J, Beaurain A, Luxem S, et al. (2010) Electrocoagulation of fermentation wastewater by low carbon steel (Fe) and 5005 aluminum (Al) electrodes. J App Electrochem 40: 1511-1517.
15. Kilic M, Hosten C (2010) A comparative study of electrocoagulation and coagulation of aqueous suspensions cf kaolinite powders. J Hazard Mater 176: 735-740.
16. Holt P, Barton G, Mitchell C (2005) The future for electrocoagulation as a localized water treatment technology. Chemosphere 59: 355-367.
17. Holt P, Barton G, Wark M, Mitchell C (2002) A quantitative comparison between chemical dosing and electrocoagulation. Colloids Surf A 211: 233-248.
18. Merzouk B, Gourich B, Sekki A, Madani K, Chibane M (2009) Removal turbidity and separation of heavy metals using electrocoagulation–electroflotation technique A case study. J Hazard Mater 164: 215-222.
19. Fouad YOA, Konsowa AH, Farag HA, Sedahmed GH (2008) Performance of an electrocoagulation cell with horizontally oriented electrodes in oil separation compared to a cell with vertical electrodes. Chem Eng J 145: 436-440.
20. Chou WL, Wang CT, Chang SY (2009) Study of COD and turbidity removal from real oxide-CMP wastewater by iron electrocoagulation and the evaluation of specific energy consumption. J Hazard Mater 168: 1200-1207.
21. Can OT, Bayramoglu M, Kobya M (2003) Decolourization of reactive dye solutions by electrocoagulation using aluminium electrodes. Ind Eng Chem Res 42: 3391-3396.
22. Emamjomeh M, Sivakumar M (2009) Fluoride removal by a continuous flow electrocoagulation reactor. J Envir Manag 90: 1204-1212.
23. Merzouk B, Madani K, Sekki A (2010) Using electrocoagulation electroflotation technology to treat synthetic solution and textile wastewater, two case studies. Desalin 250: 573-577.
24. Solak M, Kilic M, Yazici H, Sencan A (2009) Removal of suspended solids and turbidity from marble processing wastewaters by electrocoagulation: Comparison of electrode materials and electrode connection systems. J Hazard Mater 172: 345-352.
25. Matteson M, Dobson R, Glenn R, Kukunoor N, Waits WH, et al. (1995) Electrocoagulation and separation of aqueous suspensions of ultrafine particles. Colloids Surf A 104: 101-109.

Simulation and Optimization of a Continuous Biochemical Reactor

Ghanim M. Alwan*

Chemical Engineering, University of Technology, Baghdad, Iraq

Abstract

The present work focused on the dynamic and optimization of a continuous biochemical reactor using the glucose as a substrate. Simulated model provides the development of the process and reducing the risk of experimental runs. The selected process variables are; dilution rate (D), feed substrate concentration (Si), pH and temperature (T).The effect of D was observed at Si below 20 g/L. pH and T are affecting within Si of 60 g/L. Si is the effective process variable on the dynamic characteristics of the reactor. Reasonable agreement has found when compared the simulated results with that obtained by the previous work .Optimization technique guides the decision maker to select the best operating conditions. Stochastic genetic algorithm has found suitable for the nonlinear reactor. Optimal results indicate that the maximum biomass concentration (X) is 80.57 g/L at Si of 197.56 g/L and low D of 0.1(1/hr). Si was the sensitive variable for changing of the objective X.

Keywords: Biochemical reactor; Dynamic model; Optimization; Simulation

Nomenclatures: D: Dilution rate, [1/hr]; F: Flow rate, [L/hr]; Km: Saturation constant, [g/L]; PH: Acidity [-]; r1: Rate of cell generation [g/L.hr]; r2: Rate of substrate consumption [g/L.hr]; S: Substrate concentration in the reactor [g/L]; Si: Feed substrate concentration [g/L]; T: Temperature [C°]; t: Time [hr]; V: Volume of the reactor [L]; X: Biomass concentration in the reactor [g/L]; Y: Yield [-]; Greek Symbols μmax: Maximum specific growth rate coefficient [1/hr]; μ(s): Local specific growth rate coefficient, [1/hr]

Introduction

Lee [1], and Kapadia et al. [2] described the concept and applications of the biochemical reactors. The stirred-tank bioreactor is one of the most commonly used types for large scale production in industrial applications such as food, pharmaceuticals, various commodity and specialty chemicals. It is used mainly in two modes: the continuous mode and the fed-batch mode. In the continuous mode, the limiting substrates are constantly added to the reactor, while the output stream is simultaneously removed at the same rate, to keep the reactor volume constant. The continuous stirred biochemical reactor is widely used in the treatment of liquid wastes. Its process kinetic can be characterized by the following reaction scheme:

Substrate → Biomass +Gas

Henson [3] explained that as compared to conventional chemical reactors, bioreactors present unique modeling and control challenges due to complexity of the underlying biochemical reactions.

Karadag and Puhakka [4] and Garhyan et al. [5] studied the bioreactor performance using mixed cultures influenced by several operational parameters which affect its static and dynamic behavior such as: dilution rate, feed substrate concentration pH, hydraulic retention time, organic loading rate and temperature. In particular, the role of pH seems the most important parameter in the regulation of enzymes pool production. Ruggeri et al. [6] indicated that the pH adjustments validated the dynamics of the system. Charoenchai et al. [7] concluded that the temperature is a variable that directly affects the growth rate of organisms.

Annamalai and Doble [8] had found the mathematical modeling of fermentation process helps to; elucidate the mechanism of production process, estimate kinetic parameters such as specific growth rate of biomass and product formation rate develop the understanding between effects of operational conditions on production, and reduce laboratory experiments thereby saving time and resources.

Alhumaizi and Ajbar [9], and Shimizu [10] proved the biological processes are inherently very nonlinear and had frequently been changing optimum operating conditions. Many available mathematical models for biological reactions were not suitable for a control design since no accurate biological law had been proposed.

Kapadia et al. [2] proposed a novel robust controller for a continuous stirred biochemical reactor that controls the culture dilution rate into the reactor in order to maximize a cost function representing the biomass yield.

Genetic Algorithm (GA) is global stochastic search based on mechanics of natural selection and natural genetics.GA is based on Darwin's theory of 'survival of the fittest'. There are several genetic operators such as; selection, crossover and mutation, etc. Gupta and Srivastava [11] concluded that the deterministic algorithms for function optimization are generally limited to convex regular functions. However, many functions are either not differentiable or needed a lot of difficult mathematical treatment: decomposition, sensitivity computing, etc.

Scope of the Work

The present work focuses on the simulation of the continuous biochemical reactor using glucose as the substrate. Study the effect of the process variables on the dynamic behavior of the reactor. The selected variables are; feed substrate concentration, dilution rate, pH and temperature. The reliable simulated model can be used to generate

***Corresponding author:** Ghanim M. Alwan, Chemical Engineering, University of Technology, Baghdad, Iraq the desirable data for formulating the optimization equation. The objective is to maximize the biomass concentration in the reactor. Stochastic globe genetic algorithm search will implement to select the best operating conditions of the reactor.

Dynamic Model

Dynamic modeling for optimization and control requires models that describe the essential dynamic characteristics of the process under study. In the present work, the following assumptions have been adopted for the model:

1. Homogenous liquid-phase system.
2. Non-isothermal conditions.
3. Acidity of liquid is changed.
4. First order irreversible reaction.
5. Constant hold-up.
6. Follow the Monod law.

The component material balances for biomass(X) and substrate(S) are:

$$dX/dt = r1-(F/V)X \quad (1)$$

$$dS/dt = (F/V)\,Si-(F/V)\,S- r2 \quad (2)$$

In addition, the reaction rate equations are:

$$r1=\mu(s)\,X1 \quad (3)$$

And, $Y = r1/r2$ (4)

For Monod law;

$$\mu(s) = (\mu max *S)/ (Km+S) \quad (5)$$

X(gm/l)	Si(gm/l)	D(hr^{-1})	Y(-)	μ(1/hr)
0.001	6.0	0.3	0.4	0.4

Table 1: Optimum initial operating conditions.

$$\mu max = -40.5 + 11.78\,pH - 0.0691\,pH^2 + 1.65T + 0.003T^2 - 0.468\,pH.T \quad (6)$$

Equations (1&2) can be simplified to:

$$dX/dt = (\mu(s) - D)X \quad (7)$$

$$dS/dt = D\,(Si - S) - (X/Y)\mu(s) \quad (8)$$

where $D=F/V$

Equation 6 was correlated depended on the experimental data of Lopez et al. [12].

The simulated model will implement for the wastewater contains glucose with different concentrations from 6.0 to 200.0 gm/L. Temperature of water are varied from 20 to 30 °C and the acidity are from pH 2 to pH 4. The kinetic parameters of the biological reaction are; maximum specific growth rate coeficient (μmax=0.3 hr^{-1}), saturation constant (Km=1.0 g/L) and yield (Y=0.4) regarding to Lopez et al. [12], Cutlip and Shacham [13].

Results and Discussion

Optimal operating conditions

The initial optimal operating conditions of the system (Table 1) were estimated by the nonlinear Levenberg-Marquardt method with the aid of the MATLAB computer program.

Unsteady state conditions

The present bioreactor can be viewed as nonlinear dynamic system and the simulation is very useful tool for model validation .The unsteady- state model equation 7 and equation 8 were solved numerically using 5th order Runge-kutta method with the aid of the MATLAB program, starting from steady- state operating conditions (Table 1). Figures 1-7 explain the behaviors of the biochemical process under different values of variables dilution rate (D), feed substrate concentration (Si) at operating conditions of pH 2-4 and temperature of 20-30°C.

Dynamically, the system behaves as the first- order lag system. The

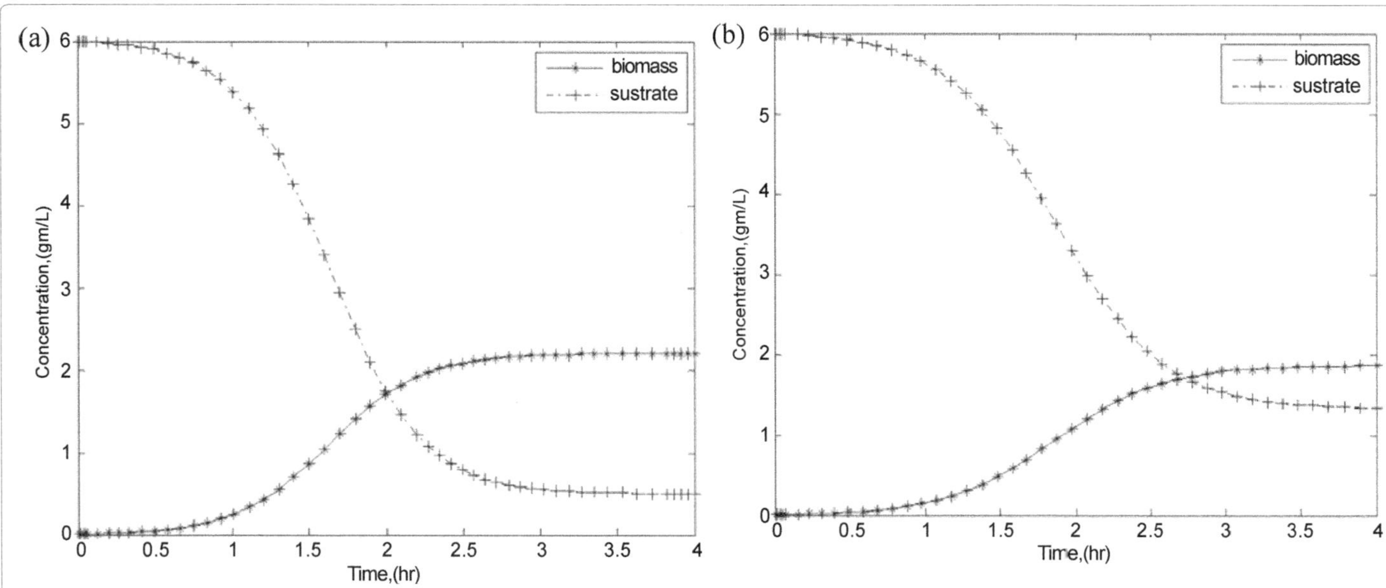

Figure 1: Unsteady state concentration of biomass and substrate at Si=6 for :(a)D=0.3, (b) D=0.8.

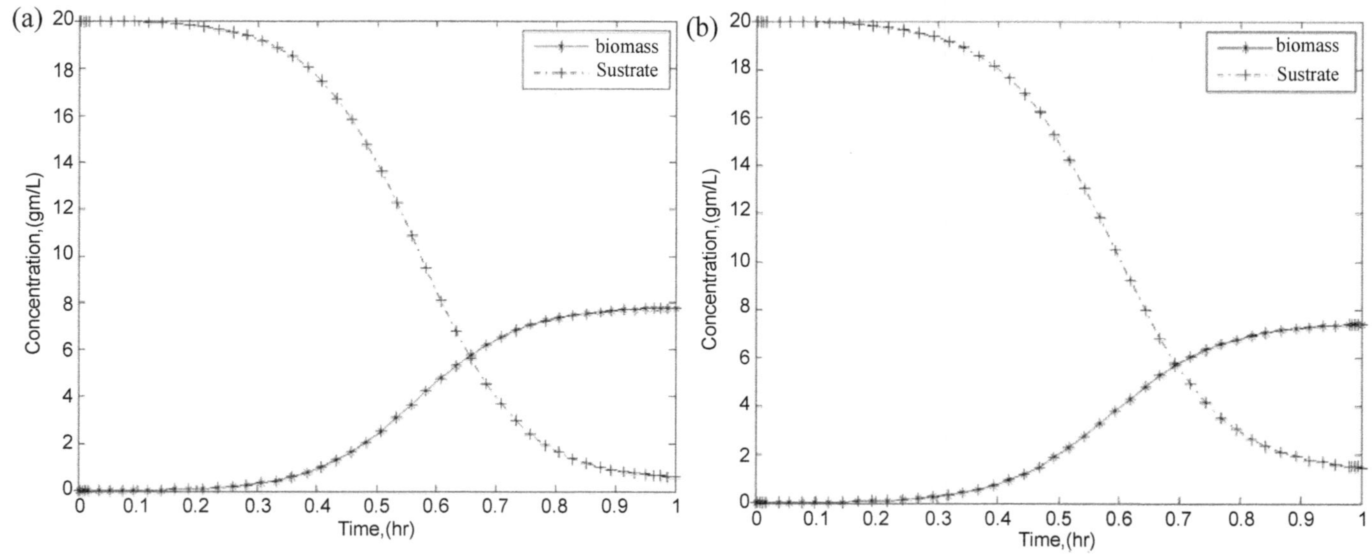

Figure 2: Unsteady state concentration of biomass and substrate at Si=20 for ;(a) D=0.3, (b) D=0.8.

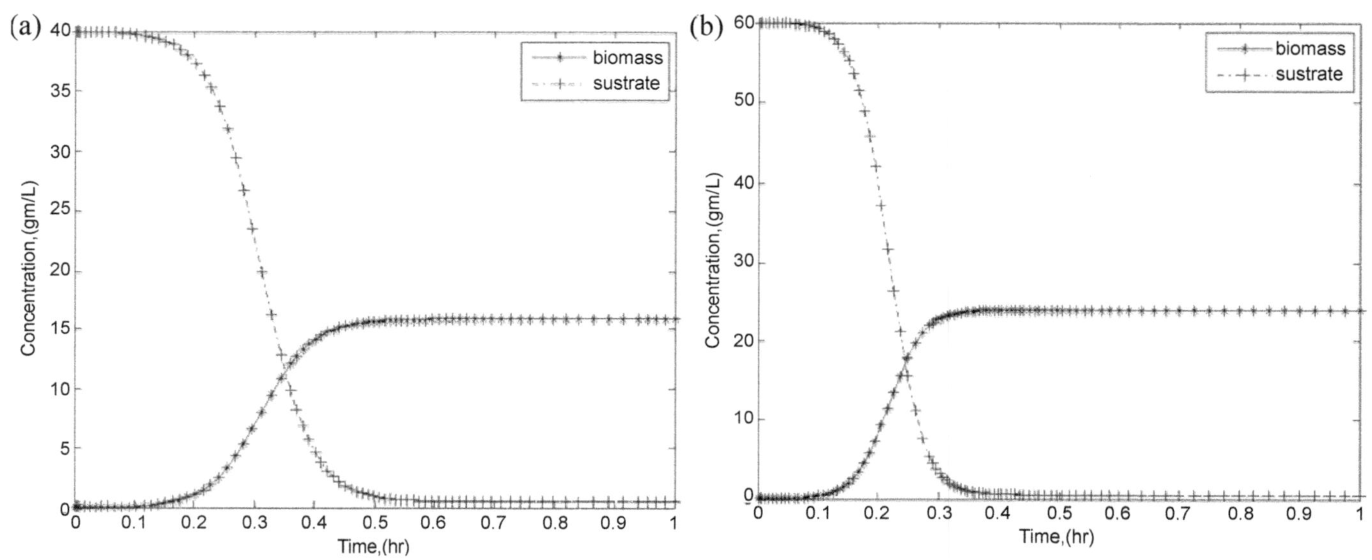

Figure 3: Unsteady state concentration of biomass and substrate at D=0.3for; (a) Si=40, (b) Si=60.

dynamic model appears that the biomass concentration curves have S-shape and more sluggish when compared with the substrate curves, which have an exponential shape because of the rate of consumed substrate is more than the rate of biomass cell generation in the reactor. The response speed of the biomass and substrate curves increase with Si and decreases with D as shown in figures1-3. The intersection point between two curves indicates to the local optimal point of the system, where the concentration of the biomass equal to that of the substrate. These points are various depended on the operating conditions.

The concentration of the biomass in the reactor decreases with increasing D (Figure 4a) for low and high Si. In the contrast, the increasing of Si increases the concentration of biomass in the reactor as shown in the figure 4b.This is due to the fact; that Si has a positive effect on the specific growth rate constant (μ) regarding to the Monod law (equation 5).While the increasing of D tends to increase the dilution of the substrate which could moderate the growth rate then reduces the concentration of the biomass in the biochemical reactor. The sensitivity of the process (steady-state gain) against Si (Figure 4b) is more than that with D (Figure 4a). The effect of Si is more pronounced at low D as shown in Figure 4. These behaviors were also concluded by Jarzebski [14].

The effect of temperature on the biomass growth rate appears in figure 5 and 6 at the temperature range from 15 to 30°C. The simulated results explain that the increasing of temperature will increase the growth rate of the biomass at low and high Si. This tends to increase the response speed of the biomass concentration. The steady- state value of the biomass concentration was unaffected with the increasing of temperature as shown in figure 5 and 6.

Figure 7 and 8 explain the effect of the water acidity (pH) on biomass growth. The effect studied for the available data ranged

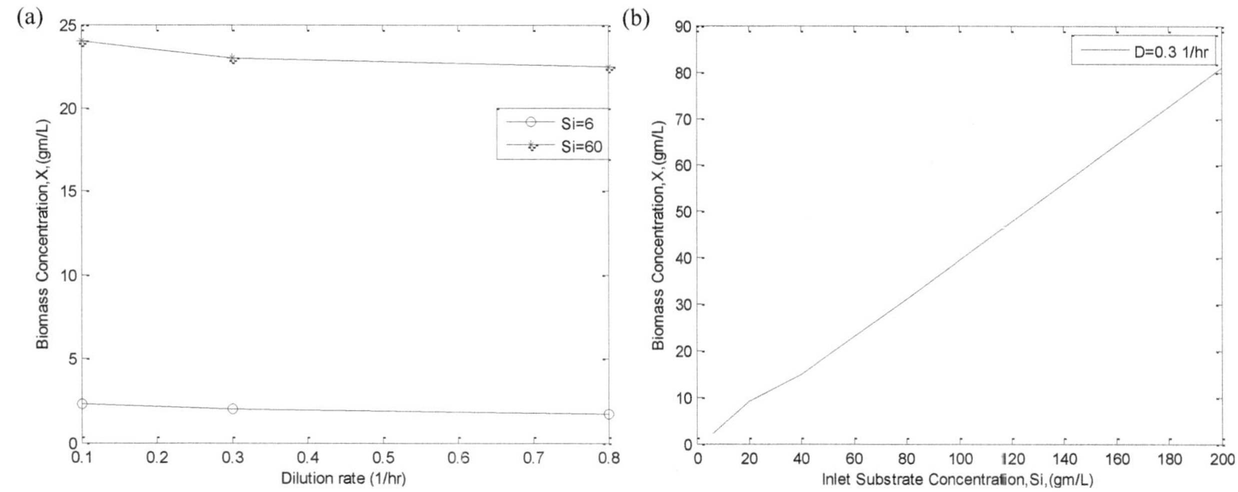

Figure 4: Biomass concentration as a function of (a) Dilution rate, (b) Inlet substrate concentration.

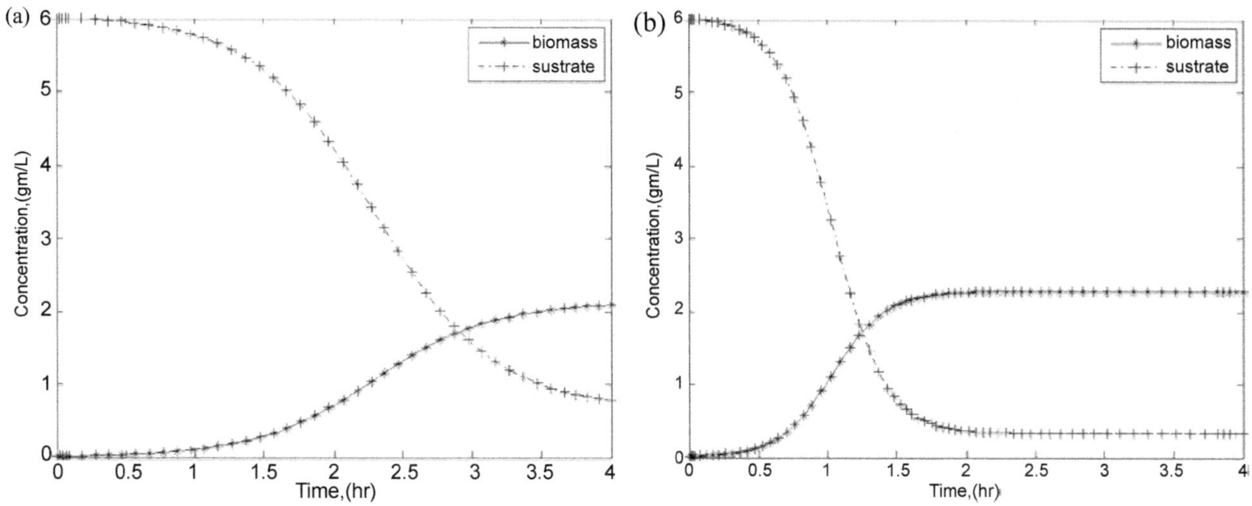

Figure 5: Effect of temperature on the process at D=0.3 and Si=6 for (a) T=20, (b) T=30.

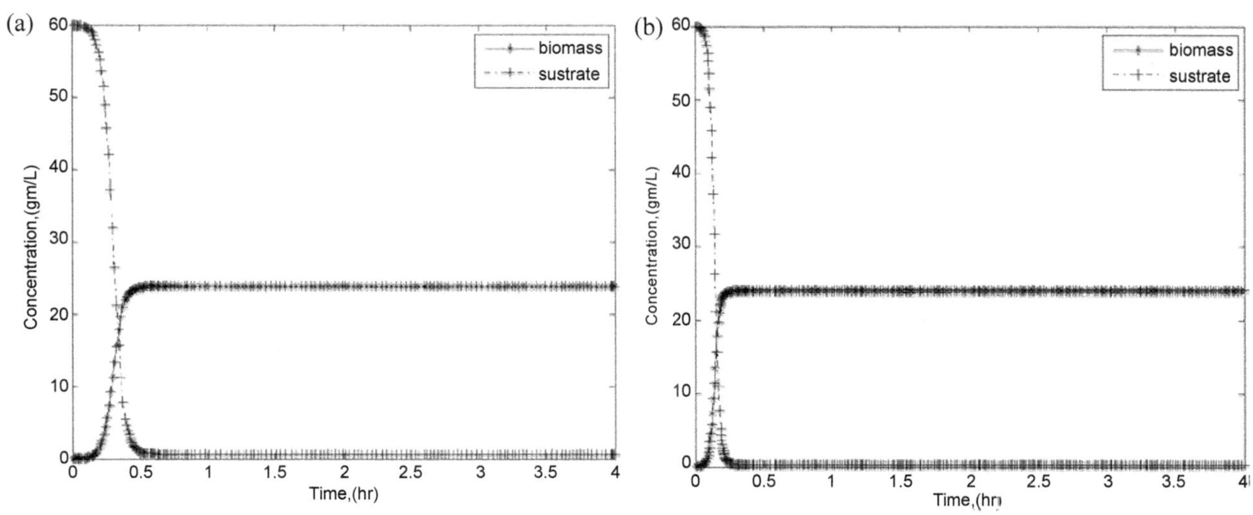

Figure 6: Effect of temperature on the process at D=0.3 and Si=60 for (a) T=20, (b) T=30.

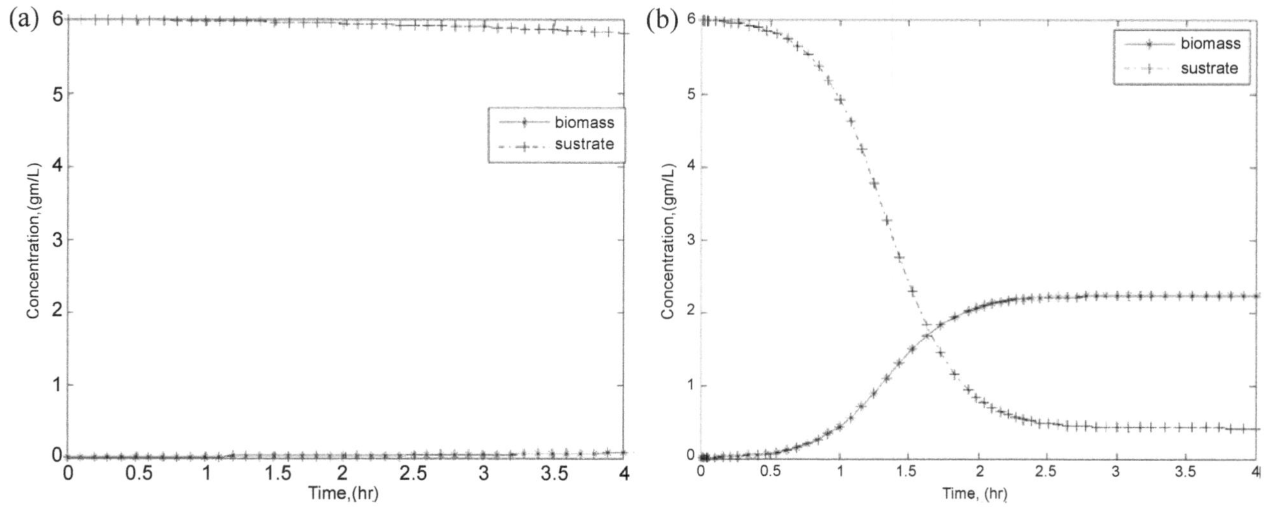

Figure 7: Effect of pH on the process at D=0.3 and Si=6 for; (a) pH=2, (b) pH=4.

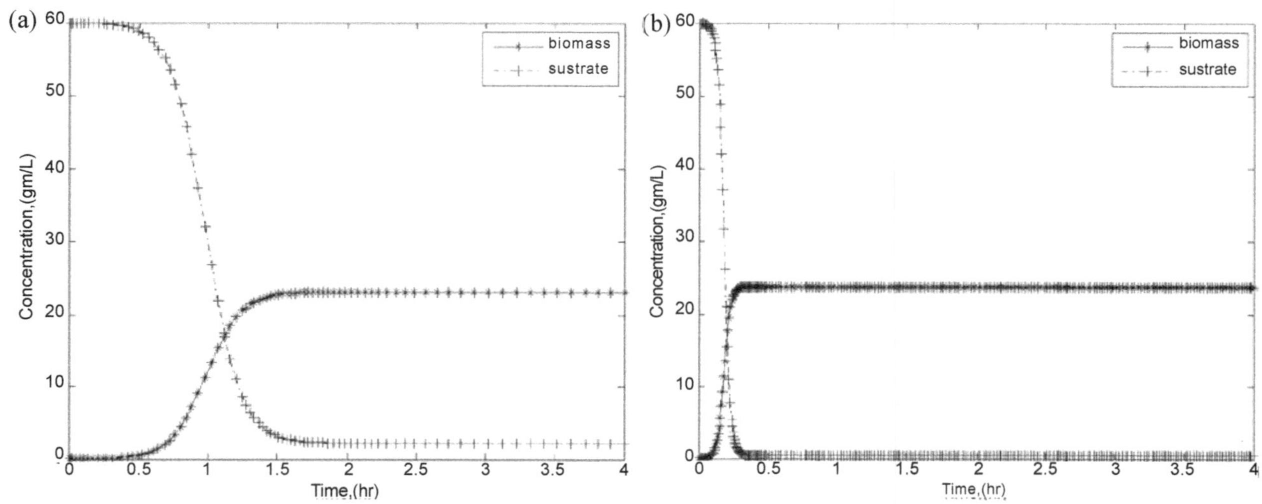

Figure 8: Effect of pH on the process at D=0.3 and Si=60 for; (a) pH=2, (b) pH=4.

between pH 2 to pH 4.The biomass growth is very slow at low acidity (pH 2) and increased with increasing to pH 4 as shown in figure 7a and 7b.The concentration of the biomass in the reactor is very low with the lower feed substrate concentration (Si=6) and pH 2 of water as shown in the figure 7a. At high substrate feed concentration (Si=60), the growth of biomass cell would enhanced at low acidity (pH=2) when compared figure 8a with figure 7a. The growth rate coefficient (μ) is directly affected by Si regarding to the Monod law. The final steady-state concentration of biomass is unaffected by the increasing of pH at high Si as shown in figure 8.

Reasonable result can observe when compared the simulated results with these obtained by Cutlip and Shacham [13] as shown in figure 9. The deviation is about 8%. This indicates that the proposed simulated model is agreed for the present biochemical reactor. Therefore, the reliable model will use to generate the desirable data for formulating the optimization equation.

Optimization problem

The available simulated data have been used to correlate the objective (concentration of biomass X) with the decision variables to facilitate the optimization scheme. The selected effective decision variables are; dilution rate (D) and inlet concentration of substrate (Si). Nonlinear regression using the Levenberg-Marquardt method is implemented with the aid of the computer program (Statistica version10).

The empirical correlation is:

$$X=0.409\,Si-0.575D-0.028DSi+0.02 \quad (9)$$

Subject to inequality constraints:

$$6.0 \leq Si \leq 200 \quad (10)$$

$$0.1 \leq D \leq 0.8$$

Equation 9 indicates that the dilution rate (D) has negative effect on the biomass concentration while the inlet concentration of substrate (Si) has positive effect. The interaction between Si and D shows that Si is more effective than D.

Optimization technique

The objective is to maximize the biomass concentration in the reactor. The optimization equation (equation 9) is interacted and nonlinear, so that the deterministic search is unsuccessful. GA has been found suitable for the present biochemical process. GA is stochastic global search based on mechanics of natural selection. Figure 9 illustrates the results/solution of the algorithm scheme. The parameters of the GA were adapted, and the selected operators are suitable for solving the problem to obtain the best optimal values. Hybrid function implemented as the combined search between genetic algorithm and pattern search to refine the values of decision variables. 51 generations occurred regarding to the nonlinearity of the process. The adapted operators of GA are explained in the table 2.

Population type	Double vector
Population size	80
Creation function	Feasible population
Scaling function	Rank
Selection function	Roulette
Crossover function	Scattered
Crossover fraction	0.8
Mutation function	Adaptive feasible
Migration direction	Forward
Migration fraction	0.1
Hybrid function	Pattern search
Number of generation	51
Function tolerance	1.0E-6

Table 2: Adapted parameters of GA.

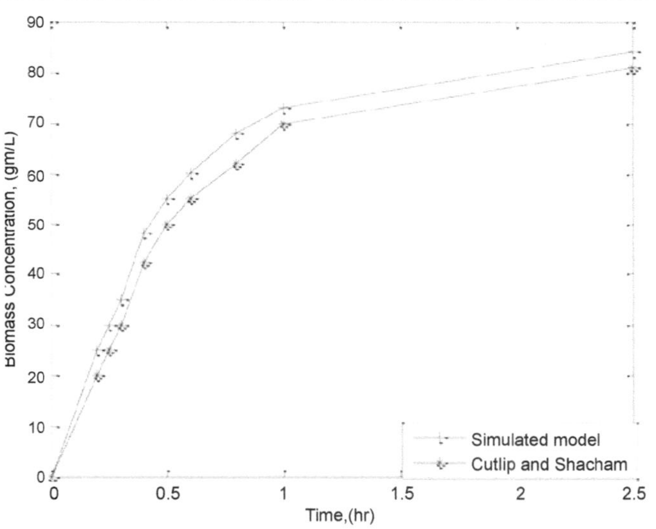

Figure 9: Comparison with previous work

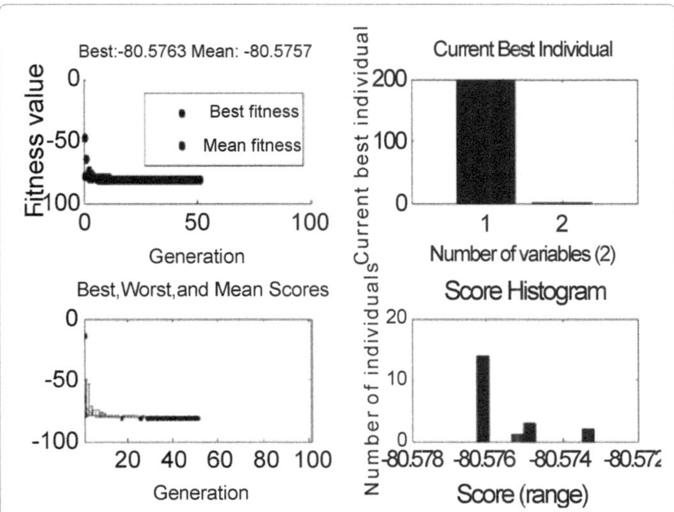

Figure 10: Solution/results of GA search.

Table 2 explains the best operators of the genetic algorithms. Figure 10 illustrates the outputs of the algorithms solutions/operators of genetic algorithm.GA is implemented with the pattern search by using the hybrid function as shown in table 2 to refine the decision variables .The best fitness, best function and score histogram as shown in figure 10 illustrate that the maximum biomass concentration is 80.57 g/L. The histogram of decision variables indicates that the optimal values are; Si=197.56 g/L (variable 1) and D=0.1hr^{-1} (variable 2), which are within the limit of inequality constraints (equation 10). The histogram of the variables in the figure 10 indicates that Si (variable 1) is the effective variable on X. Due to the nonlinearity of the bioreactor process; the optimization equation (equation 9) was solved by 51 generations as shown in figure 10.

The optimal sets of the decision variables are illustrated in figures 11a and 11b corresponding to the objective X. The scattering and stochastic of the results are appeared in these figures as a results of natural selection by GA. It is found that the optimal values of the dilution rate (D) are approximately constant within its lower bound as explained in the figure 11a. Inlet substrate concentration (Si) is more sensitive to the optimal objective change(X) as shown in figure 11b. This is due to that Si is the effective variable on X as shown in the figure 10. Si is changed within its upper bound (Figure11b).These behaviors are because of Si has positive effect while D has negative effect on X as shown in the equation 9. Optimal values of the two decision variables are stayed within optimal value of X which equal to 80.57 g/L as shown in figure 11.

Optimization technique is a powerful tool to obtain the desired operating conditions that improves the performance of the reactor. This reduces the risk of experimental runs and cost consumed for design and operation. However, the reliability of the search depends on; the best selection of decision variables, formulation of the objective function and the selection of the proper optimization technique. Palonen et al. [15] also indicated this conclusion.

Conclusions

1. Simulated model helps the study of dynamic characteristics of the biochemical reactor. Reliable model could use to generate extra data in the case of unavailable experimental results.
2. Effect of dilution rate was observed at low feed substrate concentration that is below 20 g/L. The effect of pH and temperatures were appeared within the concentration of 60 g/L.
3. Feed substrate concentration was found the effective process variable on the growth rate of the biomass cell in the reactor.
4. Maximum concentration of the biomass cell could be obtained at high concentration of substrate and low dilution rate.

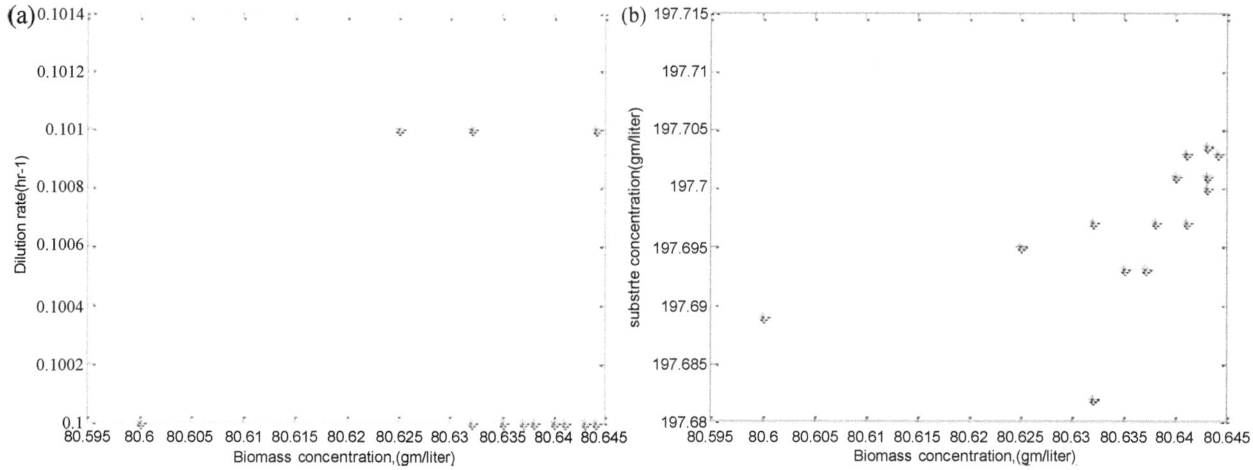

Figure 11: Optimal values of decision variables corresponding to objective X.

Optimal feed substrate was more sensitive to the variations of the objective biomass concentration.

5. Reasonable agreement was obtained when compared the simulated results with that obtained by the previous work.
6. Simulation and optimization provide the development of the process and reducing the risk of experimental runs and cost for design and operation.
7. Stochastic genetic algorithm was suitable search for the nonlinear biochemical reactor process.

References

1. Lee JM (1992) Biochemical Engineering. Prentice-Hall, UK.
2. Kapadia A, Nath N, Burg TC, Dawson DM (2010) Lyapunov-Based Continuous-Stirred Tank Bioreactor Control to Maximize Biomass Production Using the Haldane and Monod Specific Growth Model. American Control Conference, USA.
3. Henson MA (2003) Dynamic Modeling and Control of Yeast Cell Populations in Continuous Biochemical Reactors. Comput Chem Eng 27: 1185-1199.
4. Karadag D, Puhakka JA (2010) Effect of Changing Temperature on Anaerobic Hydrogen Production and Microbial Community Composition in an Open-mixed Culture Bioreactor. Int J Hydrogen Energ 35: 10954-10959.
5. Garhyan P, Elnashaie SSEH, Al-Haddal SM, Ibrahim G, Elshisini SS (2003) Exploration and Exploitation of Bifurcation/Chaotic Behavior of a Continuous Fermentor for the Production of Ethanol. Chem Eng Sci 58: 1497-1496.
6. Ruggeri B, Tommasi T, Sassi G (2009) Experimental kinetics and dynamics of hydrogen production on glucose by hydrogen forming bacteria (HFB) culture. Int J Hydrogen Energ 34: 753-763.
7. Charoenchai C, Fleet GH, Henschke PA (1998) Effects of temperature, pH and sugar concentration on the growth rates and cell biomass of wine yeasts. Am J Enol Vitic 49: 283-288.
8. Annamalai M, Doble M (2007) Modeling of D-Hydantoinase production by *Agrobacterium radiobacter* in a batch system. J Appl Sci 7: 2051-2056.
9. Alhumaizi K, Ajbar AH (2004) Optimization of an Unstructured First Order Kinetic Model of Cyclically Operated Bioreactors. J Environ Eng 132: 453-462.
10. Shimizu K (1993) An Overview on the Control System Design of Bioreactors. Measurement and Control 50: 65-84.
11. Gupta AK, Srivastava RK (2006) Integral Water Treatment Plant Design Optimization: Genetic Algorithm Based Approach. *IE Journal* 8.
12. Lopez FNA, Orlic S, Querol A, Barrio E (2009) Effects of Temperature, pH and Sugar Concentration on the Growth Parameters of *Saccharomyces Cerevisiae, S. Kudriavzevii* and their Interspecific Hybrid. Int J Food Microbiol 131: 120-127.
13. Cutlip MB, Shacham M (2007) Modular and Multilayer Modeling Application to Biological Processes. Comput Aided Chem Eng 24: 1019-1024.
14. Jarzebski AB (1992) Modelling of Oscillatory Behaviour in Continuous Ethanol Fermentation. Biotechnol Lett 14: 137-142.
15. Palonen M, Hasan A, Siren K (2009) A Genetic Algorithm for Optimization of Building Envelope and HVAC System Parameters. 11th International IBPSA Conference, Glasgow, Scotland, UK.

Batch Grinding of Dolomite Using Box-Behnken Design

Santhosh Kumar G* and Subba Rao Ch V

Department of Chemical Engineering, M.V.G.R College of Engineering, Vizianagaram, AP, India

Abstract

The aim of our research paper was to apply response surface methodology for modeling of grinding Dolomite ore using ball mill with the parameter considering;

1. Time of Grinding.
2. Ball Size.
3. Feed Quantity.
4. Feed Size.

Specific energy consumption also decreases with feed quantity at all ball sizes. Energy savings, expressed as a percentage of the energy consumption. Using these set of experimental data obtained, the models are developed to show the effect of each parameter and their interactions on the product. Maximum Specific surface is obtained at higher speed of the mill (66 rpm in our case). This also correspondence to the speed at which we have minimum energy consumption per unit surface area generated. As the grinding time increases specific surface increases. As the specific surface increases, power consumption also increases.

Keywords: Response surface methodology; Modeling; Box-Behnken design; Dolomite ore; Ball mill

Introduction

Yet grinding operations ought to be carried out for the effective processing of minerals, either to liberate mineral values or to prepare a suitable feed for separating processes like flotation, gravity concentration, classification, magnetic or electric separations etc. It is now widely understood that grinding to definite sizes must be economical [1]. The cost of energy consumption is a major factor in the crushing and grinding processes that influence the viability of a mineral processing operation [2]. In batch grinding, the particle size is continuously reduced with time but minerals of various complexities behave differently. In view of the fact, grinding studies of each mineral is to be studied to predict the product size with other influencing parameters [3]. The present study is contemplated to know the comminution parameters for Dolomite ore (Provided by RRL-Bhubaneswar).

Description

The ball mill used for the studies is an ordinary cylindrical type vessel with 8"×8" size, closed permanently at one end and with a provision to close the other end by using a lid [4,5]. The lid in turn is provided with a groove and gasket around the periphery and can be fixed to the mill my means of a suitable nut and bolt mechanism such that material does not leak out of the mill while in operation. The inside surface of the mill is smooth except that it is provided with three baffles of square cross section of 0.5"×0.5" size, along the length of the mill fixed firmly and evenly around the periphery.

The essential dimensions of the mill are given below

Inside diameter-8"=and length -8"

Thickness of the mill wall=0.38"

Theoretical critical speed of the mill–100rpm

Ball sizes and weight: 2.54 cm (1"), 64.7g.(100No.); 1.27 cm (1/2"), 28.75g.(225no.); 0.8467 cm (1/3"), 18.86g.(343no.); 0.635 cm (1/4"), 8.55g.(756no.); Width of the groove-2.54 cm.

A stepped pulley framework is connected at one end of the mill by means of the shafts through toothed wheel reduction gear system, which enables the mill to rotate on its horizontal axis. AV-belt connecting the motor and pulley frame work in turn enables the mill to rotate. The stepped pulley framework is provided with four steps and hence the mill can be run at four different speeds namely 23, 41, 66 & 82 rpm. The mill is supplied a with 100 balls of 2.54 cm diameter, 225 balls of 1.27 cm diameter, 343 balls of 0.8467 cm diameter and 756 balls of 0.635 cm diameter. With this arrangement the mill can be charged with the required amount of feed and the required size of the balls, lid placed in position, tightened to leak proof and run for a specified period of time at a selected speed. The other accessories used for conducting the experiment include a rotap sieve shaker with an automatic time switch, a set of BSS sieves with mesh numbers 10, 16, 30, 52, 60, 85 & 100 with lid and pan, a specific gravity bottle for measuring the density of Dolomite and a stop watch to note the time (Table 1).

Experimental Procedure

The feed material was prepared in the required size range to be used

Sl No	Parameters	Range of Variables
1	Time of grinding	2, 3, 4, 5 and 6
2	Mill speed	23, 41, 66 and 82 rpm
3	Size of balls	1", 1/2", 1/3" and 1/4""
4	Quantity of feed	50, 100, 150, 200, and 300
5	Feed size	–1/2" + 1/4" –1/4" + 1/8" –1/3" + 1/16" –1/16" + 1/32"

Table 1: Range of variables considered.

***Corresponding author:** Santhosh Kumar G, Department of Chemical Engineering, M.V.G.R College of Engineering, Vizianagaram-535002, AP, India in the contemplated study. The material was initially reduced to lower sizes and further crushed in crushing rolls. The crushed material was thoroughly screened using a set of sieves until the feed material of required size ranges was obtained. Totally four feed sizes (-1/2"+1/4", -3/8"+1/4", -1/4"+1/8", -1/8"+1/16") were selected for the present experimental study. The charge material was procured from the hills of sabbavaram village of vizag district, A.P. Experimental work is carried out to establish relationships for specific surface area generated and energy consumption for both the cases with and without additive and also to determine the specific breakage rate. It was also planned to establish the effect of several operating variables viz.,

1. Time of grinding in the mill (t, min.)
2. Speed of the mill (N, rpm)
3. Size of the balls (keeping the total weight of the balls constant) (B_s, "cm")
4. Quantity of feed (Q, gms)
5. Feed size (Fs, cm)
6. Dosage of Additive

Estimation of the specific surface and energy consumption

The batch grinding operation was carried out with the selected quantities of 50, 100, 150, 200 & 300 g. respectively and fed to the ball mill half filled with the grinding media of one inch steel balls counting 100 in number [6]. The open end of the mill was closed by clamping the lid tightly to the mill so that no leakage of the material takes place during grinding operation. Then the mill was run for a period of 2 min. and the reading of energy meter was noted. In the present study the whole experiment was conducted by running the mill at a constant speed of 66 rpm. The material was discharged from the mill using a brush after unclamping the lid.

The material was allowed to fall on a meshed tray placed on a cubical container. When the material was discharged on to the meshed tray, the balls would remain on the wired mesh while the grounded material will pass through the mesh and was collected in the container. Adequate measures were taken every where to prevent the loss of the material. The material was subjected to sieving as follows. It was divided into three equal fractions and then each part was subjected to sieve analysis independently on the set of seven BSS sieves with mesh numbers 10, 16, 30, 52, 60, 85 & 100 with a pan and lid. This method was adopted to prevent the overloading of the material on the sieves. A rotap sieve shaker was used for the purpose of sieving. Each part was subjected to sieving for a period of 6 min. using an automatic timer. All necessary steps for effective screening, like stopping the shaker for every 2 min. for gentle tapping and rotating the sieves by 180 degrees were taken. The material was carefully brushed from the sieves, collected and weighed on a balance and finally the material was discarded. The specific surface of the product at the end of the experiment was calculated from the sieve analysis data for each run of the mill using the formula,

$S= (Ks/\rho W)/(\Sigma wdP^{-1})$

Where, K_s=specific surface factor of the ground particles (6 for spherical particles),

w=weight of the product collected on a Particular sieve, g

W=weight of the total feed taken, g

d_p=mean size of the product retained on a sieve under consideration, cm

ρ=density of the material (g/cc).

Time of milling was studied at regular intervals of time with four different feed sizes and keeping 100 balls of 2.54 cm size, the speed of the mill at 66 rpm and weight of feed 200 gms [7].

To evaluate the effect of ball size, balls of same material of 0.635 cm, 0.8467 cm, 1.27 cm and 2.54 cm sizes were used by keeping total ball weight constant at 6470 g. The effect of the ball size was investigated at feed quantities of 50, 100, 150, 200 & 300 g. respectively, at a speed of 66 rpm and a milling time of 2 min. for four different feed sizes. The effect of feed quantity on specific surface area and energy consumption is studied with feed quantities of 100, 300, 500, 700 & 900 g. as charge material. The other parameters were maintained as 2 min. milling time at a mill speed of 66 rpm with 1.0" steel balls.

Results and Discussion

In this chapter the important variables which affect the performance of ball mill are studied [8]. The study consists in terms of the following factors: Surface area, size distribution of the product & power consumption and the following parameters:

1. Effect of time of milling
2. Effect of the size of balls
3. Effect of feed quantity
4. Effect of feed size

Effect of time of milling

The effect of time of grinding in a ball mill has been studied covering a range of 2 minutes to 6 minutes. 200 gms of feed of the size -1/2"+1/4" was fed to the mill and the mill was run at the speed of 66 rpm. The number of balls of one inch size kept as constant at 100. The specific surface are per each run is calculated and the results are shown in Figure 1. The Figure 1 shows the variation of specific surface with time, which shows that the specific surface produced, continues to increase with time. The variation of energy consumption with specific surface which reveals that power consumption increases with specific surface. The reason for increase is due to the fact that coarse particles were being introduced to the mill is easily ground. But after attaining a certain degree of fineness, further division becomes a slower due to the cushioning action of the fines formed, and may also due to small particles agglomeration, due to generated heat. Similar observations have been made by Datta and Nandi.

Effect of ball size

To study the ball size parameter on the performance of the ball mill measured in terms of the specific surface generated and energy consumed. Four sizes of ball Viz., 2.54 cm (1"), 1.27 cm (1/2"), 0.8467 cm (1/3") and 0.6350 (1/4") are chosen. The effect of ball size has been found for a feed quantity of 200 g and keeping the time of grinding at 2 min and speed of the mil at 66 rpm. Even though the sizes of the balls are changed, the total weight of the balls was kept constant at 6470 g. A graph is drawn to show the variation of specific surface with the size of the balls. From the figure it can be concluded that the specific surface area increases with ball size and shows a downtrend after a size of 1.27 cm. This is because of the increased void spaces in between the balls of larger diameter. Similar observations have been made by

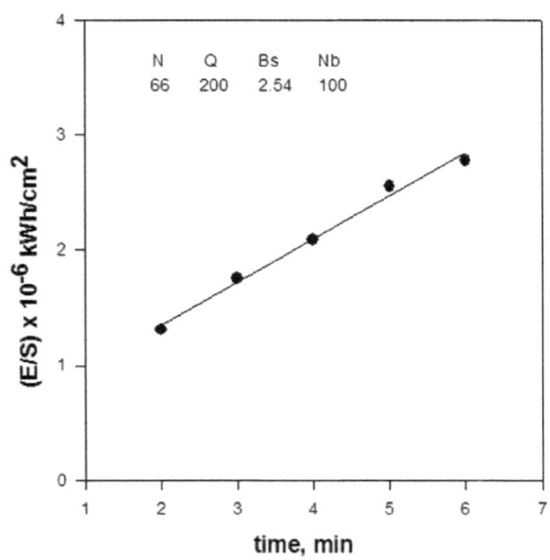

Figure 1: Variation of E/S with time.

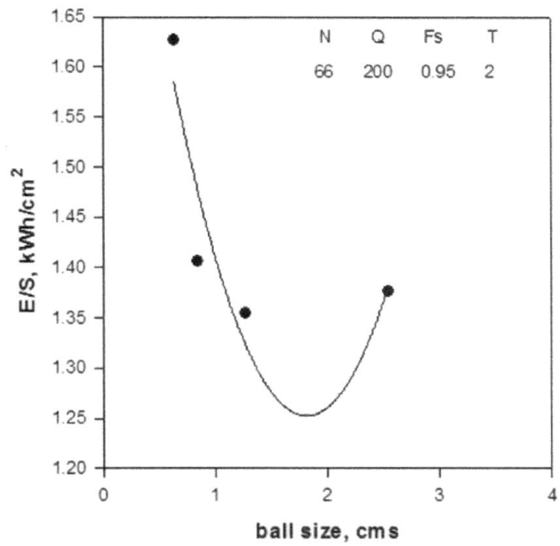

Figure 2: Variation of E/S with ball size.

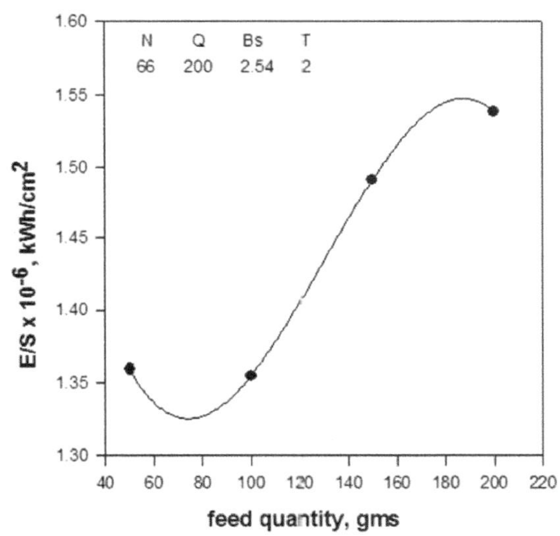

Figure 3: Variation of E'S with feed quantity.

Murthy and Arun Kumar. The variation of energy consumption with the ball size. The energy consumption is constant in the early stages and then increases with the further increase in ball size up to 1.27 cms. The variation of specific surface with energy consumption. E/S is constant in the beginning for initial small sizes of balls and then increases with increase in ball size. Figure 2 shows the variation of E/S with ball size. E/S decreases with increase in ball size.

Effect of feed quantity

To study the effect of feed quantity on the performance of the ball mill measured in terms of the surface area generated and the energy consumption, five samples of 50, 100, 150, 200 and 300 g respectively were taken. In all these systems the feed was ground for a grinding time of 2 min, at a mill speed of 66 rpm using 1" steel balls.

A graph was drawn between the quantity of feed and the specific surface; it reveals that the specific surface area decreases with increase in feed quantity. This is attributed to the choking of the mill with the increased feed quantity. At larger feed quantities, the chances of balls coming closer are meager; as a result combination by attrition and impact is greatly reduced and reveals the variation of energy consumption with the quantity of feed. The energy consumption decreased with an increase in feed quantity also reveals the variation of energy consumption with the specific surface. The energy consumption increased with an increase in specific surface. Figure 3 shows the variation of energy consumed per unit surface area produced with the quantity of feed. The E/S value is constant in the beginning upto 100 gms and then increased with the increase in feed quantities.

Effect of feed size

To study the effect of feed size on the performance of the mill measured in terms of the surface area generated and the energy consumption, five samples of sizes -1/2" + 1/4", -1/4" + 1/8", -1/8" + 1/16" and -1/16" + 1/32" respectively were taken. In all these systems the feed was ground for a grinding time of 2 min, at a mill speed of 66 rpm using 1" steel balls. A graph was drawn between the feed size and the specific surface as it reveals that the specific surface area decreases with increase in feed size and reveals the variation of energy consumption with the feed size. The energy consumption is continuously increasing and after attaining a maximum value at a feed size of 0.65 cms it starts getting decreased and it shows the variation of energy consumed with surface area. The energy gets decreased with the increase in the specific surface. Figure 4 shows the variation of energy consumed per unit surface area produced with the feed size. The E/S values increased with the increase in feed size.

Response Surface Methodology is used with Fractional Factorial Design

From the results of preliminary experiments the following factor levels were selected as Time of grinding (2-6min), feed quantity (50-200gms), ball size (.635-2.54 cm), feed size (.24-.95 cm) as listed in Table 2. 27 experimental runs were carried out according to Box-Behnken variable design with 3 additional runs at centre point to check

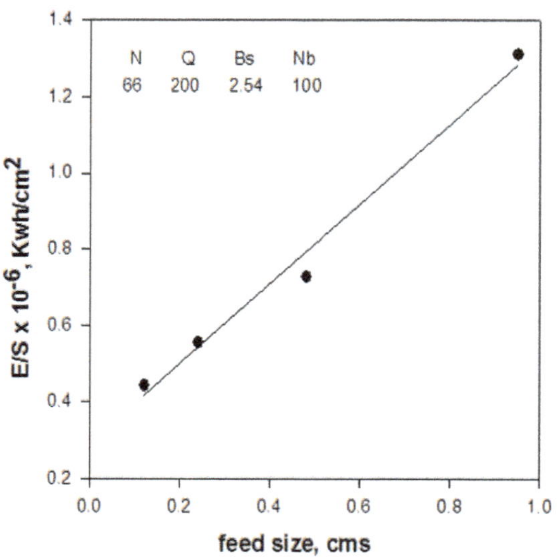

Figure 4: Variation of E/S with feed size.

the reproducibility. And the results were summarized in table.

The following quadratic equation was fitted to the above data using multiple linear regressions on statistical version 6.

$$Y= b_0 + b_1 x_1 + b_2x_2 + b_3x_3 + b_4x_4 + b_{11}x_1 2+b_{22}x_2^2+ b_{33}x_3^2 + b_{12} x_1x_2 + b_{13} x_1x_3 + b_{23} x_2x_3 + b_{44} x_4^2 + b_{14} x_1x_4 + b_{24} x_2x_4 + b_{34} x_3x_4 \quad (1)$$

The significance of each co-efficient was determined by students t-test and p-values, which are listed in Table 3. The larger the magnitude of the t-value and smaller the p-value, the more significant is the corresponding co-efficient. This data imply that first order main effects of time of grinding, feed quantity, ball size and feed size and their second order main effects of time of grinding, feed quantity and feed size are highly significant as is evident from their respective p-values ($px_1<0.00001$, $px_2<0.006$, $px_3<0.077$, $px_4<0.00001$, $px_1^2<0.006$, $px_2^2<0.002$, $px_3^2<0.02$, $px_4^2<0.00008$). The best model for minimizing energy consumption by a surface analysis was the quadratic polynomial model [9]. The fitted equation is

$$\text{Energy consumption}=103.02 + 25.82 x_1 - 10.08x_2 - 17.81x_3 - 59.97x_4 - 8.03x_1^2 - 9.18x_2^2 - 20.32x_3^2 + 0.60 x_1x_2 + 7.77x_1x_3 - 1.40x_2x_3 + 19.81x_4^2 + 2.06x_1x_4 - 0.34x_2x_4 - 3.64x_3x_4$$

The fit of the model was checked by the co-efficient of determination R^2, which was calculated to be 0.9506 indicating that 95.06% of the variability in the response could be explained by the model. The optimal conditions for the four factors were calculated to be 2.4559 min, 143.8056 gms, 1.6261 cm and 0.3407 cm for time of grinding, feed quantity, ball size and feed size respectively [10-12]. The model predicts the minimum response of 94.88 kw hr/ton for this point. The excellent correlation between predicted and measured values of these experiments justifies the validity of response model and the existence of minimum energy consumption.

Surface Area

Different surface areas are shown in the form of Response Surface Methodology (RSM) with Fractional Factorial Design (Figures 5-7).

Energy

Different energy fields are shown in the form of Response Surface Methodology (RSM) with Fractional Factorial Design (Figures 8-10).

Factors	Lower limit	Center Point	Upper Limit
Time of grinding (min)	2	4	6
Feed Quantity (gms)	50	100	200
Ball Size(cm)	0.635	1.27	2.54
Feed Size(cm)	0.24	0.48	0.95

Table 2: Range of factors.

	Std. Err.	t(21)	p	Coeff.
intercept	6.02331	17.1237	0.000000	103.1412
(1)time of grinding(L)	6.50329	7.5859	0.000000	24.6667
time of grinding(Q)	5.30991	-3.0342	0.006309	-8.0556
(2)feed quantity(L)	6.50329	-3.0412	0.006208	-9.8889
feed quantity(Q)	5.41941	-3.3966	0.002720	-9.2037
(3)ball size(L)	18.67928	-1.8559	0.077559	-17.3333
ball size(Q)	16.25822	-2.5081	0.020416	-20.3889
(4)feed size(L)	9.40588	-12.6522	0.000000	-59.5024
feed size(Q)	8.17766	4.8376	0.000088	19.7801

Table 3: Values of p and t.

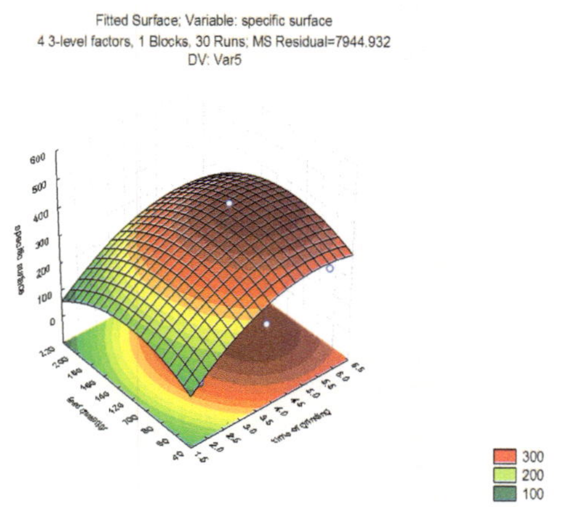

Figure 5: Response surface methodology of specific surface area with Fractional Factorial Design.

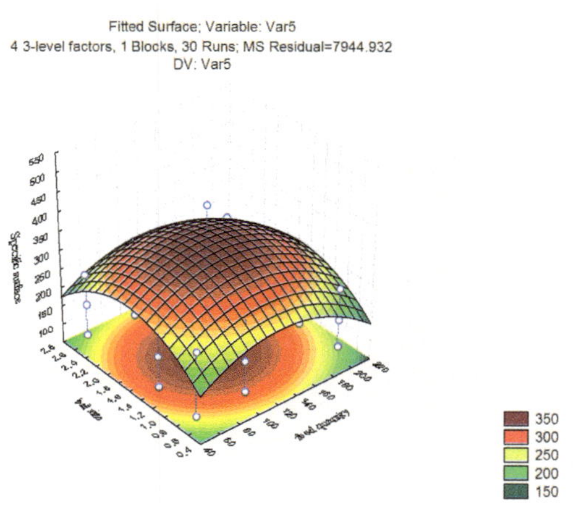

Figure 6: Response surface methodology of var5 with Fractional Factorial Design.

Conclusions

1. As the grinding time increases, specific surface increases. As the specific surface increases, power consumption also increases .Power consumption per unit specific surface increases with time.

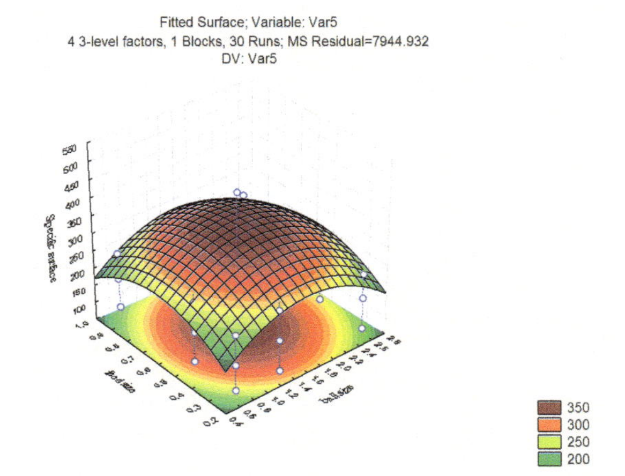

Figure 7: Response surface methodology of var5 with Fractional Factorial Design.

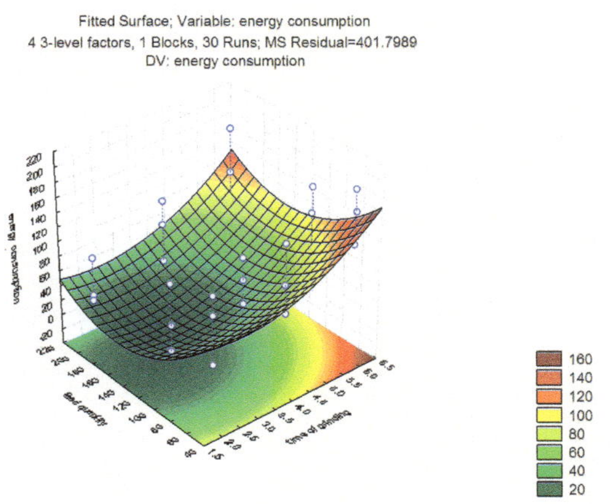

Figure 8: Response surface methodology of Energy consumption with Fractional Factorial Design.

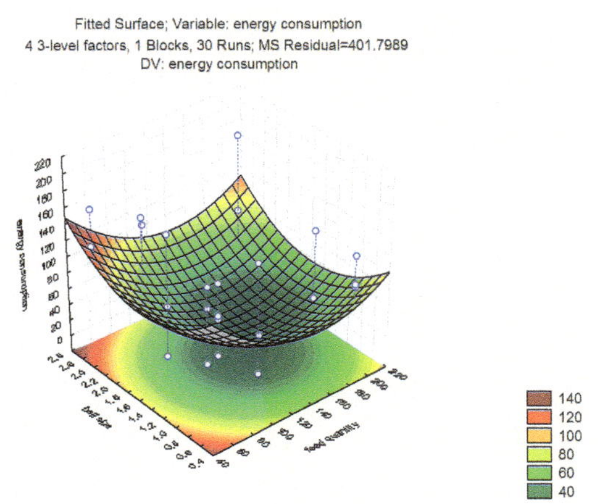

Figure 9: Response surface methodology of Energy consumption.

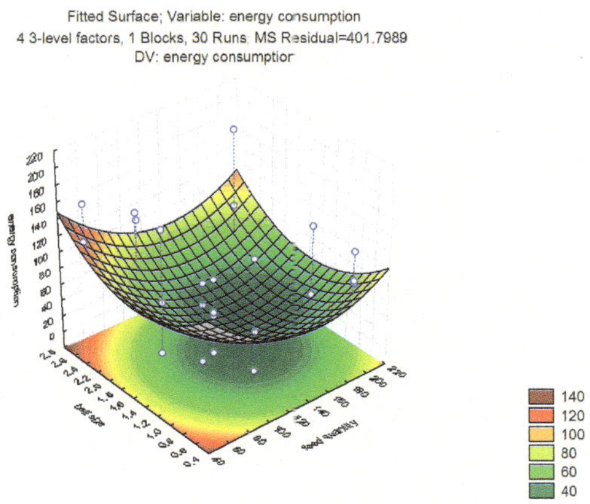

Figure 10: Response surface methodology of variable energy points.

2. As the size of the ball increases specific surface increases up to a certain value of Bs and then decreases.
3. Power consumption is almost same for all ball sizes for a particular amount of feed.
4. Power consumption per unit specific surface area of feed decreases up to certain ball size and then increases.
5. Power consumption of feed decreases with quantity of feed. The new specific surface generated also decreases with increase in feed quantity.
6. The power consumption per unit specific surface of the feed is first decreases and then increases.

References

1. Parasivam R, Vedaraman R (1992) Effects of the physical properties of liquid additives on dry grinding. Powder Technology 70: 43-50.
2. Pryor EJ (1960) Mineral processing. Mining Publications Ltd., London.
3. Taggart AF, Behre HA (1960) Hand Book of Mineral Dressing (Ores and Industrial Minerals), John Wiley and Sons, Inc Newyork.
4. Fuerstenau DW, Abouzeid AZM (1992) Energy split and kinetics of ball mill grinding of mixture feeds in heterogeneous environment. Powder Technology 72: 105-111.
5. McCabe W, Smith J, Harriott P (1996) Unit Operations of Chemical Engineering, 820, second Edition, McGraw Hill Company, New York.
6. Magdalinovic NM (1989) Calculation of energy required for grinding in a ball mill. Int J Mineral processing 25: 41-46.
7. Fuerstenau PW, Abouzeid AZM (1991) Effect of particulate environment on the kinetics and energetics of dry ball milling. Int J Miner Process 97: 152-162.
8. Zeng Y, Forssberg E (1993) Monitoring grinding parameters by signal measurements for an industrial ball mill. Int J Mineral Process 40: 1-16.
9. Nomura S, Hasoda K, Tanaka T (1991) An analysis of the selection function for mills using balls as grinding media. Powder Technology 68: 1-12.
10. Schwedes J, Bunge F (1992) Comminution and transport behaviour in agitated ball mills. Advanced Powder technology 3: 55-70.
11. Moothedath SK, Ahluwalia SC (1992) Mechanism of action of grinding aids in comminution. Powder Technology 71: 229-237.
12. Sohoni S, Sridhar R, Mandal G (1991) The effect of grinding aids on the fine grinding of limestone, quartz and Portland cement clinker. Powder Technology 67: 277-286.

Generating Electricity from Fluegas Produced By Boilers through a Thermodynamic ORC (Organic Rankine Cycle) Driven by R245fa

Omid Rowshanaie[1]*, Saari Bin Mustapha[2] and Hooman Rowshanaie[3]
[1]*University Putra Malaysia, Department of Chemical Engineering, Serdang, Malaysia*
[2]*University Putra Malaysia, Department of Chemical and Environmental Engineering, Serdang, Malaysia*
[3]*University Putra Malaysia, Serdang, Malaysia*

Abstract

In recent years, an increase in fossil fuel prices and imposing strict laws and regulations by environmental organizations have resulted in an increasing interest on the parts of researchers in oil industry to implement various projects to eliminate pollutants emitted by fluegas in order to achieved the desired standards. Today, different processes including heaters, furnaces, and boilers are used by different industries resulting in the production of a large volume of fluegas. The energy and heat wasted by these gases can be recovered even in low temperatures in order to be used in various processes such as preheating refinery fluid flows or generating electricity. ORC (Organic Rankine Cycle) is a thermodynamic cycles used for generating electricity in the majority of plants and also in this article; furthermore, its function is similar to Karno cycle. In ORC thermodynamic cycle, superhot vapor of fluegas is first produced by a boiler and then it is sent to an evaporator then sent to a turbo-expander to produce electricity. Finally, the remaining vapor is condensed by an air cooler and it is sent back to the ORC thermodynamic cycle. ORC thermodynamic cycles are usually based on vapor working fluid fluids such as NOVEC7000, R123, R134a, and R245fa (1,1,1,3,3-pentafluoropropane), (employed in the present article) as the working fluid of ORC thermodynamic cycle.

Keywords: Fluegas; ORC thermodynamic cycle; Evaporator; Turbo-expander; Air cooler

Introduction

Nowadays the oil price is dramatically increase and this phenomenon is due to the all economic obstacle in all over the world especially China, India, and Iran, On the other hand, the governments try to apply the greenhouse gases that produce from boilers such as fluegas to increase the efficiency of fossils fuels and decrease the negative aspects of these kinds of gases such as worldwide disaster as same as global warming and also air pollution; in addition, the grade of temperature of these type of gas is a little bit high, therefore we can use this type of gases in ORC (Organic Rankine Cycle) [1].

The most important obstacle in all ORC thermodynamic cycles is choosing the suitable working fluid. In this ORC thermodynamic cycle uses a number of working fluids such as: NOVEC7000, R123, R134a, and R245fa that some reason for choose one or two working fluids from these, are more important, the first and foremost is thermodynamic performance, another one is environmental behavior, and the last but not the least which in industrial world has a more benefit effect is system cost. For choosing the suitable and affordable working fluid should be focus on some point that in the below mention them.

The first and foremost is working fluid should be non-flammable, for knowing this properties would refer to the auto ignition temperature, and this is non-fiction, if auto ignition temperature of a working fluid is high, this working fluid is closed to non-flammable. Another point which causes to select a suitable working fluid for this thermodynamic cycle is volatility because when do not use turbo-expander in a hermetic and tightness container, the range of volatility of working fluid would be low for increasing the efficiency of ORC cycle. Another point is reasonable and low boiling point temperature in a normal working condition because even though boiling point temperature of working fluid is low; therefore, the evaporator need to consume lower heat energy to change the phase of working fluid from liquid to gas; although, ORC thermodynamic cycle need a little evaporator instead of big one, and the system cost is decreasing. Other points which can consider together and get the same result to select an affordable working fluid for this thermodynamic cycle are labeling pictograms, hazard statement, and toxicity of these working fluids that can see in Table 1. The last but not the least point that more influence for choosing the suitable working fluid is high heat of vaporization ($\Delta h_{Vap.}$) and high density in vapor state ($\rho_{Vap.}$) in reference temperature. Because when these properties are high for working fluid, it means the ORC thermodynamic cycle need a smaller turbo-expander; furthermore, the cost of system is decreasing slightly.

All in all, after considering all these properties which show in Table 1, the best choice for working fluid which using in this ORC thermodynamic cycle is R245fa [2].

Material and Methods

ORC Process

The process of this ORC thermodynamic cycle which show in Figure 1 is very marvelous and interest. In this ORC thermodynamic cycle, liquid R245fa as a working fluid is not completely pure and has 0.1 mole fraction H_2O. For purification or if better say, for increasing the mole fraction of R245fa and decreasing the mole fraction of H_2O, first of all the working fluid with 30×10^6 kg/h mass flow rate, 14.69°C temperature, and 101.3 KPa pressure transfer to a separator (S-100).

***Corresponding author:** Omid Rowshanaie, University Putra Malaysia, Department of Chemical Engineering, Serdang, Malaysia

	R245fa	NOVEC 7000	R123	R134a
Autoignition Temperature (°C)	412	415	770	>750
Volatility Range	Low	Medium	High	Low
Boiling Point @ 1atm (°C)	15.3	34	27.9	-26.2
Labeling Pictograms				
Hazard Statement	Contains gas under pressure; may explode if heated. Asphyxiate in high concentrations.	Contains gas under pressure. Acidity.	Acidity. Colorless. Volatile liquid with ethereal and faint sweetish odor.	Has Halogens acids and possibly carbonyl halides. Hazardous polymerization will not occur.
$\Delta h_{Vap.}$ (kj/kg) @ 100 °C	134.4	105.9	Not Available	Not Available
$P_{Vap.}$ (kg/m^3) @ 100 °C	73.15	53.03	Not Available	Not Available
Toxicity	Oral Inhalation Dermal	Skin Oral Inhalation Eyes Ingestion	Skin Eyes Inhalation Ingestion	Skin Eyes Inhalation Ingestion

Table 1: Choosing the suitable working fluid in ORC thermodynamic cycle.

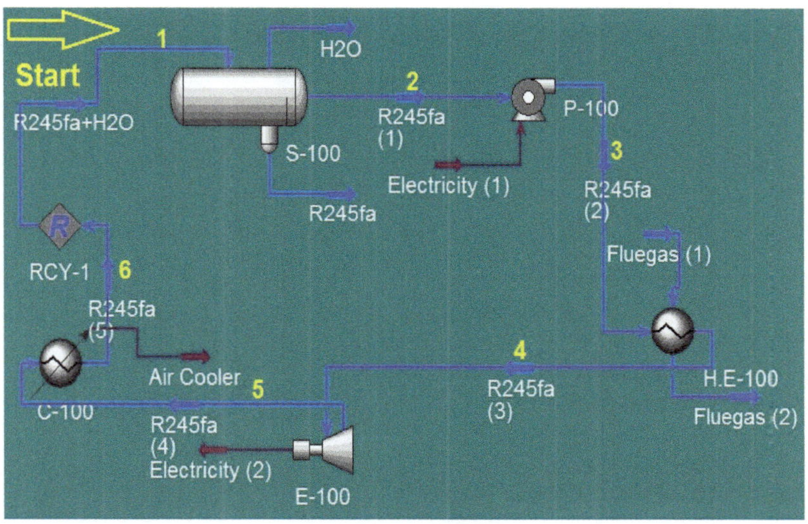

Figure 1: Scheme of ORC.

After completing the purification process of working fluid in separator, and decreasing the mole fraction of R245fa and H2O to 0.9998 and 0.0002 respectively, R245fa enter to the electrical pump (P-100) with 1651.86 hp for adjust the fluid flow and increasing the pressure, then R245fa enter to the heat exchanger (H.E-100) for changing the phase from liquid to gas because after this equipment, the working fluid should be enter to the expander (E-100) as a feed for producing the electricity. This heat exchanger (H.E-100) for changing the phase of working fluid from liquid to gas, need a heat source which named fluegas. Fluegas has 180-220 °C temperature, but in Hysys simulation software because this software cannot accept the temperature range for fluegas. Unwillingly and reluctantly using the average of this temperature range it means 200°C, 10980 KPa pressure and 67340 m^3/h volume flow rate for inlet and after exit the fluegas from heat exchanger (H.E-100), the temperature and pressure decrease to 80°C and 253.6 KPa respectively. After exit the R245fa from this heat exchanger (H.E-100) the temperature is increasing and pressure is decreasing. Then R245fa in gas phase enter to the expander (E-100) which working fluid causes to rotate the seal shaft and producing the electricity energy at about 6408 KW under best working condition and state. After exit the R245fa from expander, for recovery of R245fa, it means changing the phase of this working fluid from gas to liquid and can use again in ORC thermodynamic cycle should enter to cooler (C-100) which working by cool drops water with cool air. So with changing the phase of R245fa as a working fluid of this ORC thermodynamic cycle, R245fa can come back to this thermodynamic cycle and using again for producing electricity.

The most important point for adding a preheater as a modification is in this ORC thermodynamic cycle cannot add a preheater such as: a heat exchanger or a heater, before the evaporator (E-100), because if add a preheater before evaporator (E-100), the negative pressure drop

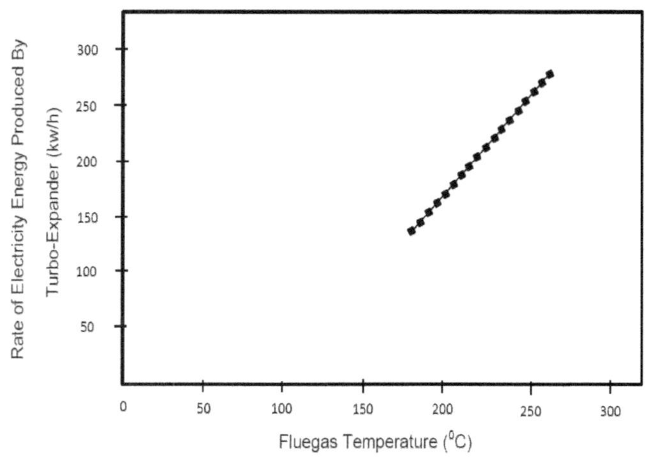

Figure 2: Increase in the usage of exhaust heat the efficiency and the outlet of net power of ORC thermodynamic cycle.

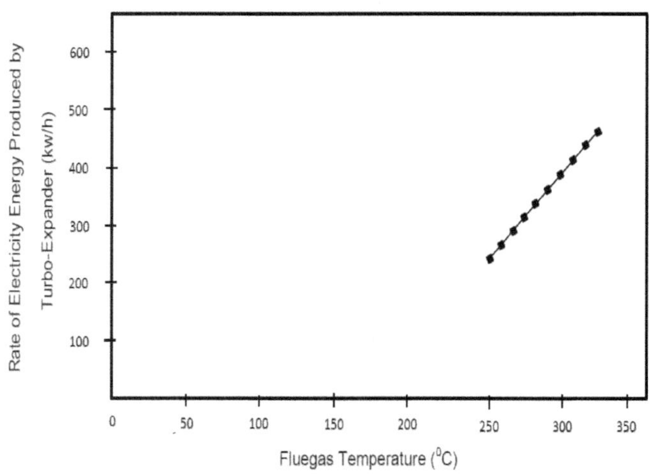

Figure 3: low temperature range of fluegas and the amount of electricity energy produced is equal to high efficiency of ORC thermodynamic cycle.

occur at next equipment; in other word, in evaporator the pressure is decreasing dramatically, and cannot use two evaporator (first one as a preheater and second one as a phase changer) together in this ORC thermodynamic cycle.

The following equation developed by Bourji et al. [3] which can estimate the amount of electricity generated by a simple ORC thermodynamic cycle:

$P = Q_{fg}[AT_{fg}T_a+BT_a+CT_{fg}+D]$

P: amount of electricity energy [kw]

Q_{fg}: volume flow rate of fluegas [m^3/h]

T_{fg}: temperature of inlet fluegas [°C]

T_a: ambient temperature (=21) [°C]

A: -0.00411 [without dimension]

B: 0.775 [without dimension]

C: 1.122 [without dimension]

D: -211.63 [without dimension]

This equation has application to 177-260°C, and defining an area which a working fluid is becoming this area to a super critical area. The temperature and pressure range of this super critical area is lower than critical point of working fluid.

Conclusion

In a nutshell, statistical estimation of other researches reveals this fact that add or remove and increase or decrease some equipment or parameters can cause to increase the performance and efficiency of ORC thermodynamic cycles. In other word, with increasing the usage of exhaust heat the efficiency and the outlet of net power of ORC thermodynamic cycle are increasing as well as possible. The most important thing that impact on net power and efficiency is selecting a sufficient working condition for ORC thermodynamic cycle. If want to have the appropriate efficiency and net power output in ORC thermodynamic cycles, the temperature of industrial process as the exhaust heat should not more than 370°C; furthermore, more than this temperature can cause heat pollution in environment and increasing the global warming. All in all, the organic Rankine cycle (ORC) has a number of advantages such as: great flexibility, high safety, and low maintenance [4].

Bourji et al. [3] doing similar research in this area, but different heat source temperature (fluegas temperature) which show in Figure 2. They using low temperature range of fluegas for this ORC process, and the amount of electricity energy produced in their research are equal to 8815 KW under high efficiency of ORC thermodynamic cycle. But present research has a different temperature range of fluegas which show in Figure 3. The amount of electricity energy produced in present research is equal to 6408 KW under high efficiency of ORC thermodynamic cycle and without any negative pressure drop in equipment.

After compare above figures to each other, Bourji et al. [3] and present research claimed that the rate of electricity energy produced by turbo-expander of ORC thermodynamic cycle with fluegas temperature have a linear relative to each other; otherwise, with increasing the fluegas temperature, the rate of electricity energy is also increasing in both above figures.

Another result that achieve in present research is relating between mass flow rate of R245fa as a working fluid of this ORC thermodynamic

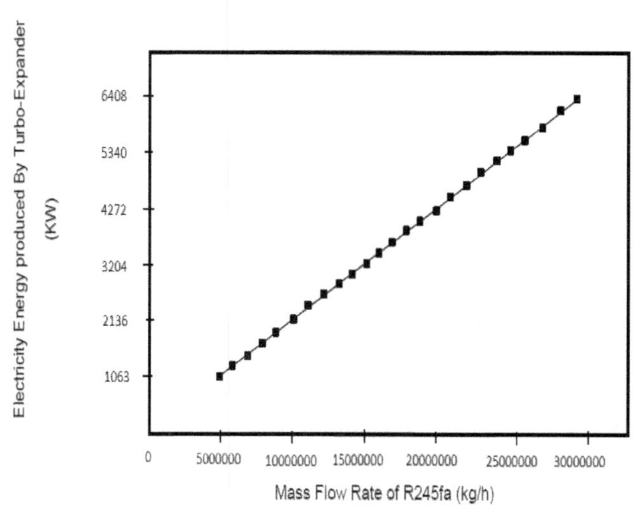

Figure 4: Comparison between mass flow rate of working fluid and electricity energy produced by turbo-expander of ORC thermodynamic cycle.

cycle and electricity energy produced by turbo-expander of ORC thermodynamic cycle which show in Figure 4.

In these results which show in Figure 4, with increasing the mass flow rate of R245fa, the electricity energy produced by turbo-expander is also increasing. Because when the mass flow rate of R245fa is increasing, it means the potential energy of R245fa for rotating the seal shaft of turbo-expander is increasing, and as a result, the amount of electricity energy is increasing.

At the end, if refinery employers using ORC process which simulated in present research, they can become the refineries to independent from electricity energy producing, and decreasing the cost of refineries for buying the required electricity energy of refinery equipment from electricity power plants.

References

1. Kang SH (2012) Design and experimental study of Organic Rankine Cycle (ORC) and radial turbine using R245fa working fluid. Energy 41: 514-524.
2. Declaye S, Quoilin S, Lemort V (2010) Design and experimental Investigation of a small scale Organic Rankine Cycle using a Scroll Expander. Refrigeration and Air conditioning Int. Conference, Purdue University, USA.
3. Ali B, John B, Jimmy W, Alan W (2010) Recover Waste Heat from Fluegas. Chemical engineering magazine.
4. Wei D, Lu X, Lu Z, Gu J (2007) Performance analysis and optimization of Organic Rankine Cycle (ORC) for waste heat recovery. Energy Conversion & Management 48: 1113-1119.

Study the Remove and Transport Mechanisms of Seleniums in Membrane (UF and NF) Processes

Mustapha Chabane* and Benamar Dahmani

Spectrochemistry and structural pharmacology laboratory, Department of chemistry, Science faculty, University of Tlemcen, Algeria

Abstract

The removal and transport of ionized and non-ionized selenium's were systemically investigated in this study. The charge repulsion between the membrane surface and selenium ions was an important mechanism for the rejection of ions by a charged membrane. In addition, selenium removal was significantly affected by the J_0/k ratio, but was not affected by the cross-flow velocity. Both diffusion and convection are proven to affect the transport of selenium ions transports. The reflection coefficients of both UF and NF membranes increased with increasing the pH, and with the reflection coefficients of selenate are being higher than those of selenite at under the same operating conditions. The spiral-wound type module exhibited slightly higher selenate removal compared to than the flat-sheet type module at under the same operation conditions.

Keywords: Selenium; Diffusion; Convection; J_0/k ratio; Membrane

Introduction

Selenium is an essential element for animal nutrition but there is narrow range of diatery [1]. Selenium has been regulated is regarded as a highly toxic chemical by World health organization(WHO) and EU to recommend maximum selenium concentration in drinking water of 10 ppb while EPA sets limit of 50ppb [2]. There are many processes to for treating selenium, including coagulation, ion exchange resins, and membranes [3]. However, for the removal of seleniums compounds down up to the MCL level of 10 µg/lL, membrane processes is believed to be the only alternative. Selenium exhibits broad toxic properties and complex chemistry characteristics, that which govern the speciation and distribution of its two major oxidation states. Both selenate (Se (VI)) and selenite (Se (IV)) may exhibit different transport phenomena in there natural system, depending on the pH, ionic strength, and oxidation state [4].

Two different mechanisms for the removal of a solute; size and electrostatic exclusions, can be resulted from the solute charge and the charged membrane surface. To demonstrate removal mechanisms in NF and UF membranes, transport phenomena for a solute through the membrane pores should be rigorously investigated, along with membrane characterizations in terms of relating to the surface charge and pore size distribution.

Therefore, the objectives of this study were to determine the important transport mechanisms of NF and UF membrane filtrations in terms of the mass transfer coefficients, and diffusivities, of both selenate and selenite ions, to determine the optimum operating conditions for their maximum removal, and to predict the seleniums removal performance of selenium compounds by membranes.

Theories

The J_0/k ratio can be used to control hydrodynamic operating conditions. For the same J_0/k ratio, it is hypothesized that solute removal, and flux decline trends, for a given solute-membrane pair are almost identical[3,5]. J_0 (the initial pure water permeation flux (equation. (1)) and k (the back-diffusional mass transfer coefficient) can be calculated from the following equations (6).

$$J_0 = \frac{Q_p}{A_m} \tag{1}$$

$$k = 1.62\left(\frac{UD^2}{d_h L}\right)^{0.33} \quad \text{(thin-channel-type module)} \tag{2}$$

Here, Q_p and A_m are the permeate flow rate (cm^3/sec) and membrane surface area (cm^2), respectively. Equation (2) is derived from the Sherwood number (Sh=kdh/D) with consideration of module configuration (flow regime), where U is the average velocity of the feed fluid (cross-flow velocity (cm/sec));, D is the diffusion coefficient of the solute (cm^2/sec) estimated by the Stokes-Einstein relationship, dh is the equivalent hydraulic diameter (cm), and L is the channel length (cm).

The non-equilibrium thermodynamic equation takes into account two driving forces (convection and diffusion). To describe this transport characteristics, the solute flux can be expressed by the equation (3) suggested by Kedem and Katchalsky [6], which is based on irreversible thermodynamics considering the membrane is considered as a black box, since the transport mechanisms are not clearly understood.

$$J_s = P_m\left(C_m - C_p\right) + J_v(1-\sigma)\cdot C^* \tag{3}$$

Here, J_s, P_m, and C* represent the solute flux, the solute permeability (the diffusive transport of the solute through the membrane pores), the reflection coefficient (which is a measure of the selectivity of a membrane for a solute), and the logarithmic average concentration ($C^* = (C_m - C_p)/\ln(C_m/C_p)$), respectively. J_v, C_m, and C_p represent solvent flux, solute concentration near membrane, and permeate side solute

***Corresponding author:** Mustapha Chabane, Spectrochemistry and structural pharmacology laboratory, Department of chemistry, Science faculty, University of Tlemcen, Algeria concentration, respectively. The values of C_m and J_v can be obtained from the concentration polarization (CP) relationship, and Js equals the product of J_v and C_p. Using experimentally determined parameters at different J_0/k ratios, including (C_m-C_p), J_s, J_v, and C*, P_m and can then be determined by a nonlinear estimation with a three dimensional plot of J_s versus J_v versus C*.

Materials and Methods

Two different types of bench-scale units; including a spiral-wound and a flat sheet membrane module were tested under various hydrodynamic operating conditions in terms of determined from J_0/k ratios. The initial test J_0/k ratios tests were performed obtained with deionized (DI) water, prior to the actual tests with selenium containing feed waters containing selenium. A particular J_0/k ratio was adjusted by controlling varying J_0 (controllable with trans-membrane pressure valve) and a fixed value of k (changeable by cross-flow velocity control), or vice versa. Based on these pre-determined J_0/k ratios, membrane filtration tests were performed with feed water and membranes. All of the filtration experiments were conducted in the recycle mode.

Selenium removal measurements were conducted on one NF membrane (ESNA (Hydraunautics)) and two UF membranes (PW (Desal.) and GM (Desal.)). The membrane properties, in terms of MWCO, charge, and pure water permeability, are listed in Table 1.

Synthetic solutions were used to perform bench-scale membrane tests. These solutions were prepared with pure water, and spiked with either 100 µg/L of selenate (Se (IV)) or selenite (Se (VI)) was spiked as a pure component, at various pHs conditions (4, 7, and 10). Sodium hydroxide (NaOH) and hydrochloric acid (HCl) were added to the synthetic solutions to adjust solution the pH. Selenium concentrations were measured with continuous hydride generation (PERKIN ELMER FIAS 400) using an Atomic Adsorption Spectroscopy (AAS, PERKIN ELMER 5100ZL).

Membrane	Material	MWCO (Dalton)	Zeta Potential (mV) @ pH 7.0	PWP (l/day-m²-kPa)	Contact angle (°)
GM (UF)	Polyamide TFC	8000	-45.09	4.34	61.0
PW(UF)	Polyethersulfonate	10000	-29.11	28.22	73.9
ESNA (NF)	Polyamide TFC	250	-9.90	2.35	63.0

Table 1: Membrane properties.

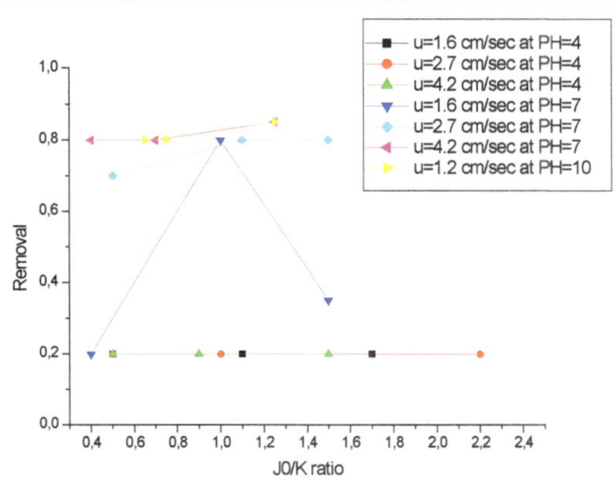

Figure 1: The results of selenium ions removal by the spiral wound type module with the NF membrane.

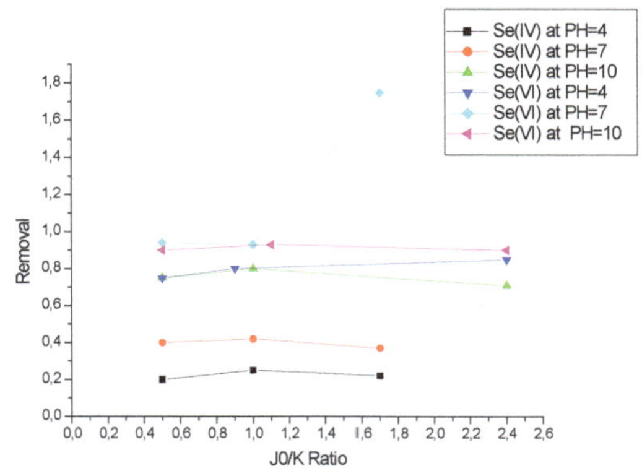

Figure 2: The removal behaviors of selinite and selenate ions by the NF membrane.

This study was comprised of three main experiments. Experiment I included the characterization of the membrane characterizations in terms of surface charge and pore size distribution. Experiment II included the evaluation of the ion removal efficiencies of ions by changing various chemical and hydrodynamic operating conditions. Experiment III included the determinations of transport coefficients, including the solute permeability (P_m) and the reflection coefficient (σ), with using the modified Kedem and Katalsky [6] models; the modified models used implied convection and diffusion transport terms, and transports coefficients (k), and with the reflection coefficient were being determined at the constant cross-flow velocity mode.

Hypothesis

The removal of seleniums compounds removal by NF and UF membranes can be substantially affected by electrostatic exclusion. It is hypothesized that selenate and a selenite exhibit different transport characteristics, which are pH and oxidation state depending on pH and oxidation state. Selenium rejection is hypothesized to increase at higher pHs and in the higher oxidation state. Transports of selenium compounds during the operation can be affected by both membrane pore sizes and charges. In addition, operating conditions, in terms of J_0/k ratios, is also important to selenium transportations. It is also hypothesized that the transport characteristics of selenium compounds may exhibit differently with different types of membrane type and J_0/k ratios.

Results and Discussions

Selenium removal by the spiral-wound module of the NF membrane

Figure 1 presents shows the results of selenite ions rejection by the spiral-wound type module of the NF membrane.

The removal of selenite increased with increasing pH value due to the charge repulsion. The relationship between selenite removal and pH was also investigated. The removals of selenite ions in the pH range of 7 to 10 were much higher than for those in the pH range of 4 to 7. This is because the pKa value of selenite is approximately 9.2, so in the latter pH range selenite ions are still are non-charged species in pH range of 4 ~ 7.

The J_0/k ratio also affected selinite removal. Without charge repulsion (pH 4 and 7), the maximum value of selinite removal was found for a J_0/k ratio of approximately 1.0. That means the major transport mechanism changed from diffusion to convection at a J_0/k ratio of 1.0. But tThe removal of selinite increased with increasing values of J_0/k ratio at pH 10. Unlike pH and J_0/k ratio, the cross-flow velocity (U) did not affect significantly the removal of selinite. The same experiments were performed for selenate. However, selenate exhibited high removal values (above 0.92) due to electrostatic interaction between the negatively-charge membrane and ionized selenate ions.

Selenium removal by flat-sheet types module with the NF and UF membranes

Figures 2 and 3 show the removal behaviors of selinite and selenate ions conducted by using the flat-sheet type module of with both the NF (ESNA) and UF (GM) membranes. At the same pH and concentration of ions, the removal of selenium compounds removals by the NF membrane was higher than those by the UF membrane due to size exclusion. However, the UF membrane has the potential to remove selenium removal up to 0.87. The Diffusion is always the major transport mechanism for the removal of selenate and charged selinite ions by the NF membrane was always diffusion, but the major transport mechanism of uncharged selinite changed from diffusion to convection at J_0/k ratio of approximately 1. On the other hand conversely, the major transport mechanism of selinate and selenite

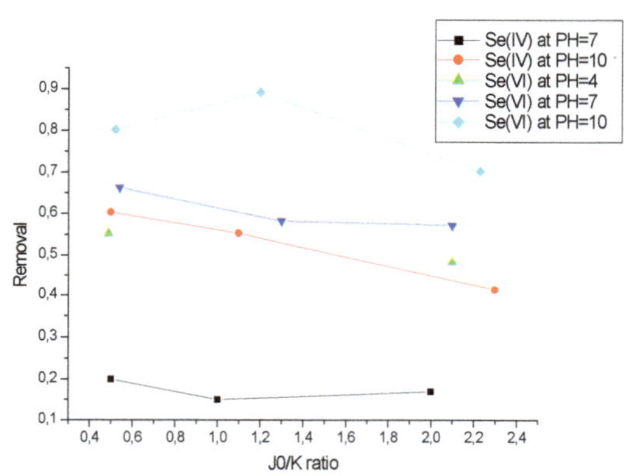

Figure 3: The removal behaviors of selinite and selenate ions by the UF membrane.

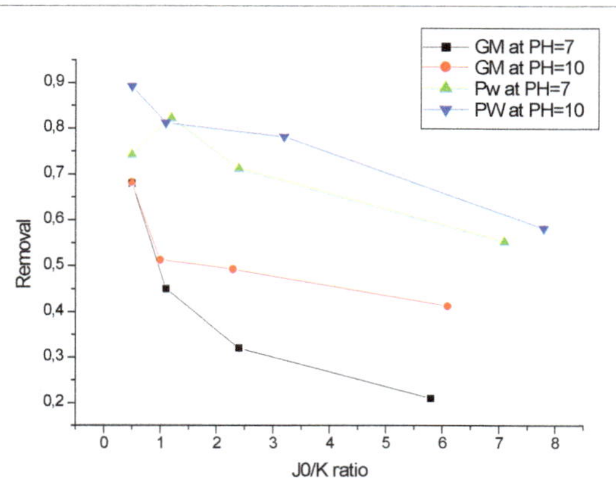

Figure 4: Selenate removal at pH value 7 and 10 with two different UF membranes (PW and GM).

	Membrane	Module	pH	σ	P_m (cm/sec)
Selenate	ESNA	Spiral-wound	10	0.85	7.1×10^{-6}
	ESNA	Spiral-wound	7	0.82	6.8×10^{-6}
	ESNA	Spiral-wound	4	0.84	6.9×10^{-6}
	GM	Flat-sheet	10	0.62	2.3×10^{-4}
	GM	Flat-sheet	7	0.43	2.2×10^{-4}
	GM	Flat-sheet	4	0.19	2.0×10^{-4}
	PW	Flat-sheet	10	0.51	1.7×10^{-3}
Selenite	ESNA	Spiral-wound	10	0.80	7.0×10^{-6}
	ESNA	Spiral-wound	7	0.61	6.9×10^{-6}
	ESNA	Spiral-wound	4	0.31	6.9×10^{-6}
	GM	Flat-sheet	10	0.31	2.2×10^{-4}
	GM	Flat-sheet	7	0.13	2.2×10^{-4}
	PW	Flat-sheet	10	0.11	1.7×10^{-3}

Table 2: Estimated transport coefficients.

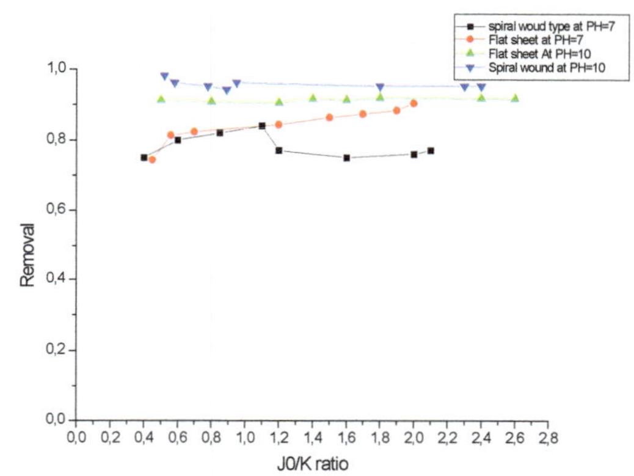

Figure 5: Effects of Module type on selenium Removal.

with the UF membrane was convection, with the exception for highly-charged selenate (i.e. selenate at pH of 10).

Selenium removal by flat-sheet types of two different UF membranes

Figure 4 shows the removal behaviors of selenium ions removal at two different pHs conditions (pH =of 7 and 10,) by two different UF membranes (GM and PW). The major transport mechanism of the PW membrane is always convection in spite of even in conditions of high pH condition (pH = 10), but the major transports mechanism of the GM membrane, for highly charged selenate, changed from diffusion to convection at J_0/k ratio above approximately 1.0 It is believed that pore size of the GM membrane (MWCO of 8000) is slight larger than that of the PW membrane (MWCO of 1000).

In addition, the selenate removal by the GM membrane was slightly higher than that of the PW membrane. However, the water permeability of the PW membrane is quite significantly higher than that of the GM membrane, thus the PW membrane is believed to be more efficient in the case of selenate at a higher pH conditions than the GM membrane based on the operation condition (especially low J_0/K ratio value)

Determination of Transport Coefficients from the Bench-scale Membrane Tests

Using equation (3), transports coefficients were estimated. Table 2 lists the detail transport coefficients

From table 2, it was found that membrane pore sizes, as well as charge solute properties y solute, and the feed solution chemistry were affected membrane transport. Both diffusion and convection are proven to affect the ion transports of ions,. However, in the case of the NF membrane, diffusion is the most dominant transport mechanism over convection, but whereas convection is more dominant mechanism in case for the UF membrane, convection is more dominant transport mechanism over diffusion. The reflection coefficients (σ) for both the UF and NF membranes were increased with increasing the pH values, and the reflection coefficients (σ) of selenate were being higher than those for selenites,.which can both results can be explained by the charge repulsion.

Effect of Module Type on selenium Removal (Spiral-Wound Type vs. Flat-Sheet Type)

Figure 5 shows the effects of module type on selenium removal. Selinate removal, under both neutral and basic conditions (pH 7 and 10), was evaluated. From the results, the spiral-wound type module showed slightly higher selinate removal than the flat sheet module, type membrane under for both pHs conditions.

Conclusions

From bench-scale membrane tests, electrostatic interactions are an important mechanism for the removal selenium ions by both UF and NF membranes. Membrane transports were affected by membrane pore size, and the charge repulsion between selenium ions and the charged membrane. The reflection coefficients (σ) of the NF membrane were higher than those of UF membrane. However, some experiments (NF membrane: uncharged selenium ion, UF membrane: highly charged a selenium ions) exhibited indicate that the major transport mechanism was changed by the J_0/k ratio. Therefore, the J_0/k ratio can be used as an important operating parameter during the operation of both UF and NF membranes operations. In addition, it is possible that the UF membrane can be applied to remove selenium ions efficiently, through control of operating conditions without any additional treatment.

References

1. Sandy T (2010) Review of available technologies for the removal of selenium from water. North Amerian metals council.
2. Bleiman N, Mishael YG (2010) Selenium removal from drinking water by adsorption to chitosan–clay composites and oxides: Batch and columns tests. J Hazard Mater 183: 590-595.
3. Gloder associate Inc (2009) Literrature review of treatement technologies to remove selenium for mining influenced water 08-1421-0034.
4. Hem JD (1992) Study and interpretation of the chemical characteristics of natural water. USGE Water-Supply Paper 2254.
5. Amy CJG, Pellegrino J (2000) Membrane filtration of natural organic matter: comparison of flux decline, NOM rejection, and foulants during filtration with three UF membranes. Desalination 127: 283-298.
6. Kedem O, Katchalsky A (1958) Thermodynamic analysis of the permeability of biological membranes to non-electrolytes. Biochem Biophs Acta 27: 229-246.
7. Amy CJG, Pellegrino J (2000) Membrane filtration of natural organic matter: factors and mechanisms affecting rejection and flux decline with charged ultra filtration (UF) membrane. J Member Sci 164: 89-110.
8. Mulder M (1996) Basic principles of membrane technology. Kluwer Academic Publishers 210-279.

Liquid-Liquid Equilibria of Water + Lacticacid + Butylacetate

Afolabi, Tinuade Jolaade and Alao Adeyinka Idowu*

Department of Chemical Engineering, Ladoke Akintola University of Technology, Ogbomoso, Oyo State, Nigeria

Abstract

Experimental liquid-liquid equilibria of the ternary system of (water + lactic acid + butyl acetate) were studied at T=(298.15, 303.15 and 308.15) K under atmospheric conditions. Complete phase diagrams were obtained by determining solubility and tie-line data. The ternary system exhibits type-1 behavior of LLE. The reliability of the experimental tie-line data was ascertained by using both Hand and Othmer-Tobias correlations. The UNIFAC group contribution method was used to predict the observed ternary liquid-liquid equilibrium data. Distribution coefficients and separation factors were evaluated to evaluate the extracting capability of the solvent. Separation factors were more than unity and the UNIFAC prediction was satisfactory.

Keywords: Lactic acid; Liquid-liquid equilibria; Butyl acetate; Separation factor; UNIFAC

Introduction

Fermentation is one of the viable processes in the production of some chemicals in manufacturing industries. Two of the major obstacles of fermentation are purity of the product and efficiency of the medium. Industrially, lactic acid is produced through the fermentation of sugar, starch, molasses or cheese whey in the presence of micro-organisms such as *Lactobaccilus delbrueckii* [1-8]. The pure lactic acid is used and applied; in the food industry (as food additives and preservatives), in pharmaceutical industry (as calcium source in calcium lactate), as solvent for lacquers and moisturizers in hand skin lotion or for the production of biodegradable polymers [1,2]. The main obstacle or problem of its industrial production, just like other fermentation processes, is the accumulation of the acid. During the fermentation process, the lactic acid accumulation leads to the simultaneous decrease in the pH of the medium and this causes the process inhibition. Therefore there is need to either remove the acid in situ or add a neutralizer that will optimize the fermentation process, in order to increase the productivity of the acid and the performance of the bioreactor [3-8]. This has therefore caused an increase in interest towards the acid recovery from fermentation broth. Different processes have been applied for this purpose which includes precipitation with calcium hydroxide, use of ion exchange resins and electrodialysis [9-14]. Some of the methods are either complicated, yield low product, have low purity, cause disposal and environmental problems, or costly, which at the end makes about 50% of the total cost spent on separation and purification [15,16]. In lieu of all these separation processes, distillation and extraction processes can also be used to separate water-based products, the former being non-applicable in the case of lactic acid because the acid can be deformed at high temperature [2,17]. Research has been carried out on the probable use of some groups of extractants (such as alcohols, acetates, ketones, etc.) for the latter process; long chain alcohols were found suitable for this purpose and their liquid–liquid equilibrium data were also determined [18,19]. Also, the equilibrium condition between the organic and aqueous phases was found to be a very important aspect of the liquid-liquid extraction process and variables most generally considered for the process are: physical properties (density difference between liquid phases and boiling point), chemical properties (distribution coefficient, selectivity and flammability), availability and cost [17,20,21].

This study is a part of research program on the recovery of lactic acid from fermentation broth using organic solvents. Several studies have been carried out to obtain LLE data on the extraction of lactic acid from its dilute aqueous solution [19,22,23]. Ternary liquid-liquid equilibria are usually determined by either the titration method, continuous measurement or cell equilibration method [21,24]. In the cell equilibration method, the solubility data is first established using the cloud point method and the tie lines are determined with the aid of a gas chromatograph (GC) [25,26] or a refractometer, which has been initially calibrated [20]. Based on hydrogen bonding interactions, butyl acetate is considered as a probable solvent for the removal of lactic acid from fermentation broth [18]. The LLE results for (water + lactic acid + butyl acetate) at three different temperatures (T = 298.15K, 303.15K and 308.15K) are reported here. The reliability of the experimental data was ascertained by using both the Othmer-Tobias and Hand correlating equations [27].

The liquid-liquid equilibrium data were predicted by the UNIFAC (universal functional group activity coefficient) method [24-29]. The UNIFAC method uses the activity coefficient, γ_i for the prediction. At equilibrium, the activities of the component ion in both phases are equal and the mole fractions x_i^E and x_i^R of the conjugate phases are calculated using the following equation:

$$\gamma_i{}^E X_i^E = \gamma_i{}^R X_i^R \qquad (1)$$

where $\gamma_i{}^E$ and $\gamma_i{}^R$ are the corresponding activity coefficients of component i in extract and raffinate phases. The interaction parameters between each of the main groups were used to calculate activity coefficients of component i. The values of the UNIFAC parameters for LLE predictions were taken from a UNIFAC table [29].

Experimental Section

Materials and equipment's

Analytical grade butyl acetate and lactic acid (purities >90%) were purchased and used without further purification. Deionized

***Corresponding author:** Alao Adeyinka Idowu, Department of Chemical Engineering, Ladoke Akintola University of Technology, Ogbomoso, Oyo State, Nigeria

W_1	W_2	W_3	Brix
1.0000	0.0000	0.0000	0.0
0.9950	0.0000	0.0050	0.7
0.9400	0.0525	0.0075	3.5
0.9150	0.0740	0.0110	5.0
0.8685	0.1085	0.0230	8.0
0.7575	0.1825	0.0600	15.0
0.6800	0.2100	0.1100	20.0
0.5900	0.2400	0.1700	25.9
0.5100	0.2600	0.2300	30.3
0.4500	0.2700	0.2800	32.8
0.4000	0.2700	0.3300	34.2
0.1200	0.0000	0.8800	33.3
0.1250	0.0150	0.8600	33.4
0.1300	0.0300	0.8400	33.5
0.1245	0.0475	0.8100	33.6
0.1580	0.0620	0.7800	33.8
0.1785	0.0915	0.7300	34.0
0.2025	0.1375	0.6600	34.2
0.2200	0.1800	0.6000	34.4
0.2550	0.2150	0.5300	34.6
0.2700	0.2400	0.4900	34.7

Table 1: Solubility Data for Water + Lactic acid + Butyl acetate at 298.15K.

W_1	W_2	W_3	Brix
1.0000	0.0000	0.0000	0.0
0.9950	0.0000	0.0050	0.7
0.9425	0.0475	0.0070	3.9
0.9190	0.0725	0.0110	5.4
0.8850	0.0975	0.0175	7.9
0.8550	0.1195	0.0255	10.2
0.7700	0.1800	0.0500	15.1
0.6900	0.2200	0.0900	20.2
0.5950	0.2750	0.1300	25.1
0.5000	0.3000	0.2000	29.2
0.3800	0.3080	0.3120	33.0
0.1050	0.0000	0.8950	33.4
0.1080	0.0220	0.8700	33.5
0.1165	0.0285	0.8550	33.6
0.1325	0.0425	0.8250	33.8
0.1480	0.0700	0.7820	34.0
0.1550	0.0938	0.7512	34.2
0.1640	0.1160	0.7200	34.4
0.2015	0.1485	0.6500	34.6
0.2200	0.1780	0.6020	34.9
0.2550	0.2110	0.5340	35.0

Table 2: Solubility Data for Water + Lactic acid + Butyl acetate at 303.15K.

and redistilled water was used throughout the experiment. Setra digital balance (410S model) was used to measure the masses of the materials. The stated accuracy of the balance is 10^{-3} g. Brix values were measured with Digital Refractometer (203X model) by MISCO Limited London. The surface was cleaned with water before and after every measurement. The refractometer has the variation of ± 0.1 brix value. Rotary evaporator (52A model) was used to stir the mixture at a constant temperature. Variation of the rotary evaporator is ± 0.1 °C. Boiling points were measured with Fischer boiling point apparatus.

Experimental procedure

The binodal curves for the (water + lactic acid + butyl acetate) system at T=(298.15, 303.15 and 308.15) K were determined by the cloud point method [14,24,25]. The major central part of the solubility curve was obtained by titrating the heterogeneous mixture of (water + butyl acetate) with lactic acid until the turbidity disappeared. For the water and butyl acetate-side limited regions, binary mixtures of (water + lactic acid) were titrated with butyl acetate and binary mixture of (butyl acetate + lactic acid) titrated with water until the transition from homogeneity to turbidity was observed. The corresponding brix values were recorded. Mutual solubility values of (water + butyl acetate) binaries were also measured using the method based on cloud point detection. The transition point between homogeneity and heterogeneity zones was also determined visually. The titration was performed by adding in minute drops the material and the visual inspection of the transition was achieved by waiting approximately 5 minutes at the transition point and observed the heterogeneity. The graphs of variation of brix value with corresponding mass fractions of water (W_1) and butyl acetate (W_3) were plotted at (T=298.15, 303.15 and 308.15) K for calibration.

End point determinations of tie-lines were based upon the independent analysis of the conjugate phases which were regarded as being in equilibrium. Mixtures of known masses of water, lactic acid and butyl acetate lying within the heterogeneous gap were introduced into rounded bottom flask attached to a rotary evaporator and were stirred rigorously for about 2 hours. The mixture was then transferred into a vertical settler to settle down into water and butyl acetate layers. The brix values of the resulting two phases were measured using the refractometer [20]. The composition of each phase was determined from the calibration graph.

Results and Discussion

Experimental LLE data

The experimental solubility data of the phase equilibria of (water + lactic acid + butyl acetate) at T=(298.15, 303.15 and 308.15) K (Tables 1-3), while the corresponding binodal curves (Figures 1-3). The ternary system exhibits type-1 behavior of LLE [21]. This type is characterized by the total miscibility between two pairs and partial miscibility in the third pair as shown in the figures. The heterogeneous area (that

W_1	W_2	W_3	Brix
1.0000	0.0000	0.0000	0.0
0.9950	0.0000	0.0050	0.7
0.9700	0.0245	0.0055	1.7
0.9500	0.0440	0.0060	3.2
0.9200	0.0685	0.0115	4.5
0.8840	0.0980	0.0180	6.8
0.8300	0.1350	0.0350	10.6
0.7900	0.1600	0.0500	13.1
0.6800	0.2200	0.1000	20.1
0.5400	0.2700	0.1900	30.0
0.4500	0.2800	0.2700	34.3
0.1050	0.0000	0.8950	33.1
0.1225	0.0200	0.8575	33.4
0.1265	0.0250	0.8485	33.5
0.1310	0.0310	0.8380	33.6
0.1475	0.0500	0.8025	33.8
0.1825	0.0875	0.7300	34.3
0.1950	0.1050	0.7000	34.7
0.2275	0.1425	0.6300	35.2
0.2550	0.1850	0.5600	35.7
0.2600	0.2350	0.5050	35.8
0.2700	0.2600	0.4700	36.0

Table 3: Solubility Data for Water + Lactic acid + Butyl acetate at 308.15K.

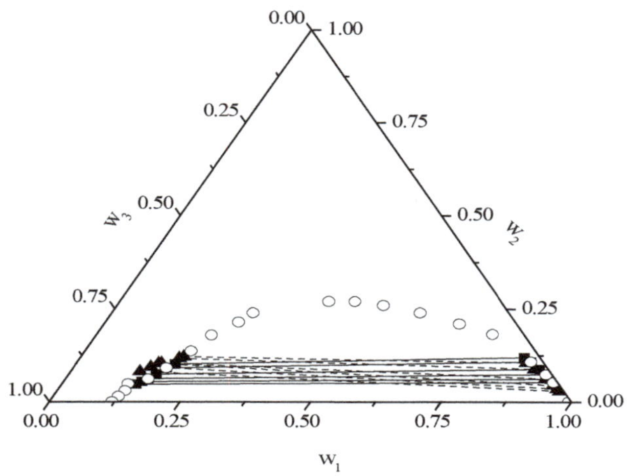

Figure 1: Solubility curve (circles),experimental tie lines (squares) and predicted tie-lines (triangles) for water (1) + lactic acid (2) + butyl acetate (3) at T = 298.15K.

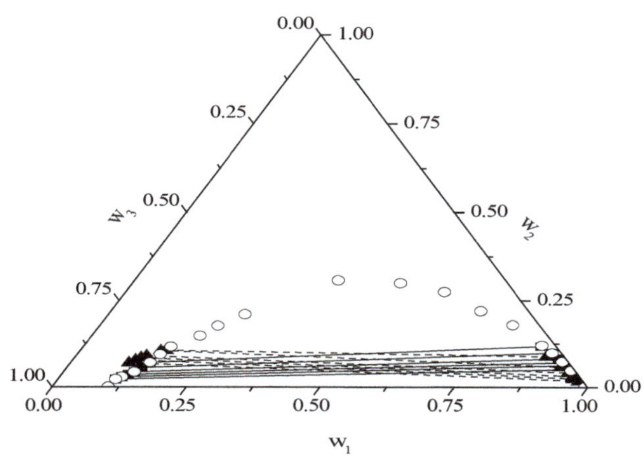

Figure 2: Solubility curve (circles),experimental tie lines (squares) and predicted tie-lines (triangles) for water (1) + lactic acid (2) + butyl acetate (3) at T = 303.15K.

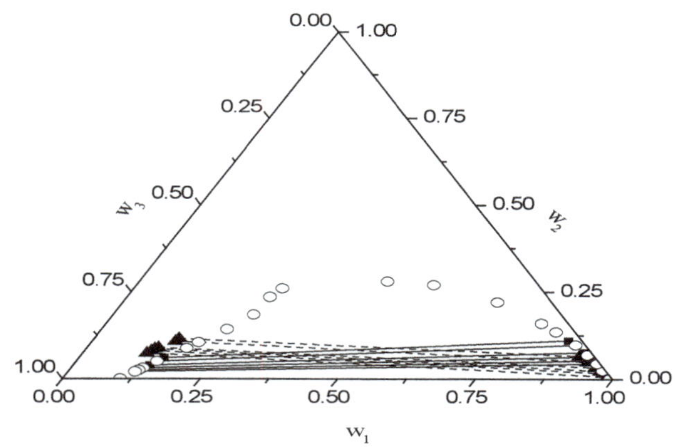

Figure 3: Solubility curve (circles),experimental tie lines (squares) and predicted tie-lines (triangles) for water (1) + lactic acid (2) + butyl acetate (3) at T = 308.15K.

is area under the curve) increased with temperature in the order of 298.15K<303.15K<308.15K. This due to the fact that solubility increases with temperature which has in turn increased the heterogeneous area [2].

Calibration curves were determined for the ternary system at T=(298.15, 303.15 and 308.15) K. The resulting curves are reported in Figures 4-6 respectively. The tie-lines data of the ternary system at T = (298.15, 303.15 and 308.15) K are eventually evaluated from the resulting curves (Tables 4-6). The tie-lines for the system at the three temperatures are shown in Figures 1-3, respectively. The slope of the tie-lines obtained in this work shows that lactic acid is more soluble in water than in butyl acetate.

Distribution coefficient and separation factor

To indicate the ability of the solvent in the recovery of lactic acid, distribution coefficients for water ($D_1 = w_{13}/w_{11}$) and lactic acid ($D_2 = w_{23}/w_{21}$) and separation factors (S) were calculated from experimental data. w_{13} and w_{23} are the mass fractions of water and lactic acid in

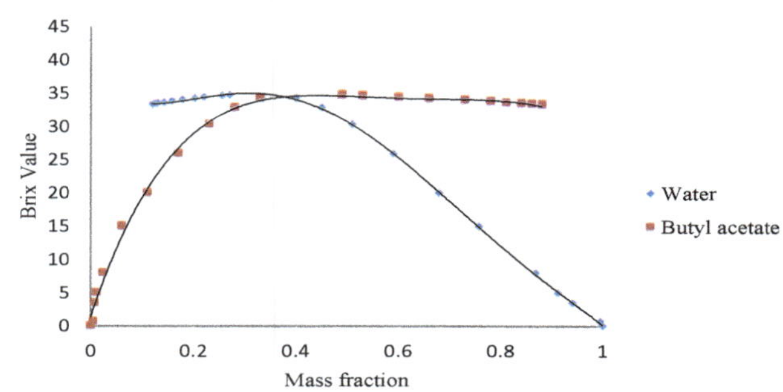

Figure 4: Calibration curve for water + lactic acid + butyl acetate at 298.15K.

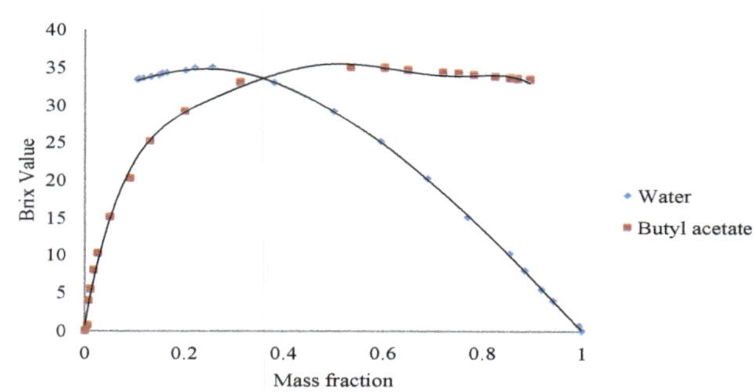

Figure 5: Calibration curve for water + lactic acid + butyl acetate at 303.15K.

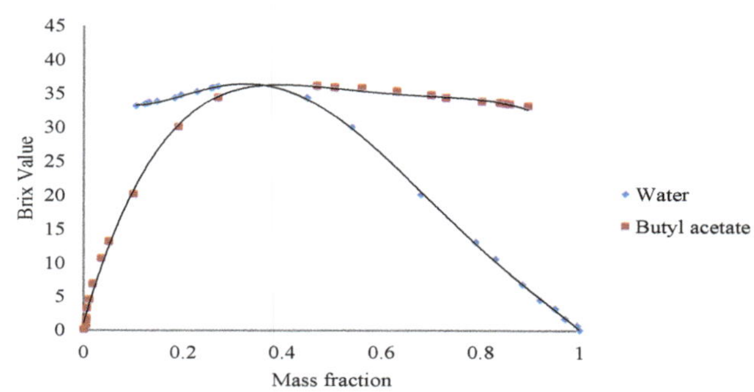

Figure 6: Calibration curve for water + lactic acid + butyl acetate at 308.15K.

Water phase				Butyl acetate phase			
Brix	W_{11}	W_{21}	W_{31}	Brix	W_{13}	W_{23}	W_{33}
3.5	0.9400	0.0525	0.0075	33.6	0.1425	0.0475	0.8100
3.9	0.9300	0.0615	0.0085	33.7	0.1500	0.0550	0.7950
5.0	0.9150	0.0740	0.0110	33.8	0.1580	0.0620	0.7800
6.4	0.8950	0.0885	0.0165	33.9	0.1675	0.0750	0.7575
8.0	0.8685	0.1085	0.0230	34.0	0.1785	0.0915	0.7300
10.1	0.8500	0.1200	0.0300	34.1	0.1900	0.1025	0.7075

Table 4: Experimental Tie-line Data for Water (1) + Lactic acid (2) + Butyl acetate (3) at 298.15K.

Water phase				Butyl acetate phase			
Brix	W_{11}	W_{21}	W_{31}	**Brix**	W_{13}	W_{23}	W_{33}
3.8	0.9425	0.0505	0.0070	33.5	0.1080	0.0220	0.8700
4.8	0.9300	0.0610	0.0090	33.6	0.1165	0.0285	0.8550
5.4	0.9190	0.0700	0.0110	33.7	0.1250	0.0350	0.8400
6.5	0.9050	0.0810	0.0140	33.8	0.1325	0.0425	0.8250
7.9	0.8850	0.0975	0.0175	33.9	0.1405	0.0545	0.8050
10.2	0.8550	0.1170	0.0280	34.0	0.1480	0.0700	0.7820

Table 5: Experimental Tie-line Data for Water (1) + Lactic acid (2) + Butyl acetate (3) at 303.15K.

Water phase				Butyl acetate phase			
Brix	W_{11}	W_{21}	W_{31}	**Brix**	W_{13}	W_{23}	W_{33}
3.2	0.9500	0.0440	0.0060	33.9	0.1225	0.0200	0.8575
3.6	0.9390	0.0535	0.0075	34.0	0.1265	0.0250	0.8485
4.2	0.9275	0.0640	0.0085	34.1	0.1310	0.0310	0.8380
5.6	0.9100	0.0765	0.0135	34.2	0.1395	0.0390	0.8215
6.8	0.8840	0.0980	0.0180	34.3	0.1475	0.0500	0.8025
8.7	0.8650	0.1100	0.0250	34.4	0.1550	0.0600	0.7850

Table 6: Experimental Tie-line Data for Water (1) + Lactic acid (2) + Butyl acetate (3) at 308.15K.

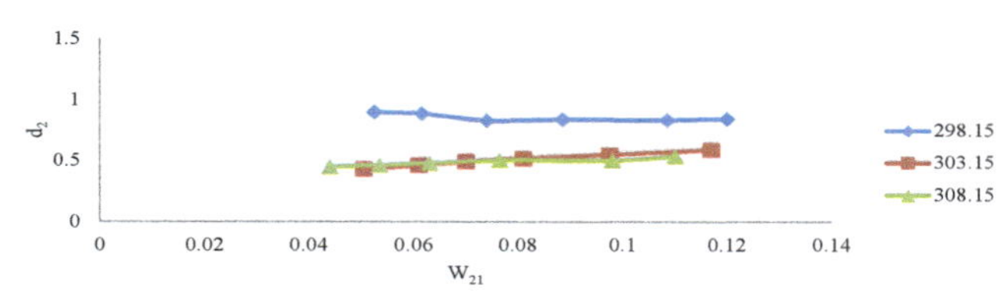

Figure 7: Variation of distribution coefficient of lactic acid with mass fraction of lactic acid in water-rich phase for water + lactic acid + butyl acetate ternary system at 298.15, 303.15 and 308.15K.

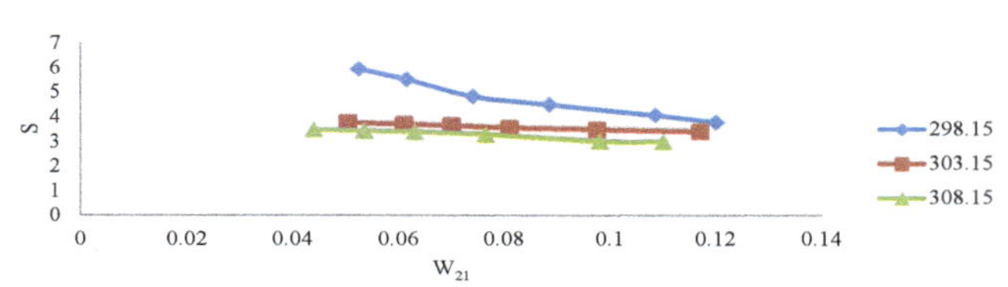

Figure 8: Variation of separation factor with mass fraction of lactic acid in water-rich phase for water + lactic acid + butyl acetate ternary system at 298.15, 303.15 and 308.15K.

organic-rich phase, respectively. w_{11} and w_{21} are the mass fractions of water and lactic acid in aqueous phase, respectively. The separation factor is defined as the ratio of distribution coefficients of lactic acid (1) to water (2), $S = D_2/D_1$.

The variation of distribution coefficient and separation factor for lactic acid as a function of the solute mass fraction in the aqueous phase for the ternary system at different temperatures are shown in Figures 7 and 8.

The experimental results indicate that butyl acetate has a high separation factor, which shows the ability of this solvent to extract lactic acid from water. This quantity is greater than one (separation factors varying between 3.0438 and 5.9683) for the system reported in this work, which means the extraction of lactic acid by butyl acetate is possible. The separation factors and distribution coefficients are not constant over the whole two-phase region.

In order to compare the effectiveness of extraction of lactic acid by butyl acetate to those of long-chain alcohols, the separation values measured at 298.15 K in this work are compared with those of octanol, nonanol and decanol. The separation values for the three long-chain alcohols (varying between 4.43 and 9.19) were higher than that of butyl acetate, which shows that long-chain alcohols were better extractants compared to butyl acetate [20], though the efficiency of the acetates could be better off using those of long chains [2].

Reliability of the tie-line data

The reliability of experimentally measured tie-line data can be ascertained by applying Othmer-Tobias and Hand correlations [27] at each temperature as shown in equations 4 and 5 respectively.

$$\ln\left(\frac{1-w_{11}}{w_{11}}\right) = a_1 + b_1 \ln\left(\frac{1-w_{33}}{w_{33}}\right) \quad (4)$$

$$\ln\left(\frac{w_{21}}{w_{11}}\right) = a_2 + b_2 \ln\left(\frac{w_{23}}{w_{33}}\right) \quad (5)$$

Where w_{11} the mass fraction of water in water-rich is phase; w_{21} is the mass fraction of lactic acid in water-rich phase; w_{23} is the mass fraction of lactic acid in solvent-rich phase; and w_{33} is the mass fraction of butyl acetate in solvent-rich phase.

Othmer-Tobias plots were constructed of $\ln\left[(1-w_{33})/w_{33}\right]$ *vs* $\ln\left[(1-w_{11})/w_{11}\right]$; and Hand plots were construcred of $\ln(w_{23}/w_{33})$ *vs* $\ln(w_{21}/w_{11})$. The correlations are shown in Figures 9 and 10 at the three different temperatures.

The linear plots of the correlations were used to determine the coefficients a_i and b_i shown in Table 7. Standard deviations were calculated to evaluate the deviations of the experimental values from the linear plots, according to equation 6. These are reported in Table 8.

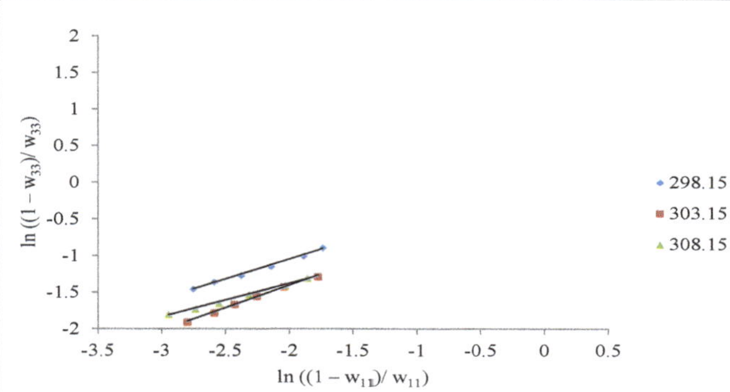

Figure 9: Othmer-Tobias correlation for water + lactic acid + butyl acetate at 298.15 , 303.15 and 308.15K.

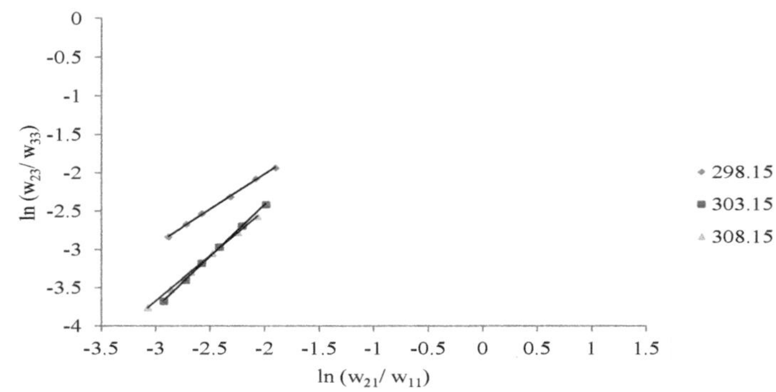

Figure 10: Hand correlation for water + lactic acid + butyl acetate at 298.15 , 303.15 and 308.15K.

298.15			303.15			308.15		
D_1	D_2	S	D_1	D_2	S	D_1	D_2	S
0.1516	0.9048	5.9683	0.1146	0.4356	3.8010	0.1289	0.4545	3.5260
0.1613	0.8943	5.5443	0.1253	0.4672	3.7296	0.1347	0.4673	3.4692
0.1727	0.8378	4.8512	0.1360	0.5000	3.6765	0.1412	0.4844	3.4306
0.1872	0.8475	4.5270	0.1463	0.5247	3.5840	0.1533	0.5098	3.3255
0.2055	0.8433	4.1036	0.1588	0.5590	3.5209	0.1669	0.5102	3.0569
0.2235	0.8542	3.8219	0.1731	0.5983	3.4564	0.1792	0.5455	3.0438

D1 and D2 are the distribution coefficients of water and lactic acid respectively.

Table 7: Distribution coefficients (Di) and Separation factors (S) for Water + Lactic acid + Butyl acetate at 298.15K, 303.15K and 308.15K.

Temperature (in K)	298.15				303.15				308.15			
Coefficients	a_i	b_i	δ_1	δ_2	a_i	b_i	δ_1	δ_2	a_i	b_i	δ_1	δ_2
Othmer – Tobias	0.042	0.544	0.003	0.002	–0.167	0.617	0.002	0.002	–0.457	0.460	0.002	0.002
Hand	–0.176	0.919	0.025	0.019	0.283	1.350	0.007	0.008	–0.124	1.184	0.016	0.016

Table 8: Coefficients of Othmer–Tobias and Hand Equations and Standard Deviations for water + lactic acid + butyl acetate at 298.15, 303.15 and 308.15K.

Group	Group Parameters	
	R_k	Q_k
CH_3	0.9011	0.8480
CH_2	0.6744	0.5400
H_2O	0.9200	1.4000
CH_3CO	1.6724	1.4880
CH_3COO	1.9031	1.7280
CH_2O	0.9183	0.7800
COOH	1.3013	1.2240

Table 9: UNIFAC Group Parameters.

$$\delta_i = \left(\frac{\sum_{k=1}^{n} \left(w_{ji}^{calc} - w_{ji}^{exp} \right)^2}{N} \right)^{1/2} \quad (6)$$

Where N is the number of the tie lines and i=j=1and 3. δ_1 and δ_2 (i = 1 or 2) are the standard deviations of the mass fractions of water in the raffinate phase and butyl acetate in the extract phase respectively. The standard deviations calculated were low. These give credence to the experimental data obtained.

Prediction of the experimental tie-line data

The UNIFAC method was also used to predict the equilibrium data of the ternary mixtures; the method using the group parameters and group-group interaction parameters of CH, CH_2, CH_3, OH, COOH and H_2O functional groups obtained by Poling et al. [29], given in Tables 9 and 10, respectively.

A FORTRAN programme was written to predict the LLE data according to UNIFAC equations 6, 7 and 8. The

Predicted values (W_{ij}^{pre}; i=component, j=phase) are compared with the experimental values at the respective temperatures (Tables 11-13).

$$\ln \gamma_i = \ln \gamma_i^C + \ln \gamma_i^R \quad (6)$$

$$\ln \gamma_i^C = \ln(\varphi_i / x_i) + 1 - \varphi_i / x_i - (z/2) q_i [\ln(\varphi_i / \theta_i) + 1 - \varphi_i / \theta_i] \quad (7)$$

$$\ln \gamma_i^R = \sum_k v_{ki} (\ln \Gamma_k - \ln \Gamma_{ki}) \quad (8)$$

The quality of the prediction was evaluated with the residual function (F) calculated from the difference between the experimental value and predicted value for each ternary system according to equation 9; where n is the number of tie-lines, w_{iexp} is the experimental mass fraction and w_{ical} is the predicted mass fraction of component i. j refers to phases and k=1,2,3,4.....n(tie lines).

$$F = \left\{ \sum_k \min \left[\sum_i \sum_j \left(w_{i_{exp}} - w_{i_{cal}} \right)^2 \right] / 6n \right\}^{1/2} \quad (9)$$

The residual function (F) calculated for (water + lactic acid + butyl acetate) system at T = 298.15K, 303.15K and 308.15K were 0.0271, 0.0387and 0.0332, respectively.

Main group	$-CH_3$	$-CH_2$	H_2O	$-CH_3CO$	$-CH_3COO$	$-CH_2O$	$-COOH$
$-CH_3$	0	0	1318	476.4	232.1	251.5	663.5
$-CH_2$	0	0	1318	476.4	232.1	251.5	663.5
H_2O	300	300	0	−195.4	72.87	540.5	−14.09
$-CH_3CO$	26.76	26.76	472.5	0	−37.36	−103.6	669.4
$-CH_3COO$	114.8	114.8	200.8	372.2	0	−235.7	660.2
$-CH_2O$	83.36	83.36	−314.7	191.1	461.3	0	664.6
$-COOH$	315.3	315.3	−66.17	−297.8	−256.3	−338.5	0

Table 10: UNIFAC group-group interaction parameters (a_{mn}) [29].

Water phase			Butyl acetate phase		
W_{11}^{pre}	W_{21}^{pre}	W_{31}^{pre}	W_{13}^{pre}	W_{23}^{pre}	W_{33}^{pre}
0.9231	0.0276	0.0071	0.1451	0.0903	0.8632
0.9119	0.0342	0.0090	0.1530	0.0989	0.7891
0.8991	0.0430	0.0113	0.1608	0.1070	0.7572
0.8623	0.0590	0.0160	0.1738	0.1125	0.7833
0.8139	0.0815	0.0234	0.1963	0.1218	0.7157
0.7910	0.0972	0.0300	0.2042	0.1265	0.7086

Table 11: The predicted liquid-liquid equilibrium data for water + lactic acid + butyl acetate at 25 °C.

Water phase			Butyl acetate phase		
W_{11}^{pre}	W_{21}^{pre}	W_{31}^{pre}	W_{13}^{pre}	W_{23}^{pre}	W_{33}^{pre}
0.8667	0.0151	0.0071	0.1174	0.0737	0.8813
0.8682	0.0213	0.0133	0.1243	0.0815	0.8533
0.8747	0.0278	0.0111	0.1313	0.0881	0.8355
0.8647	0.0366	0.0141	0.1387	0.0940	0.8183
0.8318	0.0515	0.0191	0.1495	0.1031	0.7370
0.7746	0.0768	0.0292	0.1634	0.1067	0.7505

Table 12: The predicted liquid-liquid equilibrium data for water + lactic acid + butyl acetate at 30 °C.

Water phase			Butyl acetate phase		
W_{11}^{pre}	W_{21}^{pre}	W_{31}^{pre}	W_{13}^{pre}	W_{23}^{pre}	W_{33}^{pre}
0.9471	0.0064	0.0060	0.1228	0.0775	0.8539
0.9347	0.0159	0.0076	0.1271	0.0842	0.8347
0.9187	0.0216	0.0100	0.1323	0.0917	0.8366
0.9047	0.0310	0.0138	0.1403	0.0961	0.8394
0.8673	0.0458	0.0197	0.1504	0.1070	0.7329
0.8378	0.0603	0.0259	0.1600	0.1200	0.7568

Table 13: The predicted liquid-liquid equilibrium data for water + lactic acid + butyl acetate at 35 °C.

Conclusion

The binodal and tie-line data for (water +lactic acid + butyl acetate) ternary systems were determined at temperatures of 298.15K, 303.15K and 308.15K. The tie-line data of the system were satisfactorily ascertained by the Othmer-Tobias and Hand correlating equations. Butyl acetate was able to recover lactic acid from aqueous lactic acid solution. The UNIFAC method used predicted the experimental data satisfactorily.

References

1. Nanninga GL (1983) Fermentative production of lactic acid. Antonie van Leeuwenhoek 49: 86.
2. Wong YC, Wong CT, Onyiruka SO, Akpanisi LE (2002) University Organic Chemistry; The Fundamentals, Manhattan Publishers, Hong Kong.
3. Buyukkileci AO, Harsa S (2004) Batch production of L(+) lactic acid from whey by Lactobacillus casei (NRRL B-441). J Chem Technol Biotechnol 79: 1036-1040.
4. Calabia BP1, Tokiwa Y (2007) Production of D-lactic acid from sugarcane molasses, sugarcane juice and sugar beet juice by Lactobacillus delbrueckii. Biotechnol Lett 29: 1329-1332.
5. Dumbrepatil A1, Adsul M, Chaudhari S, Khire J, Gokhale D (2008) Utilization of molasses sugar for lactic acid production by Lactobacillus delbrueckii subsp. delbrueckii mutant Uc-3 in batch fermentation. Appl Environ Microbiol 74: 333-335.
6. Hofvendahl K, Hahn–Hagerdal BL (1997) l-lactic acid production from whole wheat flour hydrolysate using strains of Lactobacilli and Lactococci. Enzyme Microb Technol 20: 301-307.
7. Kertes AS1, King CJ (1986) Extraction chemistry of fermentation product carboxylic acids. Biotechnol Bioeng 28 269-282.
8. Tejayadi S, Cheryan M (1995) Lactic acid from cheese whey permeate Productivity and economics of a continuous membrane bioreactor. Appl Microbiol Biotechnol 43: 242-248.
9. Cao X, Yun HS, Koo YM (2002) Recovery of l-(+)-lactic acid by anion exchange resin Amberlite IRA-400. Biochemical Engineering Journal 11: 189-196.
10. Inci I, Hasdemir M, Bilgin M, Aydin A (2000) Investigation of Extraction of Lactic Acid With Alamine -336 Using Different Diluting Solvents. Turkish J Eng Env Sci 24: 47-54.
11. Choi JI, Hong WH (1999) Recovery of Lactic Acid by Batch Distillation with Chemical Reactions Using Ion Exchange Resin. Journal of Chemical Engineering of Japan 32: 184-189.
12. Kim YH, Moon SH (2001) Lactic acid recovery from fermentation broth using one-stage electrodialysis. Journal of Chemical Technology and Biotechnology 76: 169-178.
13. Hongo M1, Nomura Y, Iwahara M (1986) Novel method of lactic Acid production by electrodialysis fermentation. Appl Environ Microbiol 52: 314-319.
14. Wee YJ1, Yun JS, Lee YY, Zeng AP, Ryu HW (2005) Recovery of lactic acid by repeated batch electrodialysis and lactic acid production using electrodialysis wastewater. J Biosci Bioeng 99: 104-108.
15. Moresi M, Parente E (2005) Fermentation (Industrial) | Production of Organic acid. Encyclopedia of Food Microbiology 705-717.
16. US Department of Energy (1999) Advanced electrodeionisation technology for product purification, waste recovery and water recycling, Office of Industrial Technologies, Washington DC.
17. Perry RH, Green DW (1984) Perry's Chemical Engineers Handbook, 7th Edn, McGraw Hill Publishers, New York.
18. Robbins LA (1980) Principles of Unit Operations, 2nd Edition, Wiley Publishers, New York.
19. Sahin S, Kirbaslar SI, Bilgin M (2009) (Liquid plus liquid) equilibria of (water plus lactic acid plus alcohol) ternary systems. Journal of Chemical Thermodynamics 41: 97-102.
20. Afolabi TJ, Edewor TI (2011) World Academy of Science, Engineering and Technology 60: 1501-1504.
21. Thornton JD (1992) Science and Practice of Liquid-liquid Extraction: Phase equilibria, mass transfer and interfacial phenomena, extractor hydrodynamics, selection, and design. Clarendon Publishers, Oxford.
22. Petritis VE, Petritis CJ (1953) Equilibrium Data of Water–Lactic acid–Solvents. Journal of Chemical Engineering Data 4: 197-198.
23. Weiser RB, Geankoplis CJ (1955) Lactic Acid Purification by Extraction. Ind Eng Chem 47: 858-862.
24. Bilgin M (2006) Phase equilibria of liquid (water + butyric acid + oleyl alcohol) ternary system. Journal of Chemical Thermodynamics 38: 1634-1639.
25. Senol A, Sevgili L (2006) Phase equilibria for ternary liquid systems of (water + tetrahydrofuran + nonprotic aromatic solvent) at T = 298.2 K. Journal of Chemical Thermodynamics 38: 578-584.
26. Ghanadzadeh A, Ghanadzadeh H, Abbasnejad S, Bahrpaima Kh (2010) Phase equilibria in ternary aqueous mixtures of 1,3-butanediol with 2-ethyl-1-hexanol at T = (298.2, 303.2 and 308.2) K. Fluid Phase Equilibria 291: 90-94.
27. Othmer DF, Tobias PE (1942) Liquid-Liquid Extraction Data - The Line Correlation. Ind Eng Chem 34: 693-696.

28. Frendeslund A, Jones RL, Prausnitz JM (1975) Group-contribution estimation of activity coefficients in nonideal liquid mixtures. American Institute of Chemical Engineers Journal 21: 1086-1099.

29. Poling BE, Prausnitz JM, O'Connell JP (2004) The Properties of Gases and Liquids, 5th Edition, McGraw Hill Publishers, New York.

Effect of Non-Solvent Additive on Effectiveness of Polyvinylidene Fluoride Membrane Fabricated With Thermal Induced Phase Separation Method for Carbon Dioxide Absorption

Nayef Ghasem* and Mohamed Al-Marzouqi

Department of Chemical & Petroleum Engineering, UAE University, Alain, UAE

Abstract

In the present work porous poly (vinylidene fluoride) (PVDF) hollow fiber membranes were fabricated via thermally induced phase separation (TIPS) method for the application in gas-liquid membrane contactor. For this purpose long air gap distance was used (90 mm). Scanning electron microscopy (SEM) was used for membrane characterization. Gas permeation test was performed using carbon dioxide as test gas. It was observed that the effective surface porosity and membrane pore size increased with increased glycerol concentration. CO_2 absorption using the fabricated hollow fiber membranes were measured in a gas-liquid hollow fiber membrane contactor. The results of the CO_2 absorption rate of the tested fibers revealed that complete removal of CO_2 was achieved using 7% glycerol added to the casted solution at normal operating conditions for equal gas to liquid volumetric flow rates using sodium hydroxide as absorbent liquid.

Keywords: Hollow Fiber Membrane; Membrane Contactor; CO_2 Separation, PVDF

Introduction

Packed bed absorbers have been extensively used in many industries and it is successful in removing undesirable gases and impurities from gas streams. In spite of the success of the packed bed absorbers, there are number of weak points such as flooding at high flow rates and foaming Gabelman and Huang [1]. To overcome this drawback in gas absorbers, microporous hollow fiber membrane contactors for gas separation have been attracting the attention of many researchers as an alternative technology, which overcomes the weaknesses of conventional gas absorption equipments Rangwala [2]. In gas membrane contactor gas–liquid interfacial area remains undisturbed by the change of gas and liquid flow rates. In hollow fiber membrane contactor, phase separation after absorption operation is not necessary because one phase is not dispersed into other phase in the module, by contrast, membrane can introduce resistance to mass transfer Li and cheng [3]. There are several important issues to be considered in the selection of the membrane contactor such that; membrane must be highly hydrophobic to prevent the membrane pores from being filled with aqueous absorbent solutions. Membrane pores must be small to prevent the penetration of absorbents into the pores since the smaller the pore radius, the larger the liquid entry pressure. Membrane porosity must be large so that a large gas–liquid interface is available for gas absorption. The membrane must have enough mechanical strength Kim and Harriot [4]. Several hydrophobic microporous membranes have been used for membrane gas absorption, which include polyethylene (PE), polypropylene (PP), poly (vinylidene fluoride) (PVDF) and (poly(tetrafluoroethylene) (PTFE). Microporous polyethylene and polypropylene membranes have been often used in the membrane contactors since they are cheap and their modules are available commercially. Microporous polyethylene and polypropylene membranes are wetted by some absorbents with low interfacial tensions, which drastically decrease the module efficiency Lawler [5] Wang et al. [6]. Microporous PTFE membrane is showed good gas absorption performance as well as good stability without membrane wetting Nishikawa et al. [7]. Different PVDF hollow fiber membranes were fabricated via non-solvent induced phase separation (NIPS) method. The effect of membrane structure on gas separation was performed and found that the mass transfer coefficients of the membranes estimated from the membrane structure parameters such as pore size, pore size distribution and effective pore length agreed with those observed in the experiments on dilute H_2S absorption in an aqueous NaOH solution under the condition of membrane resistance controlling Kong and Li [8]. Yeon et al. [9] carried out a sequence of tests on the absorption of CO_2 from a CO_2/N_2 mixture into aqueous amine solutions using hollow fiber membrane contactor prepared by using commercially available PVDF membranes. They found that when an aqueous monoethanolamine (MEA) solution was used as absorbent, the membrane was wetted after few hours of operation; by contrast, a high CO_2 removal was maintained for longer period when a mixed solution of MEA and triethanolamine (TEA) was used. Atchariyawut et al. [10-12] studied absorption of CO_2 into water, NaOH and MEA solutions using a commercial microporous PVDF membrane and found that when a 2 molar NaOH solution was used, the membrane mass transfer resistance occupied almost the total resistance. Xu et al. [13] uses PVDF hollow fiber membranes with an inner skinless surface prepared via the NIPS process in the removal of CO_2 from gas mixture. The experiments showed a higher CO_2 absorption rate than other PVDF and polyethylene membranes and attained performance levels equivalent to a PTFE membrane in the experiments of pure CO_2

***Corresponding author:** Nayef Ghasem, Department of Chemical & Petroleum Engineering, UAE University, Alain, UAE absorption into water. As the PVDF membranes, which have been used so far in membrane contactors, were prepared by the NIPS method, their mechanical strength was not so high due to the macrovoid formation. In addition, the formation of skin layer near the inner surface of the membrane is not suitable for membrane contactor application. Khaisri et al. [14] compared the membrane resistances and absorption performances of three different commercial hydrophobic microporous membranes, including PVDF, PP and PTFE membranes, for the CO_2 absorption into aqueous MEA solutions. They found that the CO_2 absorption performance using PTFE membrane had excellent stability compared to other membranes. An alternative method for fabricating microporous membranes without formation of the macrovoids is the thermally induced phase separation (TIPS) method [15,16]. In the TIPS process, the membrane structure was controlled by changing the dope composition and spinning parameters Rajabzadeh et al. [17-19]. Several different PVDF membranes with different structures were prepared and used for CO_2 absorption into aqueous MEA solutions and their efficiencies were compared with that of a commercial PTFE hollow fiber membrane. The effect of the reaction products of the CO_2-MEA reaction on the solubility of CO_2 and diffusivities of CO_2 and MEA in liquid was also discussed [20,21] In the present work porous PVDF hollow fiber membrane is fabricated using predefined concentrations of PVDF/Triacetin/glycerol dope solutions, fabrication of PVDF polymers can be easily performed using TIPS method, by contrast, PTFE is fabricated using stretching and melting method which is not easy method compared to TIPS method, also PVDF is polymer is cheaper than PTFE. The fabricated hollow fibers are used to construct gas liquid membrane contactor module suitable for the removal of CO_2 from CO_2/CH_4 gas mixture using aqueous sodium hydroxide as the absorbent liquid.

Experimental

Materials

PVDF polymer (Solef 6020) was purchased from Solvay, glycerol triacetate (Triacetin) and ethanol from Sigma Aldrich and glycerol from Croda Company, the Netherlands. All chemical were used as received without further treatments.

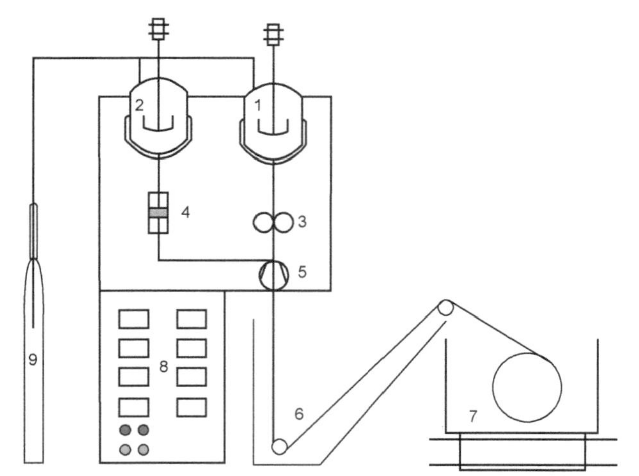

Figure 1: Schematic of thermally induced phase separation (TIPS) process (1) dope vessel, (2) bore liquid vessel, (3) gear pump, (4) bore liquid flow meter, (5) spinneret, (6) quench tank, (7) water trough, (8) control panel, (9) N_2 cylinder

Spinning parameters	Values
Dope composition	30%PVDF/(0,2,5,7)%Glycerol/balance triacetin
Bore liquid	triacetin
Spinneret ID/OD	0.9/2.3 mm
Dope solution flow rate	7 gram/min
Bore fluid flow rate	10 gram/min
Take up winder	10 rpm
Quenching bath temperature	25°C
Air gap	9 cm

Table 1: Hollow fiber spinning conditions.

Preparation of hollow fiber membranes

Hollow fiber membranes were prepared by the device shown in Figure 1. The apparatus consists mainly of two jacked vessels (2 liters each); one vessel is used for dope solution the other vessel is for the bore liquid. The dope vessel is equipped with mixer and jacket filled with high boiling point silicon oil (300°C). The bore liquid is heated with heating mental. The polymer, solvent and additive is stirred manually by spatula for 5 minutes before transferred to the dope vessel. The dope solution is composed of PVDF, triacetin as solvent and glycerol as non solvent additive. The casting solution was heated to 160°C, a temperature of 30°C above the dope solution cloud point to ensure homogeneous solution. The dope solution is heated for 1 hour with continuous gentle mixing. After one hour of the stirring, the stirrer is stopped and the heating continues for 2 more hours without mixing to liberate possible air bubbles. Then, the dope solution is introduced to a spinneret by a gear pump under nitrogen pressure. The spinneret outer and inner diameters were 2.3 and 0.9 mm, respectively. The solvent was introduced into the inner tube to make the lumen of the hollow fiber. The hollow fiber membrane was extruded from spinneret and then immersed in a water quenching bath where the temperature is monitoring at 25°C to induce the separation. Finally the fibers were wounded on a take-up winder. In the spinning process, the variable was the concentration of glycerol. The mass flow rate of the polymer solution and solvent in the spinneret were fixed at 0.9 and 0.7 gram/min, respectively. The take-up speed was fixed at 10 rpm. The fiber membranes were kept in ethanol to extract the solvent remained in the hollow fiber membranes, and then ethanol was detached by immersing the membranes in pure water. The fabrication spinning conditions and parameters are shown in Table 1.

SEM observation

The prepared membranes were freeze-dried with a freeze dryer (EYELA, FD-1000, Japan) to obtain dry hollow fiber membranes without affecting the membrane structure. The dry hollow fiber membranes were fractured in liquid nitrogen and sputtered with gold. The structure of the cross section, the outer and inner surfaces of the hollow fiber membranes were examined through a scanning electron microscopy (SEM, Jeol, JCM-5000 NeoScope, Japan) with an accelerating voltage of 10kV.

Membrane contact angle

The water contact angle of the outer surface of hollow fiber membranes were measured with a drop master (Kyowa, drop master series, Japan). A droplet of 0.5μl was overthrown on the outer surface of the fibers, and then the contact angle was measured.

Membrane mechanical strength

The tensile strength of the hollow fiber membranes were measured by a tensile apparatus (Shimadzu Co, Autograph AG-X, Japan). The hollow fibers were fixed vertically between two pairs of tweezers with the length of 90 mm. The membrane was extended at a constant elongation velocity of 20 mm/min until it was broken. Each data of tensile stress at break was repeated five times.

Gas permeation test

Gas permeation through the hollow fiber membranes were measured using 6 hollow fibers with 17 cm length were glued with epoxy resin at one end and the other side was potted to plastic tubing (Figure 2). The shell inside diameter was 12 mm and length is 200 mm. The CO_2 gas is introduced to the shell side and penetrates through the skin of hollow fiber membranes and leaves from the top of the hollow fibers membrane module. The permeate gas flow rate is measured using digital gas flow meter. The equations used for the calculation of the mean pore size and the effective surface porosity are described below. It was assumed that membrane pores are straight and cylindrical in shape and permeation of gas is under the Poiseuille and Knudsen flow regimes in parallel connection Bakeri et al. [22]. The gas permeance under the Poiselle flow regime is given by:

$$P_p = \frac{1}{8\mu}\frac{r_{p.m}^2}{RT}\frac{\xi}{L_p}\bar{p} \quad (1)$$

The gas permeance under Knudsen flow regime is:

$$P_k = \frac{2}{3}\left(\frac{8RT}{\pi M}\right)^{0.5}\frac{r_{p.m}}{RT}\frac{\xi}{L_p} \quad (2)$$

Hence, the total permeance of permeated gas is:

$$\bar{P} = P_p + P_k = \frac{2}{3}\left(\frac{8RT}{\pi M}\right)^{0.5}\frac{r_{p.m}}{RT}\frac{\xi}{L_p} + \frac{1}{8\mu}\frac{r_{p.m}^2}{RT}\frac{\xi}{L_p}\bar{p} \quad (3)$$

Linearization of equation (3)

$$\bar{P} = A + B\bar{p} \quad (4)$$

Where,

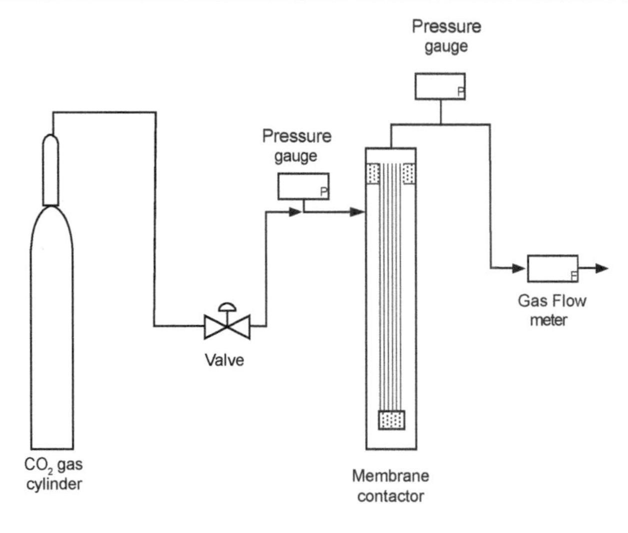

Figure 2: Schematic of the experimental setup used in gas permeation test.

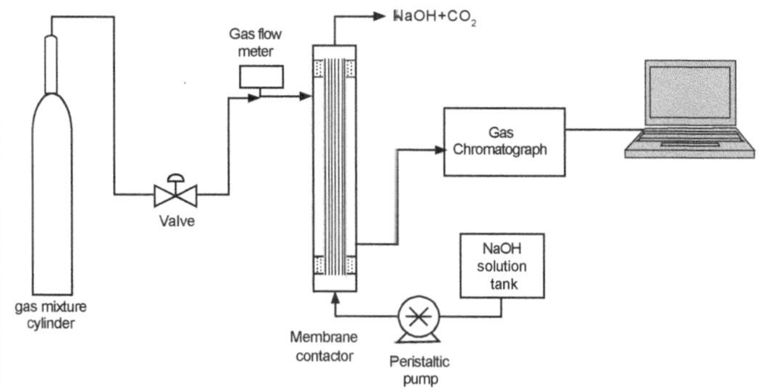

Figure 3: Schematic of the experimental setup used in gas separation test.

$$A = \frac{2}{3}\left(\frac{8RT}{\pi M}\right)^{0.5}\frac{r_{p,m}}{RT}\frac{\xi}{L_p} \quad (5)$$

$$B = \frac{1}{8\mu}\frac{r_{p,m}^2}{RT}\frac{\xi}{L_p} \quad (6)$$

In the above equations, $\bar{P}$ is total gas permeance ($mol.m^{-2}.Pa^{-1}.s^{-1}$), R is universal gas constant (8.314 $J.mol^{1}.K^{1}$), T is absolute temperature (K), M is molecular weight of gas (g/mol), $r_{p,m}$ is mean pore radius (m), μ is viscosity of gas (Pa.s), ξ is the surface porosity (Ap/A_T) where A_P is area of pores and A_T is total area of membrane, L_P is effective pore length (m) and $\bar{p}$ is mean pressure (Pa):

$$\bar{p} = \left(\frac{Pu + Pd}{2}\right)$$

where P_u is upstream pressure and P_d is downstream pressure. Experimentally obtained gas permeance is plotted versus mean pressure, which should result in a straight line. The intercept (A) of the straight line with the ordinate and slope (B) are used in the following equations to calculate the mean pore size and the effective surface porosity (ξ/L_p).

$$r_{p,m} = \frac{16}{3}\frac{B}{A}\left(\frac{8RT}{\pi M}\right)^{0.5}\mu \quad (7)$$

$$\frac{\xi}{L_p} = \left(\frac{8\mu RTB}{r_{p,m}^2}\right) \quad (8)$$

Gas separation

CO_2 was separated from a gas mixture containing 91% of CH_4 and 9% of CO_2 and 0.5M NaOH aqueous solution was used as liquid absorbent. Five hollow fiber membranes, length of 20 cm were introduced to a plastic tube, the inside plastic tube diameter is 12 mm. The ends of the tube were blocked with epoxy resin. The absorbent liquid was introduced to the tube side and the gas mixture in the shell side of the hollow fiber membranes contactor as shown in Figure 3. The inlet gas volumetric flow rate is adjusted by mass flow controller (Alicat Scientific, USA). The liquid absorbent volumetric flow rate is controlled by peristaltic pump. The CO_2 composition in the exit gas stream is measured using gas chromatography (GC-2014 Shimadzu, Japan).

Results and Discussion

Membrane morphology

Figure 4 is the SEM micrograph of the hollow fiber membrane prepared with 30%PVDF, 70%Triacetin solvent. The SEM micrographs of the cross section, outer surface and inner surface are shown in Figure 4a,4b,4c, respectively. The membrane structure revealed that the porosity at the inner surface is higher than that at the outer surface. This is attributed to two reasons, the long air gap distance (90 mm) compared to previous study which was between 0 to 5 mm Rajabzadeh S et al. [18] and the bore liquid temperature which was 160°C (the same temperature of the dope solution).

Figure 5 shows the membrane structure of the PVDF/triacetin/glycerol system. In this case, 5% Glycerol is added to the dope solution as additive. A difference structure has been observed on the outer surface. The outer surface of the membrane prepared with PVDF/triacetin/glycerol system is more porous than that prepared with PVDF/triacetin system without glycerol additives. This is confirmed in Figure 6, the results of the gas permeability tests as explained in the following section.

Gas permeation test

Figure 6 shows gas permeability versus inlet gas pressure for hollow fiber membrane fabricated from 30% PVDF and (0%, 2%, 5% and 7% glycerol), the balanced is triacetin. At specific gas pressure, the figure revealed that gas permeability increased as the percent glycerol added to the cast solution increased. Slight increase in gas permeability is observed with increasing inlet gas pressure at fixed percentage of glycerol added to the dope solution. After analysis of gas permeability, it can be disclosed that addition of glycerol increased the porosity of the hollow fiber membrane, that's why the permeability of gas increased with the when the glycerol concentration in the dope solution increased. The addition glycerol was effective to increase gas permeability. This trend is explained by the result shown in Table 2. The table shows the mean pore radius and effective surface porosity for hollow fiber membrane prepared with different concentration of glycerol additive. It is clear from the data disclosed in Table 2 that the pore size increased and effective surface porosity also increased as the percent glycerol added to the dope solution increased. It should be noted that in measuring those values, the assumption is made that the skin layer governs the gas transport. Hence the mean pore size and the effective surface porosity obtained by approach mentioned in section 2.4 are those of the skin layer.

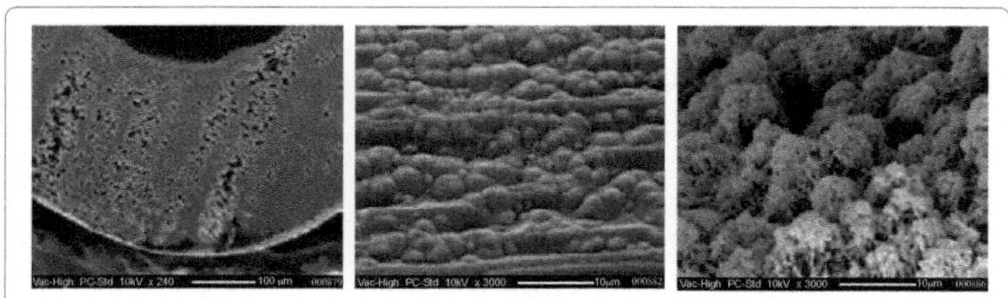

(a) Cross section (b) External surface (c) Inner surface

Figure 4: SEM micrographs for hollow fiber membrane prepared in 30%PVDF/70%triacetin dope solution. (a) Cross-section; (b) External surface; (c) inner surface.

(a) Cross section (b) External surface (c)Inner surface

Figure 5: SEM micrographs for hollow fiber membrane prepared in 30%PVDF/65 %/Triacetin/5% Glycerol cast solution. (a) Cross-section; (b) External surface; (c) inner surface.

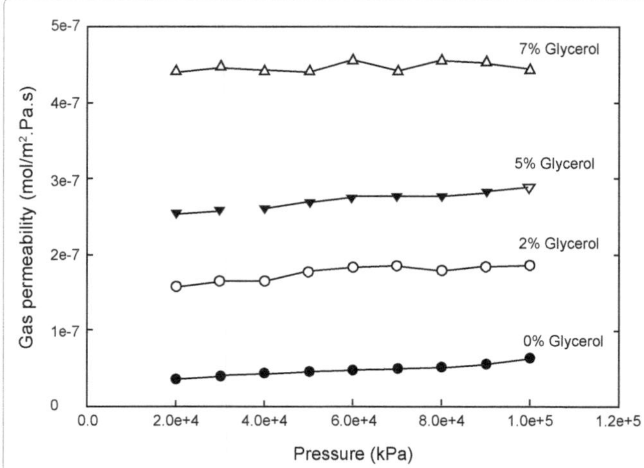

Figure 6: Gas permeability for several hollow fiber membranes prepared with different concentration of glycerol with TIPS method at Bath temperature of 25°C.

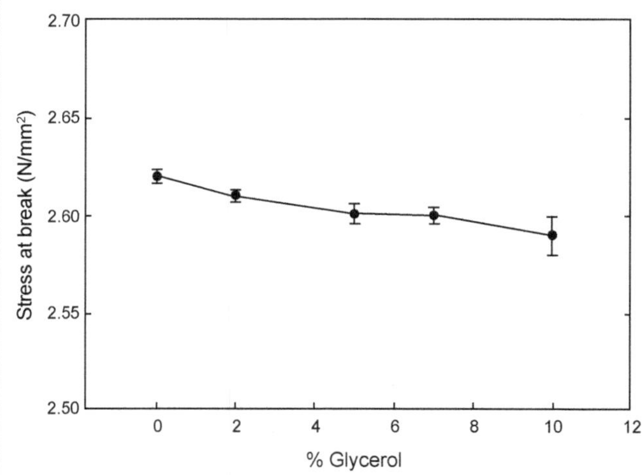

Figure 7: Membrane strength for hollow fiber membrane prepared with different glycerol concentration.

Membrane strength

Figure 7 shows the tensile stress at break for PVDF hollow fiber membranes prepared with various glycerol concentrations. The figure revealed that, tensile strength decreased slightly as glycerol concentration increased. This is attributed to the increase in the membrane porosity of the hollow fiber membranes with glycerol concentrations in the dope solutions. These results agreed with previous studies Rajabzadeh et al. [17-19].

Membrane water contact angle

The effect of percent glycerol added to the dope solution on the average value of the membrane water contact angle is shown in Figure 8. The figure demonstrated sharp increase in the membrane contact angle up to 5% glycerol, further increase in the percent glycerol shows slight increase in the membrane water contact angle.

CO_2 separation

In a membrane contactor for gas absorption, a gas mixture flows on one side of a hydrophobic microporous membrane (in this case shell side) while a liquid absorbent flows on the other side (tube side). The gas–liquid interface is formed at the pore openings adjacent to the liquid. In the membrane contactor module, gas diffuses from the gas side across the membrane and reaches the gas–liquid interface, where gas is absorbed in the liquid. To avoid wetting the membrane pores should not be filled with liquid but with gas; since the liquid phase diffusivity is much lower than gas phase diffusivity, the mass transfer resistance through liquid-filled pores is much larger than that through gas-filled pores. Figure 9 shows the percent removal of CO_2 from CO_2/CH_4 gas mixture versus the inlet gas volumetric flow rate. The gas was inserted to the shell side of the hollow fiber membrane contactor at fixed liquid absorbent (0.5M NaOH) volumetric flow rate (10 ml/min) flows in the tube side of hollow fiber membrane contactor fabricated at different percentage of glycerol additives (0%, 2%, 5% and 7%). The figure shows that, at equal inlet ratio of gas volumetric flow rate to absorbent liquid volumetric flow rate, complete removal of carbon dioxide was achieved. Increasing ratio of gas flow rate at fixed absorbent liquid flow rate decreases the percent CO_2 removal. In other word, when the ratio of gas to liquid absorbent flow rate increased, percent removal of CO_2 decreased. For fiber fabricated with 7% glycerol, it was possible to achieve complete removal for gas to liquid volumetric flow rate up to 3 times, further increase in the gas to absorber ratio resulted in lower percent removal of carbon dioxide, as it is clear from the decline in the curve shown in Figure 9. The percent removal of CO_2 for 0% glycerol was very low. Therefore, in order to investigate the effect of surface property on CO_2 absorption efficiency, we performed experiments for two cases; operation where absorbent flows in the tube and gas mixture flows in the shell side and vice versa. Experimental results revealed that, when gas flowed in the lumen side of the membrane, the percent removal of CO_2 was insignificant and bubbles was observed in the liquid absorbent. That leads to pressure build up of the gas in the tube side and consequently penetrated to the low pressure side and caused bubble formation.

Glycerol concentration (wt %)	Pore radius (μm)	Effective surface porosity (m^{-1})
0	0.005	20
2	0.011	165
5	0.051	585
7	0.077	907

Table 2: Mean pore radius for hollow fiber membrane prepared with different concentration of glycerol.

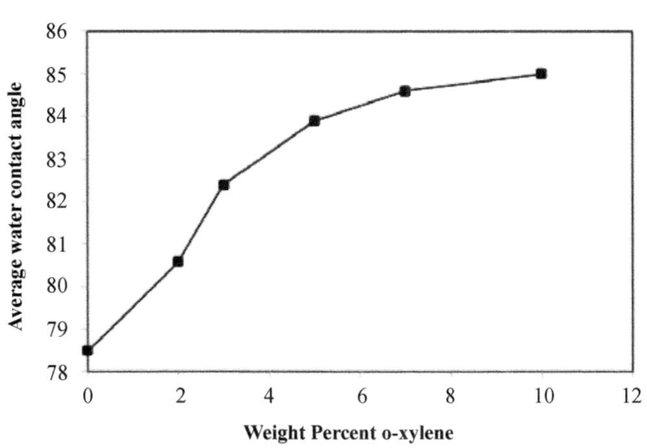

Figure 8: Effect of glycerol concentration in the dope solution on membrane water contact angle.

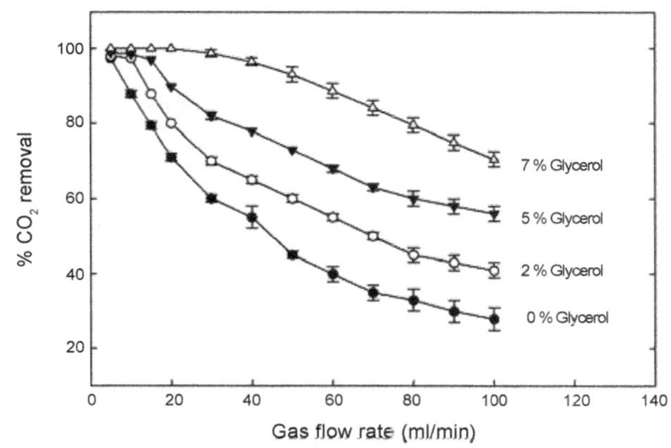

Figure 9: Percent CO_2 removal versus inlet gas flow rate of fabricated hollow fiber membrane prepared with different glycerol concentration.

Conclusions

Microporous hollow fiber membranes were prepared via TIPS method using PVDF polymer, triacetin as solvent and varied quantities of glycerol as non solvent additives gas-liquid phase separation. In the case of zero glycerol content in the cast solution; solid-liquid separation covers the structure of the fabricated fibers and liquid-liquid separations governed the cases with glycerol addition to the dope solution. Hollow fiber membranes prepared with PVDF/triacetin/glycerol system, were successfully used to remove CO_2 from CO_2/CH_4 gas mixture. It was found that as the percentage of glycerol increased in the dope solution the removal rate of CO_2 increased where complete removal of CO_2 is achieved at equal ratio of gas to liquid volumetric flow rate. It can be concluded that addition of glycerol to dope solution was efficient to increase the effective surface porosity and pore radius. The effect of glycerol additive to the cast solution on the membrane tensile stress was insignificant.

References

1. Gabelman A, Huang ST (1999) Hollow fiber membrane contactors. J Memb Sci 159: 61-106.
2. Rangwala HA (1996) Absorption of carbon dioxide into aqueous solutions using hollow fiber membrane contactors. J Memb Sci 112: 229-240.
3. Li JL, Cheng BH (2005) Review of CO_2 absorption using chemical solvents in hollow fiber membrane contactors. Sep Purif Technol 41: 109-122.
4. Kim BS, Harriot P (1987) Critical entry pressure for liquid in hydrophobic membranes, J Colloid Interface Sci 115: 1639-1648.

5. Lawler D (2004) Integrated Water Treatment: Softening and Ultrafiltration. Awwarf Publisher.

6. Wang R, Zhang HY, Feron PHM, Liang DT (2005) Influence of membrane wetting on CO_2 capture in microporous hollow fiber membrane contactors. Separation and Purification Technology 46: 33-40.

7. Nishikawa N, Ishibashi M, Ohta H, Akutsu N, Matsumoto H, et al. (1995) CO_2 removal by hollow fiber gas-liquid contactor. Energy Conversion and Management 36: 415-418.

8. Kong J, Li K (2001) An improved gas permeation method for characterizing and predicting the performance of microporous asymmetric hollow fiber membranes used in gas absorption. J Memb Sci 182: 271-281.

9. Yeon SH, Lee KS, Sea B, Park YI, Lee KH (2005) Application of pilot-scale membrane contactor hybrid system for removal of carbon dioxide from flue gas. J Memb Sci 257: 156-160.

10. Atchariyawut S, Feng C, Wang R, Jiraratananon R., Liang DT (2006) Effect of membrane structure on mass transfer in the membrane gas-liquid contacting process using microporous PVDF hollow fibers. J Memb Sci 285: 272-281.

11. Atchariyawut S, Jiraratananon R, Wang R (2007) Separation of CO_2 from CH_4 by using gas-liquid membrane contacting process. J Memb Sci 304: 163-172.

12. Atchariyawut S, Jiraratananon R, Wang R (2008) Mass transfer study and modeling of gas-liquid membrane contacting process by multistage cascade model for CO_2 absorption. Separation and Purification Technology 63: 15-22.

13. Xu A, Yang A, Young S, deMontigny D, Tontiwachwuthikul P (2008) Effect of internal coagulant on effectiveness of polyvinylidene fluoride membrane for carbon dioxide separation and absorption. J Memb Sci 311: 153-158.

14. Khaisri S, deMontigny D, Tontiwachwuthikul P, Jiraratananon R (2009) Comparing membrane resistance and absorption performance of three different membranes in a gas absorption membrane contractor. Separation and Purification Technology 65: 290-297.

15. Mulder M (1996) Basic Principles of Membrane Technology. (2nd edn), Kluwer Academic Publishers.

16. Lloyd DR, Kinzer KE, Tseng HS (1990) Microporous membrane formation via thermally induced phase separation. I. Solid-liquid phase separation. J Memb Sci 52: 239-261.

17. Rajabzadeh S, Maruyama T, Ohmukai Y, Sotani T, Matsuyama H (2009) Preparation of PVDF/PMMA blend hollow fiber membrane via thermally induced phase separation (TIPS) method. Separation and Purification Technology 66: 76-83.

18. Rajabzadeh S, Maruyama T, Sotani T, Matsuyama H (2008) Preparation of PVDF hollow fiber membrane from a ternary polymer/solvent/nonsolvent system via thermally induced phase separation (TIPS) method. Separation and Purification Technology 63: 415-423.

19. Rajabzadeha S, Yoshimotoa S, Teramotoa M, Al-Marzouqib M, Matsuyama H (2009) CO_2 absorption by using PVDF hollow fiber membrane contactors with various membrane structures. Separation and Purification Technology 69: 210-220.

20. Park HH, Deshwal BR, Kim IW, Lee HK (2008) Absorption of SO_2 from flue gas using PVDF hollow fiber membranes in a gas-liquid contactor. J Memb Sci 319: 29-37.

21. Ren J, Wang R, Zhang HY, Li Z, Liang DT, et al. (2006) Effect of PVDF dope rheology on the structure of hollow fiber membranes used for CO_2 capture. J Memb Sci 281: 334-344.

22. Bakeri Gh, Ismail AF, Shariaty-Niassar M, Matsuura T (2010) Effect of polymer concentration on the structure and performance of polyetherimide hollow fiber membranes. J Memb Sci 363: 103-111.

23. Wang D, Teo WK, Li K (2004) Selective removal of trace H_2S from gas streams containing CO_2 using hollow fiber modules/contactors. Separation and Purification Technology 35: 125-131.

Catalytic Dehydration of Methanol to Dimethyl Ether (DME) Using the $A_{162,2}Cu_{25,3}Fe_{12,5}$ Quasicrystalline Alloy

Jamshidi LCLA[1]*, Barbosa CMBM[2], Nascimento L[3] and Rodbari JR[4]

[1]*Post-Graduate Program in Chemical Engineering-PPCE/Technology and Geosciences Centre-TGC-UFPE. Av. Moraes Rego, 1235 - University City, CEP: 50670-901,Recife-PE, Brazil.
[2]Post-Graduate Program in Chemical Engineering-PPCE/Technology and Geosciences Centre-TGC-UFPE, Brazil. Av. Moraes Rego, 1235 - University City, CEP: 50670-901,Recife, PE, Brazil
[3]Post-Graduate Program in Chemical Engineering-PPCE/Technology and Geosciences Centre-TGC-UFPE. Av. Moraes Rego, 1235 - University City, CEP: 50670-901,Recife, PE, Brazil
[4]Department of Sociology-DS/Islamic Azad University of Roudehen Branch, Roudehen City-Complex of University Roudehen, Tehran, Ira

Abstract

Dimethyl ether (DME) has been considered a potential and promising energy alternative for petroleum subproducts due to its good burning characteristics, and to its high cetana content which is superior to that of diesel. Furthermore, DME can be considered a cleaner fuel than diesel. DME can be produced by dehydration reaction of methanol by using solid catalysts in catalytic reactions. This study shows an analysis of the performance of $Al_{62,2}Cu_{25,5}Fe_{12,3}$ quasicrystalline alloy as catalyst for dehydrating methanol to produce DME. These quasicrystalline alloys are stable at high temperatures, show a low thermal conductivity and exhibit a fragile nature, which turn them to be easily crushed. Also, their activity is not affected by water. In this research it were used the following special measurements: (i) X-Ray Diffratometry (XRD) for analyzing the phases evolution of the alloys; (ii) Scanning Electron Microscopy (SEM) in order to study the surface microstructure and (iii) Transmission Electron Microscopy-TEM for studying internal phases; quasicrystal nuclei morphologies, initial defects and for testing methanol catalytic conversion and selectivity. The latter characteristics were also analyzed for DME and for other subproducts formed in the catalyst. The study showed a good performance of the $Al_{62,2}Cu_{25,5}Fe_{12,3}$ quasicrystalline alloy used as catalyst for DME production.

Keywords: $Al_{62,2}Cu_{25,5}Fe_{12,3}$ quasicrystalline alloy; Dimethyl ether (DME); Methanol dehydration

Introduction

Dimethyl ether (DME) has attracted a worldwide attention because of its potential as an alternative for substituting petroleum. Its use in diesel motors causes low emissions of soot particles and of NO_X [1]. Hence, it can be considered an environmentally compatible fuel. DME production has been investigated by several researchers in many parts of the planet.

At environmental conditions; i.e., at 1 atmosphere pressure and temperature of 25°C, DME is present in gaseous state. When it is submitted to higher pressures or to lower temperatures it liquefies easily in the same way as PLG (Petroleum Liquefied Gas) [2].

Probably in next future, DME can be stocked and distributed by using almost the same technology applied to PLG. Hence, DME can be applied as a PLG substitute although it presents a thermal power lesser than that of propane gas (C_3H_8) and butane gas (C_4H_{10}) which are the main constituents of this fuel [3]. Furthermore, it is important to state that there is an increase in the use of DME as thermoelectric energy.

DME is generally produced when methanol is dehydrated according to the following chemical reaction [4]:

$$2CH_3OH \rightarrow CH_3OCH_3 + H_2O \quad (1)$$

$$(\Delta G = -12,1 kcal.mol^{-1}, 25°C)$$

where ΔG refers to the standard free energy for the reaction given by equation (1).

Several researches were performed in order to formulate better the use of catalyst solids such as crystalline solids, amorphous, zeolites, and quasicrystalline solids. The principal aims of these researches were to ensure not only a higher stability during the catalytic reaction but also to optimize DME production [5].

Recently, quasicrystals are been applied in catalyst reactions due to its stable equilibrium phases even at high temperatures. The quasicrystalline materials occupy a position between crystal and amorphous materials. They exhibit a complex structure showing a repetition of quasiperiodicity in the atoms arrangement together with rotational symmetries which are not observed in crystals [5].

The catalytic behavior of the AlPdMn quasicrystalline alloy was investigated in comparison with that one of pure AlPd crystals, pure copper and pure iron crystals. In the research, particles with size less than 200nm of each sample were mixed with magnesium oxide (MgO) and were calcined at 775K during 5 hours [6]. In the reaction of methanol decomposition, the quasicrystal catalyst achieved a great amount of hydrogen gas at low temperature in the beginning of the reaction. Several palladium quasicrystals alloys were tested and they showed to be highly active for methanol decomposition.

The first vapor reform reaction by using $Al_{62,2}Cu_{25,5}Fe_{12,3}$ quasicrystal was obtained after lixiviation of the alloy giving an H_2 production

***Corresponding author:** Agostinho L.C.L, Post-Graduate Program in Chemical Engineering-PPCE/Technology and Geosciences Centre-TGC-UFPE. Av. Moraes Rego, 1235 - University City, CEP: 50670-901, Recife-PE, Brazil of 235 L. $kg^{-1}.min^{-1}$ at 553°C [7]. In catalytic reactions there was also formation of other metalorganic products such as dimethyl ether (DME); methanoic acid and others [8].

This study shows the behavior of $Al_{62,2}Cu_{25,5}Fe_{12,3}$ quasicrystal alloy in catalytic reactions and its applicability for obtain DME from the methanol dehydration.

Theoretical Considerations

Methanol dehydration Synthesis to DME

DME is obtained from the synthesis gas which is a mixture of carbon monoxide and hydrogen gas originated from natural gas reform, from petroleum subproducts, from gaseous coal and from biomass (Bio-DME). In this process, the synthesis is achieved from two distinct catalytic functions: hydrogenate and dehydration functions, respectively [9]. The first one is based in the initial methanol formation and produces the synthesis gas. The second one is related to DME production from condensation of two molecules of methanol. In both processes, reactions take place through a catalyst which can be metallic solids, amorphous solids, crystals, semiconductors, zeolites and quasicrystalline alloys.

The main reactions observed in DME synthesis are a reaction limited by the equilibrium of methanol syntheses and a reaction not limited by the equilibrium of the methanol dehydration in catalysts [10].

Reactions involved in this process are as follows [11]:

- Reaction for methanol synthesis defined as:

$$4H_2 + 2CO \rightleftarrows 2CH_3OH \tag{2}$$

with enthalpy value given by ΔH = - 43,2 $kcal.mol^{-1}$.

- Reaction for methanol dehydration given by:

$$2CH_3OH \rightleftarrows CH_3OCH_3 + H_2O \tag{3}$$

in which enthalpy value is H = - 5,6 $kcal.mol^{-1}$.

- Water – gas reaction:

$$CO + H_2O \rightleftarrows H_2 + CO_2 \tag{4}$$

With enthalpy value given by ΔH = - 9,8 $kcal.mol^{-1}$.

The overall reaction is given by:

$$3H_2 + 3CO \rightleftarrows CH_3OCH_3 + CO_3 \tag{5}$$

In which enthalpy value is ΔH = - 58,6 $kcal.mol^{-1}$ and the product of each reaction is the reactant of the reaction given by equation (5).

The hydrogen formed in the reaction of equation (4) is the reactant for syntherizing methanol. This allows a high conversion of the synthesis gas in the overall reaction according to equation (5). It is important to state that the combination if these reactions results in a synergic effect making easier not only the thermodynamics of methanol synthesis bit also the reactions of displacement associated to water-gas. Normally, the latter reactions are observed in solid catalysts such as quasicrystals, zeolites, crystal solids, amorphous solids, metallic solids and semiconductors because the catalysts can support high temperatures around 500°C [12].

Regarding the use of zeolites as catalysts, the studies have showed that when methanol is catalyzed on the surface of acid zeolites firstly it dehydrates producing DME and secondly the mixture is converted in olefins and aromatic aliphatic compounds until C_{10} [13].

The mechanism of methanol adsorption, its dehydration, the C - C first liaison formation as well as the nature of intermediaries compounds involved in the reactions are not entirely understood [14]. Kasaie and Sohrabi suggested that methanol reaction occurs in Brønsted site and its adjacent locals in the Lewis basic sites forming both $[CH_3 - OH_2] + [CH_3O]$ species in the quasicrystal surface. Afterwards, when it condensates DME and water are produced [15]. This kinetic mechanism utilizes Langmuir-Hinshelwood to explain the heterogeneity of the catalytic reaction on the quasicrystal surface and is based in the studies of Bandiera et al. [16].

Methanol molecules are adsorbed in two different types of sites: acidic site (Lewis) and basic site (Brønsted) [17]. Initially, in the catalyst surface occurs the adsorption o one proton of H^+ according to the following equation [18]:

$$CH_3OH + H^+ \rightleftarrows [CH_3OH_2]^+ \tag{6}$$

Simultaneously, methanol molecules are adsorbed in the basic site giving:

$$CH_3OH + O^{2-} \rightleftarrows [CH_3O]^- + [OH]^- \tag{7}$$

The above species are then absorbed on surface producing DME by condensation according to the following equation:

$$[CH_3OH_2]^+ + [CH_3O]^- \rightleftarrows CH_3 - O - CH_3 + H_2O \tag{8}$$

The catalyst surface is regenerated by the following reactions:

$$H_2O + [OH]^- \rightleftarrows [H_3O]^+ + O^{2-} \tag{9}$$

$$[H_3O]^+ \rightleftarrows H_2O + H^+ \tag{10}$$

According to Kubelková et al. [19], only one molecule of methanol is adsorbed on the surface with one transition of protonated H^+. After dehydration, one methoxy group remains in the surface. Another molecule of methanol reacts with DME methoxy group according to the reaction given by:

$$CH_3OH_{(ads)} \rightleftarrows CH_3OH_2{}^+ \rightleftarrows -CH_3 + H_2O \tag{11}$$

$$-CH_3 + CH_3OH \rightleftarrows CH_3 - O - CH_3 + -H \tag{12}$$

Methanol dehydration to DME in quasicrystal catalytic reaction

When transition metals are used as catalysts, especially in steam reform reaction, being or not supported, they must show a very well defined structure. In the active phase dispersion, the metallic ions oxidation states define the "d" (electronic sublevel) character and the atomic radium [20]. These factors have a great influence in the selectivity and activity of catalysts. The metallic ions used as dopants show a lesser size and valence than these of the supported ion used. When the ions of low valence are substituted by ions of high valence, spaces in the structure of the crystalline solids are formed. In this case, it is observed the presence of gaps for the oxygen atoms which are interlinked in the SiO_2 net [21]. These vacancies exhibit an important role in the stabilization of catalytic systems.

The catalytic dehydration of methanol on the quasicrystalline alloy $Al_{62,2}Cu_{25,5}Fe_{12,3}$ catalyst is due to oxidation reaction of methanol + methanoic acid, DME and intermediary products. One of the advantages of this type of catalyst is that in the beginning of the water reaction in gaseous state there is no blockage in the local active sites.

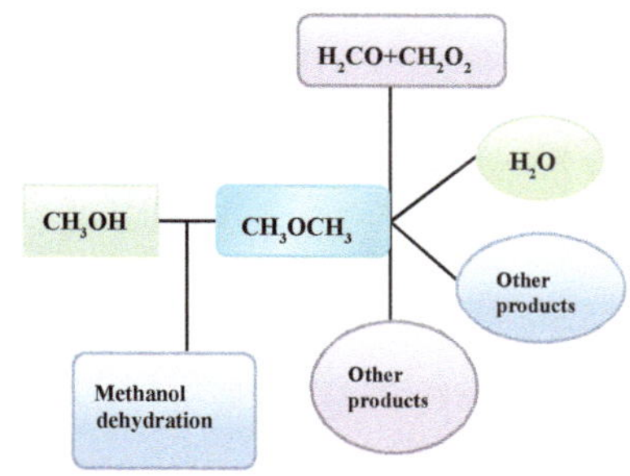

Figure 1: Fluxogram of methanol dehydration reaction to DME.

Therefore, the reaction products show a better selectivity due to a lesser formation of weak acid sites. Figure 1 below illustrates the dehydration reaction of methanol in the $Al_{62,2}Cu_{25,5}Fe_{12,3}$ quasicrystalline alloy.

Experimental results show that the methoxy radical ($-O-CH_3$) is the intermediary more stable product formed after methanol adsorption in the quasicrystal surface. With increase in temperature this species is decomposed into formaldehyde (H_2CO). It is important to state that in the iron (Fe) surface occurs methane formation (CH_4). As to copper (Cu) surface, the methoxy radical is adsorbed much easier in gaps of energy. This process is observed in the top of surface in the (111) crystallographic plane formation or near the first saturation layer of the adsorbed atoms. It was observed later that the methoxy radical is adsorbed in all adsorption sites of the copper surface in $Al_{62,2}Cu_{25,5}Fe_{12,3}$ quasicrystalline alloy. This is due to the layer that exhibits valence3d, 4s and 4p sublevels which effectively influence the catalytic sites of the catalyst [22]. In this process, part of the methanol formed comes from the copper interaction with reaction substrate in the quasicrystal surface. Copper has a tendency to oxidize by using the oxygen present in the catalyst surface and thus producing copper oxide (CuO) and copper hydroxide ($Cu(OH)_2$) species [23]. There is also a possibility for ion pairing between copper and magnetite (Fe_3O_4) forming copper ferrite ($CuFe_2O_4$) and between copper and iron oxide forming delafossite ($CuFeO_2$). All these substances show good electrical properties for metallic conduction [24]. In the quasicrystal surface an initial oxide layer is observed when temperature reaches 670°C [25]. This is a passivation layer that is comprised of a thin oxide film. The mass of aluminum atoms formed in the quasicrystal surface is due to a directional force supplied by oxide exothermicity which is greater than all the others species of the alloy. It was also observed that $Al(CuFe)+AlFe_3$ quasicrystalline alloy when submitted to temperature of 350°C produces, during the reaction process, the $Al_7Cu_2Fe_2$ chemical that exhibits a tetragonal form [26]. Hence, it is observed oxidation in both aluminum and copper. In the range of temperature 425ºC to 550ºC this chemical prevails, but associated to ferric aluminum traces (Al_2Fe_3). This fact shows that an increase of 125°C in temperature is sufficient for oxidizing the quasicrystalline alloy. Aluminum of the qc surface also acts as an acid function for catalyzing the DME hydrolysis to MeOH (CH_3OH). On the quasicrystal surface, aluminum forms aluminum hydroxide (AlOH). These agglomerates together with the reaction elements (hydrogen, oxygen and carbon) make arrangements with linking of clusters and represent the quasicrystal acid site. The reaction of one molecule methanol is initially absorbed in the qc local site. Then, methanol is dehydrated leaving a methyl group linked to the oxygen of the qc structure of this arrangement.

Experimental Methodology

Preparation of the $Al_{62,2}Cu_{25,5}Fe_{12,3}$ quasicrystalline alloy

In the preparation of the precursory alloy to be applied as catalyst, it was utilized $Al_{62,2}Cu_{25,5}Fe_{12,3}$ ternary system. Powders with nominal composition of $Al_{62,2}Cu_{25,5}Fe_{12,3}$ were classified in terms of its granulometry and prepared with purity superior to 99,90%. The alloy was obtained by applying air fusion to the pure elements. Induction oven controlled by Argonium 5, 0 atmospheres was utilized. The alloy was then submitted to several foundries in order to ensure the complete dissociation of its components and to achieve a good homogeneous quasicrystalline phase. The samples were prepared at the Fast Solidification Laboratory, - LSR of the Federal University of Paraiba, UFPB by utilizing a high frequency POLITRON generator (40kVA). For the preparation of the alloy it was weighed 10g of the samples by using a SHIMADZU AY220 model balance (precision around 10^{-4}g). The alloys were formed by applying the solidification method in a special oven and, thus, results a mixture of quasicrystalline and crystalline league. For this purpose, it was used a NABERTHERM resistance oven where the samples were submitted to a temperature of 750°C during 8 hs and 24 hs, respectively. After this procedure, samples were placed in induction oven at argonium rare-effect atmosphere. Samples were heated in a rate of 30ºC.min^{-1}. For catalytic reactions and to avoid decrease in the oxidation reaction, samples were encapsulated in a quartz tube under continuous vacuity during thermal treatment.

X-Ray Diffratometry (XRD) tests were applied in order to follow the phase evolution and the samples stability during the melting process. It was used a SIEMENS DIFFRATOGRAM D 5000 and it was applied CuKα radiation with wave length of $\lambda = 1,5406\mathring{A}$. The tests were performed at a temperature of 298K, 40Kv tension , 30mA electric current, step of 0,01°, time by step of 3s and angle of 2θ (2-theta) with variation of 20° to 120°.

Scanning Electron Microscopy (SEM) tests were applied for analyzing quasicrystalline surface morphology by using a LEO MODEL 1430 scanning electronic microscope coupled to an OXFORD probe. Samples after been melted and submitted to catalytic tests were immersed in isopropyl alcohol solution. Afterwards they were placed in a DABIATLANTE-3L Ultrasonic Cable in order to promote the powder non agglomeration. After this procedure, samples were analyzed for SEM.

Dispersive Energy Spectroscopy (DES) in samples was obtained in the micrographic of dispersive energy analysis after thermal treatment of 24 hs coupled to SEM analysis. The samples were previously metalized with gold in a medium thickness of 12 nm.

The experimental analysis performed in the Transmission Electronic Microscope (TEM) used a TECNAC 20 Model with tension between 20 to 200 kV, 1.9 Å and point resolution of 0,2 nm. The tests helped to describe not only the quasicrystal structure but also the samples morphology and thus propitiating informations such as surface defects, discordances and particles size.

Catalytic tests on the $Al_{62,2}Cu_{25,5}Fe_{12,3}$ quasicrystalline alloy

The catalytic tests were performed in the catalytic evaluator reactor, TCAT-1 Model, at atmospheric pressure. In general, methanol concentration in the admission gas has always increased to 18.6% (in molar unity). The rest was a mixture of air, N_2 and methanol, generally of argonium (37%) + N_2 (18.4%) + methanol (26%). For the tests it was weighted 300 mg which was introduced into a U reactor of Pyrex glass. This vessel was heated at environmental temperature until 450°C in argonium in a heat rate of 20°C.min^{-1}, 50 mL.min^{-1} flux [27].

After reaching the temperature of 450°C, the sample stayed at this condition during 2 hours in order to remove the physically fissorved water.

The following procedure was to transfer the methanol from a saturator at environmental temperature to the catalytic bed through a heated line at 120°C with air flow of 50 mL.min^{-1}. During the reactive process, the catalytic bed was maintained at a constant temperature of 450°C by using a COELHW1500 temperature controller. The effluent products of the reactor were successfully "online" injected by means of a 10 ways valve in a VARIANCP 3800 gas chromatograph. This experiment has a thermal conductivity detector in intervals of 15 minutes until it reaches the pseudo - stationary state. Experimental tests were made with six consecutive injections at intervals of 15 minutes and, thus, totalizing 90 minutes.

Results and Discussions

X-Ray Diffratometry results

X-Ray Diffratometry spectra of $Al_{62,2}Cu_{25,5}Fe_{12,3}$ quasicrystal samples are shown in Figure 2 and 3, respectively, for thermal treatments of 8 hs and 24 hs. This phase shows clearly that the sizes of the grains are critical to initiate the already existing transformation. In these peaks can be noticed the quasicrystalline icosaedral phase (i) and the crystalline phase (β) with $Al_{50-x}(Cu, Fe)_{50+x}$ composition [28].

In these figures it can be recognized the quasicrystalline icosahedral i-phase, β-$Al_{0,5}Fe_{0,5}$ phase, tetragonal phases such as $\theta - AlCu_3$, $\theta - CuAl_2$ which can be identified in the peaks. Generally, $\beta -$ phase coexists with the quasicrystalline phase when the applied process does not propitiate the thermodynamic conditions for the alloy to be completely quasicrystalline. Hence, it is formed the icosaedral phase. By comparing Figures 2 and 3 it can be noticed that the peaks for $\beta -$ phase suffer a displacement to the left. When thermal treatment is accomplished (in 24 hs), the quasicrystalline alloy is homogeneous and the alloy is practically monophasic; i.e., this is the required time to ensure the sample saturation. This technique allows characterizing the purity of the phase which is termed "bulk". The spectra are in accordance to ROSAS and PEREZ [29], once they define this phase as a solid solution for controlling the formation of quasicrystalline phase.

Scanning electronic microscopy (SEM) and energy dispersive spectroscopy (EDS)

Figures 4a, 4b, 5 and 6, respectively show the results obtained by using Scanning Electronic Microscopy and EDS in the samples before and after the catalytic reactions. Figures 4a and 4b show, respectively,

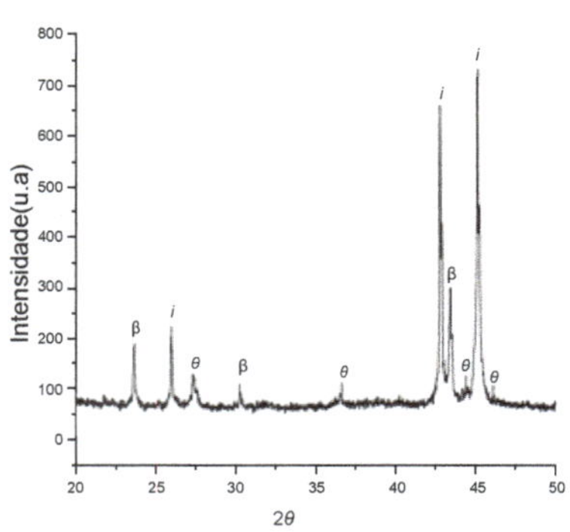

Figure 2: X-Ray Diffratometry for $Al_{62,2}Cu_{25,3}Fe_{12,5}$ sample with thermal treatment in 8 hs.

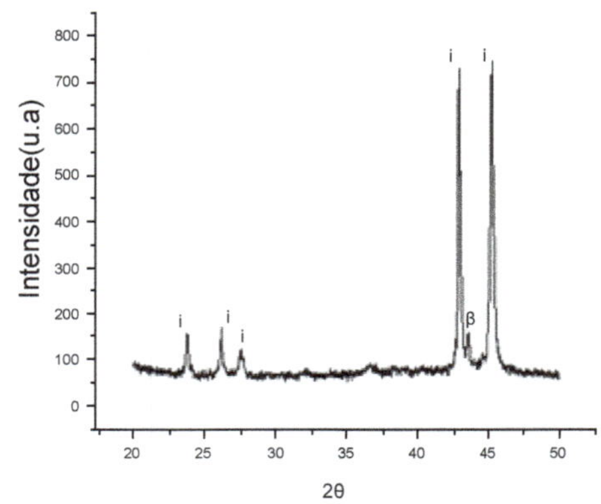

Figure 3: X-Ray Diffratometry for $Al_{62,2}Cu_{25,3}Fe_{12,5}$ sample with thermal treatment in 24hs.

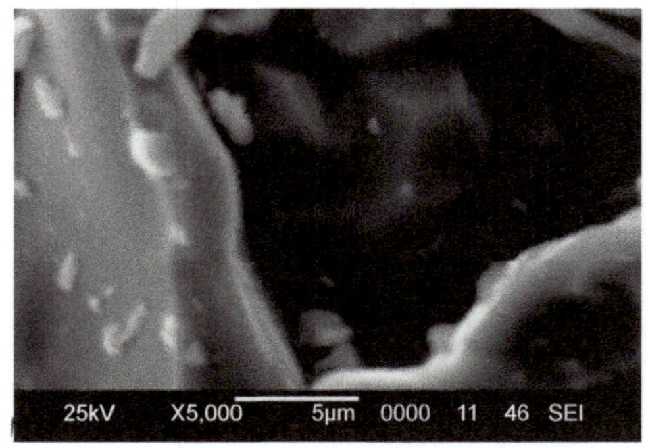

Figure 4a: Image obtained of the $Al_{62,2}Cu_{25,3}Fe_{12,5}$ quasicrystalline alloy showing the icosahedral phase in 24 hs.

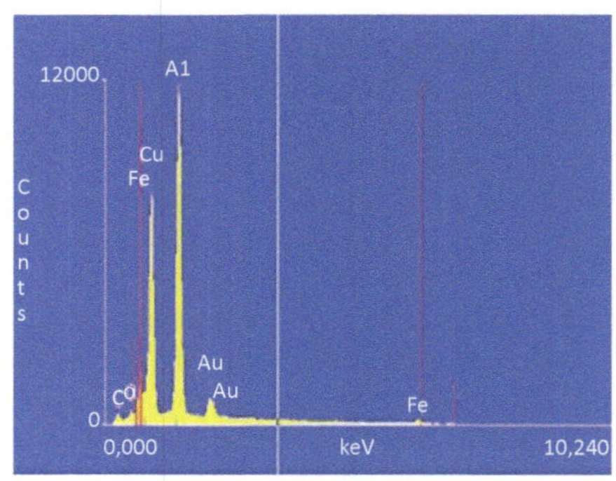

Figure 4b: SDE elementary analysis of $Al_{62,2}Cu_{25,3}Fe_{12,5}$ quasicrystalline alloy showing icosahedral phase after thermal treatment in 24 hs.

the results of SEM and EDS applications in samples after thermal treatment of 24 hs.

In Figure 4a it is observed a very good geometrical uniformity which is a result of cleavage fractures, a bit rough, with icosaedral symmetry, but with some porosity. On the surface it is possible to notice quasicrystalline grains in the form of dodecahedral polyhedrons. In Figure 4b, it is observed a major predominance of aluminum in relation to others quasicrystalline alloy components.

Figures 5 and 6, respectively, show the results of the SEM measurements applied in the samples after catalytic reaction.

Figure 5: $Al_{62,2}Cu_{25,3}Fe_{12,5}$ quasicrystalline alloy with 8hs of thermal treatment and submitted to catalytic reaction.

Figure 6: $Al_{62,2}Cu_{25,3}Fe_{12,5}$ quasicrystalline alloy with 24 hs of thermal treatment and submitted to catalytic reaction.

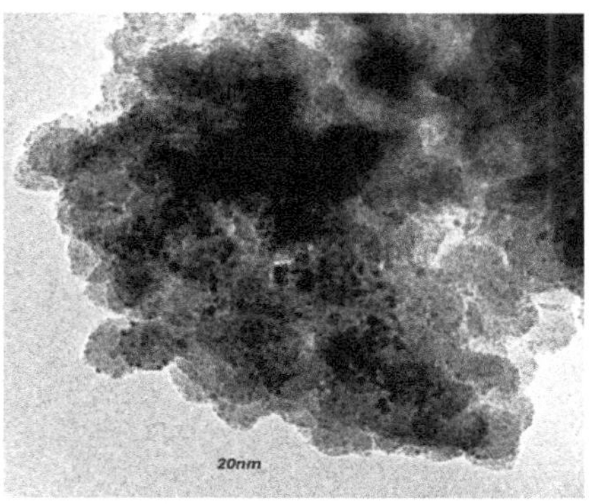

Figure 7: Image of TEM-Transmission Electronic Microscopy for $Al_{62,2}Cu_{25,5}Fe_{12,3}$ quasicrystalline alloy sample after thermal treatment of 24 hs.

The microstructure of the sample which is illustrated in Figure 5 shows nodules formation in the catalyst surface due to an oxide θ-Al_2O_3 layer formation which is non homogeneous. This layer penetrated the $Al_{62,2}Cu_{25,5}Fe_{12,3}$ alloy surface and allowed α-Al_2O_3 formation. Hence, it is evidentiated a superficial nodule morphology that has almost 5% of copper. In this figure, it is observed a decomposition of quasicrystalline phase to crystalline phase due to the increase in temperature of catalytic reaction. This reaction has occurred in temperature above 400°C and formed initially on the whole qc surface, an amorphous oxide layer. This process is due to nucleation and oxide crystallites growth in the γ-Al_2O_3 phase which is distributed in aleatoric form in the interface between metal oxide and the amorphous. This growth is fed by diffusion to the metal/ oxide interface. The atomic oxygen of the amorphous layer is generated through fissures of the amorphous layer.

Naturally, it has occurred the reduction in quasicrystalline phase and an oxidation in intermetallic phase which, in its turn, acted as nuclei or catalytic sites.

Transmission electronic microscopy-TEM

Due to the existence of icosaedral quasicrystalline phase (crystalline and intermetallic phases) it was applied TEM-Transmission Electronic Microscopy for analyzing the structure characterization of the samples. Figure 7 shows the quasicrystalline alloy after thermal treatment of 24 hs and at temperature of 750°C.

In Figure 8 it can be observed the quasicrystalline alloy microstructure after thermal treatment and after catalytic reaction.

In both figures, it is possible to visualize particles agglomerates morphology in a range from 8.10^{-1} nm to 20nm after catalytic reaction. Also, in those figures, it can be seen an image of dark areas in which occurred various morphologies of the phases.

Nature of metallic particles was confirmed by EDS-Energy Dispersive Spectroscopy and showed preferential localization according to the metal. Regarding Cu, its metallic particles are located in the quasicrystal edges. In these regions, it is possible to find big copper crystals indicating enrichment in the layer of the quasicrystal catalyst. In other words, there was a copper particle syntherizing during the catalytic reaction.

Faungnawakij et al. [30] stated that the iron addition to Cu/ZnO/Al_2O_3 contributed significantly to conduct highly dispersed Cu particles formation. Fang et al. [31] Said that the addition of FeO_x to Cu/Al_2O_3 gives rise to highly dispersed copper formation. As to iron, Fe it is important to notice its beneficial effect regarding the catalytic behavior through the interaction between copper and iron oxide (FeO) and its oxides in catalysts of type (CuFe)-Cu/FeO_x or supported catalysts. This benefit refers to the increase in catalytic activity and to its good thermal stability. Other researchers showed a good relation of Cu single orientation to Fe_3O_4 in which Cu took the same direction of Fe_3O_4 reducing to a ferrite such as $CuFe_2O_4$ (delaphossite). In this case, its magnetic and electric properties are function not only of its ionic radium and valence but also from its chemical, morphological and stoichiometric properties and size of particles that propitiate the catalytic properties in chemical reactions [32].

Bergstein [33] related that the corresponding net in the interface between Cu and Fe_3O_4 which is the reduced form $CuFe_2O_4$ (delaphossite) facilitates the formation of new metallic phases. Chen et al. [34] showed that the direct interaction between copper species and iron oxide

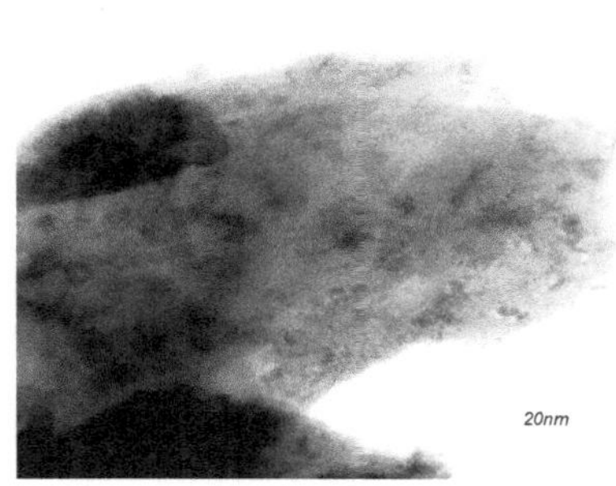

Figure 8: Image of TEM-Transmission Electronic Microscopy for $Al_{62,2}Cu_{25,3}Fe_{12,5}$ quasicrystalline alloy sample after a catalytic reaction.

agglomerates could form new species to inhibit syntherization and increases the catalytic activity. In resume, dehydration reaction for methanol in copper surface originates the formation of DME and other intermediary products [35].

Catalytic processes evaluation

Tables 1 and 2 show that $Al_{62,2}Cu_{25,5}Fe_{12,3}$ quasicrystalline alloy has a very good performance. The synthesis of methanol to DME reached a good and satisfactory level of selectivity. Also, it occurs a good methanol conversion rate. From the results of these tables it was made the graph of Figure 9 with selectivity curves (%) related to time (minutes). These curves allow observing some behavior of the reactions products. Water curve remained constant and showed that there was no increase during vapor reaction. Generally, in this type of reaction and in the presence of Al_2O_3, it is observed a preferential tendency for water to be adsorbed in quasicrystal surface instead of methanol. This implies in a reduction in the methanol activity. However, in this case, there was no decrease in the methanol context (CH_3OH). On the contrary, there was an increase in its conversion [36].

Acidity on the catalyst surface influences the products distribution as well as the catalytic activity. Hence, in the dehydration synthesis of methanol to DME, the selectivity is an important factor in methanol conversion. Figures 9 and 10 show, respectively, as a function of selectivity and conversion rate that DME efficiency rate has a good selectivity in this reaction. Two main factors must be considered: (i) the presence of weak acid sites and that the water curve in all reaction time was constant and contributes in a satisfactory way to DME selectivity and (ii) the decrease of selectivity rate with time was not significant because of coke formation on the surface and that it was observed olefins production.

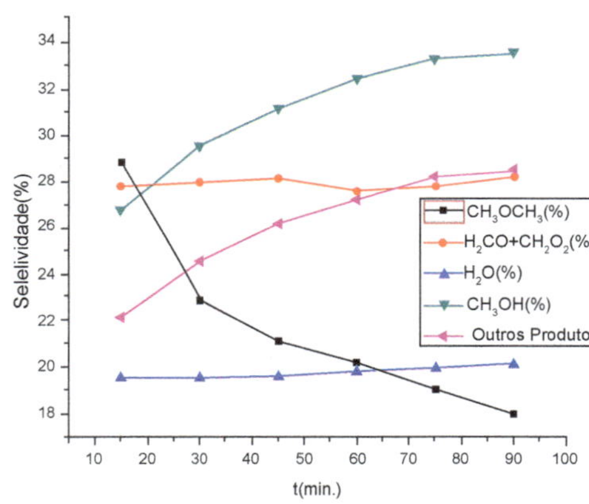

Figure 9: Catalytic behavior of $Al_{62,2}Cu_{25,5}Fe_{12,3}$ quasicrystalline sample as a function of products selectivity (%) and time (min).

t(min.)	15	30	45	60	75	90
CH_3OCH_3 (%)	25,87	22,88	21,1	20,13	19,01	18,02
$H_2CO+CH_2O_2$ (%)	27,86	28	28,15	27,61	27,82	28,29
H_2O(%)	19,54	19,56	19,6	19,85	19,93	20,12
CH_3OH (%)	26,72	29,56	31,15	32,41	33,25	33,55
Other products. (%)	22,18	24,53	26,16	27,22	28,26	28,53

Table 1: Products selectivity (%) in methanol reaction as a function of injection time (min).

t(min.)	15	30	45	60	75	90
CH_3OCH_3 (%)	25,87	22,88	21,1	20,13	19,01	18,02
CH_3OH (%)	26,72	29,56	31,15	32,41	33,25	33,55

Table 2: Conversion (%) of the catalytic reaction subproducts as a function of time (min.).

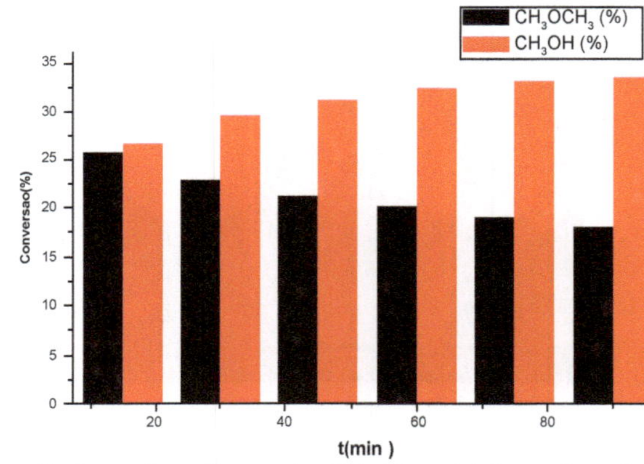

Figure 10: Comparison of conversion (%) of the catalytic reaction subproducts as a function of time (min).

Methanal + methanoic acid subproducts ($H_2CO+CH_2O_2$) showed its selectivity curve practically constant in the time intervals of the study. The analysis of methanol selectivity curve showed an increase of almost 10% during the time intervals of the research. This growth may be attributed to a total oxidation of methanol which was verified later in the reaction for methoxy radical production (CH_3O) [37]. This radical is considered the more stable intermediary product after adsorption of methanol in qc surface. In the reaction as a function of methanol there is a decomposition of formaldehyde (H_2CO). Normally, the next step after formaldehyde formation is the production in variable stages of carbon monoxide, CO or of carbon dioxide, CO_2.

In general, reactions occur in sequential stages. In the first stage, there is methanol dehydration. Next step shows that the hydrogen ion which is liberated can exothermally react with air giving rise to a reaction that is definitively an oxidative dehydrogenation. It is important to state that formaldehyde was produced as subproduct in the methanol dehydration. However, the aluminum which is present in the qc alloy gives some sites for formaldehyde production [38].

In terms of DME efficiency which is evaluated through the multiplication of methanol conversion, the results indicate that $Al_{62,2}Cu_{25,5}Fe_{12,3}$ quasicrystalline alloy as catalyst shows a good stability in catalytic reaction and can be applied for methanol dehydration to DME.

Conclusions

The main conclusions of this study are as follows:

1. $Al_{62,2}Cu_{25,5}Fe_{12,3}$ quasicrystalline alloy used in the experiments is thermodynamically stable at high temperatures satisfying the initial requisite for utilizing it as a catalyst;

2. Iron and copper from the $Al_{62,2}Cu_{25,5}Fe_{12,3}$ quasicrystalline alloy showed a direct interaction between copper species and ironed oxide agglomerates and thus making possible to produce new species in order to inhibit syntherization and to increase catalytic activity;

3. Copper particles (outside transition metals with $3d^{n-2}4s^2$ energy sublevels) are observed in the quasicrystal surface. They

constitute the transition metallic element more favorable to the methanol oxidation reaction as well as methyl radical (CH_3) formation. The latter constitutes the most stable intermediary product formed;

4. The quasicrystal $Al_{62,2}Cu_{25,5}Fe_{12,3}$ catalyst proved to be adequate for dehydrating methanol to DME due to the good reaction performance. Its activity was not affected by water once it remained constant in the reaction process. The water in the reaction has a positive effect on the methanol dehydration because it regenerates the catalyst by removing carbon deposition and, thus, limits the deactivation due to coke production;
5. Aluminum present in the surface of $Al_{62,2}Cu_{25,5}Fe_{12,3}$ quasicrystalline alloy propitiated aluminum oxide (Al_2O_3) production, These agglomerates, together with reaction elements such as hydrogen, oxygen and carbon, formed link arrangements of cluster type representing the acid sites (Lewis and Brønsted sites) in the quasicrystal;
6. The metallic cations of the $Al_{62,2}Cu_{25,5}Fe_{12,3}$ quasicrystal have the ability to form complexes (ion pairing) when react with water and can be represented by a general formulation such as $[M(H_2O)_n]^{m+}$;
7. The low cost for producing this type of alloy for catalytic reactions constitutes one advantage and makes possible its use for industrial catalysts.

Acknowledgement

To the National Council for Scientific and Technologic Development Agency-CNPq for its financial support; to the Magnetism and Magnetic Materials Laboratory of the Physics Department of Federal University of Pernambuco, to the Federal University of Paraiba and to the Fuel and Lubricant Laboratory of the Federal University of Rio Grande do Norte-UFRN.

References

1. Lima Neto EP, Almeida ELF, Bomtempo JV (2008) Restructuring the cadeias the chemical and energy: the way methanol. Revista Brasileira de Energia 14: 127-149.
2. Ogawa T, Inoue N, Shikada T, Ohno Y (2003) Direct dimethyl ether synthesis. Journal of Natural Gas Chemistry 12: 219-227.
3. Song J, Huang Z, Qiao X, Wang W (2004) Performance of controllable premixed combustion engine fueled with dimethyl ether. Energy Conversion and Management 45: 2223-2232.
4. Varisli D, Dogu T, Dogu G (2007) Ethylene and diethyl-ether production by dehydration reaction of ethanol over different heteropolyacid catalysts. Chemical Engineering Science 62: 5349-5352.
5. Huttunen-Saarivirta E (2004) Microstructure, fabrication and properties of quasicrystalline Al-Cu-Fe alloys: a review. J Alloys Compd 363: 150-174.
6. Schmithusen F, Cappello G, De Boissieu M, Boudard M, Comin F, et al. (2000) Study of the response of the fivefold i-AlPdMn surface to different temperature treatments. Surface Science 444: 113-122.
7. Tsai AP, Yoshimura M (2001) Highly active quasicrystalline Al-Cu-Fe catalyst for steam reforming of methanol. Appl Catal A: Gen 214: 237-24.
8. Yaripour F, Baghaei F, Schmidt I, Perregaard J (2005) Catalytic dehydration of methanol to dimethyl ether (DME) over solid-acid catalysts. Catal Commun 6: 147-152.
9. Erdener H, Arinan A, Orman S (2011) Future Fossil Fuel Alternative; DME (A review). IJRER 1.
10. Sun KP, Lu W, Qiu F, Liu S, Xu X (2003) Direct synthesis of DME over bifunctional catalysts: surface properties and catalytic performance. Appl Catal A: Gen 252: 243-249.
11. Shen WJ, Jun KW, Choi HS, Lee KW (2000) Thermodynamic investigation of methanol and dimethyl ether synthesis from CO2 hydrogenation. Korean J Chem Eng 17: 210-216.
12. Lenarda M, Storaro L, Frattini R, Casagrande M, Marchiori M, et al. (2007) Oxidative methanol steam reforming (OSRM) on PdZnAl hydrotalcite derived catalyst. Catal Commun 8: 467-470.
13. Kito-Borsa T, Cowley SW (2004) Kinetics, Characterization and mechanism for the selective dehydration of ethanol to diethyl ether over solid acid catalysts. Am Chem Soc Div Fuel Chem 49: 856.
14. Blaszkowski SR, Santen Van RA (1996) The mechanism of dimethyl ether formation from methanol catalyzed by zeolitic protons. J Am Chem Soc 118: 5152-5153.
15. Kasaie M, Sohrabi M (2009) Kinetic study on methanol dehydration to dimethyl ether applying clinoptilolite zeolite as the reaction catalyst. J Mex Chem Soc 53: 233-238.
16. Bandiera J, Naccache C (1991) Kinetics of Methanol Dehydration on dealuminated H-mordenite- model with acid and basic active centers. Appl Catal 69: 139-148.
17. Fu Y, Hong T, Chen J, Auroux A, Shen J (2005) Surface Acidity and the dehydration of methanol to dimethyl Ether. Thermochim Acta 434: 22-26.
18. Mas C, Dinjus E, Ederer H, Henrich E, Renk C (2006) Dehydration of methanol to dimethylether. Forschungszentrum Karlsruhe, Karlsruhe 1-21.
19. Kubelková L, Nováková J, Nedomová K (1990) Reactivity of surface species on zeolites in methanol conversion. Journal of Catalysis 124: 441-450.
20. Russel JB (1994) Química Geral. 2° Ed São Paulo: Makron Books 1.
21. Kittel C (2007) Introduction to Solid State Physics - Editora Guanabara Two.
22. Agostinho LCL (2009) Study of the Applicability of Quasicrystals AlCuFe in Catalytic Reactions in Oxidation of Methanol, Thesis (Master of Science in Materials. Federal University of Paraíba, João Pessoa, Paraíba.
23. Srinivas V, Barua P, Ghosh TB, Murty ES (2004) Oxidation behavior of AlCuFe nanoquasicrystal powders. Journal of Non-Crystalline Solids 334 & 335: 540-543.
24. Estrella M, Barrio L, Zhou G, Wang X, Wang Q, et al. (2009) In Situ Characterization of CuFe2O4 and Cu/Fe3O4 Water-Gas Shift Catalysts. J Phys Chem C 113: 14411-14417.
25. Nascimento L, Agostinho LCL, Cavalcanti BF (2009) Comportamento Da Oxidação Na Fase Icosaedral Do Quasicristal Al62,2Cu25,3Fe12,5. Acta Microscópica 18: 295-303.
26. Turquier F, Cojocaru VD, Stir M, Nicula R, Lathe C, Burkel E (2004) Formation and stability of single-phase AlCuFe quasicrystals under pressure. Rev Adv Mater Sci 8: 147-151.
27. Agostinho LCL, Nascimento L, Cavalcanti BF (2010) Aplicação da liga quasicristalina Al62,2Cu25,3Fe12,5 em reação catalítica do metanol. Revista Analytica ANO 8: 78-85.
28. Cheung Y, Chan CK, Zhu HY (2001) Characterization of Icosahedral phase in as-cast quasicrystalline Al65Cu20Fe15 alloys. Materials Characterization 47: 299-305.
29. Rosas G, Perez R (2001) On the transformations of the ψ-AlCuFe icosahedral phase. Material Letters 47: 225-230.
30. Faungnawakij K, Tanaka Y, Shimoda N, Fukunaga T, Kikuchi R, et al. (2007) Hydrogen production from dimethyl ether steam reforming over composite catalysts of copper ferrite spinel and alumina. Appl Catal B: Environ 74: 144-151.
31. Fang DR, Ren WZ, Liu ZM, Xu XF, Zhang HM, et al. (2010) Synthesis of mesoporous Cu-Mn-Al2O3 materials and their applications to preferential catalytic oxidation of CO in a hydrogen-rich stream. Chem Res 26: 105-109.
32. Krishnan V, Selvan RK, Augustin CO, Gedanken A, Bertagnolli H (2007) EXAFS and XANES investigations of CuFe2O4 Nanoparticles and CuFe^{2}O4-MO2 (M= Sn,Ce) Nanocomposites. J Phys Chem C 111: 16724-16733.
33. Bergstein A (1968) Formation of the spinel Cu0.5Fe2.5O4 from $CuFe_2O_4$ and αFe_2O_3. Materials Research Bulletin 3: 787-796.
34. Chen W-S, Lee M-D (1992) Nonoxidative dehydrogenation of cyclohexanol over copper-iron binary oxides. Appl Catal A: Gen 83: 201-211.
35. Lindstrom B, Pettersson LJ (2001) Hydrogen generation by steam reforming of

methanol over copper-based catalysts for fuel cell applications. Int J Hydrogen Energy 26: 923-933.

36. Tanabe T, Kameoka S, Tsai AnP (2010) Microstructure of leached Al-Cu-Fe quasicrystal with high catalytic performance for steam reforming of methanol. Appl Catal A: Gen 384: 241-251.

37. Wen-Chun WU, Chuang CC, Lin JL (2000) Bonding geometry and reactivity of methoxy and ethoxy groups adsorbed on powdered TiO2. J Phys Chem B 104: 8719-8724.

38. Yoshimura M, Tsai AP (2001) Quasicrystal application on catalyst. Journal of Alloys and Compound 342: 451-454.

Performance Study on Photocatalysis of Phenol Solution in a UV Irradiated Reactor

Rajarshi Kar, Oindrila Gupta, Kunal Mandol, and Sangita Bhattacharjee*

Chemical Engineering Department, Heritage Institute of Technology, Kolkata, India

Abstract

Photocatalysis is a "green" technology with potential applications in various disciplines, such as chemical synthesis and environmental technologies. In this work, heterogeneous photocatalytic degradation of phenol in aqueous solution was studied using TiO_2 photocatalyst in a UV irradiated new type of annular reactor. Degradation of phenol has been investigated against irradiation time for three different flow rates. The kinetic and equilibrium study showed that the mineralization of aqueous phenol could be modeled using Langmuir –Hinshelwood model and the rate constant has been reported.

Keywords: Photocatalysis; Annular reactor; Phenol; TiO_2; Langmuir-Hinshelwood model

Introduction

Photocatalysis is the acceleration of a photoreaction in the presence of a catalyst. In recent years, interest has been focused on the use of TiO_2 as a photocatalyst for the destruction of polluting materials. Heterogeneous photocatalysis offers an efficient technique for wastewater treatment particularly when treated for low concentration and high volume fluids. Activation of semiconductor catalyst TiO_2 is achieved through the absorption of photon of ultraviolet band gap energy resulting in the formation of electron donor (reducing) and electron acceptor (oxidising) sites. The photocatalytic activity of TiO_2 depends on the relative rates of generation and recombination of electron-hole pairs as well as the levels of adsorbed radical-forming species on TiO_2 surfaces. Hydroxyl radicals ($HO^{\cdot}$) and superoxide ion are the principal agents responsible for the oxidation of aqueous organic contaminants [1]. The carbon containing pollutants are oxidized to carbon dioxide and water, while the other elements bonded to the organic compounds are converted to anions such as nitrate, sulfate or chloride.

The purification of wastewater by heterogeneous photocatalysis is one of the most rapidly growing areas of interest to both research workers and water purification plants [2]. It is useful for cleaning biologically toxic or non-bio degradable materials such as aromatics, pesticides, petroleum constituents, dyes and volatile organic compounds in waste water [3]. Commercial application of the process is called advanced oxidation process (A.O.P). In recent years, interest has focused on the use of TiO_2 as a photocatalyst for the destruction of polluting materials due to its efficiency, non toxicity, high activity, photochemical inertness and low cost [4]. In another work [5] phenol degradation had been examined with combined photocatalysts TiO_2 and ZnO under illumination of a 500W super high pressure mercury lamp mounted axially. The results revealed that, both TiO_2 and ZnO were effective though the latter was much superior in activity.

Though photocatalysis has been proven to be a promising technology, new reactor design ideas are necessary for implementation of important parameters like light distribution inside the reactor, maintenance of the photocatalyst in suspension so as to ensure better mass transfer and subsequent conversion of the solute. This work examined the mineralisation of aqueous phenol solution in an indigenously designed and fabricated new type of annular photo reactor using TiO_2 as catalyst. Chemical, pharmaceutical industries, petrochemical industries as well as petroleum refinery use phenols to a great extent. Due to toxicity of phenols and their derivatives, their disposal in receiving water bodies causes severe problem. Presence of such compounds strongly reduces biodegradation of other components. Conventional process such as adsorption onto activated carbon is effective in treatment of phenolic water but they only transfer the contaminants from one medium to another.

This investigation was aimed to study the variation of phenol concentration in aqueous solution with reactor operating time and thus to determine the conversion in this reactor. The effects of flow rate of phenol solution on degradation of phenol were investigated and reported. The model proposed by Langmuir-Hinshelwood was employed to describe the kinetics of the photodecomposition reaction of phenol and the rate constant was evaluated under specified operating condition.

Theory

The heterogeneous photocatalysis permits total degradation (mineralization) of the most toxic organic molecules in aqueous solution in the presence of a catalyst. The hydroxyl radicals generated in this process are powerful oxidant due to high reactivity. This species is unstable and must be continuously produced in situ by means of chemical or photochemical reactions.

The heterogeneous photocatalysis is based on a semiconductor catalyst which is activated by UV light. The anatase form of Titanium dioxide is mainly used as catalyst which is activated by UV light. In this process, excitation of the of TiO_2 by a UV light wavelength of $\lambda \leq$ 400 nm generates electron-hole pair (e^-_{cb},h_{vb}^+), the hydroxyl radicals are generated by the holes which can degrade organic pollutants present in wastewater. Dissolved oxygen of the solution scavenges the electron generated, preventing the recombination of electrons and

***Corresponding author:** Sangita Bhattacharjee, Chemical Engineering Department, Heritage Institute of Technology, Kolkata, India holes. Conduction band electrons and valence band holes (Equation1) generated from UV irradiated TiO_2 interact with surface adsorbed molecular oxygen to generate superoxide radical anions, $O_2^{\cdot-}$ (Equation 2) and with water to produce the highly reactive $HO^{\cdot}$ (3) [1].

$$TiO_2 + h_v \longrightarrow e^-_{cb} + h_{vb}^+ \quad (1)$$

$$e^-_{cb} + O_2\,(ads) \longrightarrow O_2^{\cdot-} \quad (2)$$

$$h_{vb}^+ + Ti\text{-}OH_2 \longrightarrow HO^{\cdot}_{ads} + H^+ \quad (3)$$

$$HO^{\cdot}_{ads} + \text{Reactants} \longrightarrow \text{Photo oxidized products} \quad (4)$$

The photodegradation efficiency or conversion was calculated by the following relation

$$X\,(\%) = 100 \times (C_O - C)/C_O \quad (5)$$

Where X is the photodegradation efficiency or conversion.

Materials and Methods

Reagents

The catalyst TiO_2 (more than 99.9% purity) was procured from Sigma, Aldrich and was used without any pre-treatment. From XRD analysis (Figure 1), it was found that, this TiO_2 contained only anatase form. Phenol with a purity>99.9 % was obtained from Merck, India. The aqueous solutions were prepared with deionised water.

Calibration curve

Six numbers of dilute solutions of phenol were prepared for construction of standard curve. Aqueous phenol solution was scanned for maximum absorbance and was found to give peak absorbance at (λ_{max}) 270nm wavelength in a UV-visible Shimadzu spectrophotometer. The absorbance vs. concentration data was plotted. This standard curve was used extensively to determine phenol concentration during experimental runs.

Photoreactor

The photoreactor of capacity 1720 ml contained 305 mm long, four no. UV lamps each with flux 13.5 W/m^2. The sources of irradiation were placed uniformly inside the reactor to assure the maximum energy exchange between the source of irradiation and reaction mixture that would flow out continuously. A pump was used to circulate the mixture between the reactor and the reservoir. The schematic diagram of the experimental setup is furnished in figure 2.

Experimental procedure

5 litres phenol solution (50 mg/L) was prepared using deionised water and placed in the tank of the reactor. 2 gm of TiO_2 was accurately weighed in an electronic weighing balance (Sartorius) and was mixed thoroughly with the phenol solution in the tank. The suspension was kept in circulation for 1 hour in the absence of UV light in order to reach adsorption equilibrium. Photocatalysis reaction was carried out for a period of 3 hrs. The same procedure was followed in each of three flow rate of 5, 10 and 15 Litre per hour (LPH) respectively (Figure 4). The liquid coming out of the reactor was directed to the tank and allowed to mix thoroughly with the solution in the tank. A sample of 20 ml was collected from the tank at regular interval of 30 minutes. Each sample collected was centrifuged in a Research Centrifuge, Remi (R-24) at 12,000 rpm for a period of 15 minutes to remove TiO_2 particles. The clear supernatant solution was then collected and analysed in the

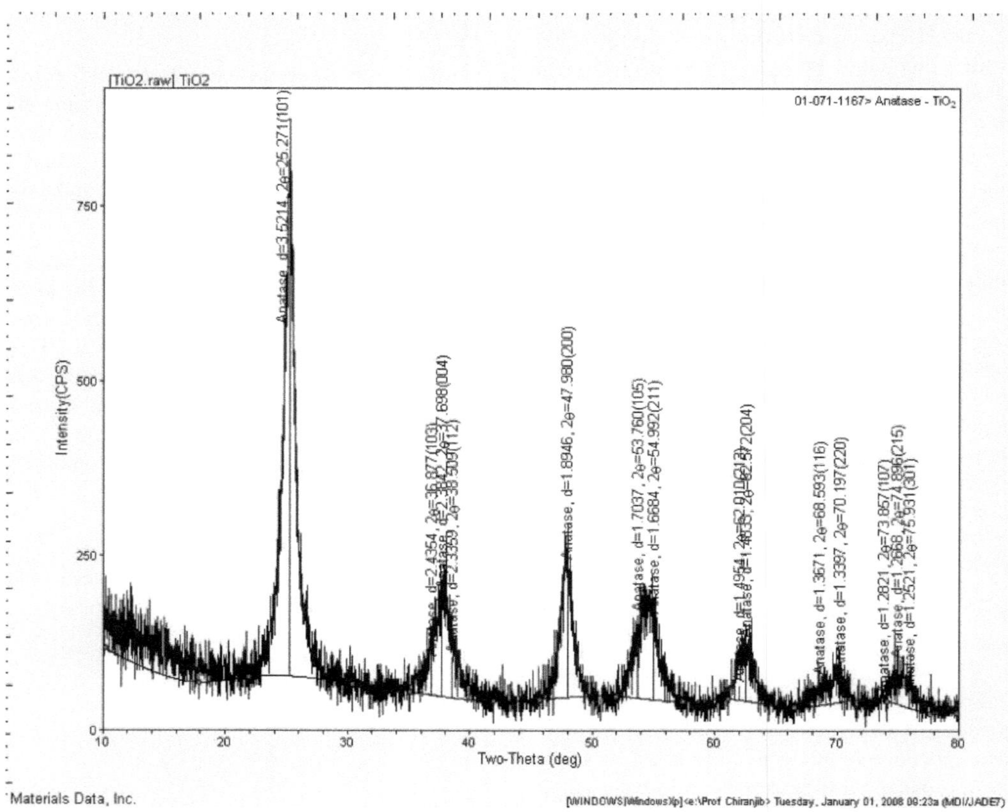

Figure 1: XRD Analysis.

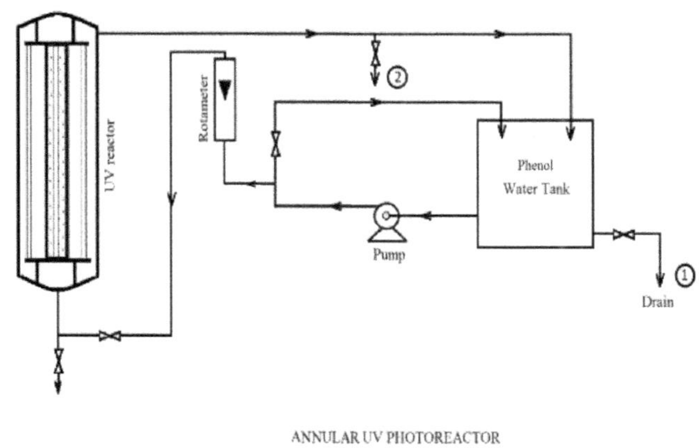

Figure 2: Experimental Setup.

Shimadzu UV spectrophotometer by measuring the absorbance at 270 nm. Using the calibration curve, the remaining phenol concentration of the samples was determined. During the reaction, the reaction temperature was maintained at 32°C using a cooling system placed between the light source and the reactor (not shown in figure 2).

Kinetic studies

The Langmuir-Hinshelwood kinetic model was used to describe the mineralization process kinetics and to assume automatically that reactions took place at the surface of the catalyst particle, without any corroborating independent evidence to support this.

The Langmuir-Hinshelwood (L-H) model was initially developed to quantitatively describe gaseous–solid reactions [4]. This model was recently employed to describe solid-liquid reactions. In this model, the rate of reaction (r) is proportional to the fraction of surface covered by the substrate (θ)

$$-r = -dC/dt = k_r\theta$$

Considering Langmuir's equation,

$$\theta = KC/(1+KC) \quad (6)$$

[θ=fractional coverage, K=equilibrium const.]

$$\text{or, } -r = -dC/dt$$

$$= k_r\theta$$

$$= k_rKC/(1+KC) \quad (7)$$

Since k_r is the true rate constant, which takes into account several parameters [5] such as the catalyst's mass, efficient photon flow, O_2 layer, etc., K is the constant of adsorption equilibrium of L-H or the Langmuir constant. C is the concentration of the organic substrate at any time t. This equation can be integrated,

Since in this case, the solution is highly diluted, C (mol/l)<10^{-3} [6] the term KC becomes<1, when the denominator of (7) is neglected and the reaction is essentially an apparent first order reaction:

$$r = -dC/dt = k_rkC = k_{app}C$$

Where k_{app} is the apparent rate constant of a pseudo first order reaction. Integrating and using boundary condition $C=C_0$ at t=0 we get,

$$\ln(C_0/C) = k_{app} \times t \quad (8)$$

Where C is phenol concentration at time t. Hence, by plotting ln (C_0/C) against time, the apparent rate constant can be found out from the slope of the line.

Results & Discussion

Effect of flowrate

Adsorption and photodegradation of phenol was carried out with varying flow rates 5, 10 and 15 LPH. The concentration of phenol was plotted against time by measuring the residual phenol concentration in solution. The conversion of phenol into carbon dioxide and water was calculated using (5). Figure 3 depicts the change in concentration of phenol against time for a reactor operating time up to 180 phenol after 3 hours UV irradiation. It was observed that highest conversion (36%) had been achieved at a flowrate of 10 LPH and lowest was 32% at 5 LPH. These results can be explained based on the fact that at lowermost flow rate (5LPH), the residence time (per pass) in the reactor was highest, but exposure time of the entire volume of phenol solution to UV irradiation was the least hence conversion was the lowest. However at highest flowrate (15 LPH), the residence time inside the UV reactor

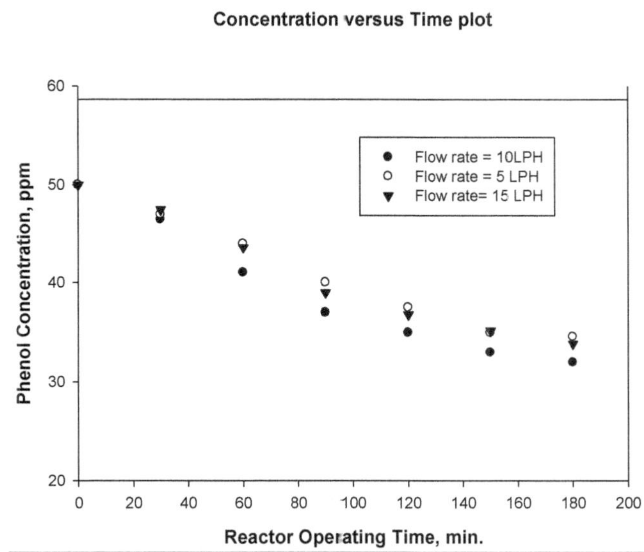

Figure 3: Phenol. Conc. versus time.

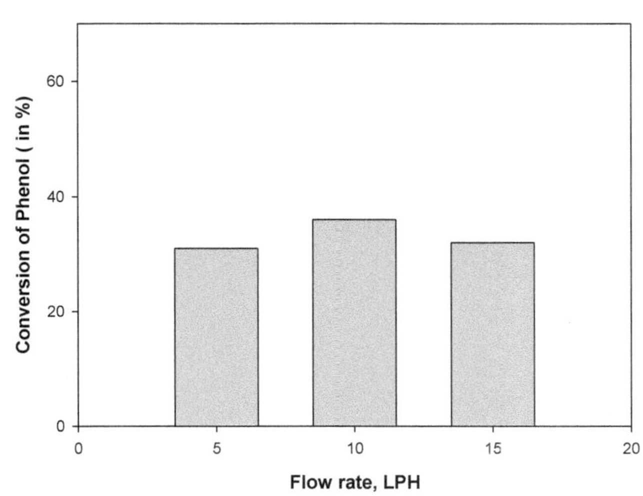

Figure 4: Conversion (%) of phenol vs. flow rate.

is least but the overall exposure of phenol solution to UV irradiation is more. At a flowrate of 10 LPH, both the residence time (per pass) within the UV reactor and overall exposure were moderate- hence, highest efficiency of degradation or conversion of phenol was achieved at this flowrate.

Kinetic study

The kinetics of the reaction was also studied using Langmuir-Hinshelwood model. Since here the phenol concentration is very dilute, much less than 5×10^{-3} mol/l [7], the reaction rate could be approximated as a pseudo-first order type (8). From the slope of $\ln(C_0/C)$ against time, the apparent rate constant was evaluated. The value of the rate constant for phenol degradation corresponding to the flowrate of 10 LPH and 0.4 g/L TiO_2 dosage was evaluated to be 0.00288 min^{-1}.

Conclusion

From the present work, it was found that the UV annular reactor was a promising technology to degrade phenol. The photodegradation of phenol in presence of TiO_2 suspension proved to be an efficient method for mineralization of phenol. Effect of flowrate on conversion of phenol was investigated by varying flowrate at 3 different values i.e. 5 LPH, 10 LPH and 15 LPH. Starting with an initial concentration of 50 ppm, 36% conversion of phenol was achieved after a reactor operating time of 3 hours with a flowrate of 10 LPH. It was also observed that photocatalysis of phenol followed a pseudo first order reaction in dilute solution.

References

1. Laoufi NA, Tassalit D, Bentahar F (2008) The degradation of phenol in water solution by TiO2 photocatalysis in a helical reactor. Global Nest Journal 10: 404-418.
2. Herrmann JM (1999) Heterogeneous photocatalysis: fundamentals and applications to the removal of various types of aqueous pollutants. Catalysis Today 53: 115-129.
3. Kurtoglu ME, Longenbach T, Gogotsi Y (2011) Preventing Sodium Poisoning of Photocatalytic TiO2 Films on Glass by Metal Doping. International Journal of Applied Glass Science 2: 108-116.
4. Matthews RW (1990) Purification of water with near UV illuminated suspensions of titanium dioxide. Water Research 24: 653-660.
5. Devipriya SP, Yesodharan S (2010) Photocatalytic Degradation of Phenol in Water using TiO_2 and ZnO. J Environ Bio 31: 247-249.
6. Satterfield CN (1999) Mass Transfer in Heterogeneous Catalysis. MIT Press, Cambridge MA, 177-179.
7. Fernandez A, Lassaletta G, Jimenez VM, Justo A, Gonzalez-Elipe AR, et al. (1995) Preparation and characterization of TiO_2 photocatalysts supported on various rigid supports(glass, quartz and stainless steel). Comparative studies of photocatalytic activity in a water purification. Applied Catalysis B: Environmental 7: 49-63.

Comparative Simulation of the Purification of Wet Phosphoric Acid by Tbp, Mibk and a Mixture (MIBK+TBP)

R. Hamdi* and A. Hannachi

Department of Chemical-Process Engineering, National Engineering School of Gabes, Omar Ibn El Khattab St., 6029 Zrig, Gabes, Tunisia

Abstract

Liquid-liquid solvent extraction technology is used in several industrial processes such as petrochemical processing, pharmaceutical production, food and hydrometallurgy. It is successfully applied for the purification of Wet Phosphoric Acid (WPA) with fairly good purification performances. The purification process is basically carried out in three steps: extraction, washing and recovery, each of these steps is a liquid-liquid extraction operation. The aim was to develop a computer code that helps predicting the overall performance of the purification process for a given set of operating conditions. The adopted approach is based on determining the necessary theoretical stages for each step in the purification process using a numerical technique. Some simulation results for the WPA purification process with Tri-Butyl Phosphate (TBP) at 45°C, Methyl Isobutyl Ketone (MIBK) at 25°C and a mixture (MIBK+TBP) at 30°C are presented.

Keywords: Wet process phosphoric acid; Purification; Solvent extraction; Simulation; NTS

Introduction

The most commonly used process for the production of phosphoric acid is: thermal and wet-process. The thermal process produces a pure acid with huge energy consumption. The wet-process involves reaction of phosphate rock with an acid (mainly sulfuric acid). This process is economic and practiced everywhere in the world [1,2]. Phosphoric acid produced by the latter process contains a variety of impurities, which vary according to the origin of the minerals. These impurities are removed from the acid by several techniques such as precipitation, adsorption, ion exchange and solvent extraction [1-12].

Liquid-liquid extraction is at the heart of many WPA purification processes. Several solvents can be used for the purification of WPA. The majority of solvents are able to extract selectively phosphoric acid (H_3PO_4), indeed alcohols, the etheroxydes, the ketones like Methyl Isobutyl Ketone (MIBK) and the phosphoric esters as Tributyle phosphates (TBP). The purification process is basically carried out in three steps: extraction, washing and recovery. All of these three steps are liquid-liquid extraction operations. Because of the complexity of such operations, it is very difficult to predict the overall performance of the purification process for a given set of operating conditions, making it very hard to find the optimal conditions for conducting the purification process [8-12].

On the basis of experiment undertaken to a scale laboratory, this study proposes to exploit the equilibriums isotherms to develop a computer code for the simulation of the operation of purification of WPA. The considered solvent is Methyl Isobutyl Ketone (MIBK).

The objective of this work is then to unfold a method for predicting the overall performance of the purification process by solvent extraction. This technique is founded on the simulation of the WPA purification process. The adopted approach is based on the determination of the necessary theoretical stages for each step in the purification process using a numerical technique. Some simulation results of a WPA purification process will be shown.

Modelling of the WPA purification

The modeling of the WPA purification was limited to three consecutive operations: extraction, washing and stripping, in order to simulate the purification of WPA by solvents with TPB, MIBK and mixtures MIBK+TBP. In this study, each step of the purification process is assumed to be conduct in a cascade of counter current theoretical stages. The phase diagram of the ternary system H_3PO_4 – water – considered solvents and the portion ratios for some trace impurities for the three systems have to be approximated. For the three solvents TBP, MIBK and mixtures MIBK+TBP, a representative H_3PO_4 distribution between the organic and the aqueous phases is shown respectively in Figure 1, 2 and 3. Also, the partition ratios for some trace impurities as function of the acid concentration in the organic phase are shown in Figure 4, 5 and 6.

For a given feed quality and an acid extraction rate in the extraction phase, the number of theoretical stages is determined as follows (Figure 7):

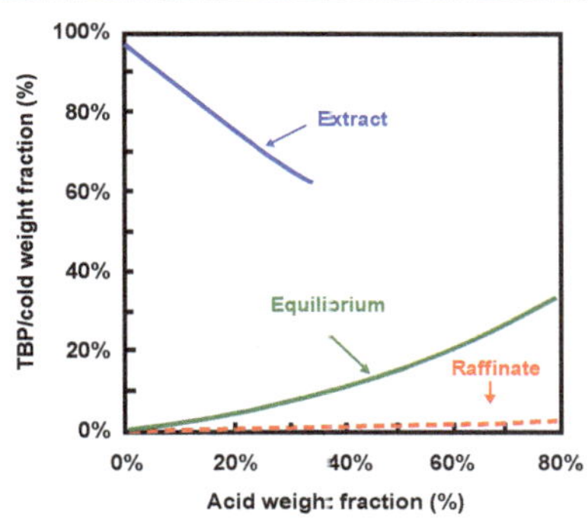

Figure 1: Equilibrium isotherms (H3PO4-H2O-TBP at 45°C).

***Corresponding author:** Rania Hamdi, Department of Chemical-Process Engineering, National Engineering School of Gabes, Omar Ibnelkhattab St., 6029 Zrig, Gabes, Tunisia

i. Assuming the raffinate quality and performing an overall balance yields the acid composition in the extract;

ii. The acid composition in the aqueous phase leaving the first stage is determined from the equilibrium curve;

iii. An overall balance between the second and the last stage yields the acid composition in the extra phase leaving the second stage and from the acid distribution isotherm the acid composition in the raffinate phase exiting the same stage;

iv. This calculation procedure is repeated until the last stage for which inlet extract phase and the outlet raffinate compositions should be consistent with the assumed extraction rate and the solvent composition. If these conditions are not satisfied, the solvent rate has to be modified and the calculation procedure has to be repeated until satisfaction of the convergence criteria.

This calculation procedure gives also the acid contents in all fluxes. A similar iterative numerical technique can be also applied to obtain impurities' contents in every flux. These iterative computation techniques have been implemented in a computer code that is capable of finding the number of theoretical stages as well as the composition of all fluxes in each step of the purification process.

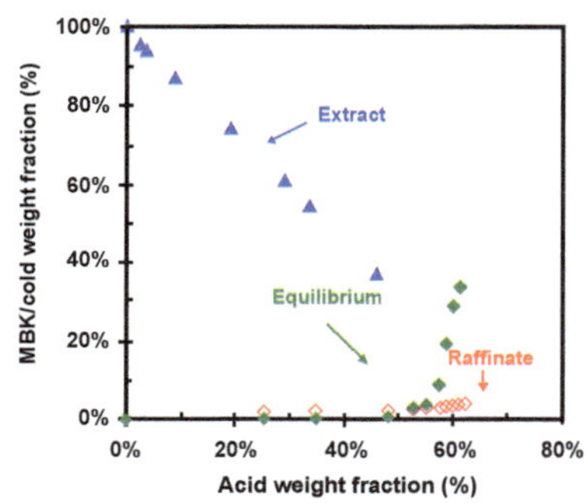

Figure 2: Equilibrium isotherms (H3PO4-H2O-MIBK at 25°C).

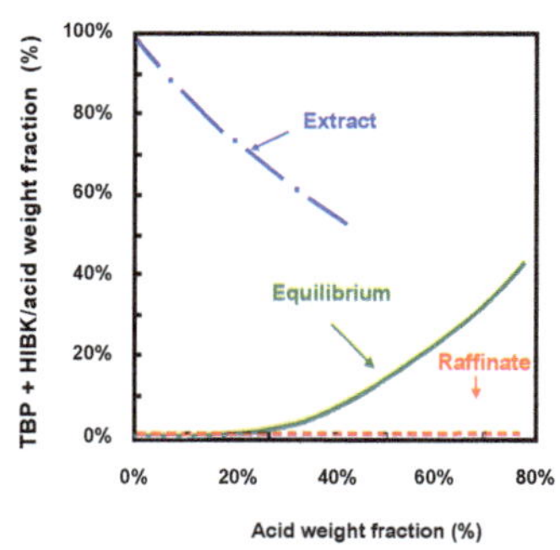

Figure 3: Equilibrium isotherms (H3PO4-H2O- TBP+MIBK at 30°C).

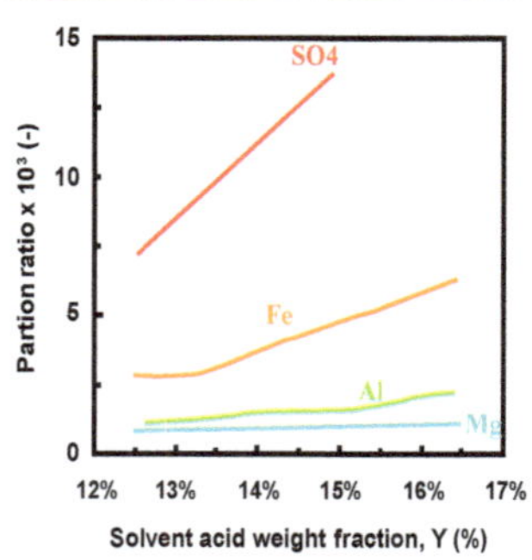

Figure 4: Some impurities' partition ratios (TBP at 45°C).

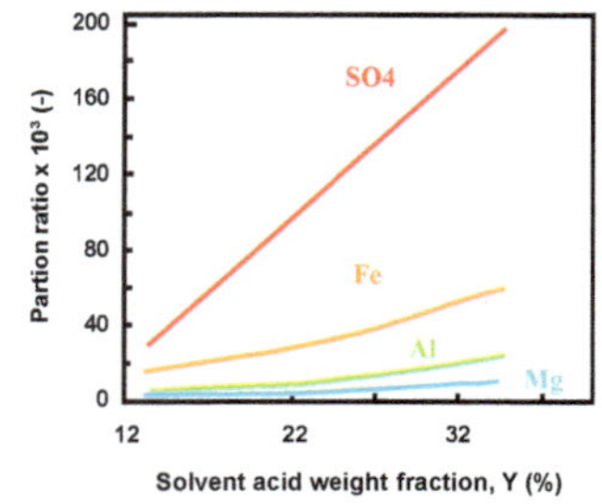

Figure 5: Some impurities' partition ratios (MIBK+TBP at 30°C).

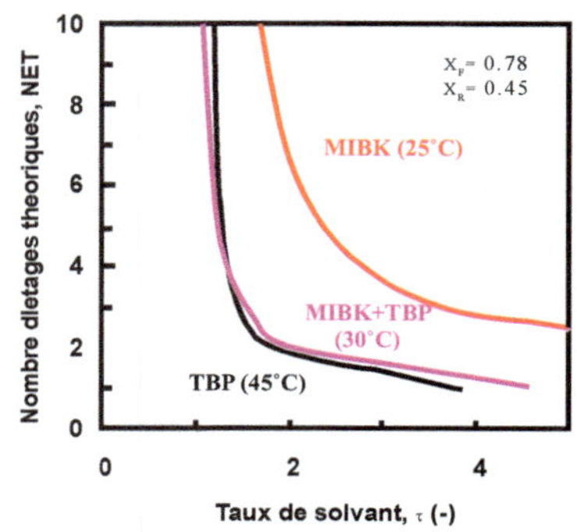

Figure 6: Number of theoretical stages for extracting H3PO4 from the WPA versus the solvent rate.

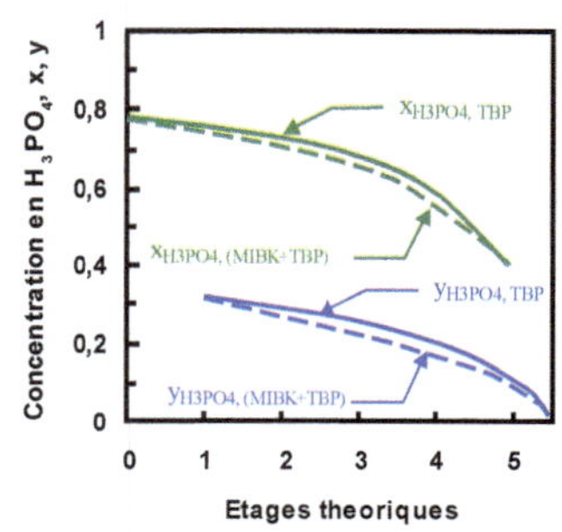

Figure 7: Raffinate and extract acid contents along the ideal extractors.

Results and Discussion

The Number of Theoretical Stages' (NTS) is an important parameter for the dimensioning of the equipments used in the purification process. A large NTS is equivalent of a difficult operation that needs huge equipments to insure adequate mass transfer between phases. For the three considered solvents, the NTS required for extracting a given amount of the acid in feed at 78% H_3PO_4 weight fraction for a residual content equal to 45% in H_3PO_4. For the same solvent rate, the operation of extraction with TPB at 45°C and mixtures MIBK+TBP require the same NTS, however MIBK the operation of extraction requires a NTS more important than foe the other solvent.

For a given amount of acid extracted by the solvent, the computer code gives the acid as well as impurities' weight fractions in all streams flowing in out of all theoretical stages in the extraction operation. As an example, for an operation that requires five stages, Figures 8, 9 and 10 give details of the composition of the acid and impurities in both phases at each stage, for the three solvents, for a WPA at 78% H_3PO_4 with a fixed H_3PO_4 extraction rate at 82%. These details provide valuable information on the intensity of solute transfer taking place in each stage. It is clear from these figures that much of the transfer is happening in the stages close to the fresh solvent feed.

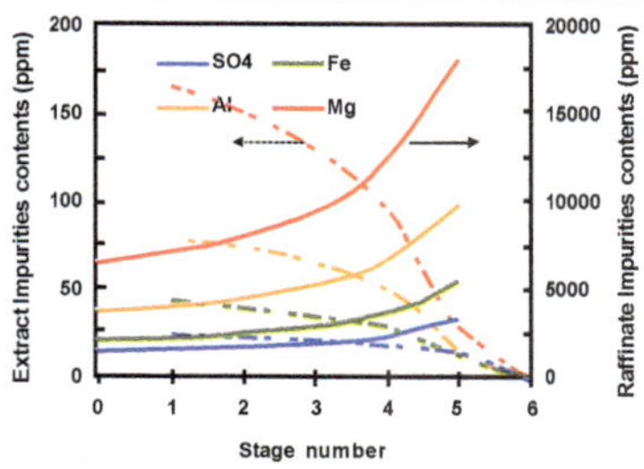

Figure 8: Raffinate and extract impurities' contents along the ideal extractors for TBP.

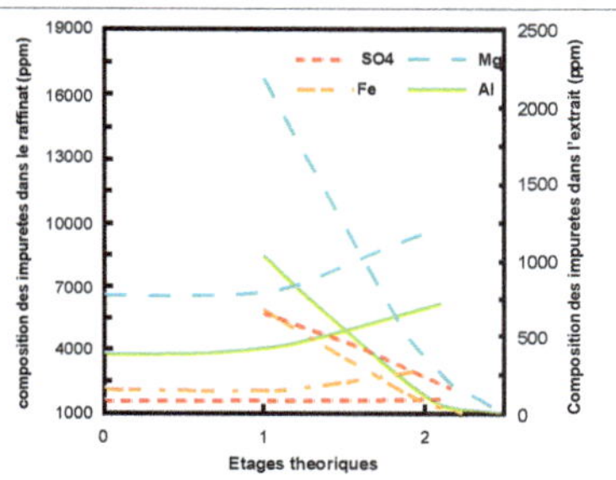

Figure 9: Raffinate and extract impurities' contents along the ideal extractors for MIBK.

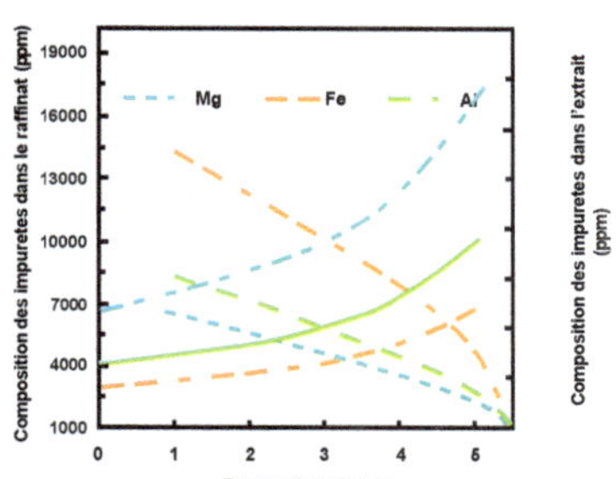

Figure 10: Raffinate and extract impurities' contents along the ideal extractors for MIBK+TBP.

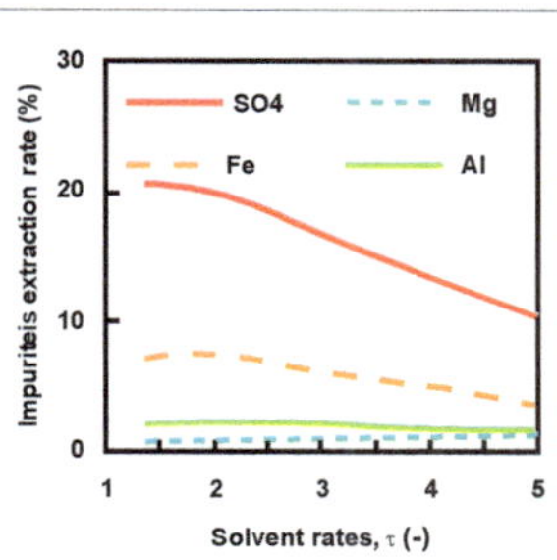

Figure 11: Purified acid impurities' contents versus acid content of the stripped solvent (TBP).

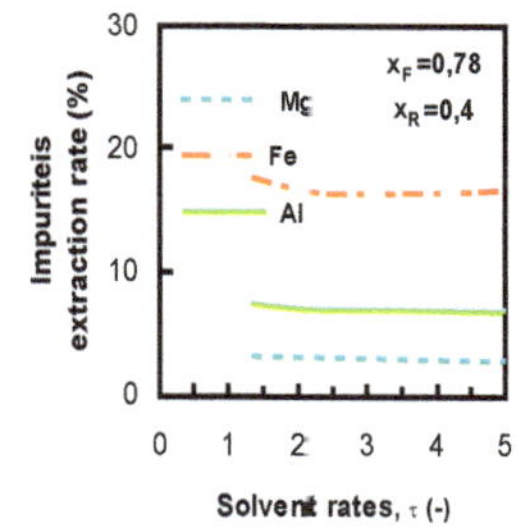

Figure 12: Purified acid impurities' contents versus acid content of the stripped solvent (MIBK+TBP).

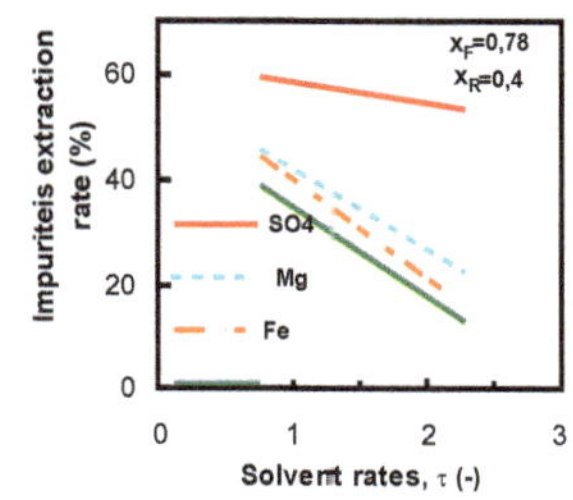

Figure 13: Purified acid impurities' contents versus acid content of the stripped solvent.

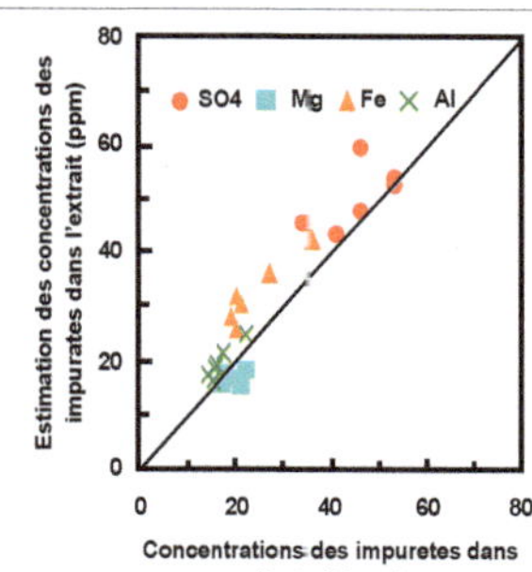

Figure 14: Comparison between the experimental results and theoretical results.

For the same operating conditions and for various solvent rates, the computed impurities' extraction rates for the TPB at 45°C, MIBK at 25°C and mixtures MIBK+TBP at 30°C are given respectively in Figure 11,12 and 13. As expected because of the increase of the required NTS, the extraction rate for the undesired matter intensifies when the solvent rate decreases. Comparison to other solvents, TBP at 45°C presents the best performances as for the reduction of the metallic impurities' content. The classification by descending order of the extraction rates of the impurities for the TBP is as follows: Mg, Al, Fe and SO_4.

The comparison between the experimental results obtained with various solvent rates for the operations of extraction with those provided by the developed code is necessary to valid our approach. Figure 14 and 15 give the extract impurities content obtained by the computer code, according to the experimental results of analysis of the impurities in the extract. It's noticed that the point are close to the first bisector, with those experimental with a light over-estimate for the TBP. For sulphates and iron, the error is higher because of the equilibrium isotherms Figure 16.

The extract exiting the first step in the H_3PO_4 purification process is then washed with pure water. This treatment is indented to eliminate the undesired impurities in the extract. The same numerical procedure as for the extraction, the result of the washing operation can also be predicted. For an extract feeding the washing operation at 31% H_3PO_4 and a fixed acid weight fraction of the washed extract at 25%, Figure 17 gives the amount of the considered impurities discharged with the washing water and their concentration levels in the cleaned extract. As expected, the impurities extraction rate in decreasing order is for Mg, Al, Fe and SO_4. The sulphates removal rates for the washing operation is important because the extract in feed is rich in sulphates. By comparing the numerical result with those experimental in Figure 18 witches shows a good agreement.

The washed extract is then feeding the stripping operation in order to re-extract the acid to obtain the final purified phosphoric acid. The last one will undergo a post treatment process. A similar approach that applied for the extraction and washing can be used for the stripping operation. For a washed extract 25% H_3PO_4 supplying the stripping operation and for three acid concentration levels of the purified acid, the concentration of some trace impurities for the TPB at 45°C, MIBK at 25°C and mixtures MIBK+TBP at 30°C. For three acid concentrations, the ratio of the impurities' content in the purified acid, rapporting at 78% H_3PO_4, with the acid feeding the first step of extraction shows that for the TBP, as an example, Aluminum content in the purified acid are lower than 0.1% of those in the feeding acid. It is clear that one has to find the optimum stripping rate that leads to high acid concentration with minimum impurities contents in the purified acid.

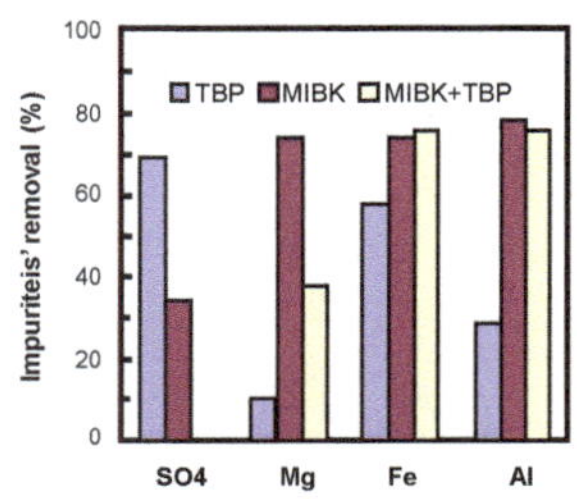

Figure 15: Impurities' removal rates for the washing operation.

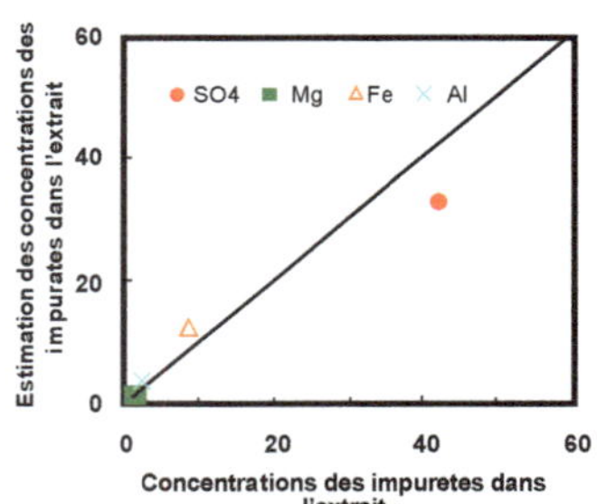

Figure 16: Comparison between the experimental results and theoretical results.

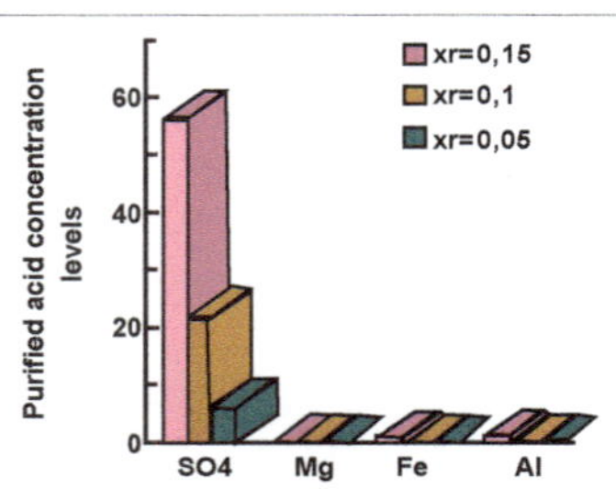

Figure 17: Purified acid impurities' contents versus acid content of the stripped solvent.

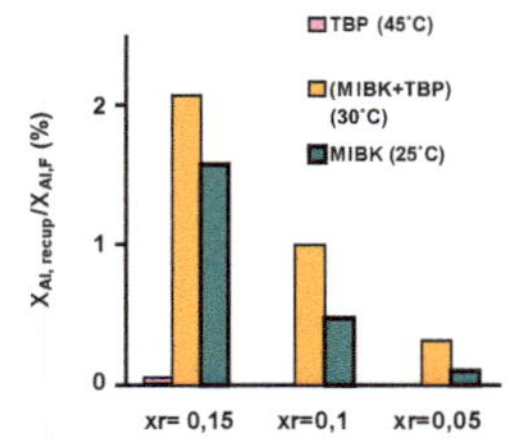

Figure 18: Ratio between Purified acid impurities' contents and feeding acid versus acid content of the stripped solvent.

Conclusion

In this study, a simulation of the purification of Wet Phosphoric Acid (WPA) was developed. The purification is based on solvent extraction. The computer code is applied for the extraction, washing and stripping the WPA. These operations are carried out in a cascade of counter current theoretical stages with three solvents witch knowing: Tri-Butyl Phosphate (TBP), Methyl Isobutyl Ketone (MIBK) and mixture (MIBK+TBP). The adopted approach is based on determining the necessary theoretical stages and the content of acid and impurities in the two phases for each step in the purification process. This approach could be used as a tool for predicting the WPA purification results in order to look for the optimal operating condition to conduct this operation.

References

1. Cuer JP, Floreancig (1980) 'Nouvelle source d'acide phosphorique très pur, procédé double solvant', Proceedings of the 2nd international congress on phosphorous compounds. Boston, USA 529-540.

2. Hignett TP (1980) Production of wet process phosphoric acid. Proceedings of the 2nd international congress on phosphorous compounds. Boston, USA 401-429.

3. Leenaerts R (1984) Purification de l'acide phosphorique de voie humide. Bull séance Acad r Sci Outre-Mer 28: 529-545.

4. Davister A, Peeterbroeck M (1982) The prayon for wet process phosphoric acid purification. CEP, Mars 35-39.

5. Ruiz F, Marcilia A, Ancheta M, Caro JA (1985) Purification of Wet Process Phosphoric Acid by Solvent Extraction with Isoamyl Alcohol. Part II. Study of the Impurities Distribution. Solvent Extraction and Ion Exchange 3: 345-355.

6. Hanna AA, Youssef NS, Ali AF (1995) Purification of wet process phosphoric acid by solvent extraction. Egypt. 38: 311-317.

7. Feki M, Ayedi HF (2000) Modélisation du traitement de purification de l'acide phosphorique de voie humide par la méthylisobutylcétone. Can J Chem Eng 78: 540-546.

8. Hannachi A, Ben Zaied T, Ouederni A, et Chtara C (2002) Simulation de l'opération de purification de l'acide phosphorique industriel par extraction. Proceedings of the 4th J TET, Hammamet, Tunisie 2: 101-106.

9. Hannachi A, Hamdi R, Abbas N, Chtara C, RATEL A (2006) Simulation of the wet process phosphoric acid purification process by solvent extraction. Proceedings of the COVAPHOS II 3: 251-260.

10. Hamdi R, khleifia N, Tlili N, Hannachi A (2009) Prediction of the concentration profiles along a continuous liquid-liquid extraction column for phosphoric acid purification. The First International Chemical and Process Engineering, Tripoli.

11. Khleifia N, Hamdi R, Hannachi A (2010) Etude de l'extraction de l'acide phosphorique industriel dans une colonne. Continue, Proceedings of the 7th J.T.E.T, Tunisia.

12. Hamdi R, Hannachi A, Chtara C, Ratel A (2009) Impact du Prétraitement de l'acide phosphorique industriel sur l'opération d'extraction de H3PO4 par un solvant organique. Proceedings of the COVAPHOS III, Maroc.

Theoretical Investigations of Methane Conversion to Heavier Hydrocarbons in a Plasma Reactor

Masoud Habibi Zare[1] and Mojtaba Nabipoor Hassankiadeh[2*]

[1]*Faculty of Engineering, Islamic Azad University, Tehran South Branch, Tehran, Iran*
[2]*Department of Chemical Engineering, Tarbiat Modares University, Tehran, Iran*

Abstract

In this study, mathematical modeling of Oxidative Coupling of Methane (OCM) to C_2 hydrocarbons (C_2H_6 and C_2H_4) over La_2O_3/CaO catalyst in a fixed-bed reactor operated under isothermal and non-isothermal conditions was investigated. The kinetic model proposed for OCM process consisted of 10 elementary reactions and 8 chemical species. In this process, methane and acetylene were the inputted feed and ethane, ethylene, propylene, propane, i- butane and n-butane were the output products. The amount of methane conversion obtained was 12.7% for the former feed, however, if pure methane was inputted this conversion rose to 13.8%. Furthermore, the plasma process would enhance the conversion, selectivity towards desired product and yield. In the present study, when methane and acetylene were fed at a molar ratio of CH_4/C_2H_2=10 to the reactor, the selectivity of C_2, C_3 and C_4 hydrocarbons was determined to be 30, 24 and 44% respectively. Concurrently, a higher yield was obtained for n-butane at about 3.9% and the minimum yield achieved for propylene was approximately 0.7%. A comparison between the thermal and the plasma process showed that the methane conversion and yield production in the plasma were higher than in the thermal process under the same operating conditions. On the other hand, product selectivity in the plasma process was determined to be lower than that of the thermal process. Finally, the results of the catalytic OCM and methane conversion processes in the plasma phase were compared with one another.

Keywords: Modeling; Heavy hydrocarbons; Methane conversion; Plasma reactor; OCM

Introduction

In recent researches, there have been lots of wide-ranging studies on the direct methane conversion method [1-3]. Oxidative coupling [4], thermal coupling and plasma reaction have been used for direct methane conversion [5-19]. Methane activation with plasma is very effective because methyl radicals could be easily formed, which could induce various kinds of chemical reactions. For plasma chemical reactions, free radicals are believed to be much more important than any other reactive particles such as electrons, ions, photons, etc. Therefore, control and manipulation of the subsequent free-radical reactions are essential to the success of plasma application in organic synthesis [15]. The electrons within the plasma serve principally to excite and decompose the gas molecules at a high rate and in a non-selective fashion. Multi-pathway chemical reaction mechanisms may not be avoidable and may lead to the production of complex product distribution. To overcome this problem, heterogeneous catalysts were introduced into the plasma reaction [8-10,13,14,18,19].

Direct conversion of methane to C_2 hydrocarbons has a large implication in the utilization of natural gas in the gas-based petrochemical and liquid fuels industries. The studies on conversion and utilization of methane are widely researched in the field of C_1 chemistry. Several technologies have been proposed to improve the efficiency and utilization of methane. From thermodynamic equilibrium studies [20], production of C_2H_6 and C_2H_4 from CH_4 and CO_2 showed low performance due to thermodynamic limitation. Almost all catalysts only achieved low methane conversion (up to 7%), higher C_2 hydrocarbon selectivity (up to 85%), and low C_2 hydrocarbon yield (up to 4.5%). The CH_4 and CO_2 reactions over these catalysts only produced C_2H_6, C_2H_4 and CO. The major difficulty involved indirect methane conversion to higher hydrocarbons and oxygenates is the intensive energy consumption required to activate the stable C-H bones in the methane molecule using conventional catalysis [21,22]. Oxidative coupling of methane to ethane requires a high reaction temperature due to the high dissociation energy of activation of most C-H bones [23]. It is expected that a Dielectric Barrier Discharge (DBD) plasma reactor is an efficient tool for converting the greenhouse gases, CH_4 and CO_2, to synthesis gas and higher hydrocarbons at low temperature and ambient pressure [21,24-26]. Pertaining to the plasma reactor, non-thermal plasma can be defined as a gas consisting of electrons, have a much higher energy than the neutral gas particles. The energetic electrons collide with molecules in the gas, resulting in excitation, ionization, electron multiplication, and formation of atoms and metastable compounds [25-27]. When the electric field in the discharge gap is high enough to cause breakdown, in most gases a large number of micro discharges are observed. The active atoms and metastable compounds subsequently collide with molecules and reactions may occur. Due to the high-energy electrons of the plasma reactor, it is expected that methane and carbon dioxide gases could be activated easily in the plasma environment and be converted into synthesis gases and higher hydrocarbons. The mathematical modeling of the catalytic process is a cost-effective alternative to conducting a large number of experiments that are needed to study the effects of various input variables, and to improve the overall process performance [28]. The oxidative coupling of methane (OCM) has been studied as a potential way to produce ethylene from methane. Due to the

***Corresponding author:** Mojtaba Nabipoor Hassankiadeh, Department of Chemical Engineering, Tarbiat Modares University, P.O. Box: 14115-143, Tehran, Iran complex oxidation and heterogeneous catalytic reaction, the process suffers from low selectivity and yield of ethylene and ethane. However, there is still research going on to improve the selectivity and yield of C_2 products, aiming towards economical industrialization. Reactor configuration and mode of operation were found to play a crucial role in producing a high yield and selectivity of ethylene production in the OCM reaction. Various types of reactors have been considered and simulated [29-31]. Since the pioneering work of Keller and Bhasin in 1982, there has been intensive research on the topic of OCM [32]. A number of experimental and theoretical studies of non-catalytic or catalytic reactors for the OCM process have been reported [29,33-35].

In this research, a comprehensive study on the performance of the catalytic fixed-bed reactor has been undertaken by modeling and simulating the OCM reaction under different conditions. A plug flow fixed-bed catalytic reactor model has been proposed. The model developed has been tested for isothermal and non-isothermal operation modes of the reactor and is useful for prediction of reactor performance for OCM reaction under different modes of operation. We use a kinetic model for plasma methane conversion to higher hydrocarbons and use mathematical modeling to simulate the process of methane conversion in the plasma reactor. Comprehensive study on the performance of the plasma reactor has been undertaken by modeling and simulating the methane conversion reaction. A plug flow plasma reactor model has been proposed. The developed model has been tested for the isothermal operation mode of the reactor and it is useful for prediction of reactor performance for methane conversion reactions under different modes of operation. By detecting the kinetic rate constant (k), the general pathway of the reactions could be drawn. The kinetic model includes the gas phase transformation of molecules that exist in the reactant and the products. The constructed model was applicable for the plasma reaction using a dielectric barrier discharge reactor at atmospheric pressure (c.a. 1 atm) and ambient temperature (c.a. 25°C).

Kinetic equations

Indarto et al. [36] procured the kinetics model to any possible compounds which existed in the products or the reactants that were involved in the process of methane conversion. The reaction rate (dC/dt) of a single hydrocarbon was then measured experimentally by injecting the pure species into the reactor [36] from which the respective expressions for the catalytic rate constants were determined. From the conducted experimental analysis, eight hydrocarbons existed in the products including methane, acetylene, ethane, ethylene, propylene, propane, i-butane and n-butane. All involved reactions were assumed to be first order and arranged as:

$$\frac{dC_i}{dt} = -\sum_{j=1}^{n} k_{ij}C_i + \sum_{j=1}^{n} k_{ji}C_j \tag{1}$$

where n is the maximum value of i and j for the k-set, which is 8, and k_{ij} denotes the reaction rate constant of i molecule that converts into a specific *j* molecule. Equation (1) expands and is given below:

$$\begin{bmatrix} \frac{dC_1}{dt} \\ \vdots \\ \frac{dC_8}{dt} \end{bmatrix} = -\begin{bmatrix} k_{11} & \cdots & k_{18} \\ \vdots & \ddots & \vdots \\ k_{81} & \cdots & k_{88} \end{bmatrix}\begin{bmatrix} C_1 & \cdots & C_8 \\ \vdots & \ddots & \vdots \\ C_1 & \cdots & C_8 \end{bmatrix}\begin{bmatrix} k_{11} & \cdots & k_{81} \\ \vdots & \ddots & \vdots \\ k_{18} & \cdots & k_{88} \end{bmatrix}\begin{bmatrix} C_1 & \cdots & C_1 \\ \vdots & \ddots & \vdots \\ C_8 & \cdots & C_8 \end{bmatrix} \tag{2}$$

The reaction rate constants (k) are presented in Reference [36].

Equation (2) expands for the eight hydrocarbons involved in this process and is given below:

For methane:

$$\begin{aligned} dC_{CH_4}/dt = &-(K_{12}+K_{13}+K_{14}+K_{15}+K_{16} \\ &+K_{17}+K_{18})C_{CH_4}+K_{21}C_{C_2H_2}+K_{31}C_{C_2H_4} \\ &+K_{41}C_{C_2H_6}+K_{51}C_{C_3H_6}+K_{61}C_{C_3H_8}+ \\ &K_{71}C_{i-C_4H_{10}}+K_{81}C_{n-C_4H_{10}} \end{aligned} \tag{3}$$

For acetylene:

$$\begin{aligned} dC_{C_2H_2}/dt = &-(K_{21}+K_{23}+K_{24}+K_{25}+ \\ &K_{26}+K_{27}+K_{28})C_{C_2H_2}+K_{12}C_{CH_4}+K_{32}C_{C_2H_4} \\ &+K_{42}C_{C_2H_6}+K_{52}C_{C_3H_6}+K_{62}C_{C_3H_8}+ \\ &K_{72}C_{i-C_4H_{10}}+K_{82}C_{n-C_4H_{10}} \end{aligned} \tag{4}$$

For ethylene:

$$\begin{aligned} dC_{C_2H_4}/dt = &-(K_{31}+K_{32}+K_{34}+K_{35}+ \\ &K_{36}+K_{37}+K_{38})C_{C_2H_4}+K_{13}C_{CH_4}+K_{23}C_{C_2H_2} \\ &+K_{43}C_{C_2H_6}+K_{53}C_{C_3H_6}+K_{63}C_{C_3H_8}+ \\ &K_{73}C_{i-C_4H_{10}}+K_{83}C_{n-C_4H_{10}} \end{aligned} \tag{5}$$

For ethane:

$$\begin{aligned} dC_{C_2H_6}/dt = &-(K_{41}+K_{42}+K_{43}+K_{45} \\ &+K_{46}+K_{47}+K_{48})C_{C_2H_6}+K_{14}C_{CH_4}+K_{24}C_{C_2H_2} \\ &+K_{34}C_{C_2H_4}+K_{54}C_{C_3H_6}+K_{64}C_{C_3H_8}+ \\ &K_{74}C_{i-C_4H_{10}}+K_{84}C_{n-C_4H_{10}} \end{aligned} \tag{6}$$

For propylene:

$$\begin{aligned} dC_{C_3H_6}/dt = &-(K_{51}+K_{52}+K_{53}+K_{54}+ \\ &K_{56}+K_{57}+K_{58})C_{C_3H_6}+K_{15}C_{CH_4}+K_{25}C_{C_2H_2} \\ &+K_{35}C_{C_2H_4}+K_{45}C_{C_2H_6}+K_{65}C_{C_3H_8}+ \\ &K_{75}C_{i-C_4H_{10}}+K_{85}C_{n-C_4H_{10}} \end{aligned} \tag{7}$$

For propane:

$$\begin{aligned} dC_{C_3H_8}/dt = &-(K_{61}+K_{62}+K_{63}+K_{64}+ \\ &K_{65}+K_{67}+K_{68})C_{C_3H_8}+K_{16}C_{CH_4}+K_{26}C_{C_2H_2} \\ &+K_{36}C_{C_2H_4}+K_{46}C_{C_2H_6}+K_{56}C_{C_3H_8}+ \\ &K_{76}C_{i-C_4H_{10}}+K_{86}C_{n-C_4H_{10}} \end{aligned} \tag{8}$$

For i-butane:

$$\begin{aligned} dC_{i-C_4H_{10}}/dt = &-(K_{71}+K_{72}+K_{73}+K_{74}+ \\ &K_{75}+K_{76}+K_{78})C_{i-C_4H_{10}}+K_{17}C_{CH_4}+K_{27}C_{C_2H_2} \\ &+K_{37}C_{C_2H_4}+K_{47}C_{C_2H_6}+K_{57}C_{C_3H_6}+ \\ &K_{67}C_{C_3H_8}+K_{87}C_{n-C_4H_{10}} \end{aligned} \tag{9}$$

For n-butane:

$$\begin{aligned} dC_{n-C_4H_{10}}/dt = &-(K_{81}+K_{82}+K_{83}+K_{84}+K_{85} \\ &+K_{86}+K_{87})C_{n-C_4H_{10}}+K_{18}C_{CH_4}+K_{28}C_{C_2H_2} \\ &+K_{38}C_{C_2H_4}+K_{48}C_{C_2H_6}+K_{58}C_{C_3H_6}+ \\ &K_{68}C_{C_3H_8}+K_{78}C_{i-C_4H_{10}} \end{aligned} \tag{10}$$

Reactor modeling

The reactor used in the experiments is a cylindrical tube (ID of 7.5 mm) with 2 parallel-straight wires (0.2 mm diameter, stainless steel) as the inner metal electrode. The effective volume and length of the reactor were 8.8 ml and 200 mm, respectively. The actual reactor dimensions and conditions used in the simulation are presented in Table 1. The schematic diagram of the DBD plasma reactor is depicted in Figure 1. The initial conditions, at the entrance of the reactor, z =0, are as follows:

$$C_n = C_n^0$$
$$C_m = C_m^0$$
$$P = P^0 \quad (11)$$
$$T = T^0$$

A model was developed to simulate the heterogeneous reactions without diffusion resistance. The system was assumed to be at steady state. Simulation was carried out for a plasma reactor. Formulation of the model equations was provided below:

$$-u_s = \frac{dC_j}{dz} + \varepsilon_b R_j = 0 \quad j = 1...8 \quad (12)$$

In which, u_s is superficial velocity (m.s^{-1}), C_j is concentration of species j (mol.m^{-3}), ε_b is catalyst bed porosity, R_j is rate of formation of component j (mol.m^{-3}.s^{-1}) and z is distance from the start of catalyst bed (m).

Based on the differential Equation (12) for eight different hydrocarbons: CH_4, C_2H_2, C_2H_4, C_2H_6, C_3H_6, C_3H_8, i-C_4H_{10} and n-C_4H_{10} were developed as:

$$-u_s \frac{dC_{CH_4}}{dz} + \varepsilon_b R_{CH_4} = 0 \quad (13)$$

Properties	Dimension
Inner diameter (mm)	7.5
Pressure (atm)	1
Length of catalyst bed (mm)	200
Flow rate (ml/min)	30
Catalyst density (kg. m^{-3})	3600
Applied Voltage(kV)	20
Frequency	AC

Table 1: Reactor parameters and constants used in this study.

$$-u_s \frac{dC_{C_2H_2}}{dz} + \varepsilon_b R_{C_2H_2} = 0 \quad (14)$$

$$-u_s \frac{dC_{C_2H_4}}{dz} + \varepsilon_b R_{C_2H_4} = 0 \quad (15)$$

$$-u_s \frac{dC_{C_2H_6}}{dz} + \varepsilon_b R_{C_2H_6} = 0 \quad (16)$$

$$-u_s \frac{dC_{C_3H_6}}{dz} + \varepsilon_b R_{C_3H_6} = 0 \quad (17)$$

$$-u_s \frac{dC_{C_3H_8}}{dz} + \varepsilon_b R_{C_3H_8} = 0 \quad (18)$$

$$-u_s \frac{dC_{i-C_4H_{10}}}{dz} + \varepsilon_b R_{i-C_4H_{10}} = 0 \quad (19)$$

$$-u_s \frac{dC_{n-C_4H_{10}}}{dz} + \varepsilon_b R_{n-C_4H_{10}} = 0 \quad (20)$$

The governing differential equations were solved utilizing the adaptive Runge-Kutta 4th-order method by MATLAB program for the eight different hydrocarbons that were available in this process.

The products' selectivity, yield and reactants' conversion were formulated as:

$$\text{Conversion of } C_xH_y = \frac{Mole\ of\ C_xH_y\ Converted}{Mole\ of\ Initial\ C_xH_y} \times 100 \quad (21)$$

$$\text{Selectivity of } C_mH_n = \frac{m \times Moles\ of\ C_mH_n\ produced}{Moles\ of\ C_xH_y\ Converted} \times 100 \quad (22)$$

$$\text{Yield } C_mH_n = \text{Selectivity of } C_mH_n \times \text{Conversion of } C_xH_y \quad (23)$$

The reaction pathways of methane conversion in a dielectric barrier discharge at flow rate of 30 ml/min and supplied power of 60 W are shown in Figure 2.

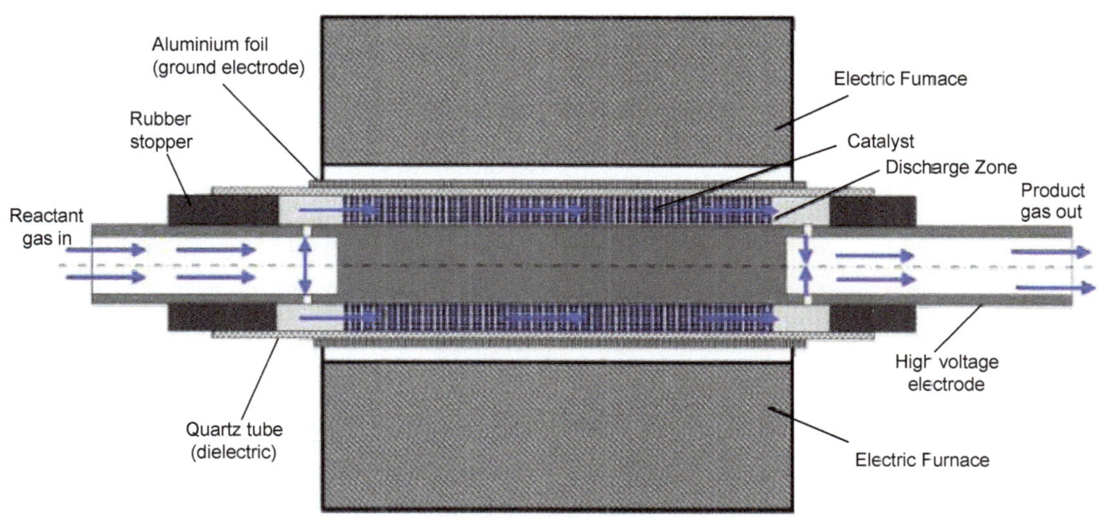

Figure 1: Schematic diagram of a catalytic DBD plasma reactor.

Results and Discussion

Simulations of methane conversion were carried out in the DBD reactor for isothermal operations. The reactor was operated at atmospheric pressure (~1 atm) and room temperature (~25°C). In order to demonstrate the obtained finding, the results were categorized as:

Validation of the reactor modeling for the OCM process

A model for the oxidative coupling of methane (OCM) over La_2O_3/CaO catalyst in a fixed-bed reactor has been developed. This model was tested for isothermal-adiabatic and non-isothermal operation modes. The methane to oxygen (CH_4/O_2) ratio in the input feed as an adjustable parameter was studied. High CH_4/O_2 ratio favored the OCM reaction but gave a low yield of ethylene due to insufficient oxygen under isothermal conditions. Conversion, selectivity and yield increased with the temperature and the maximum yield was obtained at the optimum temperature. The presence of inert N_2 gas in the feed had a positive effect on the process yield. Non-isothermal and adiabatic operations gave a temperature rise during the OCM reaction, which led to a series of reactions that affected the system conversion, yield and selectivity of the desired product. Therefore, feed temperature and CH_4/O_2 feed ratio are important parameters in order to control reactor runaway conditions.

Governing equations: This investigation benefits from a modeling performed by Thian Tye et al. [37]. They developed a one-dimensional model to simulate the reactions and also, the system was assumed to work at steady state conditions. The modeling has been carried out for a plug flow reactor, packed with La_2O_3/CaO catalyst. A developed model was applied to simulate the heterogeneous reactions without diffusional resistance. The system was assumed to be under steady state. It has been assumed that the gas flow behavior inside the reactor could be described by the ideal gas law. The differential equations set by the reactor consisted of eight mass balance equations along with an energy balance equation. These equations for each element were presented as follows:

- **Mass balance**

$$-u_s\frac{dC_j}{dz}+\varepsilon_b R_{g,j}-\rho_b R_{c,j}=0 \qquad (24)$$

- **Energy balance**

$$\frac{dT}{dz}=\frac{1}{C_{Pave}(\rho_g.u_s)}.(\rho_b\sum r_{c,j}(-\Delta H_j)+\varepsilon_b\sum r_{g,j}(-\Delta H_j)) \qquad (25)$$

- **Experimental conditions**

$$\begin{aligned}&1<P_{O_2}<20\,kPa\\&10<P_{CH_4}<95\,kPa\\&973<T<1228\,K\\&0.76<m_{cat}/\upsilon_{STP}<250\,kg.s.m^{-3}\end{aligned} \qquad (26)$$

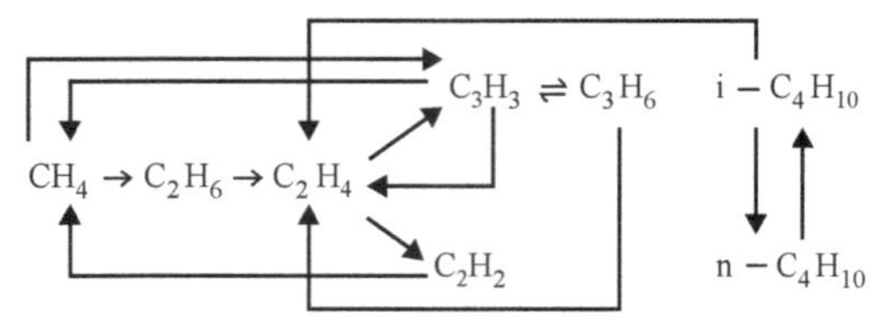

Figure 2: The pathway reaction of methane conversion.

- **Reaction scheme**

The kinetic model considered the following set of stoichiometric equations:

$$\begin{aligned}&1)CH_4+2O_2\rightarrow CO_2+2H_2O\\&2)2CH_4+0.5O_2\rightarrow C_2H_6+H_2O\\&3)CH_4+O_2\rightarrow CO+H_2O+H_2\\&4)CO+0.5O_2\rightarrow CO_2\\&5)C_2H_6+0.5O_2\rightarrow C_2H_4+H_2O\\&6)C_2H_4+2O_2\rightarrow 2CO+2H_2O\\&7)C_2H_6\rightarrow C_2H_4+H_2\\&8)C_2H_4+2H_2O\rightarrow 2CO+4H_2\\&9)CO+H_2O\rightarrow CO_2+H_2\\&10)CO_2+H_2\rightarrow CO+H_2O\end{aligned} \qquad (27)$$

The reaction rates for each step are presented below [37]:

$$r_j=\frac{K_{0,i}e^{-E_{a,j}/RT}P_C^{m_j}P_{O_2}^{n_j}}{(1+K_{j,CO_2}e^{-\Delta H_{ad,CO_{2,j}}/RT}P_{CO_2})^2},\ j=1,3,\ldots,6 \qquad (28)$$

$$r_2=\frac{K_{0,2}e^{-E_{a,2}/RT}(K_{0,O_2}e^{-\Delta H_{ad,O_2}/RT}P_{O_2})^{n_2}P_{CH_4}}{(1+(K_{0,O_2}e^{-\Delta H_{ad,O_{2,i}}/RT}P_{O_2})^{n}+K_{2,CO_2}e^{-\Delta H_{ad,O_2}/RT}P_{O_2})^2} \qquad (29)$$

$$r_7=K_{0,7}e^{-E_{a,7}/RT}P_{C_2H_6} \qquad (30)$$

$$r_8=K_{0,8}e^{-E_{a,8}/RT}P_{C_2H_6}^{m_8}P_{H_2O}^{n_8} \qquad (31)$$

$$r_9=K_{0,9}e^{-E_{a,9}/RT}P_{CO}^{m_9}P_{H_2O}^{n_9} \qquad (32)$$

$$r_{10}=K_{0,10}e^{-E_{a,10}/RT}P_{CO_6}^{m_{10}}P_{H_2}^{n_{10}} \qquad (33)$$

The reaction rates were expressed as a function of partial pressure and temperature. To convert the partial pressure of each species to concentration, the ideal gas law, $P_i=C_iRT$ was used. Moreover, the kinetic parameters' constants were presented in Reference [37]. The operating pressure at the entrance of the reactor was set equal to 110kPa. Other structural characteristics and flow conditions at the entrance of the reactor (*i.e.* boundary conditions for solving Equations 24 and 25) have been given in Table 2. The model was found to fit the experimental data plausibly. Table 3 shows a set of simulated and experimental data compared under the same velocity. In the experiment, methane and oxygen were fed with nitrogen gas as an inert into the reactor. All these runs were operated at a feed flow rate of $4cm^3/s$ over 14.8 mg of La_2O_3/CaO catalyst. It was found that for the case of high methane to oxygen ratio, the present model fits the experimental data well for run number 1, 2 and 3 compared to run 4, 5 and 6. The prediction was generally lower for low methane to oxygen ratio. When the temperature increased, the maximum error between simulated and experimental results is within 20%. The OCM kinetic equation proposed by Stansch et al. [38] predicts the experimental results reasonably. The error differences between the simulated and experimental data could be due to several reasons. The reactions of the pre- and post-catalytic zone were not taken into account in the model and the simulations [37,38]. The gas-phase reaction considered in the model is an interstitial gas-phase reaction within the catalytic bed. There could be reactions in

Properties	Dimension
	6
Pressure (atm)	110
Length of catalyst bed (mm)	2-54
Catalyst weight$_{cat}$ (g)	0.007-1.000
Flow rate v_{STP} (m^3. s_{-3}),(STP)	4×10^{-6}-13×10^{-4}
Catalyst size(mm)	0.25-0.35
Catalyst density(kg.m^{-3})	3600

Table 2: Some parameters utilized in the present work for the OCM Modeling.

the gas phase that occur before reactants enter the catalytic bed, such as gas-phase oxidation of methane; this makes a deviation from the initial conditions assumed in the simulations. The same holds for the post-catalytic zone of reactor. The gas-phase reactions like thermal decomposition and combustion contribute somehow differently to the catalytic bed performance, which were not considered in the simulation; this resulted in some differences between the simulated and experimental data.

Effect of CH_4/O_2 ratio in the OCM process: A number of simulations were carried out with varying feed composition CH_4/O_2 ratio. The results of methane conversion are plotted in Figure 3. It shows that when the feed of CH_4/O_2 ratio is decreased, conversion of methane increases. In Figures 3-5, graphs are plotted, starting from CH_4/O_2 feed ratio 2. Figure 4 shows the decreasing ratio of methane to oxygen in the inlet as a function of selectivity of the desired products decreases remarkably. Figure 5 shows the decreasing ratio of methane to oxygen in the inlet vs. yield of products increasing. This is because all the hydrocarbons in the process will be oxidized to carbon dioxides ultimately for the lesser CH_4/O_2 feed ratio (< 2). Nevertheless, reduction in the yield of C_2 product at higher selectivity, accomplished at high CH_4/O_2 feed ratio, was due to insufficient oxygen in the reaction. As a result, a higher yield was obtained with a higher oxygen concentration, in spite of a decrease in selectivity.

Validation of plasma reactor modeling

The validity of the model was tested by comparing the theoretical predictions of differential kinetics with experimental results for methane conversion in the DBD reactor reported by Indarto et al. [36]. In order to compare the simulated data with the experimental, the actual reactor dimensions and conditions were used in the simulation, presented in Table 4.

	T (°K)					
Flow Mole Ratio	1023K	1073K	1103K	973K	1023K	1103K
CH_4	0.612	0.612	0.612	0.699	0.699	0.699
O_2	0.051	0.051	0.051	0.095	0.095	0.095
N_2	0.337	0.337	0.337	0.206	0.206	0.206
CH_4 Conversion (%)						
Experimental[39]	4.9	7.9	9.9	4.1	7.1	14.4
Simulated[38]	4.9	8.5	10.9	3	5.8	13.7
Present work	6.5	8.9	11.8	7.7	9	13.76
C_2 selectivity (%)						
Experimental[39]	55.6	69.2	72.5	35.6	53.7	69.6
Simulated[38]	54.6	64.5	66.4	31.7	46.4	58.9
Present work	56.5	60	56	33.46	46	51
C_2 Yield (%)						
Experimental[39]	2.7	5.5	7.2	1.5	3.8	10
Simulated[38]	2.7	5.5	7.3	1.0	2.7	8.1
Present work	3.8	5.35	6.7	2.6	4.1	6.9

Table 3: Comparison between experimental and simulated data.

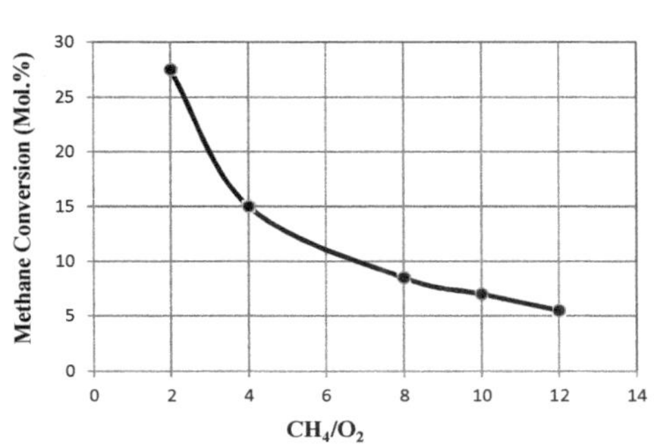

Figure 3: Influence of CH_4/O_2 molar ratio on the conversion of methane.

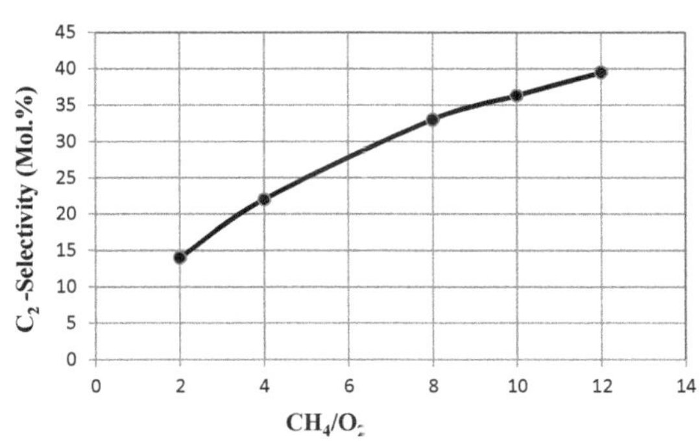

Figure 4: Influence of CH_4/O_2 molar ratio on the selectivity of the C_2 product.

Reaction	Conversion (mol%)	C_{CH4}	C_{C2H2}	C_{C2H4}	C_{C2H6}	C_{C3H6}	C_{C3H8}	$C_{i\text{-}C4H10}$	$C_{n\text{-}C4H10}$	C_{out}/C_{in}	H_{out}/H_{in}
CH_4[36]	13.8	86.2	0.17	0.45	4.7	0.00	0.93	0.26	0.43	0.95	1.01
CH_4[a]	13.8	86.2	0.1	2	4.1	0.00	2.61	1.11	2.47	0.95	1.01

[a]Present Work

Table 4: The reactant conversion and product distributions for the plasma process.

As shown in Table 5, the model calculation was found to be in good agreement with the experimental data. It also means that the proposed values of the reaction rate constant, k, as well as the model, can rationalize the plasma mechanism of methane and/or hydrocarbon conversion (Table 5).

Effect of CH_4/C_2H_2 molar ratio in a plasma reactor: The methane conversion increases with an increasing ratio of methane to acetylene in the input feed. These results show that higher conversion of methane and C_2 product yield is obtained at lower acetylene amounts (higher CH_4/C_2H_2 ratio). Table 6 shows the effect of CH_4/C_2H_2 molar ratio on the methane conversion, selectivity and product yield. Results of the simulation are presented in Figures 6 to 11. Figure 6 (a) shows the methane conversion in a length of the reactor. According to this Figure, the highest methane conversion observed was at 12.7% in CH_4/C_2H_2 =10. Figure 6 (b) shows acetylene conversion in a length of the reactor. The highest acetylene conversion observed was at 20.3% in CH_4/C_2H_2 =10. Figure 7 shows the comparison between selectivity of experimental and model simulation results with our modeling of a binary mixture of methane and acetylene. Product yield and methane conversion in the plasma process were higher than in the OCM process for the same operating conditions, but the products' selectivity in the OCM process was higher than in the process of methane conversion in the plasma reactor. It is interesting to compare experimental, simulated and present work operations under steady state in terms of the conversion, selectivity and yield obtained in the process of methane conversion in the plasma reactor. Figure 8 indicates the C_2, C_3 and C_4 selectivity vs. the reactor length. The acetylene concentration influences methane conversion and C_2, C_3 and C_4 selectivity. The larger the acetylene feed concentration, the higher the CH_4 conversion. The selectivity of all C_4 compounds decreased with increasing acetylene content, however, the selectivity of all C_2 and C_3 compounds was enhanced with increasing acetylene content. Figure 9 indicates the C_2, C_3 and C_4 yields along the reactor bed. The DBD plasma was used as an activation method for the reaction of methane and acetylene, and the products were C_2 and C_4 hydrocarbons. Methane conversion and the yield of C_2, C_3 and C_4 hydrocarbons were affected by the acetylene concentration in the feed.

Influence of the CH_4/C_2H_2 molar ratio: Figure 10 indicates the

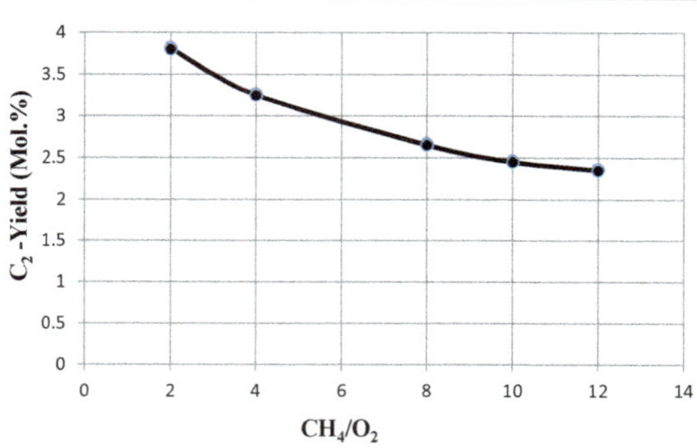

Figure 5: Influence of CH_4/O_2 molar ratio on the yield of C_2 product.

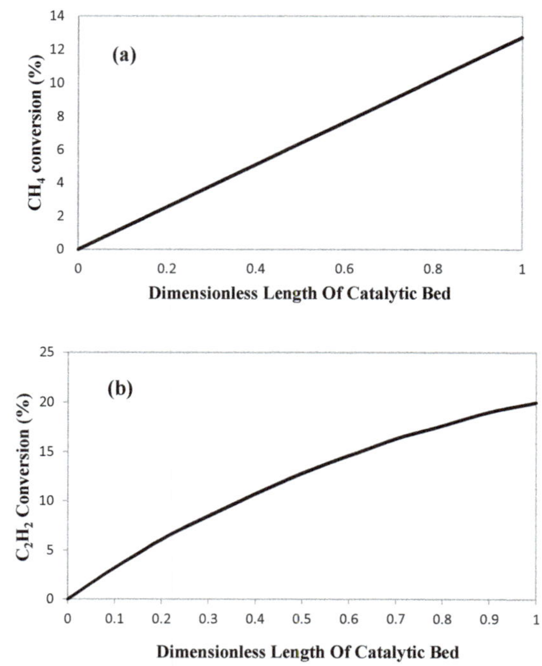

Figure 6: Conversion of methane and acetylene as a function of the dimensionless reactor length.

Selectivity	C_2H_4	C_2H_6	C_3H_6	C_3H_8	$i\text{-}C_4H_{10}$	$n\text{-}C_4H_{10}$
Experimental [36]	15	17	5	18	15	23
Simulated [36]	11	19	3	18	21	19
Present work	11	19	5	18	14	30

Table 5: The comparison between experimental and model simulation results of a binary mixture of methane+acetylene at the flow rate of 30 ml/min and supplied power of 60 W.

C_2H_4, C_2H_6, C_3H_6, C_3H_8, $i\text{-}C_4H_{10}$ and $n\text{-}C_4H_{10}$ selectivity vs. CH_4/C_2H_2 ratio. Figure 11 indicates the C_2H_4, C_2H_6, C_3H_6, C_3H_8, $i\text{-}C_4H_{10}$ and $n\text{-}C_4H_{10}$ yields vs. CH_4/C_2H_2 ratio. In Figure 10 we see the product selectivity, according to the conservation law of carbon and hydrogen, increases or decreases in the present process. In order to increase the CH_4/C_2H_2 ratio in the process, the hydrogen value has to be increased and the carbon value decreased, so that ethane and propane product selectivities increase but the n-butane selectivity decreases more than i-butane selectivity, which shows that i-butane is more stable than n-butane.

In Figure 11 we see that a product yield with a higher k value means that the reactant transformation to a certain product is easier, which increase the product's yield. Interestingly, n-butane hydrocarbon could be formed from any hydrocarbon species, confirmed by a higher value of $K_{x8} > 0.5min^{-1}$, except for methane. This result could rationalize the high yield production of C_4 compounds usually produced in a DBD using AC power supply. Higher C_3H_6 and $n\text{-}C_4H_{10}$ yield occurred

CH_4/C_2H_2	CH_4	C_2H_2	C_2H_4	C_2H_6	C_3H_6	C_3H_8	i-C_4H_{10}	n-C_4H_{10}
CH_4/C_2H_2								
Concentration (%)	88.3	14.94	2.56	4.56	1	2.81	1.42	4.32
Conversion (%)	11.7	25.29						
Selectivity (%)			10.18	18.15	6.49	16.75	14.15	34.39
Yield (%)			1.19	2.12	0.76	1.96	1.66	4.02
CH_4/C_2H_2=2.5								
Concentration (%)	90.46	28.9	3.09	4.94	1.69	3	1.74	6.18
Conversion (%)	9.54	27.75						
Selectivity (%)			9.99	15.94	8.21	14.54	14.02	39.92
Yield (%)			0.95	1.52	0.78	1.39	1.34	3.81
CH_4/C_2H_2=2								
Concentration (%)	91.54	35.88	3.36	5.12	2	3.1	1.89	7.11
Conversion (%)	8.46	28.24						
Selectivity (%)			9.92	15.12	8.85	13.72	13.97	41.94
Yield (%)			0.84	1.28	0.75	1.16	1.18	3.55
CH_4/C_2H_2=1								
Concentration (%)	96.95	70.77	4.7	6.05	3.52	3.59	2.68	11.75
Conversion (%)	3.05	29.33						
Selectivity (%)			9.7	12.5	10.9	11.11	13.82	48.85
Yield (%)			0.3	0.38	0.33	0.34	0.42	1.48

Table 6: Effect of CH_4/C_2H_2 molar ratio through the present research.

with a molar ratio of CH_4/C_2H_2 equal to 4. On the other hand, the intermolecular transformation of n-butane to i-butane occurred as i-butane is thermodynamically more stable than n-butane, as shown by K_{87} (1.38 min^{-1}). Instead of coming from the intermolecular transformation of n-butane, the formation of i-butane could be formed from C_3 compounds, K_{58}=0.54min^{-1}. K and K_{68}=0.7min^{-1}. Although i-butane will be consumed in further reactions mostly to n-butane, by K_{78}=2.54min^{-1}, we have to consider the conversion comparison between n-butane and i-butane. The conversion rate of i-butane is 10% lower than that of n-butane. In most products, an increase in the CH_4/C_2H_2 ratio leads to increased product yield and this is a good thing somewhat.

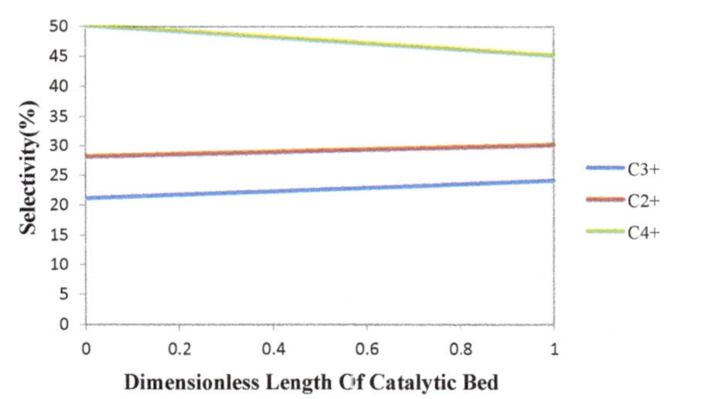

Figure 8: Variations of the C_2^+, C_3^+ and C_4^+ hydrocarbons' selectivity as a function of the reactor length.

Conclusion

The effects of plasma chemistry on the conversion of methane were studied using a dielectric barrier discharge reactor at ambient temperatures. Obtained results implied that the catalyst only enhanced the methane conversion. A dielectric barrier discharge reactor generated non-equilibrium plasma when a sufficiently high voltage was applied across the reactor's electrodes. Methane molecules were activated at this temperature and coupled to form C_2 as well as higher hydrocarbons. The mathematical modeling of methane conversion in a dielectric barrier discharge at a supplied power of 60 W showed that C_2H_4 plays an important role in the global mechanism; having a function as a branching molecule which is able to diversify such a mechanism. It will be a good advantage to predict or optimize the end products of direct methane conversion by a catalytic dielectric barrier discharge (DBD) reactor. This plasma reactor has good potential for the CO-generation of C_2^+ hydrocarbons (ethane, ethylene, acetylene, and propane) and synthesis gases (hydrogen and carbon monoxide) from methane and carbon dioxide. The synergism of the La_2O_3/CaO catalyst utilized and the plasma discharge affected the products' distribution,

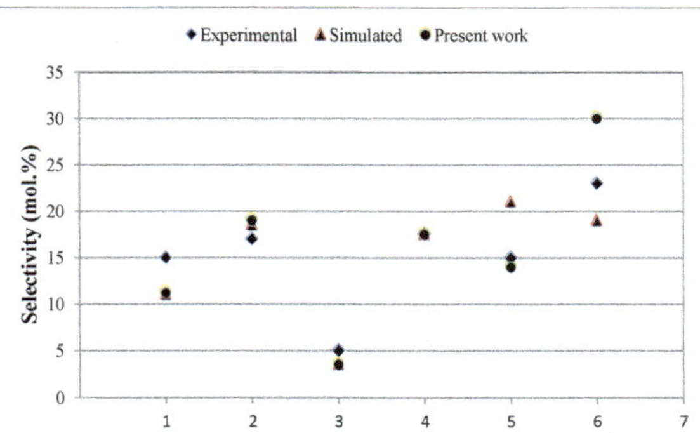

Figure 7: Comparison of product selectivities with experimental [36] and simulated data [36] (feed contains a binary mixture of methane and acetylene).

and particularly C_2^+ hydrocarbons selectivity. The CO-feed of acetylene has important effects on the coking suppression as well. A comparison between the thermal and plasma process showed that the methane conversion and yield production in the plasma process were higher than in the thermal process under the same operating conditions. However, product selectivity in the plasma process was determined to be lower than that of the thermal process.

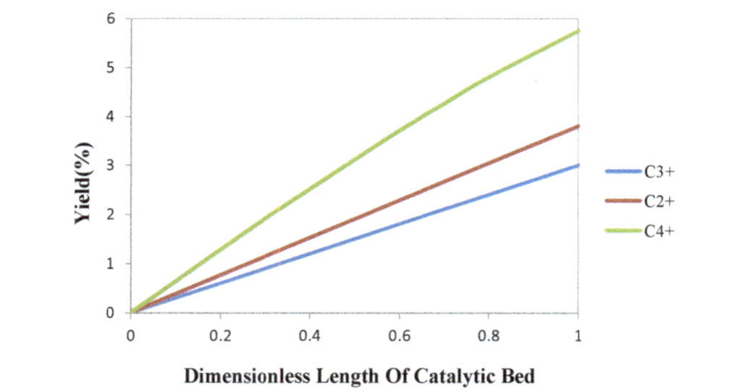

Figure 9: Variations of the C_2^+, C_3^+ and C_4^+ hydrocarbons yield along the reactor.

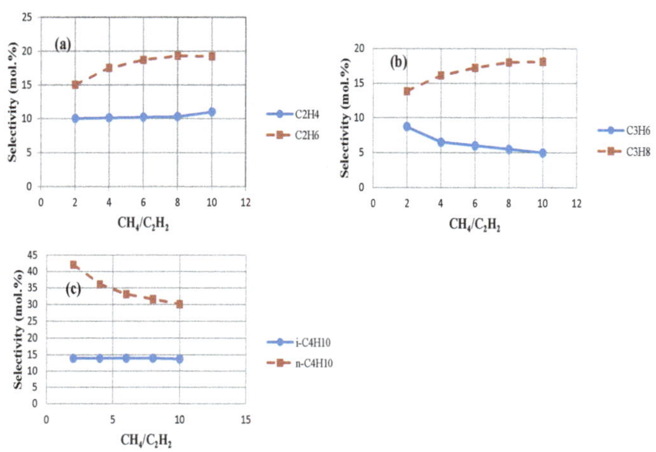

Figure 10: Influence of CH_4/C_2H_2 molar ratio on the selectivity of C_2H_4, C_2H_6, C_3H_6, C_3H_8, i-C_4H_{10} and n-C_4H_{10}

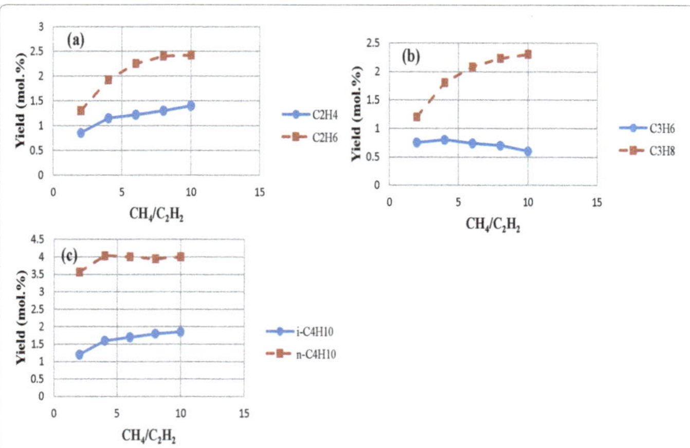

Figure 11: Influence of CH_4/C_2H_2 molar ratio on the yield of C_2H_4, C_2H_6, C_3H_6, C_3H_8, i-C_4H_{10} and n-C_4H_{10}

References

1. Sofranko JA, Leonard JJ, Jones CA (1987) The oxidative conversion of methane to higher hydrocarbons. Journal of Catalysis 103: 302-310.
2. Tan X, Thursfield A, Metcalfe IS, Li K (2009) Analysis of a perovskite ceramic hollow fiber membrane reactor for the partial oxidation of methane to syngas. Asia-Pacific Journal of Chemical Engineering 4: 251-258.
3. Munera JF, Coronel L, Faroldi B, Carrara C, Lombardo EA, Cornaglia LM (2010) Production of ultrapure hydrogen in a Pd-Ag membrane reactor using noble metal supported on La-based oxides. Modeling for the dry reforming of methane reaction. Asia-Pacific Journal of Chemical Engineering 5: 35-47.
4. Yaghobi N, Ghoreishy MHR, Eslamimanesh V (2008) An experimental kinetic model for the oxidative coupling of methane using a bench-scale reactor. Asia-Pacific Journal of Chemical Engineering 3: 202-210
5. Savinov SY, Lee H, Song HK, Na BK (1999) Decomposition of methane and carbon dioxide in a radio-frequency discharge. Ind Eng Chem Res 38: 2540-2547.
6. Liu C, Marafee A, Hill B, Xu G, Mallinson R, et al. (1996) Ind Eng Chem Res 35: 3295-3301.
7. Liu C, Marafee A, Mallinson R, Lobban L (1997) Methane conversion to higher hydrocarbons in a corona discharge over metal oxide catalysts with OH groups. Appl Catal A Gen 164: 21-33.
8. Liu Cj, Mallinson R, Lobban L (1999) Comparative investigations on plasma catalytic methane conversion to higher hydrocarbons over zeolites. Appl Catal A Gen 178: 17-27.
9. Liu Cj, Mallinson R, Lobban L (1998) Nonoxidative methane conversion to acetylene over zeolite in a low temperature plasma. Journal of Catalysis 179: 326-334.
10. Marafee A, Liu Cj, Xu G, Mallinson R, Lobban L (1997) An Experimental Study on the Oxidative Coupling of Methane in a Direct Current Corona Discharge Reactor over Sr/La2O3 Catalyst. Ind Eng Chem Res 36: 632-637.
11. Kim SS, Lee H, Na BK, Song HK (2003) Reaction pathways of methane conversion in dielectric-barrier discharge. Korean Journal of Chemical Engineering 20: 869-872.
12. Kim SS, Lee H, Choi JW, Na BK, Song HK (2003) Kinetics of the methane decomposition in a dielectric-barrier discharge. Journal of Industrial and Engineering Chemistry 9: 787-791.
13. Kim SS, Lee H, Choi JW, Na BK, Song HK (2004) Plasma-assisted reduction of supported metal catalyst using atmospheric dielectric-barrier discharge. Catalysis Today 89: 193-200.
14. Shen C, Sun D, Yang H (2011) Methane coupling in microwave plasma under atmospheric pressure. Journal of Natural Gas Chemistry 20: 449-456.
15. Huang J, Badani MV, Suib SL, Harrison JB, Kablauoi M (1994) Partial oxidation of methane to methanol through microwave plasmas. Reactor design to control free-radical reactions J Phys Chem 98: 206-210.
16. Huang L, Hari T, Nakajyo K, Ozawa S, Matsuda H (2001) Matsuda H Reduction of NO by CO in a pulsed corona reactor incorporated with CuO catalyst. Journal of Chemical Engineering of Japan 34: 1012-1016.
17. Niu J, Yang X, Zhu A, Shi L, Sun Q, et al. (2006) Plasma-assisted selective catalytic reduction of NO_x by C_2H_2 over Co-HZSM-5 catalyst. Catal Commun 7: 297-301.
18. Liu CJ, Xue B, Eliasson B, He F, Li Y, et al. (2001) Methane Conversion to Higher Hydrocarbons in the Presence of Carbon Dioxide Using Dielectric-Barrier Discharge Plasmas. Plasma Chemistry and Plasma Processing 21: 301-310.
19. Liu C, Eliasson B, Xue B, Li Y, Wang Y (2001) Zeolite-enhanced plasma methane conversion directly to higher hydrocarbons using dielectric-barrier discharges. React Kinet Catal Lett 74: 71-77.
20. Istadi A, Amin NAS (2005) Co-Generation of C2 Hydrocarbons and Synthesis Gases from Methane and Carbon Dioxide: a Thermodynamic Analysis. Journal of Natural Gas Chemistry 14: 140-150.
21. Zou JJ, Zhang Yp, Liu CJ, Li Y, Eliasson B, (2003) Starch-enhanced Synthesis of Oxygenates from Methane and Carbon Dioxide Using Dielectric-barrier Discharges. Plasma Chemistry and Plasma Processing 23: 69-82.

22. Li Y, Xu Gh, Liu Cj, Eliasson B, Xue Bz (2001) CO-generation of Syngas and Higher Hydrocarbons from CO_2 and CH_4 Using Dielectric Barrier Discharge: Effect of Electrode Materials. Energy Fuels 15: 299-302.

23. Yao SL, Ouyang F, Nakayama A, Suzuki E, Okumoto M, et al. (2000) Oxidative coupling and reforming of methane with carbon dioxide using a high frequency pulsed plasma. Energy & Fuels 14: 910-914.

24. Liu Cj, Xu Gh, Wang T (1999) Non-thermal plasma approaches in CO_2 utilization. Fuel Processing Technology 58: 119-134.

25. Wang B, Zhang X, Liu Y, XU G (2009) Conversion of CH4, steam and O2 to syngas and hydrocarbons via dielectric barrier discharge. Journal of Natural Gas Chemistry 18: 94-97.

26. Larkin DW, Zhou L, Lobban LL, Mallinson RG (2001) Product selectivity control and organic oxygenate pathways from partial oxidation of methane in a silent electric discharge reactor. Ind Eng Chem Res 40: 5496-5506.

27. Kogelschatz U (2003) Dielectric-Barrier Discharges: Their History, Discharge Physics, and Industrial Applications. Plasma Chemistry and Plasma Processing 23: 1-46.

28. Stephanopoulos G, Han C (1996) Intelligent systems in process engineering: a review. Comput Chem Eng 20: 743-791.

29. Santamaria JM, Miro EE, Wolf EE (1991) Reactor simulation studies of methane oxidative coupling on a $Na/NiTiO_3$ catalyst. Ind Eng Chem Res 30: 1157-1165.

30. Hoebink JHBJ, Couwenberg PM, Marin GB (1994) Fixed bed reactor design for gas phase chain reactions catalyzed by solids: the oxidative coupling of methane. Chemical Engineering Science 49: 5453-5463.

31. Coronas J, Gonzalo A, Lafarga D, Menendez M (1997) Effect of the membrane activity on the performance of a catalytic membrane reactor. AIChE Journal 43: 3095-3104.

32. Keller GE, Bhasin MM (1982) Synthesis of ethylene via oxidative coupling of methane: I. Determination of active catalysts. Journal of Catalysis 73: 9-19.

33. Coronas J, Menéndez M, Santamaria J (1994) Methane oxidative coupling using porous ceramic membrane reactors-II. Reaction studies. Chemical Engineering Science 49: 2015-2025.

34. Kazemeini M, Mohammadi AR (2005) Reactor Modeling of a Non-Catalytic OCM Process. Iranian Journal of Chemical Engineering 2: 3-14.

35. Seyednejadian S, Yaghobi N, Maghrebi R, Vafajoo L (2011) CFD modeling of reaction and mass transfer through a single pellet: Catalytic oxidative coupling of methane. Journal of Natural Gas Chemistry 20: 356-363.

36. Indarto A, Coowantiwong N, Choi JW, Lee H, Song HK (2008) Kinetic modeling of plasma methane conversion in a dielectric barrier discharge. Fuel Processing Technology 89: 214-219.

37. Tye CT, Mohamed AR, Bhatia S (2002) Modeling of catalytic reactor for oxidative coupling of methane using La_2O_3/CaO catalyst. Chemical Engineering Journal 87: 49-59.

38. Stansch Z, Mleczko L, Baerns M (1997) Comprehensive kinetics of oxidative coupling of methane over the La_2O_3/CaO catalyst. Ind Eng Chem Res 36: 2568-2579.

Recovery of Calcium Carbonate from Wastewater Treatment Sludge Using a Flotation Technique

Jannie P. Maree[1], Caliphs M. Zvinowanda[1]*, Munyaradzi Mujuru[1], Regina M. Matsapola[2], David J. Delport[2] and Wynand J. Louw[3]

[1]*Department of Environmental, Water & Earth Sciences, Faculty of Science, Tshwane University of Technology, Pretoria, South Africa*
[2]*Department of Chemical & Metallurgical Engineering, Faculty of Engineering & Inbuilt Environment, Tshwane University of Technology, Pretoria, South Africa*
[3]*Marlow Aquatec (Pty) Ltd, 50 Francis Street, Colbyn, Pretoria, South Africa*

Abstract

The use of flotation technique for the recovery of calcium carbonate ($CaCO_3$) from wastewater treatment sludge was investigated in this study. The parameters that were investigated included dosage of floating agents (sodium oleate and sunlight dish liquid) and the percentage solids of the slurry. The experiments were performed by floating sieved and un-sieved materials and $CaCO_3$ was determined for both conditions as well as from tailings.

Initial $CaCO_3$ analysis for the bulk material indicated that sieved and un-sieved materials had 63.4% and 32.9% $CaCO_3$ content by weight respectively. The modification of pH was effected by dosing 1g $NaCO_3$ to both 1000 g of sieved and un-sieved materials which was sufficient to raise the pH of the slurry to 9.5. A lower average recovery of 2.33% was observed on un-sieved material after using sodium oleate as a collector when compared to sunlight liquid of 31.6%. Therefore, it was concluded that for un-sieved material sunlight dishwashing liquid was a better collector compared to the latter. The results of this study proved that there is great potential of recovering commercial grade limestone from wastewater sludge.

Keywords: Collector; Conditioner; Froth; Sodium oleate; Sunlight dish liquid; Sieved

Introduction

Complete wastewater treatment does not only involve the treatment and reclamation of the liquid, but also encompasses the processing and disposal of the solids removed or generated during treatment [1]. The flotation process is widely used in industrial wastewater treatment plants, where it is used to remove fats, oil, grease and suspended solids from wastewater. These units are called Dissolved Air Flotation (DAF) units. In particular, dissolved air flotation units are used in removing oil from the wastewater effluents from oil refineries, petrochemical and chemical plants, natural gas processing plants and similar industrial facilities [2,3].

The separation process of froth flotation is a primary method of creating a high-valued concentrates which could contain usable materials [4,5]. Conventional flotation equipment from both Denver and Wemco equipment companies is available for lab and pilot plant studies. Prior to pilot plant testing, flotation parameters such as reagents, conditioning time, pH, and stages are defined at lab scale. In conjunction with the flotation studies, mineralogy, surface chemistry, and liberation determination are used to define feed parameters [6-8].

Lime treatment of wastewater produces relatively large amount of sludge by weight and for that reason, the choice of solids handling methods significantly affects capital and operation costs for the plant as well as the impact of ultimate sludge disposal on the environment [9]. A few water treatment systems have been equipped with sludge treatment facilities for recovering reusable and marketable products [10]. The procedures used at the sludge treatment facilities involve carbonation of the sludge to effect a phase separation between calcium and magnesium values [7]. The magnesium hydroxide component of the sludge is solubilised from the residual insoluble and is disposed off as a by-product waste material while reusable in the water treatment process and also available for sale on the open market. The disposal of the carbonated liquor, however, creates ecological challenges such as sludge dumps and their management.

The treatment of sludge to recover calcium carbonate has been limited to the treatment of sludges obtained from the softening of raw waters that are basically free of turbidity factors. Clay and other turbidity factors present in the raw water are separated as components of the sludge and are carried through the sludge treatment step with insoluble calcium carbonate.

Investigation has shown that froth flotation can be economically and effectively used to produce a relatively high grade calcium carbonate with good recovery from the wastewater treatment sludges when using magnesium carbonate with lime for flocculation [10]. Flotation had its beginning in mineral processing and as such it has been used for a long time in solid to solid separation using stable froths to selectively separate different minerals from each other [10]. It is believed that the cross exchange of flotation experience in the mineral flotation and water effluent treatment should lead to new and improved procedures for industrial waste treatment [11].

The flotation process has enabled the production of high brightness calcium carbonate by removing the silicate impurities from the calcium carbonate ore which would otherwise be responsible for colour imperfections in the finished product [8]. The most important criteria in evaluating the performance of calcium carbonate collector during floatation is the dosage level required and the yield of calcium

***Corresponding author:** Caliphs M. Zvinowanda, Department of Environmental, Water & Earth Sciences, Faculty of Science, Tshwane University of Technology, Pretoria, South Africa carbonate produced. In froth flotation process, calcium carbonate from the clay or any other water contaminants is conditioned with soda ash and sodium silicate to disperse the clay with the aid of pH adjustment as well. Thereafter, the slurry is conditioned with fatty acid soap which selectively coats the calcium carbonate particle with insoluble soap making it hydrophobic and collectable [8].

Flotation where possible, is carried out in an alkaline medium, as most collectors such as xanthates, are stable under these conditions [12]. The alkalinity condition minimises corrosion of cells, pipe work and other metal handling facilities. Sodium carbonate is usually added to control the alkalinity and to a lesser extend sodium hydroxide or ammonium hydroxide is used for the same purpose. Sulphuric or sulphurous acids are used where lower pH is required [13]. In most cases lime is being used to regulate pulp alkalinity as it is cheaper than most alkalis. Lime can also act as a strong depressant for pyrite and arsenopyrite when using xanthate collectors [14]. Both hydroxyl and calcium ions participate in the depressive effect on pyrite by the formation of mixed films of Fe(OH), FeO(OH), $CaSO_4$ and $CaCO_3$ on the surface [15].

The aim of this study was to recover calcium carbonate which was used after treatment of acid mine water from the sludge. Flotation was selected as a feasible technique to recover the $CaCO_3$ from the waste sludge.

Experimental

Sodium oleate and sunlight dish liquid were used as collectors for calcium carbonate in this study. The effects of collector dosage, modifiers and pH on calcium carbonate recovery were investigated.

Operational procedure for the Denver machine used in floatation studies

The following steps are the general operating principles of the Denver machine used in this study:

- The agitator mixes the slurry and the air is dispersed by an impeller stabilizer;
- The impeller is connected to a vertical hollow shaft that rotates the impeller and feeds low pressure air under the impeller plate;
- As the air flows between the rotating plate and stabilizer, it is broken into finer bubbles by the shearing action and is dispersed throughout the flotation cell;
- As the pulp contacts the impeller, it is intensely agitated and aerated;
- The flow pattern directs the bubbles to the surface and;
- The concentrates collected in the froth column at the surface is discharged at the froth overflow lips.

Figure 1 shows the floatation machine used in this study.

Experimental procedure (Flotation process)

The experiments were carried out in 5 L flotation cell at 10% solids. A mass of 500 g of dry sample was added to 4.5 L of water. It was then milled for 20 min and poured into a 5 L flotation cell. For the initial test, 1 g of Na_2CO_3 and 1 g Na_2SiO_3 were added to the slurry and allowed to condition for 3 min, then 0.5 g of fatty acid soap ($Na_2O_2H_{33}C_{18}$) was added and allowed to condition for 3 min.

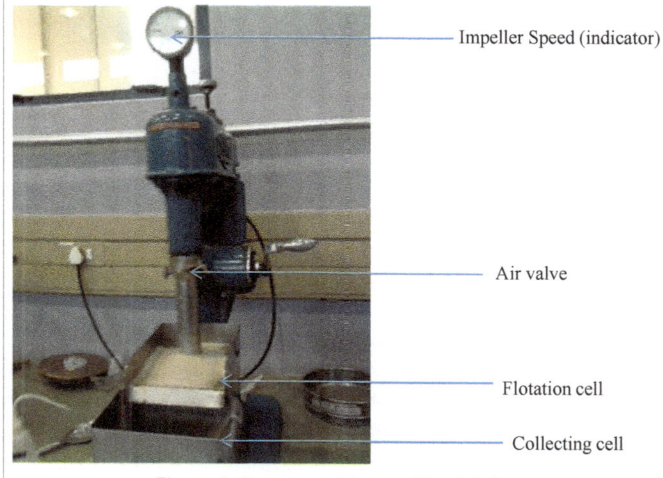

Figure 1: Denver machine used for flotation.

	Effect of sodium oleate	Effect of sunlight		
Test	1.1	1.2	1.3	1.4
Volume (L)	4500	4500	4500	4500
Feed CaCO3	500	500	500	500
Description	Unsieved	Unseived	Unseived	Unseived
Feed CaCO3 content (%)	32.9	32.9	32.9	32.9
Flotation:				
Na2CO3 (g)	1	1	1	1
Silicate (%)		30	30	30
Silicate (mL/L)	0.2	0.2	0.2	0.2
Silicate (mL)	1	1	1	1
Oleate (%)		100	100	100
Oleate (g/L)	0.4	0.1	0.2	0.4
Oleate (g)	2	0.5	1	2
pH Feed (before adding	8.51	8.6	8.6	8.6
Na2CO3)	9.45	9.54	9.54	9.54
pH Feed (after adding	650	650	650	650
Na2CO3)				
Impeller speed (rpm)				

Table 1: Conditions for the experiments on un-sieved material.

The air was then introduced to generate bubbles allowing the formation of froth. The concentrates were collected at the following times intervals, 1, 4, 7 and 20 min. The rotational velocity was set to 650 rpm. This was done on the first three tests with different dosages of $Na_2O_2H_{33}C_{18}$ (i.e. 0.5 g, 1 g and 2 g)

The preceding procedure was used for tests 4, 5 and 6 with sunlight dishwashing soap as the collector agent. The concentrates and tailing of each test were dried, weighed and then taken for analysis to determine the $CaCO_3$ concentration in them.

Experimental procedure (analysis for $CaCO_3$)

Approximately 1.065 g of sample was weighed and added to 50 mL of 1 N HCl. It was then heated to 80C for 15 min. The sample was then allowed to cool and titrated with 1 N NaOH to pH 7.

Results and Discussion

The results and discussion for un-sieved material are as follows

The experimental conditions used for un-sieved sludge are shown in (Table 1). Experiments with sodium oleate with different settings did not produce any significant recoveries from set one shown in (Table 1) and hence, these results were not reported. It was noted that the pH was almost constant for the four experiments 1.1 to 1.4 for sunlight. For each experiment, the initial pH was around 8.6 and after addition of soda ash (Na_2CO_3), the pH raised to 9.45. From literature survey, most materials responded well on the pH around 7.5 and 11.5 and calcium carbonate responded well at pH of 9.5, which was then set as target for pH modification in this study [8]. Detailed results on the recoveries are shown on (Table 2 and 3), where each frother concentrate (denoted by FC1, FC2, FC3 and FC4) and tailings concentrate (T) were analysed for $CaCO_3$ content.

Effect of collector dosage on un-sieved material

The total recoveries for the concentrates as collected at different times are shown below in (Table 2 and 3). Tables 2 and 3 show the results of un-sieved material, where the tests were conducted at the same conditions but using different collectors. It was demonstrated that the recovery increased with flotation time, however, the tests conducted with sodium oleate as fatty acid collector showed a very small increase of recovery with time.

Sodium oleate (soap) appeared to have been a weak collector when compared to sunlight dishwashing liquid (detergent). This was also shown by results obtained experiments 1.1 and 1.4 of the un-sieved sludge material, where equal dosages of fatty acids (sunlight and sodium oleate) were added, but with sodium oleate only 2.33% $CaCO_3$ recovery was obtained whereas by sunlight the recovery increased to 31.55%. Detergents have a tendency of emulsifying scum's while soaps lose their froth as they precipitate calcium as calcium salts. The failure by sodium oleate to breakdown calcium carbonated to smaller pieces easily floated in the froth might have lead to the poor recoveries observed in this study.

Figure 2 shows the difference in recovery when using sunlight and sodium oleate at the same dosages. When using dosage of 2 g sodium oleate as collector, it was found that while collecting the froth for 1 minute, approximately 1% recovery of calcium carbonate was obtained and it did not show much different when the froth was collected for 4 and 7 min respectively. When using 2 g sunlight liquid as collector, it was observed that the concentrate which was collected in the first 1 min

	Time (min)	Mass (g)	$CaCO_3$ Content (%)	$CaCO_3$ Recovery (%)
Feed		404	32.9	
FC1	1	3.3	18.97	0.99
FC2	4	3.4	21.01	1.13
FC3	7	3.7	20.57	1.20
FC4	20	0	0.00	0.00
FrotherConcentrate		**10.4**	**20.21**	**2.33**
Tails		393.6	15.520	96.67
Head (calc)		404	15.64	99.01

Table 2: Experiment 1.3 of un-sieved material (2 g sodium oleate).

	Time (min)	Mass (g)	$CaCO_3$ Content (%)	$CaCO_3$ Recovery (%)
Feed		500	32.9	
FC1	1	9.4	27.51	2.95
FC2	4	21.1	26.06	6.28
FC3	7	40.03	23.90	10.99
FC4	20	46.5	22.98	12.20
FrotherConcentrate		**117.3**	**24.21**	**31.55**
Tails		382.7	15.470	67.58
Head (calc)		500	17.52	100.00

Table 3: Experiment 1.4 of un-sieved material (2 g sunlight).

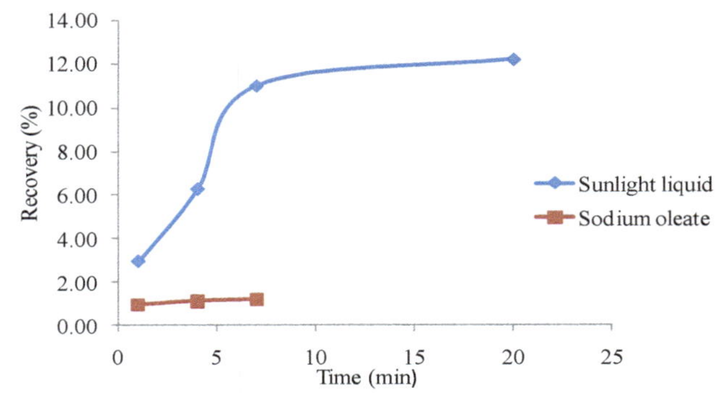

Figure 2: Graph of time versus recovery for experiments of same dosages of two different collectors used.

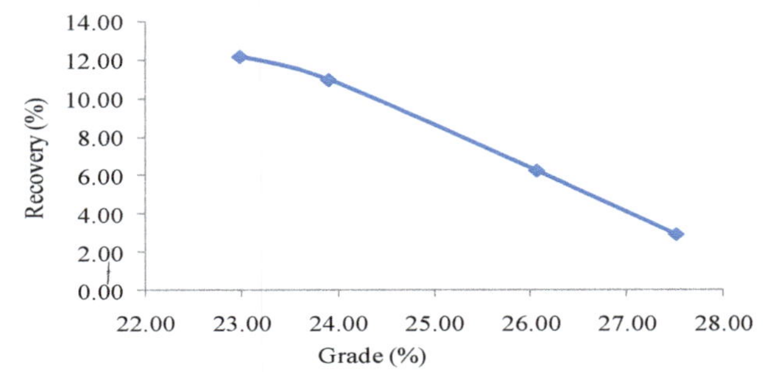

Figure 3: The graph of grade versus recovery for 2 g dosage of sunlight.

	Time (min)	Mass (g)	$CaCO_3$ Content (%)	$CaCO_3$ Recovery (%)
Feed		500	32.9	
FC1	1	1.6	28.86	0.51
FC2	4	1.7	25.84	0.49
FC3	7	3.5	21.16	0.82
FC4	20	54.1	20.00	12.02
FrotherConcentrate		**60.9**	**20.46**	**13.84**
Tails		439	17.670	86.16
Head (calc)		499.9	18.01	100.00

Table 4: Experiment 1.1 of un-sieved (0.5 g sunlight).

had a recovery of approximately 3%. The increase in the concentrate collected between 4 and 7 min was quite significant. Even after 7 min, the froth was still there and the concentrate collected for the next 20 min and the recovery increased further to reach12%.

Figure 3 shows the relationship between recovery of $CaCO_3$ and grade in the sludge. At a recovery of about 12% the grade was 23%, but at a recovery of about 2% the grade (quality of calcium carbonate produced) was higher, that is about 27.5%. These results demonstrated that with a higher recovery, a low grade $CaCO_3$ was obtained. This can be explained by the fact that when floating is performed for a long-time, some of the non- $CaCO_3$ materials will also start floating and resulting in a high recovery with a lot of impurities. Some studies have come to a conclusion that typical grade-recovery curve show a relationship of grade being inversely proportional to recovery [12]. The data of sodium oleate as collector could not be shown as the recoveries were too low when compared to the use of sunlight as collector (Figure 3).

The comparison of the effectiveness of sunlight dishwashing liquid as a collector over sodium oleate as a collector is illustrated in (Figure

4). From the graph, it is noted that with 0.5 g sodium oleate, there was no $CaCO_3$ recovered whilst sunlight managed to produce about 15% recovery.

This can be explained by the fact that impurities in the raw water vary from place to place as well as with seasons and results in the modification of the water treatment processes and the sludge produced thereafter. Hence, variations in the sludge composition would require adjustment in the flotation reagents composition as well.

The experimental conditions for the recovery of $CaCO_3$ use a dosage of 0.5 g of sunlight are shown in (Table 4). From the grade column, the initial concentrate which was collected after 1 min showed a very high grade than the other concentrates. There was grade reduction as the time and the recoveries increase.

The effects of increasing the collector concentration from 0.5 g to 1.0 g are shown in (Table 5).

From (Table 5), it is noted that the concentrates that were collected for the first three time intervals were almost equal although they were not collected for equal in mass (8.4, 8.5, 10.3). This indicates that the pattern of froth floating was not constant and $CaCO_3$ was almost coming out of the cell constantly.

The effect of collector concentration on the recovery of $CaCO_3$ from sludge is illustrated in (Figure 5). It was observed that the recoveries were almost invariant for both dosages in the first 4 minutes. This could be explained by assuming that the conditioning times were too short, therefore, giving inadequate interactions time between the collector and substance of interest $CaCO_3$. After 4 min a rapid increase in the recoveries for both doses was observed.

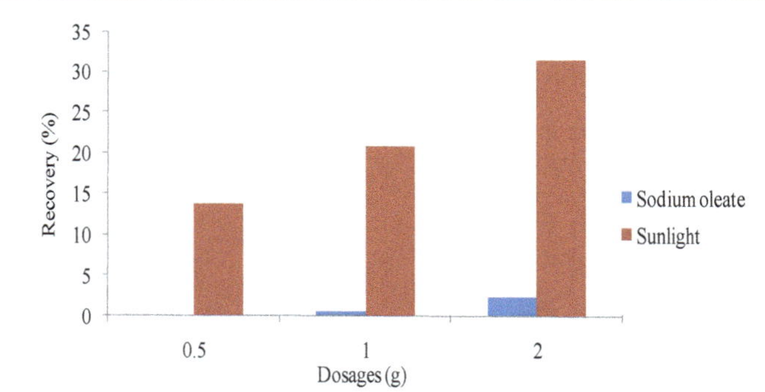

Figure 4: Bar graph of dosage versus recovery to indicate the effectiveness of sunlight over sodium oleate.

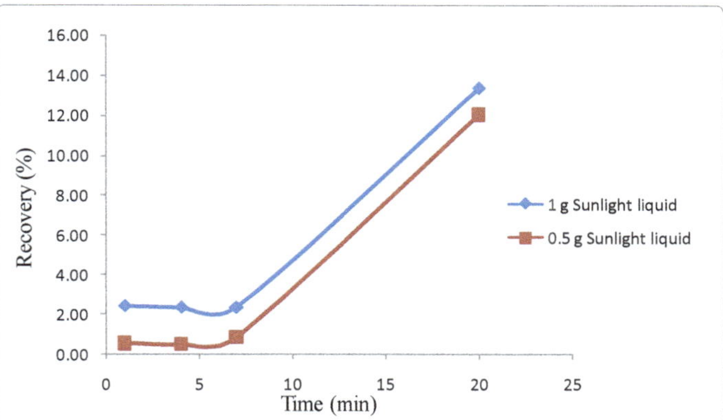

Figure 5: Graph of time versus recovery for 0.5 g and 1.0 g sunlight liquid.

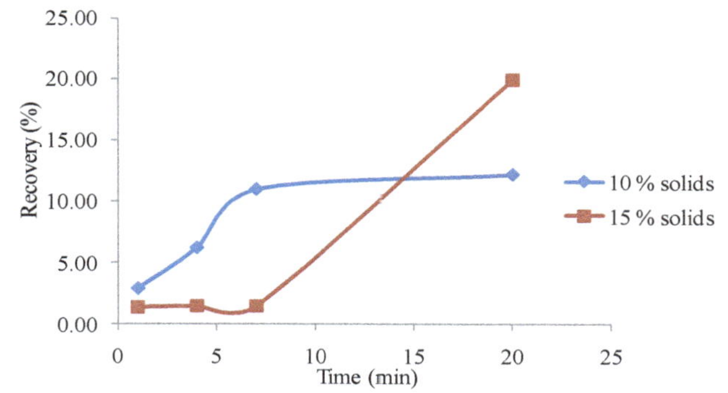

Figure 6: Graph of time versus recovery for slurry at 10 % and 15 % solids.

Products	Time (min)	Mass (g)	$CaCO_3$ Content (%)	$CaCO_3$ Recovery (%)
Feed		500	32.9	
FC1	1	8.4	26.29	2.40
FC2	4	8.5	25.38	2.35
FC3	7	10.3	21.16	2.37
FC4	20	60.9	20.16	13.35
FrotherConcentrate		**88.1**	**21.37**	**20.91**
Tails		410.8	17.800	79.53
Head (calc)		498.9	18.43	100.00

Table 5: Experiment 1.2 of un-sieved (1 g sunlight).

Products	Time (min)	Mass (g)	$CaCO_3$ Content (%)	$CaCO_3$ Recovery (%)
Feed		650	32.9	
FC1	1	3.9	23.56	1.45
FC2	4	6.4	15.01	1.52
FC3	7	7.2	13.51	1.54
FC4	20	110	11.44	19.92
FrotherConcentrate		**127.5**	**12.11**	**17.15**
Tails		522.5	10.020	82.85
Head (calc)		650	10.43	100.00

Table 6: Floatation experiment with 15% solids using sunlight.

Effect of percentage solids on the extraction of $CaCO_3$ from un-sieved material

Table 6 contains the results obtained when the slurry made from un-sieved material contained about 15% solids. From (Table 1), the overall recovery was 31.55%, while with slurry at 15 % solids a 17.15% recovery was realised. The slurry with 10% solids produced a higher grade $CaCO_3$ of 24.21% while at 15% solids it gave a lower grade of 12.11% . It was also noted that the recovery and conditioning time showed a direct proportionality relationship.The average $CaCO_3$ grade for these experiment was lower than all the other experiment conducted with different collectors.

The variation of recoveries with time when slurries of 10% and 15% solids were floated by sunlight is shown in (Figure 6). From the graph it is noted that with 10% slurry, recoveries increased more rapidly than 15% slurry until after 14 min. However, the overall recovery for slurry with 10 % solids is more than that of the slurry with 15% solids. The lag of recovery in 15% slurry could be attributed to the need for longer period to condition more concentrated slurry.

The experimental conditions that were used on sieved sludge material are shown in (Table 7). The pH was kept almost constant for these experiments. For each experiment, the initial pH was around 8.4 and after addition of soda ash (Na_2CO_3), the pH raised to 9.45. Detailed results on the recoveries are shown on (Table 8 and 9), where each frother concentrate (denoted by FC1, FC2, FC3 and FC4) and tailings concentrate (T) were analysed for $CaCO_3$ content.

Effect of collector dosage on sieved material

Table 9 indicates the results obtained when using a dosage of 2 g sodium oleate on the sieved material. It was noted that on the last two time intervals, the recovery was constant at 1.81%. The overall recovery obtained was 5.91%, which was lower than what was obtained when un-sieved material was used under similar conditions.

The effects time on recovery of $CaCO_3$ when using a dosage of 4 g sodium oleate on the sieved sludge material in shown in (Table 9). It was noted that the dosage only made a difference of 1.74% to the test performed with a dosage of 2 g. Generally sodium oleate exhibited some poor collection characteristics for $CaCO_3$.

Effect of percentage solids in sieved sludge material on recovery

Figure 7 shows the variation in recoveries with time for the 10% and 15% slurries. The curve lying below indicates that increasing percent solids from 10% to 15% did not make any improvement on recovery and instead the recovery was lowering for every time the concentrates were being collected.

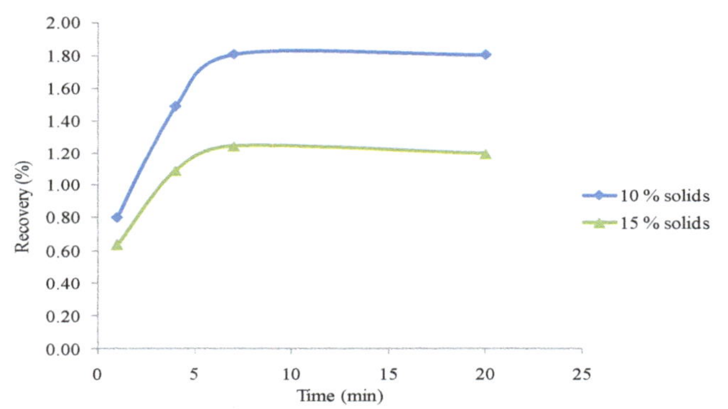

Figure 7: Graph of time versus recovery for slurry at 10 % and 15 % solids (sieved).

Effect of sodium oleate			
Test	1.1	1.2	1.3
Volume	4302	310	4500
Feed $CaCo_3$ (g)	478	500	500
Description	Sieved	Sieved	Sieved
Feed $CaCO_3$ content (%)	63.4	63.4	63.4
Flotation			
Na_2CO_3 (g)	1	1	1
Silicate (%)			
Silicate (mL/L)	0.2	0.2	0.2
Silicate (mL)	1	1	1
Oleate (%)			
Oleate (g/L)	0.4	0.4	0.8
Oleate (g)	2	2	4
pH Feed(before adding Na_2CO_3)	8.4	8.5	8.3
pH Feed(after adding Na_2CO_3)	9.45	9.5	9.43
Impeller speed (rpm)	650	650	650

Table 7: Conditions for experiments on sieved material.

	Time (min)	Mass (g)	$CaCO_3$ Content (%)	$CaCO_3$ Recovery (%)
Feed		444	63.42	
FC1	1	3.5	59.49	0.80
FC2	4	6.2	62.60	1.49
FC3	7	8	58.89	1.81
FC4	20	7.6	61.88	1.81
Frother Concentrate		**25.3**	**60.78**	5.91
Tails		418.7	58.440	94.09
Head (calc)		444	58.57	100.00

Table 8: Conditions used to float sieved sludge material with 2 g sodium oleate.

Products	Time (min)	Mass (g)	$CaCO_3$ Content (%)	$CaCO_3$ Recovery (%)
Feed		450	63.42	
FC1	1	3	55.49	0.64
FC2	4	5.6	50.70	1.09
FC3	7	7.2	45.10	1.25
FC4	20	7.8	40.00	1.20
Frother Concentrate		**23.6**	**46.06**	4.18
Tails		426.4	58.440	95.82
Head (calc)		450	57.79	100.00

Table 9: Effects of time on collection of $CaCO_3$ from sieved using sodium oleate.

Conclusions

From the experimental results obtained on flotation of sieved and un-sieved material it was concluded that:

- Recovering $CaCO_3$ from wastewater treatment sludge is definitely possible with flotation.
- Sunlight dish wash liquid showed a great potential as a collector of $CaCO_3$ from sludge than sodium oleate.
- There is a minimum conditioning time necessary to effect significant recovery of $CaCO_3$.
- pH regulation was essential to improve collection capability of collectors.
- The collector had shown a great influence amongst all the other reagents used, on the tests conducted and sunlight dishwashing liquid was successful on improving the recovery for un-sieved material throughout the experiments.

Acknowledgements

The authors would like to acknowledge financial support from the Technology for Human Resources and Industrial Programme (THRIP-SA), Tshwane University of Technology and Key Structures Holdings.

References

1. Victor G, Roger L, Larry B, Clifford R (1978) Factors affecting the design of dissolved air flotation systems. J Wat Pol Contrl Fed 50: 1835-1840.
2. Crawford RJ, Ralston J (1988) Deinking Flotation: Influence of Calcium Soap and surface- Active Substance. Min Eng 18: 59-64.
3. Collins DN, Read AD (1971) The treatment of slimes. Min Sci Eng 3: 19-31.
4. http://www.smenet.org/store/mining-books.cfm/Froth-Flotation%3A-A-Century-of Innovation/252-9
5. Sanks RL (2004) Water Treatment Plant Design For Practicing Engineers. ANN Arbor Science Publishers Inc. 105
6. Klassen VI, Mokrousov VA (1963) An Introduction to the Theory of Flotation. Butterworths, London.
7. http://en.wikepedia.org/wiki/Froth_Flotation#Waste_water_treatment

8. Bell BA, Zaferatos TM (1977) Evaluation of alternate solids handling methods for advanced waste treatment. J Wat Poltn Contrl Fed 49: 146-155.

9. Metcalf and Eddy (1979) Treatment Disposal Reuse: Wastewater Engineering. McGraw-Hill Inc, New York.

10. Rodrigues RT, Rubio J (2007) DAF–dissolved air flotation: Potential applications in the mining and mineral processing industry. International Journal of Mineral Processing 82: 1-13.

11. Finch JA, Hardie CA (1999) An example of innovation from the waste management industry: Deinking flotation cells. Minerals Engineering 12: 467-475.

12. Rubio J, Sousa ML, Smith RW (2002) Overview of Flotation as a Wastewater Treatment Technique. Minerals Engineering 15: 139-155.

13. Wills BA, Napier-Munn TJ (1997) Wills Mineral Processing Technology. Elsevier Ltd., Australia 294-295.

14. http://www.min-eng.com/pdf/willsflyer.pdf

45

Degradation Studies of Tannery Effluents using Electro Flotation Technique

P. Gomathi Priya*, V. Ramamurthi and Prabhu Anand

Department of Chemical Engineering, A.C. College of Technology, Anna University, Chennai- 600 025, India

Abstract

This study reveals that the treatment of tannery effluent samples by electro flotation. In this system, the various parameters like effect of pH, current density, COD, total solids, addition of surfactant and flocculent were investigated. It was found that maximum degradation was at the current density of 0.14A/cm^2 and at a pH of 6.0. Under the same experimental conditions the removal efficiency of chemical oxygen demand (COD) and total solids were found to be approximately 66.3% and 80% respectively. The current density was the most significant factor in the suspended solids elimination by the addition of surfactant and flocculent concentration. The addition of surfactant increases the COD removal and it also enhances the total solids removal. The flocculent also enhanced electro flotation and this technique was found to be effective in removing total solids and reducing COD from tannery effluents. These results intend an important role of these parameters in electrochemical process for removing tannery effluents.

Keywords: Electro flotation; Flocculent; Surfactant; Chemical oxygen demand; Tannery effluent

Introduction

Industrialization has its effect on environment by causing pollution, among which water pollution is contributed by the industries, the nature of effluents from industries are varied and is industry specific. Especially the leather industry has been facing a major threat regarding the ecological problems caused by their waste streams [1]. In tanneries the conversion of the raw hide into leather necessitate different mechanical and chemical treatments e.g. soaking, liming, deliming, bathing, pickling, skin degreasing and tanning, etc. The low efficiency of the chemical operations generates large quantities of effluents like 30 -35 L kg^{-1} of raw materials processed [2]. The commonly used technology for effluent treatment was physicochemical followed by units of biological treatment, and activated sludge or aerated lagoon systems [3]. Generally these conventional treatments are unable to reduce all of the polluting parameters like COD, chlorides, sulphates and ammonia [4,5]. The tanning of hides and skins to convert them into leather has been an important activity.

The tannery effluent is a mixture of biogenic matter and large variety of organic and inorganic chemicals. These may contains high concentrations of chlorides, aliphatic sulphonates, sulfates, aromatic and aliphatic ethoxylates, sulfonated poly-phenols, acrylic acid condensates, fatty acids, dyes, proteins, soluble carbohydrates and Na_2S. Conventional physio-chemical treatment of tannery effluents consists of pretreatment, coagulation/flocculation, sedimentation and sludge handling. In addition to chemical coagulation treatment, secondary and tertiary treatment methods such as biological filters, activated sludge etc. are being practiced. If nitrogen content is high, an additional step involving de nitrification is essential. The fats and proteins are partially hydrophobic they can be easily separated by floatation rather than sedimentation. Separation by flotation is relatively simple and faster compared to the other process. Hence removal of such pollutants by floatation could be more appropriate. Water treatments based on the electro flotation technique have been recently proved to circumvent most of these problems, while being also economically attractive [6-8,8-12]. In this work industrial effluent was carried out using electro floatation.

Electro flotation is a process of removing dispersed particles from liquid using gas bubbles. The gas bubbles are obtained through the electrolysis of water, which produces hydrogen and oxygen gas. The electrodes are arranged at the bottom of a tank filled with the solution containing the dispersed solids. During the electro flotation process, current is passed through the electrodes and producing bubbles of hydrogen and oxygen gas. Where the bubbles float on the tank may colloid with particles suspended in the water. Adhere to them and float them to the surface of the water. Dispersed foam was skimmed off the surface of the tank with a bladed transport system or skimmer.

Overall process of electro flotation

The basic principle of such process lies on water electrolysis subsequently. It may be determined among other variables on pH and temperature. The electrochemical reactions at the cathode and the anode are hydrogen evolution and oxygen evolution respectively.

At Anode:

$$Fe_{(s)} \rightarrow Fe^{2+} + 2^{e} \quad (1)$$

$$Fe\ (aq) + 2OH^{-}_{(aq)} \quad Fe\ (OH)_{2\,(s)} \quad (2)$$

At Cathode:

$$2\ H_2O + 2^{e-} \rightarrow H_2 + 2OH^{-}_{(aq)} \quad (3)$$

Overall:

$$Fe_{(S)} + 2H_2O_{\,(l)} \rightarrow Fe\ (OH)_{2(s)} + H_{2\,(g)} \quad (4)$$

The protonated water shows to be equivalent in both consumption and generation of positively charged species. Half reaction (first reaction) is proposed to be achieved in the anode while the second

***Corresponding author:** P. Gomathi Priya, Department of Chemical Engineering, A.C. College of Technology, Anna University, Chennai- 600 025, India at the cathode. From the literature [13] the author has performed a complete analysis about the experimental consequences of producing gas through electrolysis.

One of the gases (oxygen or hydrogen) is preferably used to perform the electro flotation, the aqueous speciation will change. The current density influences directly the number and size of bubbles [14]. If the current density is significantly low, both convective and diffusive mechanisms lead to gas to be dissolved and there may not even a bubble to occur. The latter will happen even at high current density which represents a waste of energy. Besides, this problem comes in addition to the inherent chemical irreversibility of both electrode reactions which set the voltage applied be significant. From the process point of view, during gas nucleation, other mechanisms start to work such as bubble coalescence. This mechanism makes bubbles to coarse and produce bubbles of larger size decreasing the bubbling surface available and inevitably an inefficient global process.

Materials and Methods

The tannery effluents were collected from KAS leathers, Erode, Tamil Nadu in India. These effluents were taken in a way to respect the standards of sampling and were stored in the cold storage before experimental works were started. The characterization of the effluent was carried out. The electrolytic cell is comprised of a sheet of mild steel as anode and a sheet of stainless steel as cathode each having the dimensions of 7.5cm X 4cm with the porosity of 0.5mm in both the electrodes. The Triton –X and the $FeCl_3$ was added as surfactant and flocculent during the experiment. The electrodes were connected to a digital DC regulated power supply unit (0-30V, 0-12A). A peristaltic pump was used to take away the treated effluent at different time intervals.

The number of colony forming units can be found using equation (5)

No. of colony forming units (**CFU**) (Per ml of sample) = No. of colonies X Dilution factor of the plate counted. (5)

Experimental procedure

The electro flotation was carried out in an electrolytic cell having the capacity of one liter effluent solution. The distance between the electrodes maintained was 1.5cm. Before the experiment the electrodes were washed with diluted hydrochloric acid to remove surface grease, rust, oil and other impurities on the steel surfaces and then dried. The effluent was taken in the electrolytic cell and electro flotation was carried out at different time intervals. During the electro flotation reaction the effluents were taken out and analyzed for COD and Total Solids content. The various effects of the surfactant Triton –X (0.1ml/100ml) and the flocculent $FeCl_3$ (0.5g/L) during the experiment were also discussed.

Results

Effect of current density

To carry out the optimum current density to the removal percentage of COD of tannery waste water was represented in Figure 1. The experiments were conducted using the tannery waste water at pH of 6. During the experiment, the current densities ranging from 0.035A/Cm^2, 0.07 A/Cm^2, 0.105 A/Cm^2 to 0.14 A/Cm^2 were adjusted. The current density significantly affects the efficiency of electro flotation. It showed that the COD removal efficiency was increased with increasing the current density 63% to 65.9%. The rate of reduction of COD increases with different current densities but the increase is not linear then it may be concluded that the process is controlled by diffusion.

In electro flotation, COD reduction is associated with three factors,

1. Direct oxidation of organic residue on electrode
2. Removal of material particles by gas bubbles
3. Indirect oxidation through reactive species

During the electro flotation batch experiments, the magnitude of the current density had a significant effect on COD and the total solids removal. The total solids removal efficiency was increased with increasing the current passed through the electrodes. The removal rate of COD assumed to be obeying first order kinetics; hence the rate equation is given as [15]

$$\frac{d(COD)}{dt} = -K(COD) \qquad (6)$$

The rate constant was calculated for all the current density and using the above equation the K was calculated for each value of current density and at each time interval and averaged.

From the Figure 1 it was observed that the total solids removal was reached maximum at the same current density after 60 minutes of electro flotation. The raw effluent was subjected to electro flotation and was observed for bacterial colonies by standard plate count method. The major part of the procedure deals with a series of successive dilutions of the tannery waste water in sterile bottle with sterile water. The diluted waste water samples of before and after electro flotation was taken. Then it was poured into Petri dishes containing nutrient agar. The numbers of colonies were counted after incubation for 48 hours at 37°C. The disinfection involves two mechanisms like due to oxidation and precipitation with Fe $(OH)_3$.

The chlorides ions present in the effluent also contribute disinfection. i.e. in electro flotation the anodic oxidation causes chloride ions to form hypo chlorites ions which could be used as disinfectant.

It was observed at the crude sample after series of dilution had contained 7200000 (approximately) that is 72 prominent colonies. Bacteria of mixed culture per ml. after subjecting to electro flotation for various time intervals from 5-60 minutes at 0.14A/Cm^2 of current density the colonies were found to be eliminated to the extent of 86.6% shown in the Figure 1.

Effect of pH

It has been found that the influence of pH is an important parameter for the performance of the electrochemical process [16,17]. In the electro flotation process the pH of the solution is known to play an important role in the removal efficiency of pollutants. To examine the pH on the tannery effluent was adjusted to the desired pH with diluted sodium hydroxide or hydrochloric acid. The experiments were conducted for pH ranges from 3 to 10 by keeping the current density constant at 0.035A/Cm^2 for the same electro flotation time of 60 minutes. The experiment results shows in Figure 2, and it was observed that the total solids removal efficiency and COD decreased for pH below 4 and above 9, this is due to an extent of the reduction of Fe^{3+} is favored. At the pH range of 5-9, hydrolysis and reduction takes place and it forms mononuclear or poly nuclear ferrous complexes as Fe $(OH)_3$, Fe $(OH)_2^+$. Which are effective pollutants but at high pH solutions that dominant species was Fe $(OH)_4^-$ which does not coagulate with the pollutants [18]. From this result it may be concluded that the pH of the experiment was further followed at 6.

Effect of surfactant

Total solids were removed using electro flotation and the process was further enhanced by adding the surfactant Triton –X. In the Figure 3, it was observed that the addition of surfactants enhances the total solids removal. In the process of electro flotation alone the removal percentage was found to be 66% but when this process was improved with surfactant and the removal percentage of total solids has been reached to 78%. From the same Figure, it may be concluded that, an addition of surfactant initially increases the COD value due to the release of a new organic compound in the waste water. During electro flotation, these surfactants also gets removed along with the low density colloidal particles by formation of surfactant – particulate - bubble composite and reduction of COD removal percentage is slightly decreased when surfactant is used in normal electro flotation process the removal percentage was 71% when surfactants are added the value

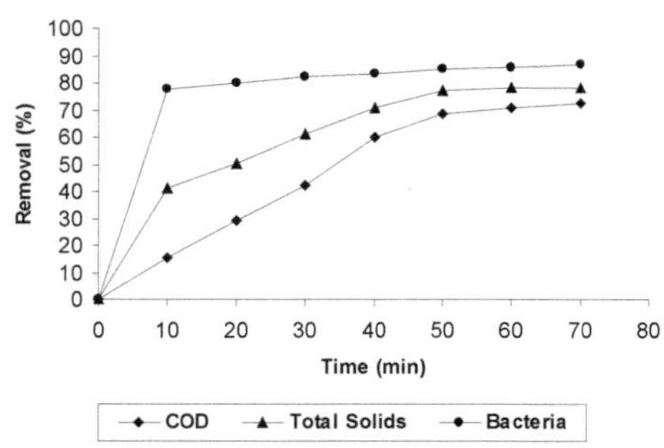

Figure 1: Effect of current density as a function of electrolysis time on COD, Total solids and Bacteria removal percentage.

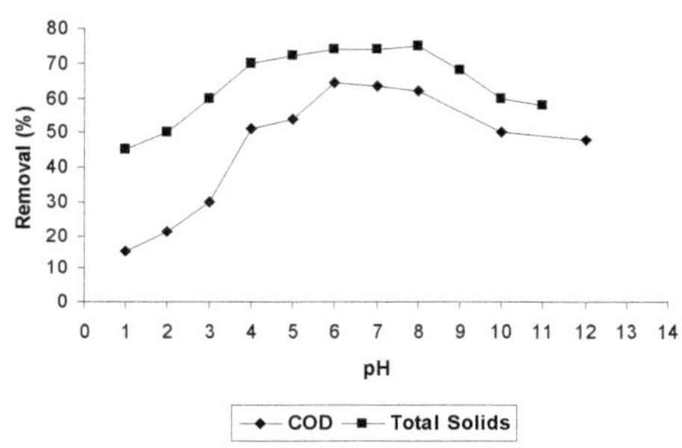

Figure 2: Effect of pH on COD and Total solids removal.

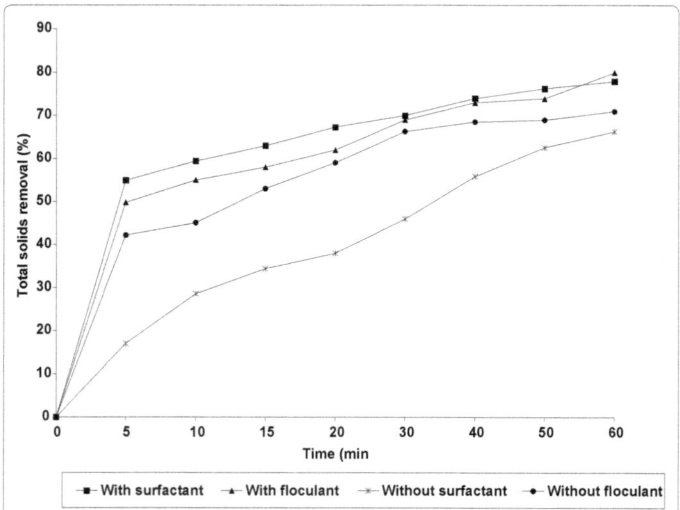

Figure 3: Effect of surfactant and Flocculants as a function of electrolysis time on Total solids removal.

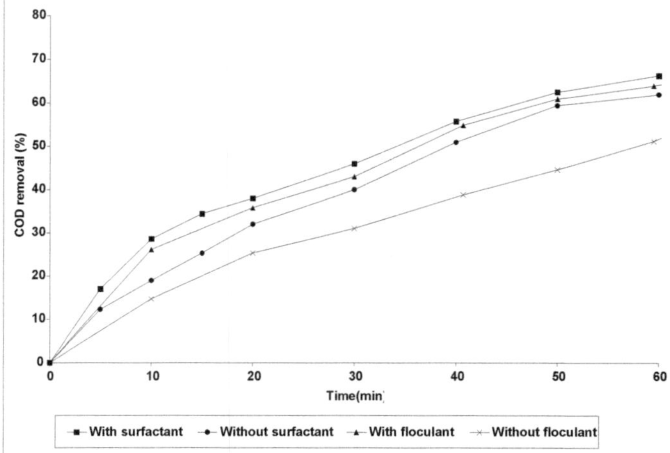

Figure 4: Effect of Flocculants and surfactant as a function of electrolysis time on COD.

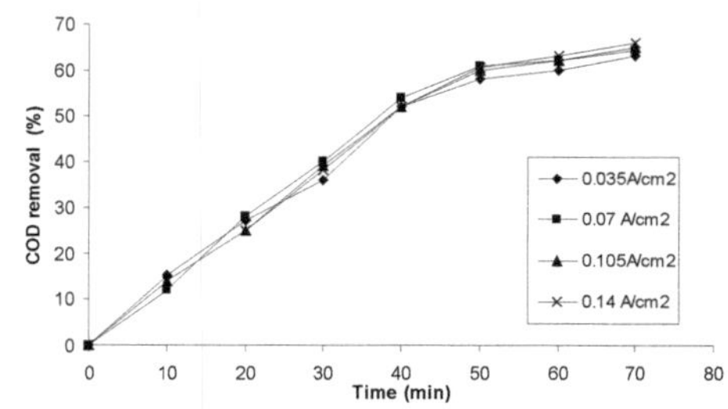

Figure 5: Effect of current density on COD removal.

of percentage removal has been increased to 80%.

Effect of flocculating agent

The well known flocculating agent known as $FeCl_3$ was added to the waste water and its effect in removal of total solids and COD has been studied. $FeCl_3$ was used in most of the tannery waste water treatment due to its efficiency and low cost [19]. The flocculating agents are polymers of high molecular weight; they have property to be adsorbed with solid –liquid or liquid –liquid interface which allows the aggregation of the colloidal particles by formation of bridges between the dispersed particles. The metal ions present in the coagulant causes coagulation of the colloidal organic matter and the suspended solids resulting in the formation of rather larger particles (flocs) which are attached to the surface of the gas bubbles and are collected at the top of the electro flotation bath [20].

The removal efficiency of Total Solids has been increased up to 80% within 60 minutes of electro flotation at the current density of 0.14A/Cm^2. The COD removal percentage has also been increased up to 64% with the addition of $FeCl_3$ this is explained by the oxidation of dissolved organic substance due to the presence of both fine oxygen bubbles and the chlorine as a strong oxidizing agent in the medium. The Chlorine was the oxidation product in the secondary electrochemical anodic reaction during electrolysis [21]. From the Figure 4, it was found that Total Solids removal efficiency and COD removal efficiency increased with the addition of flocculent to the electro flotation unit.

Discussion

Tannery effluent samples were treated by electro flotation technique and found to be effective for the removal of total solids and COD using mild steel as anode and stainless steel sheet as cathode. During the process the increase in removal of total solids and COD was found to be increasing with the different applied current densities. The addition of surfactant increases the COD removal as well as it increases the total solids removal from 66.3% to78%. The flocculants $FeCl_3$ enhanced electro flotation was found to be more effective in removing Total solids and COD from tannery effluent. After the electro flotation process COD removal was reached to about 66% and pathogenic bacteria could be eliminated to the extent of 86.6%

References

1. Swaminathan G, Muralidharan C, Rangasamy T, John Sundar V, Sivakumar V (2005) Management of total dissolved solids in tanning process through improved techniques. Journal of Cleaner Production 13: 699-703.
2. Rajamani S, Ravindranath E, Chita T, Umamaheswari B, Ramesh T, et al. (2004) Membrane application for recovery and reuse of water from treated tannery wastewater. Desalination 164: 151-156.
3. Moura Alencar MCP, Castro Dantas TN, Dantas Neto AA (2004) Evaluation and optimization of chromium removal from tannery effluent by micro emulsion in the Morris extractor. J Hazard Mater 114: 115-122.
4. Molinari R, Drioli E, Cassano A (1997) Recovery and reuse of chemical in unhearing, degreasing and chromium tanning processes by membranes. Desalination 113: 251-261.
5. Molinari R, Drioli E, Cassano A, Romano M (2001) Treatment of aqueous effluents of the leather industry by membrane processes. J Memb Sci 181: 111-126.
6. Alinsafi A, Khemis M, Pons MN, Leclerc JP, Yaacoubi A, et al. (2005) Electro coagulation of reactive textile dyes and textile waste water. Chem Eng Process 44: 461-470.
7. Daneshvar N, Oladegaragoze A, Djafarzadeh N (2006) The use of artificial neural networks (ANN) for modeling of decolorization of textile dye solution containing C. I. Basic Yellow 28 by electro coagulation process. J Hazard Mater 129: 116-122.
8. Bayramoglu M, Eyvaz M, Kobya M (2007) Treatment of the textile wastewater by electro coagulation Economical evaluation. Chemical Engineering Journal 128: 155-161.
9. Kobya M, Can OT, Bayramoglu M (2003) Treatment of textile wastewaters by electro coagulation using iron and aluminum electrodes. J Hazard Mater 100: 163-178.
10. Bayramoglu M, Kobya M, Can OT, Sozbir M (2004) Operating cost analysis of electro coagulation of textile dye wastewater, Separation and Purification Technology 37: 117-125.
11. Can OT, Kobya M, Demirbas E, Bayramoglu M (2006) Treatment of the textile wastewater by combined electro coagulation. Chemosphere 62: 181-187.
12. Kobya M, Demirbas E, Can OT, Bayramoglu M (2006) Treatment of levafix orange textile dye solution by electro coagulation. J Hazard Mater 132: 183-188.
13. Ketkar DR, Mallikajunan R, Venkatachalam S (1991) Electro flotation of quartz fins. International journal of miner Proc 31: 127-138.
14. Fukui Y, Yuu S (1985) Removal of colloidal particles in electro flotation. AIChE J 31: 201-208.
15. Laviron, Maja (1972) Influence of the adsorption of the depolarizer or of a product of the electrochemical reaction or polarographic current: XVII theoretical study of a reversible surface reaction followed by a first order chemical reaction in linear potential sweep voltammetry. J Electro anal Chem 35: 333-342.
16. Lin SH, Chen ML (1997) Treatment characteristics of textile wastewater and removal of heavy metals using the electro flotation technique. Water Res 31: 868-876.
17. Chen X, Chen G, Yue PL (2000) Separation of pollutants from restaurant wastewater by electro coagulation. Sep Purif Technol 19: 65-76.
18. Chig-Ta –Wang (2009) Removal of COD from laundry wastewater by electro coagulation/ electro flotation. J Hazard Mater 164: 81-86.
19. Aboulhassan MA, Souabi S, Yaacoubi A (2007) Pollution reduction and biodegradability index improvement of tannery effluents. Int J Environ Sci Tech 5: 11-16.
20. Jiantuan Ge, Jiuhui Qu (2004) New bipolar electro coagulation-electro flotation process for the treatment of laundry wastewater. Separation and Purification Technology 36: 33-39.
21. Ben mansour L (2007) Treatment of wastewaters of paper industry by coagulation –electro flotation. Desalination 208: 34-41.

Permissions

All chapters in this book were first published in JCEPT, by Omics International; hereby published with permission under the Creative Commons Attribution License or equivalent. Every chapter published in this book has been scrutinized by our experts. Their significance has been extensively debated. The topics covered herein carry significant findings which will fuel the growth of the discipline. They may even be implemented as practical applications or may be referred to as a beginning point for another development.

The contributors of this book come from diverse backgrounds, making this book a truly international effort. This book will bring forth new frontiers with its revolutionizing research information and detailed analysis of the nascent developments around the world.

We would like to thank all the contributing authors for lending their expertise to make the book truly unique. They have played a crucial role in the development of this book. Without their invaluable contributions this book wouldn't have been possible. They have made vital efforts to compile up to date information on the varied aspects of this subject to make this book a valuable addition to the collection of many professionals and students.

This book was conceptualized with the vision of imparting up-to-date information and advanced data in this field. To ensure the same, a matchless editorial board was set up. Every individual on the board went through rigorous rounds of assessment to prove their worth. After which they invested a large part of their time researching and compiling the most relevant data for our readers.

The editorial board has been involved in producing this book since its inception. They have spent rigorous hours researching and exploring the diverse topics which have resulted in the successful publishing of this book. They have passed on their knowledge of decades through this book. To expedite this challenging task, the publisher supported the team at every step. A small team of assistant editors was also appointed to further simplify the editing procedure and attain best results for the readers.

Apart from the editorial board, the designing team has also invested a significant amount of their time in understanding the subject and creating the most relevant covers. They scrutinized every image to scout for the most suitable representation of the subject and create an appropriate cover for the book.

The publishing team has been an ardent support to the editorial, designing and production team. Their endless efforts to recruit the best for this project, has resulted in the accomplishment of this book. They are a veteran in the field of academics and their pool of knowledge is as vast as their experience in printing. Their expertise and guidance has proved useful at every step. Their uncompromising quality standards have made this book an exceptional effort. Their encouragement from time to time has been an inspiration for everyone.

The publisher and the editorial board hope that this book will prove to be a valuable piece of knowledge for researchers, students, practitioners and scholars across the globe.

List of Contributors

Hussein K Abdel-Aal
Professor Emeritus of Chemical Engineering / Petroleum Operations, NRC, Cairo, Egypt

Rajesh Ghosh
Department of Chemical Engineering, Calcutta Institute of Technology, Uluberia, Howrah, West Bengal, India

Sounak Bhattacherjee
Department of Chemical Engineering, Calcutta Institute of Technology, Uluberia, Howrah, West Bengal, India

M Masukume
Department of Chemical and Metallurgical Engineering, Tshwane University of Technology, Pretoria, South Africa

J P Maree
Department of Environmental, Water and Earth Sciences, Tshwane University of Technology, Pretoria, South Africa

S Ruto
Department of Chemical and Metallurgical Engineering, Tshwane University of Technology, Pretoria, South Africa

H Joubert
Africa Remediation Technologies (Pty) Ltd, Cullinan, Pretoria, South Africa

N. Asiedu
Centre of Material and Process Synthesis, School of Chemical and Metallurgical Engineering, University of Witwatersrand, Johannesburg, South Africa

D. Hildebrandt
Centre of Material and Process Synthesis, School of Chemical and Metallurgical Engineering, University of Witwatersrand, Johannesburg, South Africa

D. Glasser
Centre of Material and Process Synthesis, School of Chemical and Metallurgical Engineering, University of Witwatersrand, Johannesburg, South Africa

Abrar Muslim
Department of Chemical Engineering, Engineering Faculty, Syiah Kuala University, Banda Aceh, Indonesia

Yang Zhang
Department of Chemical Engineering, Michigan Technological University, Houghton, MI49931, USA

Wen Zhou
Department of Chemical Engineering, Michigan Technological University, Houghton, MI49931, USA

Om Prakash Verma
Jacob School of Biotechnology & Bioengineering, SHIATS, Allahabad

Pankaj Kumar Ojha
Jacob School of Biotechnology & Bioengineering, SHIATS, Allahabad

Shashi B Bailey
Jacob School of Biotechnology & Bioengineering, SHIATS, Allahabad

Pankaj Singh
Jacob School of Biotechnology & Bioengineering, SHIATS, Allahabad

Rajendra Dubey
Galaxy Concepts Kolkata

B.N.Mishra
IET, Lucknow

Om Prakash Verma
Jacob School of Biotechnology & Bioengineering, SHIATS, Allahabad

Pankaj Kumar Ojha
Jacob School of Biotechnology & Bioengineering, SHIATS, Allahabad

Shashi B Bailey
Jacob School of Biotechnology & Bioengineering, SHIATS, Allahabad

Pankaj Singh
Jacob School of Biotechnology & Bioengineering, SHIATS, Allahabad

Rajendra Dubey
Galaxy Concepts Kolkata

B.N.Mishra
IET, Lucknow

Anatoly A. Shatalov
Centro de Estudos Florestais, Universidade Técnica de Lisboa, Instituto Superior de Agronomia, Tapada da Ajuda, 1349-017 Lisboa, Portugal

Helena Pereira
Centro de Estudos Florestais, Universidade Técnica de Lisboa, Instituto Superior de Agronomia, Tapada da Ajuda, 1349-017 Lisboa, Portugal

Ahmed Mohamed Al Menhali
Chemical Engineering Department, Abu Dhabi Men's College, Higher Colleges of Technology, Saada street, Abu Dhabi (Abu Dhabi), UAE

Abdulla Ali. AL Marzooqi
Chemical Engineering Department, Abu Dhabi Men's College, Higher Colleges of Technology, Saada street, Abu Dhabi (Abu Dhabi), UAE

Abdul Rahman Saleh Alameri
Chemical Engineering Department, Abu Dhabi Men's College, Higher Colleges of Technology, Saada street, Abu Dhabi (Abu Dhabi), UAE

Abdulla Mohamed Al Ameri
Chemical Engineering Department, Abu Dhabi Men's College, Higher Colleges of Technology, Saada street, Abu Dhabi (Abu Dhabi), UAE

Zin Eddine Dadach
Chemical Engineering Department, Abu Dhabi Men's College, Higher Colleges of Technology, Saada street, Abu Dhabi (Abu Dhabi), UAE

UPL Wijayarathne
Department of Chemical and Process Engineering, University of Moratuwa, Sri Lanka

KC Wasalathilake
Department of Chemical and Process Engineering, University of Moratuwa, Sri Lanka

NS Yousef
Petrochemical Department, Pharos University, Egypt

S. Darwish
Chemical Engineering Department, Faculty of Engineering, Alexandria University, Egypt

TM Zewail
Chemical Engineering Department, Faculty of Engineering, Alexandria University, Egypt

YA El-Tawail
Chemical Engineering Department, Faculty of Engineering, Alexandria University, Egypt

Mohamed Ahmed Mahmoud
Nuclear Material Authority, Cairo, Egypt
Chemical Engineering Department, Jazan University, Saudi Arabia

A.Elaziouti
DR. Moulay Tahar University, Saida, Algeria

N. Laouedj
LCPCE Laboratory, Faculty of sciences, Department of industrial Chemistry, University of the Science and the technology of Oran (USTO M.B), BP 1505 El M'naouar 31000 Oran, Algeria

Bekka Ahmed
DR. Moulay Tahar University, Saida, Algeria

Ali Seifollahzadeh
Department of Chemistry and Chemical Engineering, MAUT, Tehran, Iran

Ali Aminian
Department of Chemistry and Chemical Engineering, MAUT, Tehran, Iran

YoonKuen Chae
Department of Chemical and Biomolecular Engineering, University of Yonsei, 262 Seongsanno, Seodaemun-gu, Seoul 120-749, Korea

Shinsuke Mori
Department of Chemical Engineering, Tokyo Institute of Technology, 2-12-1 O-okayama, Meguro-ku, Tokyo, 152-855

Masaaki Suzuki
Department of Chemical Engineering, Tokyo Institute of Technology, 2-12-1 O-okayama, Meguro-ku, Tokyo, 152-855

JinWon Park
Department of Chemical and Biomolecular Engineering, University of Yonsei, 262 Seongsanno, Seodaemun-gu, Seoul 120-749, Korea

A.Reyes
Department of Chemical Engineering, University of Santiago of Chile, Av. L. B. O"Higgins 3363, Santiago, Chile

A.Mahn
Department of Chemical Engineering, University of Santiago of Chile, Av. L. B. O"Higgins 3363, Santiago, Chile

P. Huenulaf
Department of Chemical Engineering, University of Santiago of Chile, Av. L. B. O"Higgins 3363, Santiago, Chile

T. González
Department of Chemical Engineering, University of Santiago of Chile, Av. L. B. O"Higgins 3363, Santiago, Chile

A.Hattab
Research Unit: Applied Thermodynamics, National Engineering School of Gabes (ENIG)-Omar Ibn Elkhatab Street, Gabes, Tunisia

M. Bagane
Research Unit: Applied Thermodynamics, National Engineering School of Gabes (ENIG)-Omar Ibn Elkhatab Street, Gabes, Tunisia

M. Chlendi
Research Unit: Applied Thermodynamics, National Engineering School of Gabes (ENIG)-Omar Ibn Elkhatab Street, Gabes, Tunisia

M. Pakizeh
Department of Chemical Engineering, Faculty of Engineering, Ferdowsi University of Mashhad, P.O. Box 91775-1111, Mashhad, Iran

M Namvar-Mahboub
Department of Chemical Engineering, Faculty of Engineering, Ferdowsi University of Mashhad, P.O. Box 91775-1111, Mashhad, Iran

Om Prakash Verma
Jacob School of Biotechnology & Bioengineering, SHIATS, Allahabad

Abha Singh
Jacob School of Biotechnology & Bioengineering, SHIATS, Allahabad

Nirmala Singh
JDMC, KANPUR

Onkar Chaudhary
RTOU, ALLAHABAD

Mahnaz Aghajanloo
Oil and Gas Processing Centre of Excellence, School of Chemical Engineering, College of Engineering, Tehran University, Tehran, Iran

Ali Morad Rashidi
Research Institute of Petroleum Industry, Nanotechnology Research Center, West Blvd., Azadi Sport Complex, Tehran, Iran

Mohamad Ali Moosavian
Oil and Gas Processing Centre of Excellence, School of Chemical Engineering, College of Engineering, Tehran University, Tehran, Iran

R. Vijayaram
Department Of Chemical Engineering, Kongu Engineering College, Perundurai, Erode, India

Dr. K. Chandramohan
Department Of Chemical Engineering, Kongu Engineering College, Perundurai, Erode, India

A.Varasteh
Chemical Engineering, Islamic Azad University, Shahrood Branch, Semnan, Shahrood, Iran

H.A. Tahmasebi
Chemical Engineering Department, Islamic Azad University, Quchan Branch, Asian Road, Quchan, Iran

R. Gheshlaghi
Department of Chemical Engineering, Ferdowsi University of Mashhad, Mashhad, Khorasan, Iran

Patricia Guerrero-German
Department of Chemical Engineering and Metallurgy, University of Sonora, L. Blvd Encinas and Rosales, Hermosillo, Sonora, Mexico

Rosa Ma. Montesinos-Cisneros
Department of Mathematics, University of Sonora, L. Blvd Encinas and Rosales, Hermosillo, Sonora, Mexico

Armando Tejeda-Mansir
Department of Scientific and Technological Research, University of Sonora, Luis Encinas and Rosales Blvd, Hermosillo, Sonora, Mexico

A.Ahmadpour
Department of Chemical Engineering, Faculty of Engineering, Ferdowsi University of Mashhad, Khorasan, Iran

Z. Sheikhi Mehrabadi
Department of Chemical Engineering, Faculty of Engineering, Ferdowsi University of Mashhad, Khorasan, Iran

J.R. Esfandyari
The Technical teacher of saveh education office, saveh, Iran

M. Koolivand-Salooki
Department of Chemical Engineering, omideh beranch, Islamic Azad University, omideh, Iran

S. Asadi
Chemical Engineering Department, Faculty of Engineering, Ferdowsi University of Mashhad, Mashhad, P.O Box: 91775-1111, Iran

MT Hamed Mosavian
Chemical Engineering Department, Faculty of Engineering, Ferdowsi University of Mashhad, Mashhad, P.O Box: 91775-1111, Iran

A.Ahmadpour
Chemical Engineering Department, Faculty of Engineering, Ferdowsi University of Mashhad, Mashhad, P.O Box: 91775-1111, Iran

M Ferhan
Center for Biocomposites and Biomaterials Processing (CBBP), Faculty of Forestry, University of Toronto, Toronto, ON, Canada

N Yan
Center for Biocomposites and Biomaterials Processing (CBBP), Faculty of Forestry, University of Toronto, Toronto, ON, Canada

M Sain
Center for Biocomposites and Biomaterials Processing (CBBP), Faculty of Forestry, University of Toronto, Toronto, ON, Canada

Anish T. Achankunju
M.Tech Student, Department of Chemical Engineering, Coimbatore Institute of Technology, Coimbatore, Tamil Nadu, India

A.Sivalingam
Associate professor, Department of Chemical Engineering, Coimbatore Institute of Technology, Coimbatore, Tamil Nadu, India

T. Kannadasan
Head of the Department, Department of Chemical Engineering, Coimbatore Institute of Technology, Coimbatore, Tamil Nadu, India

Sunil A. Misal
Department of Chemical Engineering, Pravara Rural Engineering College, Loni-413736, India

V. S. Sathe
Department of Chemical Engineering, Dr. BATU, Lonere, Dist-Raigad, India

R.W.Gaikwad
Department of Chemical Engineering, Pravara Rural Engineering College, Loni-413736, India

Dhirendra
Department of Chemical Engineering, Pravara Rural Engineering College, Loni-413736, India

Pritish Chowdhury
Natural Products Chemistry & Chemical Engineering Divisions, CSIR North East Institute of Science & Technology, Jorhat 785 006, India

Juri Moni Borah
Natural Products Chemistry & Chemical Engineering Divisions, CSIR North East Institute of Science & Technology, Jorhat 785 006, India

Manobjyoti Bordoloi
Natural Products Chemistry & Chemical Engineering Divisions, CSIR North East Institute of Science & Technology, Jorhat 785 006, India

Pradip K. Goswami
Natural Products Chemistry & Chemical Engineering Divisions, CSIR North East Institute of Science & Technology, Jorhat 785 006, India

Aradhana Goswami
Natural Products Chemistry & Chemical Engineering Divisions, CSIR North East Institute of Science & Technology, Jorhat 785 006, India

Nabin C.Barua
Natural Products Chemistry & Chemical Engineering Divisions, CSIR North East Institute of Science & Technology, Jorhat 785 006, India

Paruchuri G. Rao
Natural Products Chemistry & Chemical Engineering Divisions, CSIR North East Institute of Science & Technology, Jorhat 785 006, India

Saba Adnan Ghani
Department of Chemical Engineering, University of Tikrit, Iraq

Rafei J. Yaqub
Department of Petroleum Engineering, University of Tikrit, Iraq

Suhaeb S. Saleh
Department of Chemical Engineering, University of Tikrit, Iraq

N. Abdennebi
Chemical Engineering Department, National engineering school of Gabes (ENIG), University of Gabes, Tunisia

M. Bagane
Chemical Engineering Department, National engineering school of Gabes (ENIG), University of Gabes, Tunisia

C. Chtara
Chemical Engineering Department, National engineering school of Gabes (ENIG), University of Gabes, Tunisia

E. M. Elsayed
Chemical Engineering Department, Faculty of Engineering, Alexandria University, Alexandria, Egypt

T. M. Zewail
Chemical Engineering Department, Faculty of Engineering, Alexandria University, Alexandria, Egypt

A.A. Zaatout
Chemical Engineering Department, Faculty of Engineering, Alexandria University, Alexandria, Egypt

Ghanim M. Alwan
Chemical Engineering, University of Technology, Baghdad, Iraq

G. Santhosh Kumar
Department of Chemical Engineering, M.V.G.R College of Engineering, Vizianagaram, AP, India

V. Subba Rao Ch
Department of Chemical Engineering, M.V.G.R College of Engineering, Vizianagaram, AP, India

Omid Rowshanaie
University Putra Malaysia, Department of Chemical Engineering, Serdang, Malaysia

Saari Bin Mustapha
University Putra Malaysia, Department of Chemical and Environmental Engineering, Serdang, Malaysia

Hooman Rowshanaie
University Putra Malaysia, Serdang, Malaysia

Mustapha Chabane
Spectrochemistry and structural pharmacology laboratory, Department of chemistry, Science faculty, University of Tlemcen, Algeria

Benamar Dahmani
Spectrochemistry and structural pharmacology laboratory, Department of chemistry, Science faculty, University of Tlemcen, Algeria

Tinuade Jolaade Afolabi
Department of Chemical Engineering, Ladoke Akintola University of Technology, Ogbomoso, Oyo State, Nigeria

Alao Adeyinka Idowu
Department of Chemical Engineering, Ladoke Akintola University of Technology, Ogbomoso, Oyo State, Nigeria

Nayef Ghasem
Department of Chemical & Petroleum Engineering, UAE University, Alain, UAE

Mohamed Al-Marzouqi
Department of Chemical & Petroleum Engineering, UAE University, Alain, UAE

Jamshidi LCLA
Post-Graduate Program in Chemical Engineering-PPCE/ Technology and Geosciences Centre-TGC-UFPE, Av. Moraes Rego, 1235 - University City, CEP: 50670-901, Recife-PE, Brazil

Barbosa CMBM
Post-Graduate Program in Chemical Engineering-PPCE/ Technology and Geosciences Centre-TGC-UFPE, Brazil Av. Moraes Rego, 1235 - University City, CEP: 50670-901, Recife, PE, Brazil

L. Nascimento
Post-Graduate Program in Chemical Engineering-PPCE/ Technology and Geosciences Centre-TGC-UFPE, Av. Moraes Rego, 1235 - University City, CEP: 50670-901, Recife, PE, Brazil

JR Rodbari
Department of Sociology-DS/Islamic Azad University of Roudehen Branch, Roudehen City-Complex of University Roudehen, Tehran, Iran

Rajarshi Kar
Chemical Engineering Department, Heritage Institute of Technology, Kolkata, India

Oindrila Gupta
Chemical Engineering Department, Heritage Institute of Technology, Kolkata, India

Kunal Mandol
Chemical Engineering Department, Heritage Institute of Technology, Kolkata, India

Sangita Bhattacharjee
Chemical Engineering Department, Heritage Institute of Technology, Kolkata, India

R. Hamdi
Department of Chemical-Process Engineering, National Engineering School of Gabes, Omar Ibn El Khattab St., 6029 Zrig, Gabes, Tunisia

A.Hannachi
Department of Chemical-Process Engineering, National Engineering School of Gabes, Omar Ibn El Khattab St., 6029 Zrig, Gabes, Tunisia

Masoud Habibi Zare
Faculty of Engineering, Islamic Azad University, Tehran South Branch, Tehran, Iran

Mojtaba Nabipoor Hassankiadeh
Department of Chemical Engineering, Tarbiat Modares University, Tehran, Iran

Jannie P. Maree
Department of Environmental, Water & Earth Sciences, Faculty of Science, Tshwane University of Technology, Pretoria, South Africa

Caliphs M. Zvinowanda
Department of Environmental, Water & Earth Sciences, Faculty of Science, Tshwane University of Technology, Pretoria, South Africa

Munyaradzi Mujuru
Department of Environmental, Water & Earth Sciences, Faculty of Science, Tshwane University of Technology, Pretoria, South Africa

Regina M. Matsapola
Department of Chemical & Metallurgical Engineering, Faculty of Engineering & Inbuilt Environment, Tshwane University of Technology, Pretoria, South Africa

David J. Delport
Department of Chemical & Metallurgical Engineering, Faculty of Engineering & Inbuilt Environment, Tshwane University of Technology, Pretoria, South Africa

Wynand J. Louw
Marlow Aquatec (Pty) Ltd, 50 Francis Street, Colbyn, Pretoria, South Africa

P. Gomathi Priya
Department of Chemical Engineering, A.C. College of Technology, Anna University, Chennai- 600 025, India

V. Ramamurthi
Department of Chemical Engineering, A.C. College of Technology, Anna University, Chennai- 600 025, India

Prabhu Anand
Department of Chemical Engineering, A.C. College of Technology, Anna University, Chennai- 600 025, India